Network Analysis

Third Edition

M. E. VAN VALKENBURG

Professor of Electrical Engineering
University of Illinois

PRENTICE-HALL, INC., Englewood Cliffs, New Jersey

Library of Congress Cataloging in Publication Data

VAN VALKENBURG, MAC ELWYN,
  Network analysis.

  Bibliography: p.
  1. Electric networks.   I. Title.
TK454.2.V36 1974   621.319'2   73-12705
ISBN 0-13-611095-9

SB75|68060|9|75

621.3 UAN

4720

PRENTICE-HALL INTERNATIONAL, INC., *London*
PRENTICE-HALL OF AUSTRALIA, PTY. LTD., *Sydney*
PRENTICE-HALL OF CANADA, LTD., *Toronto*
PRENTICE-HALL OF INDIA PRIVATE LIMITED, *New Delhi*
PRENTICE-HALL OF JAPAN, INC., *Tokyo*

Contents

Preface to the Third Edition

The objective in preparing this third edition of *Network Analysis* has been to retain the basic organization of previous editions but to make additions to reflect changes that have occurred in the teaching of the subject since the second edition appeared in 1964. The topics spanned by the subject of network analysis have increased in number and complexity. To explain my rationale in determining which topics to include and the depth to which each is treated, I must state that my perception of a course in network analysis is as a service course to the subjects that follow in the electrical engineering curriculum. The subject has long enjoyed that role, although not all practitioners are anxious to acknowledge that fact. Hence the topics to be considered should introduce material that will be useful in later courses taken by the student, and each should be studied in sufficient depth to make it easy to bridge the gap to the coverage of the next course. Thus the writing of equations to describe a network which is a model of a physical system permeates all of lumped system analysis and it should be treated in depth. The solution of the state equation, including an interpretation of the meaning of e^{At} where A is a matrix can be postponed to a later course.

One of the innovations of this third edition is Appendix E which contains a detailed listing of appropriate topics for computer homework assignments or topics for an associated software laboratory, including a detailed listing of textbook references. At the end of chapters, where appropriate, specific suggestions are made as to computer exercises which will support the topics covered in that chapter. A number of factors are involved in the decision to provide the information in this form, rather than to introduce numerical method material or to present computer printout to illustrate possible computer operations. The most important of these is that the actual use of the computer by the student is greatly influenced by the library of subroutines available to the student at a computer center. Computer center libraries continue to be in a state of rapid change and there is little uniformity from one to the next.

State variables have assumed new importance and are introduced. Those who will become proficient in the analysis of networks will learn *all* of the methods. However, it does not appear likely that the state-variable formulation of equations will replace nodal analysis, for example. It is generally good pedagogy to learn one or two methods well before adding more methods to a "bag of tricks." For the beginner, the traditional node and loop formulations are recommended.

Other additions include a treatment of Tellegen's theorem. The simple elegance of Tellegen's contribution and the wide variety of applications to which it may be applied for new insight and clarity continue to amaze engineers. A new treatment of the Nyquist criterion appears in this edition, after having been omitted from the second edition. The reason for this change is the new emphasis on active networks with their attendant stability problems, and the insight that is provided in understanding these networks through the Nyquist criterion.

This edition contains a significant number of new problems and the revision or updating of others. In addition, the number of solutions to problems given in Appendix G has been increased to make the book better suited for independent study.

The indebtedness of the author to his students, colleagues, and former teachers is great. To these and all others who have given help and encouragement in the preparation of the book, I offer an inadequate acknowledgment of appreciation. In particular, I am indebted to those users of the book who have been kind enough to send me their impressions of the first two editions and suggestions for an improved third edition. Finally, it is a pleasure to record that it has been enjoyable working with Virginia Huebner of the College Book Editorial Department of Prentice-Hall, and to express my gratitude to Evelyn, my wife, who gave invaluable assistance in proofreading the manuscript.

M. E. VAN VALKENBURG

Urbana, Illinois

Preface to the Second Edition

This book is designed for use in an introductory course or a second course in electric network analysis. The student is assumed to have the mathematical sophistication usually associated with the completion of a course in the calculus.

Teaching experience in the decade that has passed since the preparation of the first edition of the book has reinforced my conviction that the student's introduction to network analysis should begin with the so-called "transient" case and proceed to the sinusoidal steady state and related topics. I do not argue which order is the more basic or important, but rather that the beginning student will attain real understanding more readily by this route. Thus the basic organization of the second edition remains essentially that of the first.

The first three chapters of the book are concerned with an introduction of the elements that serve to model electrical devices, with definitions pertaining to networks, and to the formulation of the Kirchhoff equilibrium equations. There follows a study of the behavior of networks in terms of natural modes or the natural frequencies of response due to arbitrary excitation. The Laplace transformation provides the means by which this natural behavior of the network is unified with the characteristics of the excitation (or signal) with each represented by a transform and then studied in terms of the poles and zeros of the transform. Once this is accomplished, the case of sinusoidal excitation under steady-state operation is introduced. This, in turn, leads to the study of a number of topics important to the electrical engineer such as Bode plots, average power, insertion loss, and the various signal spectra. The early introduction of the Laplace transformation without proof has not presented any difficulties for students, in my experience. I believe that Whitehead* was correct in observing that, ". . . it is not essential that proof of the truth should constitute the first introduction to the idea."

At the time the first edition was prepared, it was felt that it was necessary to demonstrate the utility of the pole-and-zero approach by including material on a number of important applications such as LC one-port networks, image-parameter filters, amplifier-networks of the stagger-tuned variety, and some topics relating to automatic control. The pole-and-zero approach has since become firmly established in the curriculum and these chapters have accordingly been replaced. More examples are provided throughout. Some topics are treated in more depth than in the first edition including convolution, sinusoidal steady state analysis, the Routh-Hurwitz criterion, Fourier series and the Fourier integral. New topics added to this edition include the two-port parameters, complex loci and Bode plots, average power, power transfer, and insertion loss. Three appendices provide coverage of complex numbers, matrices, and magnitude and frequency scaling.

A few words are necessary concerning notation and conventions. The units for element values are assumed to be given in ohms, henrys, farads, volts, and amperes unless otherwise stated—both in the text and in the figures. I have used limits on integrals when I felt that it would add to the clarity or student understanding of the presentation; otherwise I have preferred to regard the integral sign as a symbol representing a rather detailed word statement and implying that the limits over which integration is performed must be known. Similarly, I have used $i(t)$ rather than i, and $I(s)$ rather than I when it seemed important to identify the variable. In every case, capital letters are used for variables in the frequency domain and lower-case letters for the time domain.

A number of decisions have been made painfully. I have used *order* in discussing algebraic equations relating to the time domain and then switched to *degree* when discussing algebraic equations in the frequency domain. This usage agrees with present technical practice by engineers; the use of either one or the other exclusively gives rise to strange-sounding expressions. For the symbol for transfer functions for two-port networks, I have used the subscript order 12, as in Z_{12}, knowing that some teachers have strong preferences for Z_T, Z^T, or Z_{21}. It will be clear from the text and the figures which

* Alfred North Whitehead, *The Aims of Education*, paperback edition by Mentor Books, New York, 1949, p. 15.

particular transfer function is meant so that there is little real danger of confusion. I hope that no great difficulty will arise in interchanging 1 and 2 or in replacing 12 by *T*. I have felt helpless in trying to give full credit to all contributors in a field in the *Further Reading* sections at the end of each chapter, and so have chosen to select only references that supplement the material given or provide an alternative approach that might lead to better understanding of the subject.

One of the rewards in writing a textbook such as this is that it provides an excuse for discussing and testing ideas with both colleagues and students. I must express great indebtedness to students who have, wittingly or unwittingly, helped me in fixing the order of presentation and the pattern of emphasis. Most of the revision for the second edition was accomplished while I was a member of the visiting faculties at the University of California, Berkeley and the University of Colorado. I am indebted to friends at these two schools as well as at the University of Illinois for most helpful discussions.

It is a pleasure to thank the following who have given valuable assistance in one or both of the editions: Don A. Baker of Los Alamos, Doran Baker of Utah State University, Joseph Chen of IBM, Jose B. Cruz, Jr. of Illinois, L. Dale Harris of the University of Utah, Shlomo Karni of the University of New Mexico, Wan Hee Kim of Columbia University, Jack Kobayashi of Hughes Aircraft Company, Franklin F. Kuo of Bell Telephone Laboratories, Philip C. Magnusson of Oregon State University, Wataru Mayeda of Illinois, William R. Perkins of Illinois, Ronald A. Rohrer of Illinois, Thomas M. Stout of Thompson-Ramo-Wooldridge, Glen Wade of Cornell University, and Philip Weinberg of Bradley University. Herbert M. Barnard and Edwin C. Jones, Jr. made valuable contributions in improving the text and checking the galley proofs. Editorial assistance from W. L. Everitt and Robert W. Newcomb is also acknowledged. Finally, I express my appreciation to my wife Evelyn and children for their patience and understanding during the preparation of the book.

M. E. VAN VALKENBURG

Urbana, Illinois

Network
Analysis

Development of the Circuit Concept 1

1-1. INTRODUCTION

One of the characteristics of the scientific method is the continual bringing together of a wide variety of facts to fit into a simple, understandable theory that will account for as many observations as possible. The name *conceptual scheme* has been used by the American chemist and educator James Conant for the theory or picture that results.[1] Perhaps the most familiar conceptual scheme to students of science and engineering is that of the atomic theory from which we take our pictures of the electron and of electric charge. Other important conceptual schemes are conservation of energy and conservation of charge.

Although electricity and magnetism were recognized early in the history of man—the charging of amber by friction, the use of the lodestone in navigation—it was not until the nineteenth century that significant progress was made in developing a conceptual scheme. The discovery by Galvani and Volta about 1800 that electricity could be produced by chemical means greatly simplified experimentation. Important discoveries were made in a relatively short interval of time after Volta. In 1820, Oersted identified the magnetic field with current, and Ampère measured the force caused by the current. In 1831 Faraday and, independently, Henry discovered electric induction.

[1] James B. Conant, *Science and Common Sense* (Yale University Press, New Haven, Conn. 1951).

1

These and other experiments were brought together to form a successful conceptual scheme by the English physicist James Clerk Maxwell in 1873. In Maxwell's equations, as the scheme has come to be known, all electric and magnetic phenomena are explained in terms of fields resulting from charge and current. The success of Maxwell's conceptual scheme is evidenced by the persistent agreement of results deduced from Maxwell's equations with observations for a period of over 100 years.

In view of Maxwell's success, why do we now embark upon a study of *another* conceptual scheme for the same phenomena, the electric circuit? Equally important as a question, how are the two concepts related? The answer to the first of our questions is the practical utility of the circuit concept. As a practical matter, we are not often interested in fields so much as we are in voltages and currents. The circuit concept favors analysis in terms of voltage and current from which other quantities—such as charge, fields, energy, power, etc.—can be computed if desired. The answer to our second question will require a longer answer and justification. Briefly, circuit concepts arise from the same basic experimental facts as do Maxwell's equations. However, the circuit involves approximations that are not included in the more general concept of field theory. It is important that we understand the nature of these approximations—the limitations of circuit theory—before we develop our subject.

It will be helpful to define the function of the circuit in terms of two basic building blocks: charge and energy. We regard charge and energy as the least common denominators in describing electrical phenomena, the primitive quantities in terms of which we can build our conceptual scheme of the electric circuit. A physical circuit is a system of interconnected apparatus. Here we use the word *apparatus* to include sources of energy, connecting wires, components, loads, etc. A circuit functions to transfer and transform energy. Energy transfer is accomplished by charge transfer. In the circuit, energy is transferred from a point of supply (the source) to a point of transformation or conversion called the load (or sink). In the process, the energy may be stored.

1-2. CHARGE AND ENERGY

Thales of Greece is credited with the discovery, about 600 B.C., that when briskly rubbed with a piece of silk or fur, amber becomes "electrified" and is capable of attracting small pieces of thread. This same technique for producing electricity was used centuries later by Coulomb in France (and independently by Cavendish in England) in establishing the inverse square law of attraction of charged bodies.

Our present-day understanding of the nature of charge is based on the conceptual scheme of the atomic theory. We picture the atom as composed of a positively charged nucleus surrounded by negatively charged electrons. In the neutral atom, the total charge of the nucleus is equal to the total charge of the electrons. When electrons are removed from a substance, that substance becomes positively charged. A substance with an excess of electrons is negatively charged.

The basic unit of charge is the charge of the electron. The MKS unit of charge is the *coulomb*. The electron has a charge of 1.6021×10^{-19} coulomb.

The phenomenon of transferring charge from one point in a circuit to another is described by the term *electric current*. An electric current may be defined as the time rate of net motion of electric charge across a cross-sectional boundary. A random motion of electrons in a metal does not constitute a current unless there is a net transfer of charge with time.

In equation form, the current[2] is

$$i = \frac{dq}{dt} \tag{1-1}$$

If the charge q is given in coulombs and the time t is measured in seconds, the current is measured in *amperes* (after the French physicist André Ampère). Since the electron has a charge of 1.6021×10^{-19} coulomb, it follows that a current of 1 ampere corresponds to the motion of $1/(1.6021 \times 10^{-19}) = 6.24 \times 10^{18}$ electrons past any cross section of a path in 1 second.

In terms of the atomic-theory conceptual scheme, all substances are pictured as made up of atoms. In a solid, some electrons are relatively free of the nucleus; the attractive forces on these electrons are exceedingly small. Such electrons are distinguished by the name *free electrons*. An electric current is the time rate of flow of these free electrons, passing from one atom to the next as pictured in Fig. 1-1.

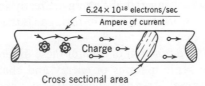

Fig. 1-1. Representation of the motion of charge in a conductor.

In some materials there are many free electrons, so that large currents are easily attained. Such materials are known as *conductors*. Most metals and some liquids are good conductors. Materials with relatively few free electrons are known as *insulators*. Common insulating materials include glass, mica, plastics, etc. Other materials

[2]The symbol i for current is taken from the French word *intensité*.

called *semiconductors* play a significant role in electronics. Two common semiconductors are germanium and silicon.

There is a common misconception that since some electric waves propagate at approximately the speed of light the electrons in a conductor travel with this same velocity. The actual mean velocity of free electron drift is but a few millimeters per second! (See Prob. 1-2 for a numerical example.)

Another conceptual scheme upon which our thinking is based is the *conservation of energy*. By our training in the methods of science, we immediately become suspicious of any scheme that claims to create energy. The law of conservation of energy states that energy cannot be created, nor destroyed, but that it can be converted in form. Electric energy is energy converted from some other form. There are a number of ways in which this is accomplished. Some of them are as follows:

(1) *Electromechanical energy conversion.* The rotating generator, patterned after the invention of Faraday in 1831, produces electrical energy from mechanical energy of rotation. Usually the mechanical energy is converted from thermal energy by a turbine and, in turn, the thermal energy is converted from chemical energy by burning fossil fuel or from nuclear fuel. Sometimes the conversion is from hydraulic energy by hydroelectric generation.

(2) *Electrochemical energy conversion.* Electric batteries produce electric energy by the conversion of chemical energy. A potentially important use of such batteries is in the electric car. Fuel cells are in this general classification.

(3) *Magnetohydrodynamics (MHD) energy conversion.* These devices generate electric energy from the mechanical energy of a high velocity ionized gas.

(4) *Photovoltaic energy conversion.* A class of devices are able to convert light energy directly into electric energy. The best known device of this type is the *solar cell.*

The function of each of these different sources of electric energy is the same in terms of energy and charge. In one form of battery, for example, two metallic electrodes—one of zinc and one of copper—are immersed in dilute sulfuric acid. The formation of zinc and copper ions causes negative charge to accumulate at the electrodes. Energy is supplied to the charge by the difference in the energy of ionization of zinc and copper in the chemical reaction. Once the battery circuit is closed by an external connection, as shown in Fig. 1-2, the chemical energy is expended as work for each unit of charge in transporting the charge around the external circuit. The quantity "energy per unit

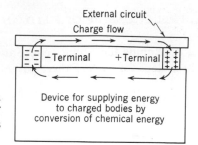

Fig. 1-2. Representation of a battery to illustrate the flow of electrons. Current direction is discussed in Section 2-1.

charge," or identically, "work per unit charge," is given the name *voltage*. In the form of an equation,

$$v = \frac{w}{q} \tag{1-2}$$

If w is the work (or energy) in joules and q is the charge in coulombs, the voltage v is in *volts* (after Alessandro Volta). The voltage of an energy source is sometimes described by the term *electromotive force*, abbreviated emf, in the electrical literature. We will avoid designating voltage as a force, because it is misleading.

If a differential amount of charge dq is given a differential increase in energy dw, the potential of the charge is increased by the amount

$$v = \frac{dw}{dq} \tag{1-3}$$

If this potential is multiplied by the current, dq/dt, as

$$\frac{dw}{dq} \times \frac{dq}{dt} = \frac{dw}{dt} = p \tag{1-4}$$

the result is seen to be a time rate of change of energy, which is *power* p. Thus power is the product of potential and current,

$$p = vi \tag{1-5}$$

Energy as a function of power is found by integrating Eq. (1-4). Thus total energy at any time t is the integral

$$w = \int_{-\infty}^{t} p \, dt \tag{1-6}$$

The change in energy from time t_1 to time t_2 may similarly be found by integrating from t_1 to t_2.[3]

---

[3] Equation (1-6) and those to follow can be written in terms of the dummy variable x as

$$w = \int_{-\infty}^{t} p(x) \, dx$$

The use of t to mean two things in one equation should not cause confusion.

1-3. THE RELATIONSHIP OF FIELD AND CIRCUIT CONCEPTS

In developing the circuit conceptual scheme, we will follow the same three steps for each of three parameters. These steps are:

(1) *The physical phenomenon.* We will discuss in a quantitative manner an electrical phenomenon which is observed by experiment. We will do this in terms of charge and energy.
(2) *Field interpretation.* We will next discuss the interpretation of the phenomenon in terms of a field quantity.
(3) *Circuit interpretation.* Finally, we will introduce a circuit parameter to relate voltage and current in place of the field relationship.

1-4. THE CAPACITANCE PARAMETER

(1) Physical phenomenon. The presence of charge on two spatially separated substances—for example, those shown in Fig. 1-3—causes an "action at a distance" in the form of a force between the

Fig. 1-3. Illustrating two spatially separated charged substances.

two substances. This phenomenon we regard as a property of nature, a basic experimental fact. Coulomb found that this force was of such nature that "like charges repel" and "unlike charges attract" and that the force varied according to the equation

$$F = \frac{q_1 q_2}{4\pi\epsilon r^2} \tag{1-7}$$

In this equation, F is the force in newtons directed from point charge to point charge, r is the separation of the point charges in meters, ϵ is the permittivity, having the free-space value of 8.854×10^{-12} farad per meter in the MKS system, and q_1 and q_2 are the charges measured in coulombs. It should be understood that this equation applies strictly to point charges only. However, the equation may be applied to any geometry of known charge distribution by vectorially adding all forces.

(2) Field interpretation. This phenomenon can be described in terms of a force on a unit charge placed between the two charged bodies. This force per unit charge, a vector quantity since force is

a vector quantity, is called an *electric field* of value

$$E = \frac{F}{q} \tag{1-8}$$

As a conceptual aid, this field may be represented by lines drawn in the direction of the force that would be exerted on the unit positive exploring charge at each point. Such lines are illustrated in Fig. 1-4. These lines are conceptual aids: They should not be thought of as actually being present. Using Eqs. (1-7) and (1-8), the electric field may be evaluated for a particular problem.

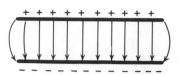

Fig. 1-4. Electric field lines or "lines of force" between two charged conductors.

(*3*) *Circuit interpretation.* The electric field E of Eq. (1-8) will exist between charged conductors of arbitrary shape. As a special case, consider the parallel-plate conductors of Fig. 1-4. Assume that the plates of area A are sufficiently large that fringing at the ends of the plates can be neglected. Let the charge on the top plate be q. The charge density on the plate will be q/A. From Gauss's law

$$q = \int_S \mathfrak{D} \cos \theta \, dS \tag{1-9}$$

where $\mathfrak{D}$ is flux density, dS is an increment of surface, and θ is the angle between $\mathfrak{D}$ and dS, we see that for the parallel plates, $q = \mathfrak{D}A$. Flux density and electric field are related by the equation $\mathfrak{D} = \epsilon E$ so that

$$E = \frac{q}{\epsilon A} \tag{1-10}$$

Now the voltage is expressed in terms of the electric field by $v_{ab} = \int_a^b E \cos \theta \, dl$. For the parallel plates, this simplifies to $v = Ed$, where d is the separation of the plates. Thus we have

$$v = \left(\frac{d}{\epsilon A}\right) q = Dq \tag{1-11}$$

where D is defined as *elastance*. The reciprocal of D is defined as the *capacitance*, C. In terms of C, Eq. (1-11) becomes

$$q = Cv \tag{1-12}$$

so that for the parallel plates

$$C = \frac{\epsilon A}{d} \tag{1-13}$$

For conductors of other shapes, C may be determined by a similar procedure. In each case, C will be a function of the geometry of the conductors and ϵ. In Eq. (1-12), if q is measured in coulombs and v in volts, then the unit of C is the *farad* (in honor of Michael Faraday), and the unit for D bears the colorful name *daraf* (farad spelled backwards). The quantity C (or the quantity D), which characterizes the system under study and permits the simple relationship between v and q to be written, is known as a *circuit parameter*, the capacitance of a system.

To reach our objective, a relationship between voltage and current in a capacitive system, there remains the task of studying the relationship of charge and current given by the equation

$$i = \frac{dq}{dt} \tag{1-14}$$

If there is an initial charge on a system, q_0, and the charge increases linearly with time, the charge at any time may be written

$$q = q_0 + kt \tag{1-15}$$

The current is found by differentiating the charge with respect to time, giving the value

$$i = \frac{dq}{dt} = k \tag{1-16}$$

Thus we see that the current in the system is independent of initial charge on that system. In going the other direction, computing charge, given the current, we integrate $dq = i\, dt$. To find the charge on the plates at time t, we integrate over all prior time.

$$q = \int_{-\infty}^{t} i\, dt \tag{1-17}$$

This equation may be written in a different form by separating the integral into two parts.

$$q = \int_{-\infty}^{0} i\, dt + \int_{0}^{t} i\, dt = q_0 + \int_{0}^{t} i\, dt \tag{1-18}$$

In arriving at this result, we note that the current for $t < 0$, or equivalent information, is ordinarily given as part of the problem, while the current for $t \geq 0$ is unknown and to be found. Actually, we are not concerned with the current in the time interval, $-\infty < t < 0$, but only in the integral of this current which is the charge accumulated on the plates of the capacitor as a result of all past current at the time $t = 0$. This constant quantity is designated q_0.

Returning once more to the relationship $q = Cv$, current and voltage are related by the equation

$$\frac{dq}{dt} = i = \frac{d}{dt}(Cv) \tag{1-19}$$

If the capacitance C does not vary with time (or with charge), then

$$i = C\frac{dv}{dt} \tag{1-20}$$

If, however, C is not constant but varies as a function of time, the current must be found from the general relationship

$$i = \frac{d}{dt}(Cv) = C\frac{dv}{dt} + v\frac{dC}{dt} \tag{1-21}$$

Similarly, starting with the equation $v = Dq$, we find that

$$v = D\int_{-\infty}^{t} i\, dt = \frac{1}{C}\int_{-\infty}^{t} i\, dt \tag{1-22}$$

Equations (1-19) and (1-22) relate the voltage and current in the capacitive system through the circuit parameter C.

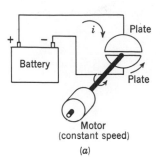

Motor
(constant speed)

(a)

EXAMPLE 1

The sketch of Fig. 1-5(a) shows two plates, one of which is driven by a constant-speed motor so that the capacitance between the two plates varies according to the equation

$$C(t) = C_0(1 - \cos \omega t) \tag{1-23}$$

If the battery potential remains constant at V volts, the current as a function of time may be found from Eq. 1-21 as

$$i = \frac{d}{dt}(Cv) = V\frac{dC}{dt} = \omega C_0 V \sin \omega t \tag{1-24}$$

This time variation of current is shown in Fig. 1-5(c).

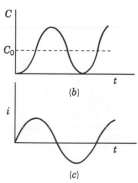

Fig. 1-5. A motor-driven capacitor produces a time-varying capacitance as shown in (b) such that the current from the battery is that shown in (c).

From the relationship $q = Cv$, we see that an instantaneous change in the product Cv implies an instantaneous change in q which, in turn, implies infinite current. In considering physical systems, we exclude the possibility of infinite current. We will revise conclusions made under this restriction in Chapter 8 when we consider the mathematical idealization of a function of extremely short duration and extremely large magnitude known as an *impulse function*. Let us now examine the consequences of our assumption of values of current of finite limit. In terms of the time interval $\Delta t = t_2 - t_1$, in which q or Cv changes a finite amount shown in Fig. 1-6, Δt cannot be zero. The instantaneous change of Cv shown as curve 1 is thus ruled out. Typical changes of Cv or q which are permitted are shown as curves 2 and 3.

Fig. 1-6. Illustrating possible time changes of the product Cv in a physical system.

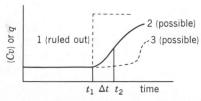

From another approach, the charge is given as

$$q(t) = q_0 + \int_0^t i \, dt \tag{1-25}$$

by Eq. (1-18). The integral portion of this equation cannot have a finite value in zero time with finite i; that is,

$$\lim_{t \to 0} \int_0^t i \, dt = 0, \qquad i \neq \infty \tag{1-26}$$

The integration process is illustrated in Fig. 1-7 as the summation of infinitesimal areas, i in height and dt in width. The interval from $t = 0$ to t_1 must be greater than zero for any area to be summed.

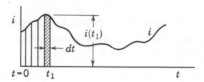

Fig. 1-7. The integration of current to give charge interpreted as the summation of infinitesimal areas.

These mathematical equations aid in visualizing the requirement that the charge in a capacitive system cannot increase or decrease in zero time. However, either capacitance or voltage can change instantaneously so long as the product of the two quantities remains constant, as

$$C_1 v_1 = C_2 v_2 \tag{1-27}$$

where the subscripts 1 and 2 refer to conditions existing at times a vanishingly small interval apart (such as before and after a switch is closed).

In most cases to be considered, the capacitance of a network does not change with time. Under this condition, the above discussion simplifies to the important conclusion that the *voltage of a capacitive system cannot change instantaneously.*

1-5. THE INDUCTANCE PARAMETER

(1) Physical phenomenon. Oersted made the important discovery in 1820 that the force between two charged substances depended on the *time rate of flow of charge* (the current). In Oersted's experiment, the needle of a compass was deflected by the presence of a current-carrying conductor, indicating that the effect was related to *magnetism.* In the same year, Ampère measured the force caused by the current and expressed the relationship in equation form. This magnetic effect is an "action at a distance" just as in the case of the force between

charged bodies. This "action at a distance" is a basic observational fact; it is not deduced from other knowledge.

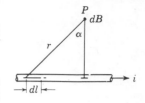

(2) *Field interpretation.* The phenomenon described above can be interpreted in terms of the force per unit magnetic pole at all points in space. Oersted discovered that this force was directed at right angles to the current-carrying conductor. In terms of the geometry of Fig. 1-8(a), Ampère described a *magnetic field density B*, the force per unit magnetic pole, of value

$$dB = \frac{\mu I \cos \alpha \, dl}{4\pi r^2} \qquad (1\text{-}28)$$

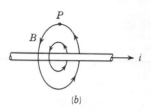

where μ is the magnetic permeability, which is a function of the medium in which the magnetic field exists, i is the current in amperes, and other quantities are defined on the figure. Figure 1-9(a) shows the cross section of a current-carrying conductor. By Eq. (1-28), the magnetic field density will be constant at a constant distance from the conductor. Continuous lines with arrows may be drawn to indicate the direction of B—as a conceptual aid. These are magnetic field density lines or "lines of force." For more complicated geometries than that shown in Fig. 1-9(a), the position of the lines can be found by integrating Eq. (1-28) or by experimentally moving a "point" magnetic pole (if one existed) from place to place in space. A magnetic compass would give an approximate measure of directions.

Fig. 1-8. Identifying quantities which determine the magnetic field at point P from Eq. 1-28.

It is sometimes convenient to replace the lines of magnetic field density by lines of *magnetic flux* defined by the integral equation

$$\phi = \int_S B \cos \theta \, dS \qquad (1\text{-}29)$$

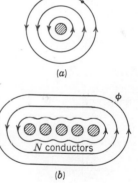

where θ is the angle between the surface of integration and the field density B. If the currents in each of N conductors, represented in Fig. 1-9(b), are in such a direction that the fluxes add, then $N\phi$ flux linkages[4] are said to exist. If, however, ϕ_1 lines of flux link N_1 conductors, ϕ_2 lines link N_2 conductors and so forth, the total number of flux linkages is found by algebraic summation as

$$\psi = \sum_{j=1}^{n} N_j \phi_j \qquad (1\text{-}30)$$

Fig. 1-9. Magnetic field and flux conventions for current directed out of the page.

Assuming that all lines link all conductors, Eq. (1-29) may be modified

[4]For a discussion of some of the problems encountered in the use (and misuse) of the concept of flux linkages, see two classics, Joseph Slepian, "Lines of force in electric and magnetic fields," *Am. J. Phys.*, **19**, 87 (1951), and L. V. Bewley, *Flux Linkages and Electromagnetic Induction* (The Macmillan Company, New York, 1952; reprinted by Dover Publications, Inc., New York, 1964).

to give flux linkages, as

$$\psi = N \int_S B \cos \theta \, dS \qquad (1\text{-}31)$$

To Faraday goes credit for the next basic experimental discovery. Faraday experimented with two conducting circuits in spatial proximity. He found that a *changing* magnetic field produced by one circuit *induced* a voltage in the other circuit. The changing magnetic field could be caused by (1) a conductor moving in space, or (2) a current changing with time.

Faraday did not envision this method of inducing voltage in terms of "action at a distance" but in terms of changes in flux linkages. A conductor moving in a magnetic field (as in the case of a generator) is thought of as "cutting flux and hence reducing the flux linkages"; the voltage induced in a stationary conductor (as in a transformer) is thought of as caused by "changing flux linkages" with time. Such pictures are valuable as conceptual aids so long as we do not attach physical significance to flux linkages which are, after all, only a means for accounting for action at a distance. Faraday's law is

$$v = k \frac{d\psi}{dt} \qquad (1\text{-}32)$$

where k is a proportionality constant. In the MKS system the units are selected to make k have unit value: When ψ is in weber-turns, t is in seconds, and $k = 1$, then v is in volts.

(3) Circuit interpretation. To derive the circuit relationship between voltage and current in the system described in (2), we begin with Faraday's law,

$$v = \frac{d\psi}{dt} \qquad (1\text{-}33)$$

or the equivalent integral form

$$\psi = \int_{-\infty}^{t} v \, dt \qquad (1\text{-}34)$$

Note, incidentally, the similarity of this expression and the one for charge in terms of current,

$$q = \int_{-\infty}^{t} i \, dt \qquad (1\text{-}35)$$

We see that ψ is to voltage as charge is to current, by comparing the two equations. Now flux linkages are related to the magnetic field by Eq. (1-31) and, in turn, the magnetic field density is related to the current by Ampère's law, Eq. (1-28). Making these substitutions, with the assumption that i can be removed from the integral,[5] we have

$$\psi = \left[N \int \left(\int \frac{\mu \cos \alpha \, dl}{4\pi r^2} \right) dS \right] i \qquad (1\text{-}36)$$

The integral term, which may be evaluated mathematically for simple geometries or may be found by measuring ψ and i, is defined as the *inductance parameter* (or the coefficient of inductance). If ψ and i refer to the same physical system, the parameter is defined as self-inductance, symbolized by the letter L as

$$\psi = Li \qquad (1\text{-}37)$$

However, if a current i_1 produces flux linkages ψ_2 in another circuit, the parameter is one of *mutual inductance*, and the letter symbol is changed to M as

$$\psi_2 = M_{21} i_1 \qquad (1\text{-}38)$$

(Again, note the similarity of these equations and the relationship $q = Cv$.) Substituting Eq. (1-37) into Faraday's law gives an equation relating voltage and current in a magnetic circuit,

$$v = \frac{d}{dt} (Li) \qquad (1\text{-}39)$$

(where M replaces L in appropriate cases). If the inductance does not vary with time, Eq. (1-39) becomes

$$v = L \frac{di}{dt} \qquad (1\text{-}40)$$

Equation (1-39) can be integrated to give

$$i = \frac{1}{L} \int_{-\infty}^{t} v \, dt \qquad (1\text{-}41)$$

The quantity $(1/L)$ is sometimes symbolized by Γ, the capital (upper-

---

[5]If the magnetically coupled system is nonlinear, containing some saturating medium, we may say that the flux linkages in circuit k is a function of the currents in all other linked circuits,

$$\psi_k = \psi_k(i_1, i_2, i_3, \ldots, i_n)$$

By Eq. (1-32), the voltage in circuit k is given by Faraday's law as

$$v_k = \frac{d\psi_k}{dt} = \frac{\partial \psi_k}{\partial i_1} \frac{di_1}{dt} + \frac{\partial \psi_k}{\partial i_2} \frac{di_2}{dt} + \cdots + \frac{\partial \psi_k}{\partial i_k} \frac{di_k}{dt} + \cdots + \frac{\partial \psi_k}{\partial i_n} \frac{di_n}{dt}$$

Each partial derivative term is evaluated with all other currents held constant. These terms may be defined as coefficients of inductance so that the voltage becomes

$$v_k = M_{k1} \frac{di_1}{dt} + M_{k2} \frac{di_2}{dt} + \cdots + L_{kk} \frac{di_k}{dt} + \cdots + M_{kn} \frac{di_n}{dt}$$

where M is used for mutual inductance and L for self-inductance. When a system is linear, this equation reduces to one which will later be written as Eq. (1-51).

case) Greek letter gamma. The henry (after the American scientist Joseph Henry) is the MKS unit for inductance.

In the case of the capacitive system, we found that charge and the product Cv could not change instantaneously. We might be led to suspect that there is a similar relationship for an inductive system in view of the analogies that have been pointed out. Indeed there is such a relationship, which may be found with the help of Eq. (1-34), in the following form:

$$\psi = \psi_0 + \int_0^t v\, dt \qquad (1\text{-}42)$$

From arguments given in the last section about capacitance, the integral in this equation has zero value for $t = 0$, assuming that v is finite. Thus, in a system altered instantaneously—say, by the closing of a switch—the flux linkages must be the same before and after the system is altered, but only for a very small interval of time. In terms of Eq. (1-42),

$$\psi = \psi_0 = \text{a constant} \qquad (1\text{-}43)$$

which is to say that the flux linkages cannot be changed instantaneously in a given system. This conclusion is described as the *principle of constant flux linkages*. If we let the subscript 1 refer to the time just before the system is altered and 2 refer to the same system after it is altered, our statements can be summarized by the equations

$$\psi_1 = \psi_2 \quad \text{or} \quad L_1 i_1 = L_2 i_2 \qquad (1\text{-}44)$$

The principle of constant flux linkages is similar to the principle of conservation of momentum in mechanics. The analogy is helpful because it is sometimes easier to visualize changes in a mechanical system than in an electric circuit. Newton's force law is

$$F = \frac{dt}{d} Mv \qquad (1\text{-}45)$$

where F is force, M is mass, and v is velocity. The product Mv is known as *momentum*; the momentum of a system cannot change instantaneously. In a system such as a missile where mass is lost as a function of time, velocity must change in such a way that momentum remains constant. We see that there are a number of analogous conservation laws:

(1) The conservation of charge:

$$q_1 = q_2 \quad \text{and} \quad C_1 v_1 = C_2 v_2, \qquad i \neq \infty$$

(2) The conservation of flux linkages:

$$\psi_1 = \psi_2 \quad \text{and} \quad L_1 i_1 = L_2 i_2, \qquad v \neq \infty$$

(3) The conservation of momentum:

$$p_1 = p_2 \quad \text{and} \quad M_1 v_1 = M_2 v_2, \qquad F \neq \infty$$

When inductance remains constant, an important specialization of the principle of constant flux linkages results. *In a fixed inductive system, the current cannot change instantaneously.*

EXAMPLE 2

In a certain inductive system, the current waveform shown in Fig. 1-10 exists. We are required to find the voltage that produces this current waveform and the associated charge, both as functions of time. We will assume that L remains constant. The relationship $v = L(di/dt)$ indicates the voltage can be found by differentiation of the current and multiplication by a constant. The result is shown in Fig. 1-11. Charge may be found by integration of the current to give the result shown in Fig. 1-12.

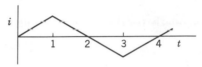

Fig. 1-10. The current variation with time considered in Example 2.

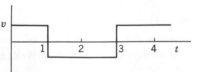

Fig. 1-11. Voltage waveform corresponding to the current of Fig. 1-10 in an inductor.

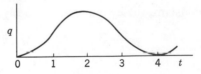

Fig. 1-12. Charge waveform corresponding to the current of Fig. 1-10.

It is important that we be able to apply the concept of inductance to several systems which are magnetically coupled. A set of three coupled coils is shown in Fig. 1-13. To simplify the system for the moment, let i_2 and i_3 be zero and consider the effect of the current i_1. The current i_1 produces ψ_1 flux linkages, found from Eq. (1-37) to be

$$\psi_1 = L_1 i_1 \qquad (i_2 = i_3 = 0) \qquad (1\text{-}46)$$

where L_1 is the self-inductance parameter (usually called just the inductance). In each of the other circuits, i_1 will produce some number of flux linkages by the proportionality of the mutual inductance parameter. For the particular system under study,

$$\psi_2 = M_{21} i_1 \quad \text{and} \quad \psi_3 = M_{31} i_1 \qquad (i_2 = i_3 = 0) \qquad (1\text{-}47)$$

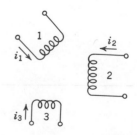

Fig. 1-13. A representation of three magnetically coupled coils.

The order of subscripts for M requires some further attention. From the two equations, it should be clear that the first subscript refers to the flux linkages and the second to the current. This particular convention is chosen to give a desired symmetry to the general equations, our next topic of study. A crutch for remembering this particular convention is that the subscripts are in the order "effect, cause," if we assume for our conceptual scheme that current produces flux.

In the general case, there will be sources or loads connected to each of the coils shown in Fig. 1-13 and no current will be zero. We will assume for the time being that the current directions and winding senses of the coils are such that all flux linkages are *additive*, postponing the more general case for Chapter 2. The total flux linkages in coil 1 will be made up of flux linkages produced by the current in coil 1 *plus* flux linkages produced by currents i_2 and i_3. In equation form,

$$\psi_1 = L_1 i_1 + M_{12} i_2 + M_{13} i_3 \tag{1-48}$$

and similarly for the other two coils,

$$\psi_2 = M_{21} i_1 + L_2 i_2 + M_{23} i_3 \tag{1-49}$$

$$\psi_3 = M_{31} i_1 + M_{32} i_2 + L_3 i_3 \tag{1-50}$$

The symmetry discussed in the preceding paragraph is now apparent. The mutual inductance coefficients have subscripts designating row and column in the above array of equations.

We are interested in flux linkages only as a stepping stone to voltage. The voltage induced in each coil is given by Faraday's law as the time rate of change of flux linkages. If the inductance parameters are constant, these voltages are readily found by differentiation to be

$$v_1 = L_1 \frac{di_1}{dt} + M_{12} \frac{di_2}{dt} + M_{13} \frac{di_3}{dt} \tag{1-51}$$

$$v_2 = M_{21} \frac{di_1}{dt} + L_2 \frac{di_2}{dt} + M_{23} \frac{di_3}{dt} \tag{1-52}$$

$$v_3 = M_{31} \frac{di_1}{dt} + M_{32} \frac{di_2}{dt} + L_3 \frac{di_3}{dt} \tag{1-53}$$

In Chapters 2 and 3, we will consider the conditions under which some terms in these equations will be negative.

1-6. THE RESISTANCE PARAMETER

(1) Physical phenomenon. The passage of electrons through a material is not accomplished without collisions of the electrons with other atomic particles. Moreover, these collisions are not elastic, and energy is lost in each collision. This loss in energy per unit charge is interpreted as a drop in potential across the material. The amount of

energy lost by the electrons is related to the physical properties of a particular substance.

(*2*) *Field interpretation.* The German physicist Georg Simon Ohm found experimentally that there is a relationship between the current in a substance and the potential drop. In terms of the field concept, the change in energy per unit of charge causes a force per unit charge—or electric field. This effect may be interpreted in terms of a field in the direction of current through the conducting substance. Ohm's experiment may be stated in terms of this field and the current per unit cross-sectional area as

$$J = \sigma E \tag{1-54}$$

where, in MKS units, J is the current density in amperes per square meter, E is the field along the conducting substance in volts per meter, and σ is the conductivity of the substance, which is a constant for each particular material.[6]

(*3*) *Circuit interpretation.* If the substance which carries the current has an idealized geometry, as that shown in Fig. 1-14(*b*), it is possible to reduce the field form of Ohm's law to relate current and voltage. If the cross section of the conductor is *uniform*, the current and current density are related by the equation

$$i = \int_S J \cos \theta \, dS = JS \tag{1-55}$$

where S is the cross-sectional area. For the same simple geometry, the electric field is uniform and directed along the length of the wire; that is,

$$v = El \tag{1-56}$$

as a special case of the more general relationship

$$v = \int_{l_1}^{l_2} E \cos \theta \, dl \tag{1-57}$$

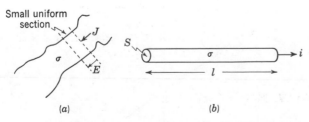

(a)   (b)

Fig. 1-14. Conductors and the identification of quantities pertaining to Ohm's law.

[6]Strictly speaking, Eq. (1-54) is a special case valid only for isotropic substances. Similarly, σ is independent of the magnitude of E only for the linear range of operation.

Substituting Eqs. (1-55) and (1-56) into the field form of Ohm's law, Eq. (1-54), gives

$$v = \left(\frac{1}{\sigma S}\right) i \tag{1-58}$$

The quantity $(l/\sigma S)$, which is a constant for constant geometry of the conductor, is given the name the *resistance parameter*—or simply the *resistance*, and is symbolized by the letter R. For geometries other than the simple one of Fig. 1-14(b), computation of the coefficient relating current and voltage for a substance will be more difficult. However, measurement of current and voltage can establish the value of the resistance parameter and bypass the computation problem. Ohm's law may be written

$$v = Ri \tag{1-59}$$

or, in terms of charge,

$$v = R\frac{dq}{dt} \tag{1-60}$$

The equation $v = Ri$ is sometimes written in the form

$$i = Gv \tag{1-61}$$

where $G = 1/R$ is known as the *conductance*. In the MKS system, the unit for resistance is the *ohm* and for conductance is the *mho*.

As well known as Ohm's law is, Ohm was ridiculed by his fellow scientists when he first announced his law in 1826, and it was some 30 years before his ideas were finally accepted. We must remember, of course, that the concepts of current and voltage were not well understood in his day, the first distinction between the two quantities having been made by Ampère in 1820. Even today, when we read newspaper

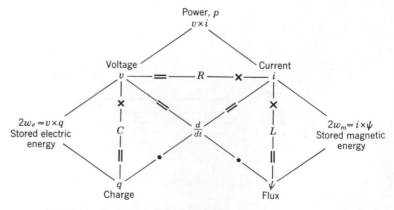

Fig. 1-15. Chart illustrating the relationships of basic quantities in terms of the circuit parameters. (From M. Kawakami, *EREM Chart*, Kyoritsu Shuppan Co., Tokyo.)

statements such as "10,000 volts passed through his body," we realize that the distinction is still not well understood by laity.

Some of the relationships discussed thus far in this chapter are summarized compactly in Fig. 1-15 and in tabular form in Table 1-1. These equations are encountered so frequently in the study of electrical engineering that they should be memorized.

Table 1-1.   SUMMARY OF RELATIONSHIPS FOR THE PARAMETERS

Parameter	Basic Relation-ship	Voltage-Current Relationships		Energy
R $G = \dfrac{1}{R}$	$v = Ri$	$v_R = Ri_R$	$i_R = Gv_R$	$w_R = \displaystyle\int_{-\infty}^{t} v_R i_R \, dt$
$L \ (\text{or } M)$	$\psi = Li$	$v_L = L\dfrac{di_L}{dt}$	$i_L = \dfrac{1}{L}\displaystyle\int_{-\infty}^{t} v_L \, dt$	$w_L = \tfrac{1}{2}Li^2$
C $D = \dfrac{1}{C}$	$q = Cv$	$v_C = \dfrac{1}{C}\displaystyle\int_{-\infty}^{t} i_C \, dt$	$i_C = C\dfrac{dv_C}{dt}$	$w_C = \tfrac{1}{2}Cv^2$

1-7. UNITS AND SCALING

The units for describing circuits introduced thus far in this chapter are summarized in Table 1-2, with prefix conventions given in Table 1-3. These units are part of a system designated by the name *International System of Units* and these are the conventional ones used in electrical engineering. In engineering applications, other units which are multiples or submultiples of these units will be used by employing the prefixes shown in Table 1-3. For example, in electronic circuits, capacitor values are usually expressed in microfarads (μF) or in picofarads (pF), inductors in millihenries (mH), etc.

For ease in computation, we will often use parameter values which are small integers or in the range of 1 to 10, rather than values which represent actual situations of engineering practice. A justification for this practice other than computational convenience will be better understood after frequency and impedance are covered. In brief, the techniques of *scaling*, discussed in Appendix C, make it possible to change practical values to the small integer range for carrying out analysis or design. We are anticipating what will later be a normal situation.

Table 1-2. SUMMARY OF SYMBOLS AND UNITS

Quantity	Symbol*	Unit	Equivalent Units	Unit Abbreviation
Charge	q	coulomb	—	C
Current	i, I	ampere	coulomb/second	amp
Flux linkages	ψ	weber-turn	—	Wb
Energy	w, W	joule	newton-meter	J
Voltage	v, V	volt	joule/coulomb	V
Power	p, P	watt	joule/second	W
Capacitance	C	farad	coulomb/volt	F
Inductance	L, M	henry	weber/ampere	H
Resistance	R	ohm	volt/ampere	Ω
Conductance	G	mho	ampere/volt	$\mho$
Time	t	second	——	sec
Frequency	f	hertz	cycles/second	Hz
Frequency	ω	radian/second	$\omega = 2\pi f$	none

*In the cases of voltage, current, energy, and power, a lower-case letter usually implies a *time-variable* quantity, an upper-case letter a *time-invariant* quantity (such as average).

Table 1-3.

Factor by Which the Unit Is Multiplied	Prefix	Symbol	Pronunciation
10^{12}	tera	T	tĕr′à
10^{9}	giga	G	jĭ′ga
10^{6}	mega	M	mĕg′à
10^{3}	kilo	k	kĭl′ō
10^{2}	hecto	h	hĕk′tō
10	deka	da	dĕk′à
10^{-1}	deci	d	dĕs′ĭ
10^{-2}	centi	c	sĕn′tĭ
10^{-3}	milli	m	mĭl′ĭ
10^{-6}	micro	μ	mĭ′krō
10^{-9}	nano	n	năn′ō
10^{-12}	pico	p	pē′cō
10^{-15}	femto	f	fĕm′tō
10^{-18}	atto	a	ăt′tō

1-8. APPROXIMATION OF A PHYSICAL SYSTEM AS A CIRCUIT

We have discussed the manner in which three electrical phenomena observed experimentally can be described in terms of circuit parameters. A problem that we must eventually face in making use of the circuit concept is that of representing a physical system in terms of these parameters. For example, can we draw a circuit that will represent an electric motor, a piezoelectric crystal, a coil of wire, an antenna, or an integrated circuit on a silicon chip?

Suppose we examine some arbitrary physical system, looking for portions of the system to be replaced by equivalent parameters. Possibly the resistive effects would be most easily recognized. A part of the system made of material of high resistivity, with small cross-sectional area and appreciable length, would be recognized as equivalent to large resistance and could easily be distinguished from another part of the system of small resistance. We have found that there is a capacitive effect between any two parts of a system. If the two parts constitute a system capable of concentration of charge, producing a high electric field—say, large area for charge storage and small distance from part to part—the capacitance of that portion of the system is large. Finally, an inductive effect is associated with every current-carrying conductor, and an effect of mutual inductance between every pair of conductors at least one of which is carrying current. If the conductors are located in space in such a way that the magnetic fields reinforce each other, then the inductance, self or mutual, of that portion of the system is large.

So much for large effects. What about smaller or secondary effects that can be recognized in much the same manner? Just how many effects must be taken into account in representing a system by equivalent parameters?

We can answer our questions only by asking another: Just how good do we expect the results to be? The accuracy of our results will be determined by how many separate electrical effects we can take into account by a parameter. We must stop somewhere. We must, at some point, make an *approximation*.

Approximation requires engineering judgment. An approximation which is valid in one case will not be in another. In many practical cases, the resistance and inductance of connecting wires are so small that they may be neglected. Likewise, in most cases of commercial capacitors, the inductive and resistive parameters may be ignored. Much less frequently the resistance and capacitance of coils can be neglected.

In the discussions to follow in other chapters, we will assume that when a schematic of a system is given, all significant parameters have been taken into account. Engineering judgment has been exercised by the individual who made up the problem. But when the student finally applies the techniques of analysis to a problem that *he* makes up, these questions associated with approximation must be answered. It is difficult to write answers to such questions in textbooks; experience is usually the best teacher.

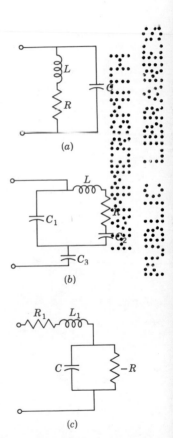

Fig. 1-16. One form of equivalent circuit or model for (*a*) an end-excited antenna, (*b*) a piezoelectric crystal, and (*c*) a tunnel diode.

Approximation is not unique to circuit analysis by any means. In solving problems by computing the electric and magnetic fields for all positions in space, there will assuredly be approximations, either in representing the physical system by mathematical equations or in solving the equations. Approximation and analysis are bound together. To ignore the problem of approximation is to lack understanding of the results of analysis.

In many cases, we do not start with an unknown system to be represented by a circuit, but instead with commercial components in combination forming a circuit. A component labeled *inductor*, however, will not behave as a pure element. It will, under some circumstances, exhibit capacitive and resistive effects. Such unwanted effects are commonly distinguished by the name *parasitic*. The decision of which effects must be taken into account involves the same engineering judgment as discussed earlier. The parasitic effects can be ignored only as long as the approximation is useful.

In all cases we have assumed that the magnetic and electric fields are isolated and that there is no interaction between the two fields. If here is such an interaction, part of the energy is lost by *radiation*.

In arriving at equations for the circuit parameters, Eqs. (1-11), (1-36), and (1-58), it was necessary to make simplifying approximations: (1) that the charge did not vary with dimensions, and (2) that the current varied with neither the length of the conductor nor the cross-sectional area. If these assumptions do not hold, the values for the parameters are different and difficult to compute.

To illustrate how current and charge might vary with space, suppose that the current is made to flow for but a brief interval of time, and that this pulsed flow is repeated at a periodic rate, a very large number of times each second. Under such conditions, the current and the charge will not be uniform throughout the system. We can imagine some portions of the system with charge and other portions without charge. This being the case, the general expressions must be used in evaluating the capacitance, inductance, and resistance parameters. These new parameter values, computed or measured, will be different from those found with uniform current and charge in the system. Must the parameters of a system be computed for every different current?

The answer to this question is, again, a practical one of engineering judgment. Certainly, there will be conditions requiring some effective value of the parameters—computed for a particular time waveform—to be used. But in many cases, the *approximation* that parameter values are equal to those found for nonvarying or static conditions gives usable results. This approximation is strictly valid only in the cases in which the variation of current and charge is slow, the so-called *quasi-stationary state*. We will assume that we are operating in this state in chapters to follow. We thus assume *constant parameters* for changing variations of current and charge.

We further assume that the parameters are constant with the variation of the *magnitude* of charge, current, or voltage. Thus, as shown in Fig. 1-17, a device having nonlinear voltage-current characteristics is represented by a model, a resistor in this case, for which voltage and current are linearly related. A system composed of such elements is said to be *linear*. We will assume that all systems to be considered (unless otherwise specified) are linear. We thereby exclude *nonlinear* elements and systems. Two such systems are illustrated by the characteristics of Fig. 1-18. The inductive system characterized by Fig. 1-18(*a*) has a hysteresis-type nonlinearity, while the capacitive system characteristic of Fig. 1-18(*b*) has a nonlinear relationship between *q* and *v*. Some nonlinear systems can be represented as linear under certain conditions. Transistors are nonlinear, but for certain analyses may be considered linear over a restricted range of operation.

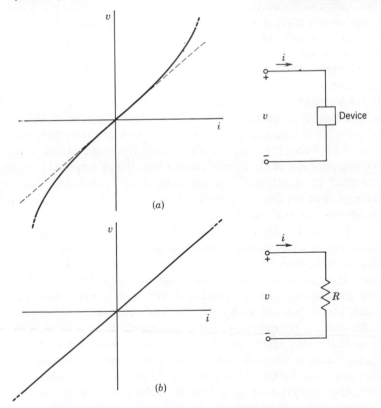

Fig. 1-17. (*a*) A device and its voltage-current relationships, and (*b*) the corresponding idealization or model which is a resistor with a linear relationship between *v* and *i*.

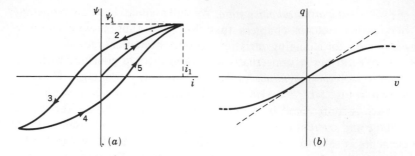

Fig. 1-18. Nonlinear characteristics: (*a*) that of an inductor with hysteresis, and (*b*) that of a capacitor. If the characteristic lines are straight and pass through the origin, then the element is said to be linear and the network parameter is a constant.

Besides the assumption of linearity we will include the requirement that all elements in a system be *bilateral*. In a bilateral system, the same relationship between current and voltage exists for current flowing in either direction. In contrast, a *unilateral* system has different laws relating current and voltage for the two possible directions of current. Examples of unilateral elements are vacuum diodes, silicon diodes, selenium rectifiers, etc.

Many electric systems are physically distributed in space. A transmission line, for example, may extend for hundreds of miles. When a source of energy is connected to the transmission line, energy is transported at nearly the velocity of light. Because of this finite velocity, all electrical effects do not take place at the same instant of time. This being the case, the restrictions discussed earlier apply in the computation of the circuit parameters. When a system is so concentrated in space that the assumption of simultaneous actions through that system is a good approximation, the system is said to be *lumped*. We will consider only lumped systems.

Our circuit approach to the approximation of a system has obscured an effect usually described in terms of the interaction of electric and magnetic fields. As an approximation, we have assumed that the magnetic field is associated only with an inductive system and that the electric field is associated only with a capacitive system. Fields cannot actually be so isolated. The consequence of interaction of the fields is the *radiation* of electromagnetic energy. Open a switch in an inductive system, and the effects will be observed as a noise in nearby radio receivers. Similarly, the ignition spark of an automobile may affect nearby television receivers. Under many conditions, however, the amount of energy lost by radiation is small, and as an approximation can be ignored. We will make this approximation.

In general, our objective is to replace physical devices or physical systems with appropriate circuit models, and then to complete the analysis of the system in terms of the overall model. The particular model we use for a device or a system will depend on many things, such as the availability of a digital computer simulation program to carry out the analysis, the refinement or accuracy needed, and the overall sensitivity of the results to the form of the model assumed.

It is important to understand that the model of a system is not unique, that the model actually used will always depend on the accuracy required. This is illustrated by Fig. 1-19: For a given physical device or system, there are many possible models, and a choice of those available must be made before analysis is made.

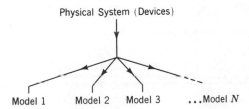

Fig. 1-19. A flowchart suggesting that a given physical device or physical system can be represented by many different models.

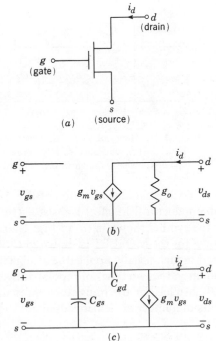

Fig. 1-20. A metal-oxide-semiconductor field effect transistor (MOSFET) with its symbol shown in (*a*) and two different models shown in (*b*) and (*c*).

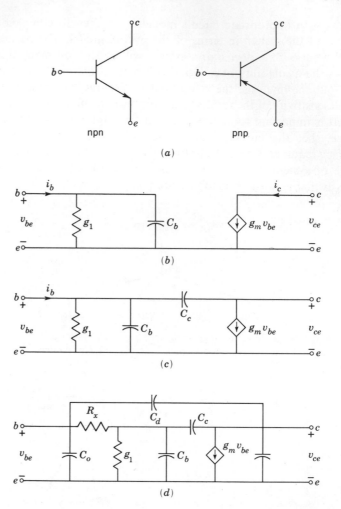

Fig. 1-21. The symbols for the *npn* and *pnp* bipolar transistors are shown in (*a*), and successively more complicated models of the bipolar transistor shown in (*b*), (*c*) and (*d*).

To illustrate the intention of Fig. 1-19, consider the metal-oxide-semiconductor field effect transistor (MOSFET) which is represented by the symbol shown in (*a*) of Fig. 1-20. A simple incremental circuit model for the MOSFET is shown in (*b*) of the figure. A more complicated model, usually applicable at higher frequencies, is shown in (*c*). Similarly, the two types of bipolar transistors are shown in (*a*) of Fig. 1-21. Models for the transistor representing increasing complexity in terms of number of elements used are shown in (*b*), (*c*), and (*d*) of Fig. 1-21.

FURTHER READING

BLACKWELL, W. A., *Mathematical Modeling of Physical Networks*, The Macmillan Company, New York, 1968. Read Chapters 1–3 for a different approach based on the linear graph.

CHIRLIAN, PAUL M., *Basic Network Theory*, McGraw-Hill Book Company, New York, 1969. Chapter 1 develops circuit concepts from a vector formulation of the field equations.

CLOSE, CHARLES M., *The Analysis of Linear Circuits*, Harcourt, Brace & World, Inc., New York, 1966. Chapter 1.

CRUZ, JOSE B., JR., AND M. E. VAN VALKENBURG, *Signals in Linear Circuits*, Houghton Mifflin Company, Boston, Mass., 1974.

DESOER, CHARLES A., AND ERNEST S. KUH, *Basic Circuit Theory*, McGraw-Hill Book Company, New York, 1969. Chapter 2.

GRAY, PAUL E., AND CAMPBELL L. SEARLE, *Electronic Principles: Physics, Models, and Circuits*, John Wiley & Sons, Inc., New York, 1969.

HAMILTON, D. J., F. A. LINDHOLM, AND A. H. MARSHAK, *Principles and Applications of Semiconductor Device Modeling*, Holt, Rinehart & Winston, Inc., New York, 1971.

LEON, BENJAMIN J., AND PAUL A. WINTZ, *Basic Linear Networks for Electrical and Electronics Engineers*, Holt, Rinehart & Winston, Inc., New York, 1970. Chapters 1 and 2 contain an especially lucid treatment of the subjects of this chapter.

LYNCH, WILLIAM A., AND JOHN G. TRUXAL, *Principles of Electronic Instrumentation*, McGraw-Hill Book Company, New York, 1962. Chapter 10 on "Determining models for physical devices" is especially recommended.

MILLMAN, JACOB, AND CHRISTOS C. HALKIAS, *Integrated Electronics: Analog and Digital Circuits and Systems*, McGraw-Hill Book Company, New York, 1972.

WEDLOCK, BRUCE D., AND JAMES K. ROBERGE, *Electronic Components and Measurements*, Prentice-Hall, Inc., Englewood Cliffs, N.J., 1969.

WING, OMAR, *Circuit Theory with Computer Methods*, Holt, Rinehart & Winston, Inc., New York, 1972. See Chapter 2.

DIGITAL COMPUTER EXERCISES

As related to the study of capacitors and inductors in this chapter, see exercises in numerical integration in Appendix E-2, especially Chapter 2 of reference 7 in Appendix E-10 by Huelsman.

PROBLEMS

1-1. A solid copper sphere 10 cm in diameter is deprived of 10^{13} electrons by a charging scheme. (a) What is the charge of the sphere in coulombs? What is the sign of the charge? (b) What is the percentage decrease in the total number of electrons in the sphere?

1-2. In a certain copper conductor, the current density is 250 amp/cm^2, and there are 5×10^{22} free electrons in 1 cm^3 of the copper. What is the mean electron drift in centimeters per second?

1-3. The current in a circuit varies according to the equation $i = 2e^{-t}$ amperes for t greater than zero, and is zero for t less than zero. Find the total charge that passes through the circuit in coulombs.

1-4. The system shown in Fig. 1-5(a) is constructed so that the capacitance of the plates has the time variation shown in Fig. P1-4. Sketch the

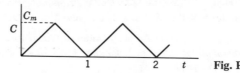

Fig. P1-4.

waveform of the variation of current in the circuit with time. On your sketch, superimpose the waveform of capacitance variation to show the relationship of the two variables.

1-5. For the system considered in Prob. 1-4, shown in Fig. 1-5(a), suppose both the voltage and the capacitance vary with time according to the equations $v = V_0 \sin \omega t$ and $C = C_0(1 - \cos \omega t)$. What is the equation for the current in the circuit under these circumstances? Sketch this current as a function of time.

1-6. A simple capacitor is constructed from two parallel plates of metal separated by a dielectric material. Assuming no fringing of the electric field at the edges of the plates, show that the capacitance of this capacitor is

$$C = \frac{\epsilon A}{d} \text{F}$$

where A is the area of either plate in meters^2, d is the distance of separation of the plates in meters, and

$$\epsilon = K\epsilon_0 = 8.854 \times 10^{-12}K$$

where K is the relative dielectric constant (which is 1 for air).

1-7. Consider a capacitor of the type described in Prob. 1-6. Upon measurement, it is found that the capacitance per inch^2 is 5×10^{-12} F. What will be the capacitance of a sheet of the dielectric separated plates of area 10 ft^2?

1-8. Two strips of tinfoil are separated by a strip of mica (for which $K = 5$ in Prob. 1-6) by a distance of 0.01 mm. A second strip of mica is placed over one of the sheets of tinfoil and the combination is wound into a roll such that in the roll strips of tinfoil and mica alternate. If the strips are 10 cm wide, what must be the total length of the strip in order that the total capacitance be 1 μF?

1-9. The tuning capacitor that is located in radio receivers is represented in Fig. P1-9. The plates are separated by air by a distance d; the

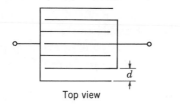

Top view   Side view

Fig. P1-9.

movable plates and the fixed plates are each connected together as shown by the figure. Neglecting fringing at the edges, determine the maximum capacitance of the tuning capacitor.

1-10. From the defining equation for energy, $w = \int_{-\infty}^{t} vi\, dt$, show that, for the inductance, $w_L = \frac{1}{2}Li^2$ and $w_L = \frac{1}{2}\psi^2/L$.

1-11. From the equation for energy in Prob. 1-10, show that for the capacitance, $w_C = \frac{1}{2}Cv^2$ and $w_C = \frac{1}{2}Dq^2$.

1-12. Assume that the inductance parameter is defined as the constant relating stored energy and the current squared by the equation $w_L = \frac{1}{2}Li^2$. Making use of the relationship $p = vi$, show that for constant inductance the voltage across the inductor is $v_L = L(di/dt)$.

1-13. Carry out a similar derivation to the one suggested in Prob. 1-12 starting with energy for a capacitive system, $w_C = \frac{1}{2}Dq^2$, to show that for constant D,

$$v_C = Dq = D \int_{-\infty}^{t} i\, dt$$

1-14. A number of devices and systems make use of banks of capacitors, such as power transmission systems, nuclear particle accelerators, and electronic flash units used in photography. Capacitors for these devices are designed using a dielectric material selected to prevent breakdown at the operating voltage. For a given dielectric constant, show that the energy stored is directly proportional to the volume of capacitors. From this, show that for a given dielectric material the cost per joule storage of capacitors is approximately constant and independent of voltage rating or capacitance value per individual capacitor.

1-15. The voltage and charge in a certain nonlinear capacitor are related by the equation $q = Kv^{1/3}$. For this capacitor, compute the energy stored as a function of capacitor charge.

1-16. The voltage and charge of a nonlinear capacitor are related by the equation

$$v(q) = \frac{S_0}{a} \sinh(aq)$$

where S_0 and a are constants. (a) Sketch v as a function of q. (b) Compute the energy stored as a function of q.

1-17. A 12-V car battery is connected to a 1-μF capacitor. Compute the energy which will be stored in the capacitor.

1-18. A capacitor of capacitance 1-μF is charged to 200 V. If the stored energy is used with 100 per cent efficiency to lift a 100-lb boy, through what distance will he be lifted?

1-19. (a) A voltage waveform identical to the current waveform of Fig. 1-10 is applied to a capacitive system. Sketch the waveforms for current and charge. (b) The current waveform shown in Fig. 1-10 is applied to a capacitive system. Sketch the waveforms for voltage and charge.

1-20. A current of 1 amp is supplied by a source to an inductor of value $\frac{1}{2}$ H. Compute the energy stored in the inductor. What happens to this energy if the source is replaced by a short circuit?

1-21. The current in a 1-H inductor follows the variation shown in the accompanying figure. The current increases from 0 at $t = 0$ at the rate of 1 amp/sec (for several seconds, at least). Find: (a) the flux linkages in the system after 1 sec, (b) the time rate of change of flux linkages in the system after 2 sec, and (c) the quantity of charge having passed through the inductor after 1 sec.

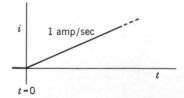

Fig. P1-21.

1-22. A time-variable inductor changes with time as shown in Fig. P1-4 with C replaced by L in henrys and C_m replaced by L_m. This inductor is connected to a current source of constant value I_0 amperes. Determine $v_L(t)$, the voltage across the inductor.

1-23. The variation of a time-variable inductor is shown in Fig. P1-32 where $x = L(t)$ rather than i_C. This inductor is connected to a constant current source of value I_0 amperes. Determine $v_L(t)$, the voltage across the inductor.

1-24. In the circuit shown the switch K is closed at $t = 0$ (the reference time). The current flowing in the circuit is given by the equation $i(t) = (1 - e^{-t})$ amp, $t > 0$. At a certain time the current has a value of 0.63 amp. (a) At what rate is the current changing? (b) What is the value of the total flux linkages? (c) What is the rate of change of flux

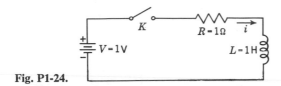

Fig. P1-24.

linkages? (d) What is the voltage across the inductor? (e) How much energy is stored in the magnetic field of the inductor? (f) What is the voltage across the resistor? (g) At what rate is energy being stored in the magnetic field of the inductor? (h) At what rate is energy being dissipated as heat? (i) At what rate is the battery supplying energy?

1-25. In the circuit shown the capacitor is charged to a voltage of 1 V, and at $t = 0$ the switch K is closed. The current in the circuit is known to be of the form $i(t) = e^{-t}$ amp, $t > 0$. At a certain time the current has a value of 0.37 amp. (a) At what rate is the voltage across the capacitor changing? (b) What is the value of the charge on the capacitor?

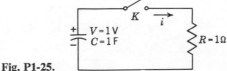

Fig. P1-25.

(c) What is the time rate of change of the product Cv? (d) What is the voltage across the capacitor? (e) How much energy is stored in the electric field of the capacitor? (f) What is the voltage across the resistor? (g) At what rate is energy being taken from the electric field of the capacitor? (h) At what rate is energy being dissipated as heat?

1-26. Show that the following quantities all have the dimension of time: (a) RC; (b) L/R; (c) $\sqrt{LC}$. Show that (d) R^2C has the dimension of inductance, (e) $\sqrt{L/C}$ has the dimension of resistance, (f) L/R^2 has the dimension of capacitance.

1-27–1-38. The following set of problems refers to the elements and the waveforms shown in the accompanying figure. For each part of this problem, sketch the required quantity, carefully making the time scale, significant amplitudes, slopes, and so on. Give enough detail to permit the waveform to be constructed from the data alone.

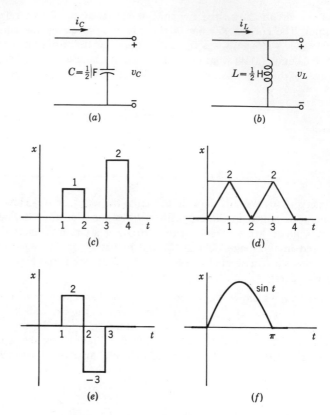

Fig. P1-27 to 38. (a, b, c, d, e, f)

	Network of:	Given that x is:	Shown in:	Sketch:	Initial Condition:
1-27.	a	v_C	d	i_C	none
1-28.	a	v_C	f	i_C	none
1-29.	a	i_C	c	v_C	$v_C(0) = 0$
1-30.	a	i_C	d	v_C	$v_C(0) = 0$
1-31.	a	i_C	e	v_C	$v_C(0) = 0$
1-32.	a	i_C	f	v_C	$v_C(0) = 0$
1-33.	b	v_L	c	i_L	$i_L(0) = 0$
1-34.	b	v_L	d	i_L	$i_L(0) = 0$
1-35.	b	v_L	e	i_L	$i_L(0) = 0$
1-36.	b	v_L	f	i_L	$i_L(0) = 0$
1-37.	b	i_L	d	v_L	none
1-38.	b	i_L	f	v_L	none

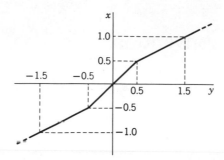

Fig. P1-39.

1-39. The figure shows a piecewise linear characteristic. Let $x = q_C$ and $y = v_C$ so that the characteristic represents a nonlinear capacitor. If the voltage applied to the capacitor is that shown in Fig. P1-32, plot the corresponding $i_C(t)$.

1-40. Repeat Prob. 1-39 if the voltage waveform is that shown in Fig. P1-30.

1-41. The piecewise linear characteristic shown as Fig. P1-39 represents a nonlinear inductor with $x = \psi_L$ and $y = i_L$. If the current in the inductor is given in Fig. P1-32, plot the corresponding $u_L(t)$.

1-42. Repeat Prob. 1-41 if the inductor current waveform is that shown in Fig. P1-30.

2 Conventions for Describing Networks

2-1. REFERENCE DIRECTIONS FOR CURRENT AND VOLTAGE

In Section 1-2, we discussed the battery as a source of energy and described the direction of electron flow from and into the terminals of the battery. In terms of this battery, the direction of current is opposite to that for charge—or out of the positive terminal and into the negative terminal. This particular convention follows a decision made by Benjamin Franklin in 1752. Franklin's choice was made before electricity was identified with the electron, before the electron or the nature of charge were known. Actually, electrons flow from the negative terminal to the positive terminal, which is in the opposite direction to that established by Franklin. To distinguish the two conventions, the flow of electrons is termed *electron current* and current assumed positive in the direction of Franklin's convention is called *conventional current* (or simply *current*, since this is the current we will use).

In discussing direction of current, as we did in the last paragraph, it is first necessary to establish a *reference direction*. For our example, we did this in terms of the terminals of the battery. For more complicated systems, the reference direction for current is conveniently indicated by the direction of an arrow. We may then describe current direction as in the reference direction or in a direction opposite to the reference direction.

Consider next a source of energy having the property that the

34

terminal voltage changes polarity as a function of time. In order that we may describe the voltage from such a source—in the form of an equation or a table of values—we need a voltage reference scheme. The scheme we will use is identical to that used for the battery in the previous discussion. One terminal of the source is marked plus, the other minus. (This is overspecification, of course; one mark may be omitted.) When the polarity of the source coincides with the reference marks, the voltage is described by a positive number. When the polarity is opposite to the reference marks, the voltage is designated by a negative number. Conversely, a positive value of voltage implies that the polarity is that of the reference; a negative value implies polarity opposite to the reference.

Some books and journals that you will study use an arrow in place of the plus and minus signs with the head of the arrow corresponding to the plus sign and the tail to the negative sign.

The same reference scheme applies to the voltage of the passive elements introduced in Chapter 1. These passive elements have voltage-current relationships which are written in terms of the reference directions shown in Fig. 2-1. Thus the equation

$$v_R(t) = Ri_R(t) \tag{2-1}$$

Fig. 2-1. Reference directions for voltage and current for all passive elements.

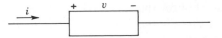

implies the reference scheme for voltage and current shown in Fig. 2-2(a). Similarly, the equations[1]

$$v_L(t) = L\frac{di_L(t)}{dt} \tag{2-2}$$

and

$$v_C(t) = \frac{1}{C} \int_{-\infty}^{t} i_C(t)\, dt \tag{2-3}$$

imply the reference schemes for voltage and current of Figs. 2-2(b) and (c).

In making use of the reference schemes just described, we arbitrarily select a reference direction for the current and then assign the corresponding voltage signs, as indicated by Fig. 2-1. For example, a network completely marked with reference directions is shown in Fig. 2-3. If analysis of the network for a specified source shows that v_2 is negative at some instant of time, then we know that the actual polarity of the voltage v_2 is opposite to that indicated by the reference signs for that particular instant of time.

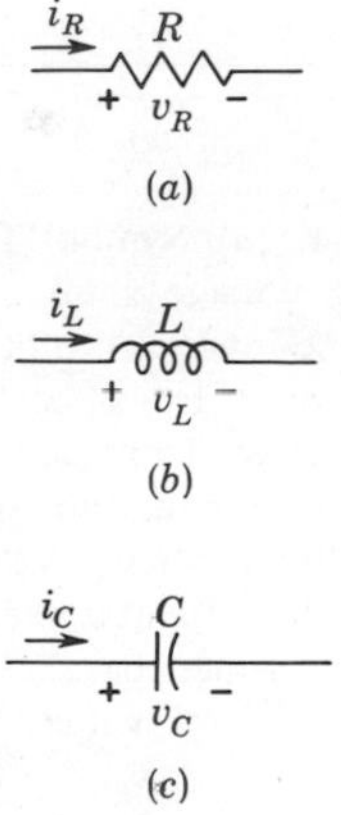

Fig. 2-2. The reference directions of Fig. 2-1 applied to (a) the resistor, (b) the inductor, and (c) the capacitor.

---

[1] As in Chapter 1, we are using t as both the variable and the limit, rather than

$$v_C(t) = \frac{1}{C} \int_{-\infty}^{t} i_C(x)\, dx$$

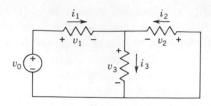

Fig. 2-3. Network with reference directions assigned for all branch voltages and currents.

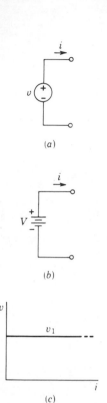

Fig. 2-4. (a) Symbol for time-varying voltage source, and (b) for a time-invariant voltage source (represented by a battery). (c) The voltage from any voltage source is not a function of current from the source. Voltage v_1 is the value of v at a fixed instant of time. Lines parallel to that marked v_1 are values at other times.

2-2. ACTIVE ELEMENT CONVENTIONS

Several kinds of sources of electric energy were described in Section 1-4. Models or idealizations of such sources are called *active elements* which are further classified by their voltage-current characteristics. Two basic models are the *voltage source* and the *current source.*

The voltage source. The voltage source is assumed to deliver energy with a specified terminal voltage, $v(t)$, which is independent of the current from the source. The symbols and reference conventions for this source are shown in Fig. 2-4, together with the voltage-current characteristic. When the voltage generated does not vary with time, the symbol for the battery shown in Fig. 2-4(b) is used; for all other cases we employ the symbol of Fig. 2-4(a). The source is said to be *idle* if the output terminals are open such that $i(t) = 0$. Observe also that when $v(t) = 0$ with the source turned off, the source is equivalent to a short circuit.

Many generators such as those used by utility companies may be represented by the model shown in Fig. 2-5(a), consisting of a source in series with a resistor. For this combination the terminal voltage does vary with the magnitude of the output current, as shown in Fig. 2-5(b). For some calculations, such as system stability, it is necessary to include inductive and sometimes capacitive effects in the models of actual generating devices.

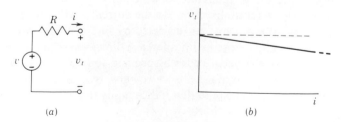

Fig. 2-5. One model for a voltage source in which R represents source resistance. For the model of (a), the terminal voltage depends on source current as shown in (b) where $v_t = v - Ri$.

The current source. The current source is assumed to deliver energy with a specified current through the terminals, $i(t)$. The symbol and reference conventions for the current source differ from those for the voltage source and are shown in Fig. 2-6. Also shown in the figure is the voltage-current characteristic of the current source. A current source is said to be *idle* when the output terminals are shorted together such that $v(t) = 0$. When the source is not idle, it delivers the same current independent of the network to which it is connected, including no network at all. Observe that when the source is turned off, so that $i(t) = 0$, it is equivalent to an open circuit.

Certain devices may be represented by the model shown in Fig. 2-7, consisting of a source in parallel with a resistor. The corresponding variation of terminal current with voltage at the terminals is shown in (b) of the figure. Such devices as transistors, vacuum tubes, and photo-electric cells make use of current sources in their model representations.

The reader will observe that the descriptions for the voltage and current sources are *duals* in the sense that the roles for current and voltage are interchanged in the two sources. As a result, other dual quantities will be recognized, such as open circuit and short circuit, series and parallel, etc.

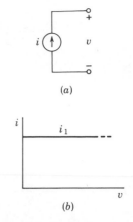

Fig. 2-6. (a) Symbol for the current source for which i does not depend on v as shown in (b). Other lines may be drawn parallel to that shown for a specific current, i_1.

The controlled source (or dependent source). The voltage and current sources we have just described are models for which voltage and current respectively are fixed and so are not adjustable. In a controlled source, the source voltage or current (depending on the type of source) is not fixed, but is dependent on a voltage or current at some other location in the network. The controlled voltage source, for example, may be visualized as a voltage source for which the magnitude of the voltage of the source is determined by the setting of a dial which, in turn, is controlled by some measured voltage or current. One kind of controlled source is shown in Fig. 2-8. Here the control variable is the voltage v_1 which controls the source voltage μv_1, where μ is a constant. For such a controlled source, the voltage-current characteristics become a set of curves, as shown in Fig. 2-8(b).

The controlled source of Fig. 2-8 is different from the other two sources we have described in that it is a three-terminal model. The three terminals are paired, with one common terminal, and one pair is described as the input, the other pair as the output. Observe also that the controlled source is *unilateral* in the sense that the input variable v_1 in Fig. 2-8(a) controls the output v_2, but that conditions at the output, such as the magnitude of the current i_2, have no influence on the input.

The controlled source of Fig. 2-8 is only one of four kinds, depending on whether the control variable is voltage or current, and

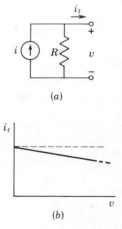

Fig. 2-7. A model for a current source in which R represents shunt resistance. For the model of (a), the terminal current in (b) is given by $i_t = i - (1/R)v$.

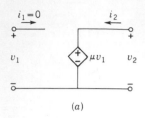

(a)

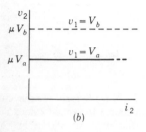

(b)

Fig. 2-8. The voltage-controlled voltage source for which v_2 depends on v_1 but not on i_2.

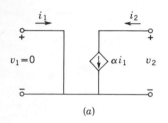

(a)

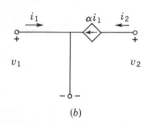

(b)

Fig. 2-9. A current-controlled current source with two identical representations.

the source controlled is a voltage source or a current source. Another combination is shown in Fig. 2-9(a) for which the control variable is the current i_1 and the controlled source is a current source. An equivalent and simplified representation of this model is shown in Fig. 2-9(b). The two remaining forms of controlled sources are shown in Fig. 2-10. Controlled sources are used in many models of devices such as the transistor and the vacuum tube.

2-3. THE DOT CONVENTION FOR COUPLED CIRCUITS

When the magnetic field produced by a changing current in one coil induces a voltage in other coils, the coils are said to be *coupled*, and the windings constitute a *transformer*. If the details of transformer construction are known, then for a current changing in one coil it is possible to compute the magnitude and direction of the voltages induced in all other windings. The necessity for cumbersome blueprints showing construction is eliminated by two characterizing factors. The value of the coefficient of mutual inductance M (discussed in Chapter 1) is equivalent to details of construction in computing *magnitude* of induced voltage. Most manufacturers mark one end of each transformer winding with a dot (or some such symbol). The dot is equivalent to details of construction as far as *voltage direction* is concerned. In this section, we will discuss the meaning of dot markings, how they are experimentally established, and their significance in network analysis.

Two windings are shown on a magnetic core in Fig. 2-11. In this figure, the winding sense is indicated for two windings, winding 1-1 (which might be called the *primary* winding) and winding 2-2 (the *secondary* winding). A time-varying source of voltage $v_g(t)$ is connected to winding 1-1 in series with resistor R_1. At a given instant, the voltage source has the polarity shown, and the current $i_1(t)$ is in the direction shown by the arrow and is increasing with time. The $+$ end of the winding is shown marked with a *dot*. Let us assume that the current flows into this dot. We will outline, step-by-step, our conceptual scheme of what happens as a consequence of this current.

(1) Current in winding 1-1 causes a magnetic field ("action at a distance") to exist, which is concentrated along the axis of the coil. The magnitude of the field can be computed from Ampère's law

$$dB = \frac{\mu i_1 dl \cos \alpha}{4\pi r^2} \tag{2-4}$$

(These symbols are defined in Chapter 1.)

(2) There is a magnetic flux ϕ associated with the magnetic field having a value

$$\phi = \int_S B \cos \theta \, dS \qquad (2-5)$$

and having a direction determined experimentally and given by the right-hand rule: If fingers wrap around the coil with the fingers pointed in the direction of the current, the thumb of the right hand indicates the direction of flux inside the coil. This flux is assumed to be confined to the magnetic core, which has the property of being a preferred path for the flux. Applying the right-hand rule, the flux is seen to have the direction indicated by the arrow (clockwise).

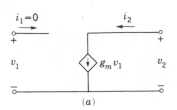

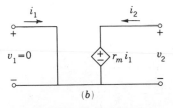

Fig. 2-10. The figure shows the voltage-controlled source and the current-controlled voltage source. These together with those shown in Figs. 2-8 and 2-9 constitute the four basic types of controlled sources.

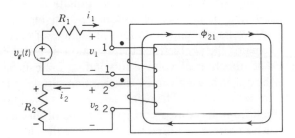

Fig. 2-11. A two-winding transformer used to describe the dot convention.

(3) Since winding 2-2 is on the same magnetic core as winding 1-1, the flux produced in winding 1-1 *links* winding 2-2. This linking flux can be described as ϕ_{21}, where the subscripts have the order "effect, cause." The number of flux linkages in winding 1-1 is

$$\psi_1 = N_1 \phi_{21} \qquad (2-6)$$

In terms of Faraday's law, ψ_1 can be computed from the voltage at terminals 1-1 as

$$\psi_1 = \int_{-\infty}^{t} v_1 \, dt \qquad (2-7)$$

Combining Eqs. (2-6) and (2-7) gives the value of flux in terms of the voltage v_1.

$$\phi_{21} = \frac{1}{N_1} \int_{-\infty}^{t} v_1 \, dt \qquad (2-8)$$

(4) Because ϕ_{21} is changing with time, a voltage is induced in winding 2-2 according to Faraday's law. The flux linkages in winding 2-2 are

$$\psi_2 = N_2 \phi_{21} \qquad (2-9)$$

and v_2 has the magnitude

$$v_2 = \frac{d\psi_2}{dt} \qquad (2\text{-}10)$$

In the discussion of Chapter 1, the coefficient of mutual inductance was introduced to relate flux linkages with current as $\psi = Mi$. For the system under study

$$\psi_2 = M_{21}i_1 \qquad (2\text{-}11)$$

and Eq. (2-10) may be written in equivalent (but more useful) form as

$$v_2 = M_{21}\frac{di_1}{dt} \qquad (2\text{-}12)$$

if M_{21} does not vary with time. Equation (2-12) tells us that a voltage is induced in winding 2-2 having a magnitude of M_{21} volts per unit time rate of change of current i_1. There remains the problem of the direction of this voltage.

(5) The direction of voltage in winding 2-2 can be found with the aid of a law given by the German physicist Lenz in 1834. In terms of the transformer, *Lenz's law* states that the voltage induced in a coil by a change of flux tends to establish a current in a direction to oppose the change in flux that produced the voltage. The flux ϕ_{21} is directed upward in Fig. 2-11 and is increasing. To produce a flux ϕ_{12} to oppose this increase in ϕ_{21} requires (by the right-hand rule) that the current flow in the direction shown by the arrow (right to left). Lenz's law is really an application of conservation of energy, since if i_2 produced a flux to aid ϕ_{21}, another increasing current would be induced in 1-1 and so on in a vicious cycle to produce infinite current.

Now that the direction of current in winding 2-2 is established, the top end of the winding is seen to be positive and so is marked with a dot. With a time-varying voltage, the dotted terminals are positive at the same time (and, of course, negative at the same time). This action is illustrated in Fig. 2-12. As shown, $v_g(t)$ increases from zero to a constant value at time t_1. The current i_1 and so flux ϕ_{21} increase with time as shown in (*b*). Note, incidentally, that Eq. (2-8) does not apply directly, since it gives ϕ_{21} in terms of v_1 rather than v_g. The induced voltage v_2 is proportional to the time rate of change of i_1 and so has the time variation shown in (*c*). This example suggests a simple experimental method for establishing the dotted ends of transformer windings. On the winding selected as 1-1, arbitrarily mark one end of the winding with a dot and to this terminal connect the positive terminal of a battery, connecting the negative terminal to the remaining end of

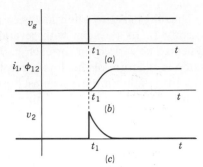

Fig. 2-12. Voltage and current waveforms in the magnetically coupled network of Fig. 2-10.

the winding. The end of winding 2-2 that momentarily goes positive, as measured with a voltmeter, is the terminal to be dotted in winding 2-2.

Of what value are the dots, which we can now establish, in network analysis? Figure 2-13 shows the transformer of Fig. 2-11, including dots, with the generator and resistor load interchanged. The positive terminal of the voltage source is connected to the dotted end of winding 2-2. A step-by-step analysis of this transformer will show that an increasing current flowing into the dotted terminal of winding 2-2 causes the upper end of winding 1-1 to be positive and so to be the dotted terminal. We would expect, after all, that dots established from 1-1 to 2-2 should agree with those established from 2-2 to 1-1.

Now suppose that the voltage source of Fig. 2-13 has reverse polarity to that shown and that an increasing current flows *out of the dot*. Another step-by-step analysis or simply intuitive reasoning will show that the dotted terminal of winding 1-1 becomes negative under such conditions.

We conclude that, for a transformer with polarity markings (dots), increasing current *into* the dotted terminal on one winding induces a voltage in the second winding which is positive at the dotted terminal; conversely, increasing current *out* of a dotted terminal induces a voltage in the second winding which is positive at the undotted terminal. This important rule will be applied in Chapter 3 in formulating circuit equations.

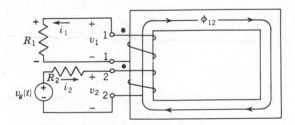

Fig. 2-13. The network of Fig. 2-10 with the voltage source moved from winding 1 to winding 2.

Thus far our discussion has been limited to a transformer with two windings. In a system with several windings, the same type of analysis can be carried on for each pair of windings, provided some variation in the form of the dots is employed (such as ● ■ ▲ ◆) to identify the relationship between each pair of windings. In Chapter 3, it will be shown that the information given by the pair of dots can be given in the sign of the coefficient of mutual inductance. For a system with many windings, this scheme avoids the confusion of a large number of similar dots. Both schemes have advantages for particular problems, and both are used in electrical engineering literature.

In the system shown in Fig. 2-14, for example, the winding sense of each coil of the transformer is indicated. The polarity markings for each set of coils are shown on the figure. In each case, one of the dots for each winding-pair was arbitrarily selected and the position of the other dot was then determined.

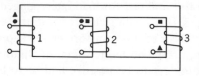

Fig. 2-14. Three windings on a magnetic core with dots of different shapes needed to describe the network.

2-4. TOPOLOGICAL DESCRIPTION OF NETWORKS

To construct the *graph* corresponding to a given schematic of a network, we replace all elements[2] of the network with lines, constructing a skeleton of the network. An example of the construction of a graph is shown in Fig. 2-15(*a*) and (*b*). If we also indicate a reference direction by an arrow for each of the lines of the graph, then it is known as an *oriented graph*, as shown in (*c*) of the figure. The lines in the

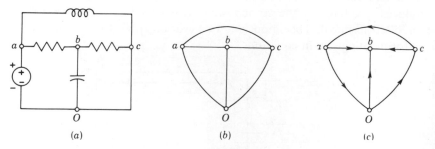

Fig. 2-15. (*a*) A network and (*b*) its graph. (*c*) An oriented graph.

[2]In the next chapter we will describe a procedure in which branches that contain only active elements are excluded from the graph to simplify analysis and computation.

graph are identified as *branches*. The junction of two or more branches is known as a *node* (and also as a *vertex*). Thus graphs are composed of nodes and branches, or sometimes oriented branches.

We make use of the graph in describing the *topological* properties of networks. Topology deals with properties of networks which are unaffected when we stretch, twist, or otherwise distort the size and shape of the network—for example, by shortening or lengthening the connection wires between elements. Figure 2-16 shows three graphs.[3] Although these graphs appear to be different, they are actually identical topologically in that the relationship of the branches and nodes is identical. Topological properties of networks are well understood by repairmen who compare the schematic diagram for a radio or television set with the actual wiring in the chassis. The two do not look the same by any means, but they must be topologically identical if the set is to operate properly.

Some topological structures occur so frequently in electrical engineering that they are given special names. Several of these are shown in Fig. 2-17, and the names given are: (*a*) T-network, (*b*) π-network, (*c*) ladder network, (*d*) bridged-T network, (*e*) bridge network (so-called because it is employed in making measurements by the Wheatstone bridge), (*f*) the lattice network. Notice that when the lattice has an element connected to the right-hand pair of terminals, it is equivalent topologically to the bridge network of (*e*). Structures (*a*) through (*d*) are said to be *grounded* or *unbalanced*.

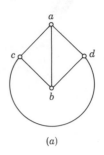

(*a*)

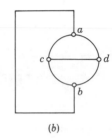

(*b*)

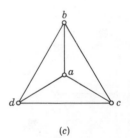

(*c*)

Fig. 2-16. Three topologically-equivalent graphs.

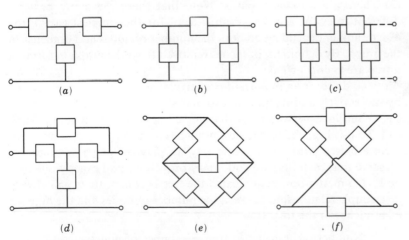

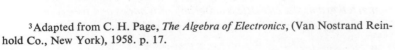

Fig. 2-17. The topological structures shown are known as (*a*) the *T*-network, (*b*) π-network, (*c*) ladder network, (*d*) bridged-*T* network, (*e*) bridge network, and (*f*) the lattice network.

[3]Adapted from C. H. Page, *The Algebra of Electronics*, (Van Nostrand Reinhold Co., New York), 1958. p. 17.

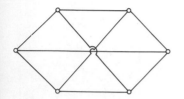

Fig. 2-18. A nonplanar graph.

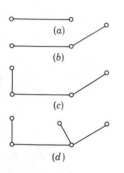

Fig. 2-19. A sequence of trees generated by adding one branch to the previous tree, illustrating the relationship between the number of nodes and branches in a tree.

Topology as a branch of geometry dates back at least to Euler who, in 1735, made use of topology in solving the famous Königsberg bridge problem. It was first applied to the study of electric networks by Kirchhoff in 1847, at about the same time that the first systematic treatise on the subject was published in German by Listing. Today topology is an important branch of mathematics and has applications in many fields of science and engineering.

Several topological quantities of importance in network analysis will be outlined in this section for application in later chapters. The graph of a network may have more than one separate *part* in the case of networks with magnetic coupling. Although the graphs shown in Fig. 2-16 are *planar* in that they may be drawn on a sheet of paper without crossing lines, there are graphs that are *nonplanar*, an example of which is shown in Fig. 2-18.

In our discussion in the next chapter, we will be interested in *node pairs* and *loops*. A node pair is simply two nodes which we identify for specifying a voltage variable. A loop (or *mesh*) is a closed path in a graph (or a network) formed by a number of connected branches.

A *subgraph* of a given graph is formed by removing branches from the original graph. A subgraph of importance in our study is the *tree* (a term used at least since the time of Kirchhoff). A tree of a connected graph (one part) of n nodes has the following properties: (1) It contains all of the nodes of the graph; nodes are not left in isolated positions. (2) It contains $n - 1$ branches, as we shall see soon. (3) There are no closed paths. Note that there are many possible different trees for a given graph (except for the simplest cases), the exact number depending on both the number of nodes and branches in the graph. By definition, branches removed from the graph in forming a tree are *chords* or *links*.

To discover an important property of a tree, let us construct one by successively adding branches so that at each step we have a tree. As shown in Fig. 2-19, we start with a single branch having a node at each end. To this branch, we add new branches in such a way that we never form a closed path, in the same way we play the game of dominos. Observe that each time we add a new branch we add exactly one new node. No matter how complicated the tree becomes, there will always be a simple relationship between the number of nodes and the number of branches in the tree. It is

$$\text{Number of branches in tree} = \text{number of nodes} - 1 \qquad (2\text{-}13)$$

Now the number of nodes in a tree is exactly the same as the number of nodes in the corresponding graph. Hence we may say that *a tree of a connected graph is a circuitless subgraph of* n *nodes and* n − 1 *branches*. Two examples of trees for a given graph are shown in Fig. 2-20.

We will also be concerned with the number of chords in a graph. This number is the difference between the total number of branches in the graph and the number of branches in a tree given by Eq. 2-13. For a graph with b branches and n nodes, this number is

$$\text{Number of chords} = b - (n - 1) = b - n + 1 \qquad (2\text{-}14)$$

For a graph with p separate parts, p replaces 1 in Eqs. (2-13) and (2-14).

Kirchhoff made use of the concept of trees in his 1847 paper in describing the proper choice of current variables for the analysis of a network. We will describe his procedure in the next chapter.

FURTHER READING

CHAN, SHU-PARK, *Introductory Topological Analysis of Electrical Networks*, Holt, Rinehart & Winston, Inc., New York, 1969. Chapter 1 amplifies the concepts introduced here.

CHIRLIAN, PAUL M., *Basic Network Theory*, McGraw-Hill Book Company, New York, 1969. Chapter 2.

CRUZ, JOSE B., JR., AND M. E. VAN VALKENBURG, *Signals in Linear Circuits*, Houghton Mifflin Company, Boston, Mass., 1974.

DESOER, CHARLES A. AND ERNEST S. KUH, *Basic Circuit Theory*, McGraw-Hill Book Company, New York, 1969. Chapters 2 and 3.

GUILLEMIN, ERNST A., *Introductory Circuit Theory*, John Wiley & Sons, Inc., New York, 1953. This has become the classic in the elementary treatment of network topology.

KARNI, SHLOMO, *Intermediate Network Analysis*, Allyn and Bacon, Inc., Boston, 1971. See Chapter 5.

KIRCHHOFF, G., *IRE Transactions on Circuit Theory*, Vol. CT-5, No. 1, pp. 4–7, March, 1958. This is a translation of the original Kirchhoff paper published in 1847, upon which modern graph theory is based.

LYNCH, WILLIAM A., AND JOHN G. TRUXAL, *Introductory System Analysis*, McGraw-Hill Book Company, 1961. Conventions for mechanical elements are discussed as well as those for electrical elements.

NEWMAN, JAMES R., *The World of Mathematics*, Simon and Schuster, Inc., New York, 1956, pp. 570ff. The Königsberg bridge problem which is mentioned in the chapter is described.

SESHU, SUNDARAM AND M. B. REED, *Linear Graphs and Electrical Networks*, Addison-Wesley Publishing Co., Inc., Reading, Mass., 1961. This is the standard treatise on graph theory, but is very advanced compared to our discussion.

WING, OMAR, *Circuit Theory with Computer Methods*, Holt, Rinehart & Winston, Inc., New York, 1972. See Chapter 1.

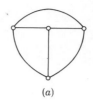

(a)

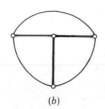

(b)

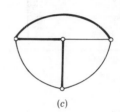

(c)

Fig. 2-20. For the graph of (*a*), the heavy lines of (*b*) and (*c*) are trees.

DIGITAL COMPUTER EXERCISES

In studying the topological descriptions of networks, write an algorithm which can be programmed for the computer which will determine all trees of a given graph. Test your algorithm with a simple example. In Appendix E-10, see reference 14 by Steiglitz and reference 16, Chapter 1, by Wing.

PROBLEMS

2-1. For the controlled source shown in the figure, prepare a plot similar to that given in Fig. 2-8(*b*).

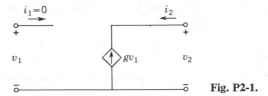

Fig. P2-1.

2-2. Repeat Prob. 2-1 for the controlled source given in the accompanying figure.

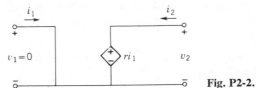

Fig. P2-2.

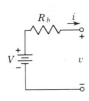

Fig. P2-3.

2-3. The network of the accompanying figure is a model for a battery of open-circuit terminal voltage V and internal resistance R_b. For this network, plot i as a function v. Identify features of the plot such as slopes, intercepts, and so on.

2-4. The magnetic system shown in the figure has three windings marked 1-1′, 2-2′, and 3-3′. Using three different forms of dots, establish polarity markings for these windings.

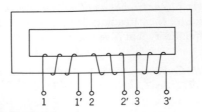

Fig. P2-4.

2-5. Place three windings on the core shown for Prob. 2-4 with winding senses selected such that the following terminals (placed in the order shown in the figure for Prob. 2-4) have the same mark: (a) 1 and 2, 2 and 3, 3 and 1, (b) 1′ and 2′, 2′ and 3′, 3′ and 1′.

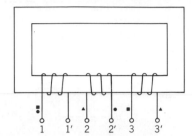

Fig. P2-5.

2-6. The figure shows four windings on a magnetic flux-conducting core. Using different shaped dots, establish polarity markings for the windings.

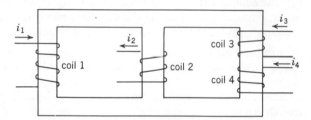

Fig. P2-6.

2-7. The accompanying schematic shows the equivalent circuit of a system with polarity marks on the three coupled coils. Draw a transformer with a core similar to that shown for Prob. 2-6 and place windings on the legs of the core in such a way as to be equivalent to the schematic. Show connections between the elements in the same drawing.

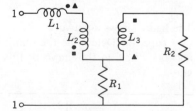

Fig. P2-7.

2-8. The accompanying schematics each show two inductors with coupling but with different dot markings. For each of the two systems, determine the equivalent inductance of the system at terminals 1-1′ by combining inductances.

2-9. A transformer has 100 turns on the primary (terminals 1-1′) and 200 turns on the secondary (terminals 2-2′). A current in the primary

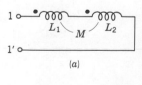

(a)

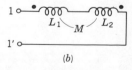

(b)

Fig. P2-8

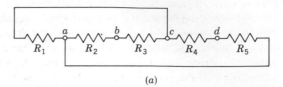

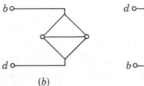

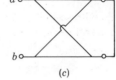

(a)

(b)   (c)

Fig. P2-10.

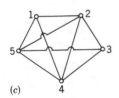

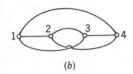

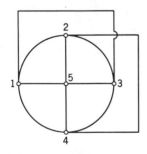

Fig. P2-11

causes a magnetic flux which links all turns of both the primary and the secondary. The flux decreases according to the law $\phi = e^{-t}$ weber, when $t \geqq 0$. Find: (a) the flux linkages of the primary and secondary, (b) the voltage induced in the secondary.

2-10. In (a) of the figure is shown a resistive network. In (b) and (c) are shown graphs with two of the four nodes identified. For these two graphs, assign resistors to the branches and identify the two remaining nodes such that the resulting networks are topologically identical to that shown in (a).

2-11. Three graphs are shown in the figure. Classify each of the graphs as *planar or nonplanar*.

2-12. For the graph of the figure, classify as planar or nonplanar, and determine the quantities specified in Eqs. (2-13) and (2-14).

2-13. In (a) and (b) of the figure for Prob. 2-11 are shown two graphs which may be equivalent. If they are equivalent, what must be the identification of nodes a, b, c, d in terms of nodes 1, 2, 3, 4 if a is identical with 1?

2-14. The figure shows a network with elements arranged along the edges of a cube. (a) Determine the number of nodes and branches in the network. (b) Can the graph of this network be drawn as a planar graph?

Fig. P2-12

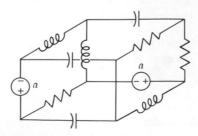

Fig. P2-14.

2-15. The figure shows a graph of six nodes and connecting branches. You are to add nonparallel branches to this basic structure in order to accomplish the following different objectives: (a) What is the minimum number of branches that may be added to make the resulting structure nonplanar? (The structure will then be identified as a Kuratowski basic nonplanar graph.) (b) What is the maximum number of branches you may add before the resulting structure becomes nonplanar?

2-16. (a) Display five different trees for the graph shown in the figure. Show branches with solid lines and chords with dotted lines. (b) Repeat (a) for the graph of (c) in Prob. 2-11.

2-17. Determine *all* trees of the graphs shown in (a) of Prob. 2-11 and (b) of Prob. 2-10. Use solid lines for tree branches and dotted lines for chords.

2-18. For the graphs shown in (c) of Prob. 2-11 and in the figure for Prob. 2-16, the number of different trees is large. For which of the two structures will the number of different trees be greater? Give reasons for your answer. (To discourage a tabulation in determining your answer, the total number for one of the structures is 125.)

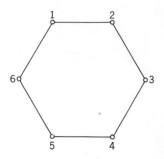

Fig. P2-15

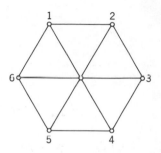

Fig. P2-16

3 Network Equations

3-1. KIRCHHOFF'S LAWS

Network equations are formulated from two simple laws that were first expressed by Kirchhoff in 1845.[1] These laws concern the algebraic sum of voltages around a loop and currents entering or leaving a node. In describing these laws, we rely on concepts introduced in the last chapter. The word *algebraic* is used to indicate that we take into account reference polarities and reference directions in the summations. We will also introduce other references in the form of a positive direction for traversing a loop and a choice of entering or leaving as positive for the current summation at a node.

Kirchhoff's voltage law states that the algebraic sum of all branch voltages around any closed loop of a network is zero at all instants of time. It is a consequence of the law of conservation of energy, voltage being the energy (or work) per unit charge. Consider a unit charge placed at node A in a network. We will move this unit charge from node A to node B, from B to C, C to D, etc., and each time we will

[1]Historically, the work of Kirchhoff closely followed the pioneer contributions of Faraday in describing electric induction, of Oersted in relating magnetism and electricity in 1820, of Ampère in relating force and current in 1820–25, and of Ohm in relating voltage and current in 1826. Kirchhoff's result was first published as an appendix to a paper in 1845, and then in detail in a paper published in 1847. It is interesting to note that Kirchhoff was a 23-year-old student at the time of the first publication of these laws.

determine the energy lost or gained. If we tabulate these energy changes for the various moves, assigning a positive number for a gain and a negative one for a loss, then we know that when we return to node A the summation of the changes must equal zero. An increase in energy in going from A to B is identified as a voltage *rise*, while a decrease in energy is a voltage *drop*. We may state Kirchhoff's voltage law in a different form in terms of voltage drops and rises: Around any closed loop, at any instant of time, the sum of the voltage drops must equal the sum of the voltage rises.

Figure 3-1 shows a simple resistive network with voltage reference directions assigned for the elements and a clockwise loop direction selected for the application of the Kirchhoff voltage law. Starting at node A, we assign a positive sign to the voltage if the polarity marks occur in the order $+$ to $-$, a negative sign for the opposite order. Thus we write

$$-v_g + v_1 + v_2 + v_3 = 0 \qquad (3\text{-}1)$$

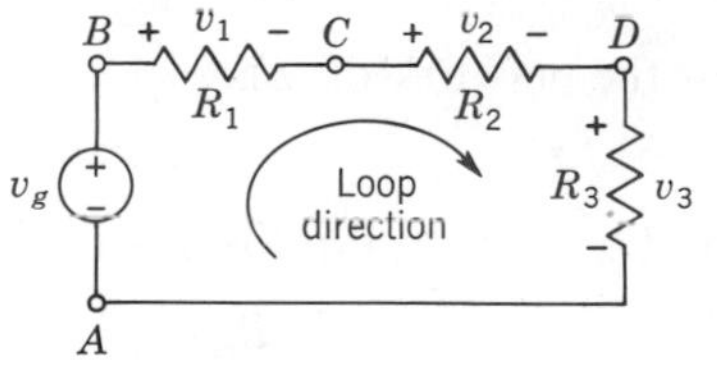

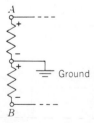

Fig. 3-1. One-loop resistive network to which Kirchhoff's voltage law is applied.

This equation may be manipulated either by multiplying by -1 or by moving terms from one side of equation to the other. In the form

$$v_1 + v_2 + v_3 = v_g \qquad (3\text{-}2)$$

the equation is interpreted as the sum of voltage drops being equal to the sum of the voltage rises.

In talking about rises and drops, it is convenient to have a *reference* node (also called a *datum* node). In the same way that we talk about sea level being the reference for elevations, so in talking about voltages we select a reference called *ground* which is indicated by the symbol shown in Fig. 3-2. In the figure, the voltage at node A is higher than ground potential, while that at node B is lower than (or below) ground potential.

Kirchhoff's current law states that the algebraic sum of all branch currents leaving a node is zero at all instants of time. The law is a consequence of conservation of charge. Charge which enters a node must leave that node because it cannot be stored there. Since the algebraic summation of charge must be zero, the time derivative of this summation must also equal zero. Just as in the voltage law, the Kirchhoff current law may be stated in other ways. For example, we may say that the sum of currents entering a node must equal the sum of the currents leaving the node. Thus, in terms of the assigned

Fig. 3-2. Part of a network in which one node is identified as the ground (or reference or datum) node.

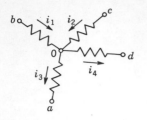

Fig. 3-3. Part of network which illustrates at node O the Kirchhoff current law.

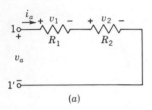

(a)

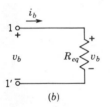

(b)

Fig. 3-4. Networks which are equivalent when $R_{eq} = R_1 + R_2$.

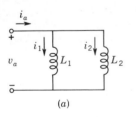

(a)

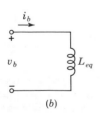

(b)

Fig. 3-6. Equivalent inductive one-port networks when Eq. 3-11 is satisfied.

reference directions shown in Fig. 3-3, we write

$$i_1 + i_2 = i_3 + i_4 \tag{3-3}$$

Alternatively, this equation may be manipulated to the following equivalent form:

$$-i_1 - i_2 + i_3 + i_4 = 0 \tag{3-4}$$

which may be interpreted in terms of the choice of a positive direction for currents leaving the node.

As an example of the application of the Kirchhoff voltage law, consider the network shown in Fig. 3-4. We wish to determine the conditions under which the two networks of the figure are equivalent. Two networks are said to be *equivalent* at a pair of terminals if the voltage-current relationships for the two networks are identical at these terminals. For the networks of (a) and (b) of the figure to be equivalent, we must find the conditions under which $i_a = i_b$ when $v_a = v_b$. For the network of (a),

$$v_a = v_1 + v_2 = R_1 i_a + R_2 i_a \tag{3-5}$$

while for the network of (b),

$$v_b = R_{eq} i_b \tag{3-6}$$

Equating Eq. (3-5) to Eq. (3-6) with $i_a = i_b$, we obtain

$$R_{eq} = R_1 + R_2 \tag{3-7}$$

Thus the summation of resistance for resistors connected in series is equal to the equivalent resistance of the combination. Generalizing the result given by Eq. (3-7) for the network shown in Fig. 3-5, we obtain for a series connection,

$$R_{eq} = R_1 + R_2 + \ldots + R_n = \sum_{j=1}^{n} R_j \tag{3-8}$$

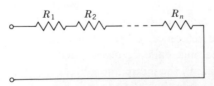

Fig. 3-5. Series-connected resistive network.

As another example, the Kirchhoff current law may be applied to the network shown in Fig. 3-6(a) to determine the condition under which it is equivalent to the network of Fig. 3-6(b). Thus,

$$i_a = i_1 + i_2 = \frac{1}{L_1} \int v_a \, dt + \frac{1}{L_2} \int v_a \, dt \tag{3-9}$$

from Eq. (1-41). Similarly, for the network of (b), we have

$$i_b = \frac{1}{L_{eq}} \int v_b \, dt \tag{3-10}$$

For $v_a = v_b$ and $i_a = i_b$, the conditions for the networks to be equivalent, we see that it is necessary that

$$\frac{1}{L_{eq}} = \frac{1}{L_1} + \frac{1}{L_2} \quad \text{or} \quad L_{eq} = \frac{L_1 L_2}{L_1 + L_2} \tag{3-11}$$

3-2. THE NUMBER OF NETWORK EQUATIONS

An important problem in analyzing a network concerns the number of equations that must be written in order to describe completely the voltages and currents in the network. The answer may appear to be obvious since we must always write the same number of equations as we have unknown quantities or variables. It will turn out, however, that a smaller number may be chosen for simultaneous solution. Two questions arise which we will answer in this section. How may we properly choose our variables so that we have a minimum number of them? How can we be sure that the equations we write are independent?

We first make a number of restrictions for the discussion of this section. We consider a branch to be the same as an element, or one branch to represent a single element. Thus the part of the network shown in Fig. 3-7 has two branches, marked 1 and 2. We will show later that under some conditions the two branches may be replaced by an equivalent branch for some calculations. We will also assume that we have voltage sources in series with other elements, and current sources in parallel with other elements. In the next section, this restriction will be removed and it will be shown that we have not lost generality. We also exclude from consideration two classes of untenable situations in networks: We assume that there are no loops consisting only of voltage sources, nor can the network be separated into two parts joined only by current sources.

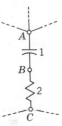

Fig. 3-7. Two branches of a network.

Consider a network composed of b branches excited by active elements for which we are to find responses in the network. The unknown quantities of interest are the branch voltages and the branch currents, making a total of $2b$ unknowns for the b branches. Since the voltage-current relationships are known for each of the elements by equations like $v = Ri$, $v = L\,di/dt$, and $v = (1/C) \int i\,dt$, we may reduce the number of unknown quantities from $2b$ to b. In other words, if we know the branch currents, then we may routinely determine the branch voltages, and vice versa.

We make use of the Kirchhoff laws discussed in the last section to write the b equations in b unknowns. In preparation for writing these equations, we first select the datum or reference node. For the remaining nodes, we then write equations using Kirchhoff's current law. This accomplished, it is then necessary that we write

$$b - (n - 1) = b - n + 1 \tag{3-12}$$

equations using Kirchhoff's voltage law in order that we have a total of b equations. We may then solve for the unknown voltages and currents, provided, of course, that the equations we have written are *independent*. What do we mean by independent?

A set of equations is said to be linearly dependent if at least one of the equations can be expressed as a linear combination of the others. Thus, if we obtain an equation by adding or subtracting two other equations, one equation is dependent on the other two, and may not be used in finding a solution. Thus, given the equations

$$3i_1 + 2i_2 - i_3 = 4$$
$$-i_1 + 5i_2 + 3i_3 = -2 \tag{3-13}$$
$$i_1 + 12i_2 + 5i_3 = 0$$

we note that the third equation may be obtained by multiplying the second by two and adding it to the first. Then the equations are dependent and no unique solution may be found for i_1, i_2, and i_3. A geometrical interpretation of this kind of situation may be given with the aid of Fig. 3-8 showing a three-dimensional current space. Two equations of the form of those in Eqs. (3-13) each represent a plane in this current space. If these planes intersect, they define a line shown by

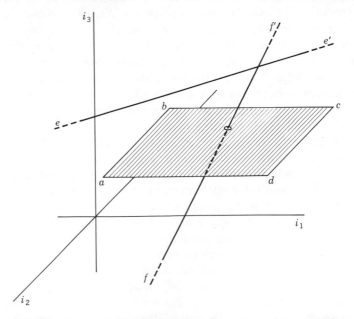

Fig. 3-8. Current space representation of equations of the general form of Eqs. (3-13.) Note that the line e-e' may be in the plane *abcda*.

e-e' or f-f' in the figure. If the third equation represents the surface shown as a-b-c-d-a, then (i) the line intersects the surface, as f-f' does, (ii) the line is parallel to the plane as is the case for e-e', or (iii) the plane includes the line e-e'. In case (i), the intersection of the line with the plane defines the solution i_1, i_2, i_3. For (ii) there is no solution, and for (iii) there is no unique solution, but rather a family of solutions defined by the line.

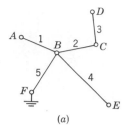

(a)

The equations we write must be independent if we are to be able to solve the equations. In addition, we desire that the set of equations be written in terms of the smallest number of independent variables. The reason for this requirement will be clear to you after you have solved a set of five simultaneous equations and compare the number of operations required in comparison to the number for solving three simultaneous equations. In the interest of saving time and effort, we select the smallest number of variables, consistent with the requirement that they be independent.

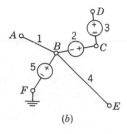

(b)

A justification for our selection of independent voltage variables is based on Fig. 3-9. In (a) of that figure, a tree of a graph is shown. Node F is selected as the datum node; there are then as many other nodes in the tree as there are tree branches. Observe in (b) of the figure that the voltage of node D is defined by voltage sources inserted in three tree branches. Since there are no closed paths in the tree, all branch currents are zero and the branch voltage sources determine the voltage at node D. Similarly, by inserting voltage sources in every branch, all node voltages are determined, as in (c) of the figure. For the example, five sources are sufficient to determine all five node voltages with respect to the datum. Conversely, if the node-to-datum voltages are known, then the tree branch voltages may be determined. The chord voltages are also determined by the node-to-datum voltages, each chord voltage being the difference of two node-to-datum voltages. Thus it seems plausible that the number of independent voltages needed is equal to the number of branches in a tree. For an n node network, this number is $n - 1$ (as discussed in Chapter 2 and earlier in this section) which is the number of node-to-datum voltages in the network. There are other choices of sets of variables from which the branch voltages can be determined,[2] but we will make use of only the node-to-datum voltage variables.

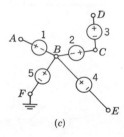

(c)

Fig. 3-9. Figures pertinent to the number of independent voltage variations in an n-node network.

Now $n - 1$ is clearly less than b, since there are $n - 1$ branches in a tree and there are always fewer branches in a tree than in the graph from which it is derived unless, of course, the graph is a tree to start with. Hence there are fewer node-to-datum variables than node-to-node variables for the b individual branches.

[2]These choices are explained in detail by E. A. Guillemin in *Introductory Circuit Theory*, (John Wiley & Sons, Inc., New York, 1953), Chapter 1.

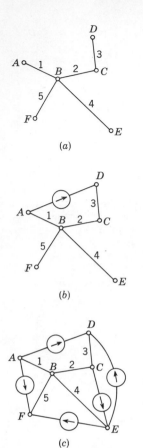

(a)

(b)

(c)

Fig. 3-10. Figures pertinent to the number of independent current variables.

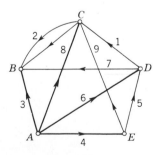

Fig. 3-11. Graph which illustrates the identification of one set of independent current variables. Intersecting lines are not connected except at the marked nodes.

We now turn to the matter of selecting current variables fewer in number than the b branch currents. The discussion to follow is the dual to that just given for voltage variables, and begins by considering a tree of a graph. To the tree shown in Fig. 3-10(a), we connect a current source in place of a chord of the graph as in (b) of the figure. For this connection, we see that this current will be identical to that in branches 1, 2, and 3. If we now place current sources in all chord positions of the graph, as shown in (c), we see that the current in each branch will be a linear combination of currents from the sources in the chord positions. If the chord currents are known, then we may solve algebraically for the branch currents; hence the number of variables necessary to fix the tree branch current is equal to the number of currents in the chords. From Chapter 2, we know that the number of chords is

$$\text{Chords} = b - n + 1 \tag{3-14}$$

and this is the minimum number of current variables we require. Note that this number is always less than b because there are always fewer chords than branches. That these current variables do indeed form an independent set was shown by Kirchhoff in 1847.

These current variables are called *loop (or mesh)* currents and their paths may be determined by replacing the chords upon the tree one at a time. It is conventional to assign the reference direction to the loop current which is the reference direction of the chord with which it is identified. Since the tree of a graph is not unique, there are many possible choices of loop current variables.

As an example of the identification of the independent loop currents, consider the graph of Fig. 3-11. The tree selected is identified by the heavy lines; reference directions are shown for all branches of the graph. The paths of the loop currents are seen to include the following branches: 1-8-6 (i_a), 2-3-8 (i_b), 7-3-6 (i_c), 9-8-4 (i_d), and 5-6-4 (i_e). The nine branch currents may be written in terms of the five loop currents we have just identified. For example, one of the nine equations is

$$i_6 = i_a + i_c - i_e \tag{3-15}$$

Using the same graph with node A identified as the datum node, we may illustrate the relationship of branch voltages to the set of node-to-datum voltages, v_3, v_8, v_6, and v_4. Branch voltage 9, for example, is

$$v_9 = v_8 - v_4 \tag{3-16}$$

Using these equations, we may compute the node-to-node branch voltages if we know the node-to-datum voltages.

As a second example, see the graph of Fig. 3-12. This graph is less complicated than that of Fig. 3-11 in that it is planar. The choice of the tree indicated by the heavy lines and the chord reference directions is convenient in that the loops involve the same number of branches and have the same counterclockwise direction. The four sectors bounded by the branches of the four loops are spoken of as *window panes* (in this case, for a circular window) in formulating the rule of thumb that loop currents are assigned to window panes when the graph is so simple that the window panes can be identified by inspection. The graph shown in Fig. 3-13 presents no difficulty in

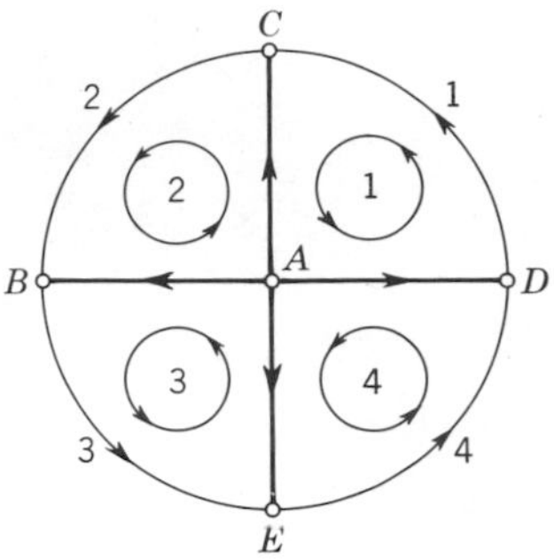

Fig. 3-12. Graph which illustrates the use of "window panes" in selecting current variables.

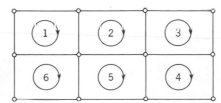

Fig. 3-13. Graph which contains six "window panes" with six loop current variables identified.

assigning the six loop currents to the window panes as shown. For this graph, it is easy to work backwards and from the assigned loops identify the tree and the chords. For simple, planar networks, the window-pane method is conveniently used. When networks to be analyzed are complex and nonplanar, the tree-and-chord-set method may be used to advantage.

3-3. SOURCE TRANSFORMATIONS

Equivalent networks were introduced earlier in this chapter in our study of networks with only one kind of element. In this section, we will extend our study of equivalent networks with emphasis on active sources. Voltage sources may be transformed into equivalent current sources and vice versa; the position of sources in the network may be shifted. Our objective in this network manipulation is to prepare the network for an analysis that is simple and direct.

Two elementary operations with sources are illustrated in Fig. 3-14. Two voltage sources, v_1 and v_2, connected in series, with reference polarities as indicated, are equivalent to a single voltage source, $v_1 + v_2$. Similarly, two current sources in parallel, i_1 and i_2, are equivalent to a single source, $i_1 + i_2$, as illustrated in (b) of Fig. 3-14. The other two equivalent networks of Fig. 3-14 remind us that voltage sources cannot be connected in parallel unless the two sources have identical voltages, and, similarly, the current sources cannot be connected in series unless identical. The paralleling of generators with nonsimilar voltage waveforms, for example, results in heavy currents and equipment damage.

Figure 3-15 shows a resistor in parallel with a voltage source. The current through this resistor is determined only by the voltage

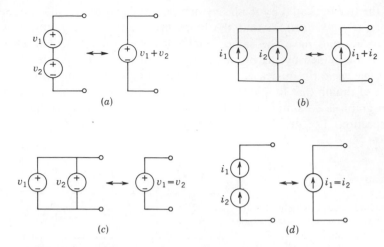

Fig. 3-14. Illustrating rules under which sources may be combined.

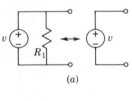

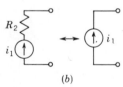

Fig. 3-15. Two examples of extraneous elements so far as terminal behavior is concerned.

source and not by the remainder of the network. As far as computations in the remainder of the network are concerned, a resistor in parallel with a voltage source may be ignored or omitted entirely from the network representation. The same situation applies to a resistor in series with a current source, as shown in (b) of Fig. 3-15. This resistor in no way affects the current from the source. As far as computations in the remainder of the network are concerned, a resistor in series with a current source may be omitted from the network representation.

Let us next turn to the matter of voltage and current source equivalents. In Fig. 3-16(a), let $v(t)$ be the voltage of the source and $v_1(t)$ the voltage at the node located between resistor R_1 and the rest of the network. Kirchhoff's voltage law for the circuit of Fig. 3-16(a) is

$$v(t) = R_1 i(t) + v_1(t) \tag{3-17}$$

Solving this equation for $i(t)$ gives

$$i(t) = \frac{v(t)}{R_1} - \frac{v_1(t)}{R_1} \tag{3-18}$$

In this current equation, we will identify each of the individual terms in relationship to the network of Fig. 3-16(b). The term $v(t)/R_1$

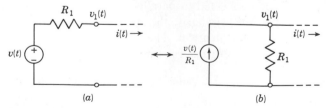

Fig. 3-16. Source transformation involving one resistor.

represents the current from the source, the term $v_1(t)/R_1$ represents the current in the resistor R_1 connected in parallel with the current source. The difference of these two currents is the current to the external network, $i(t)$. Note that $i(t)$ and $v_1(t)$ are the same for the two networks. With respect to the remainder of the network, the source-resistor combinations of (a) and (b) of Fig. 3-16 are described by the same equations, Eqs. (3-17) and (3-18), and so are equivalent.

The reasoning we have just applied to the resistor in series with a voltage source may be extended to either an inductor or a capacitor in the same series position. Writing equations analogous to those of the last paragraph, we conclude that the current source equivalent for the voltage source-inductor combination is that shown in Fig. 3-17, while that for the voltage-source capacitor is given in Fig. 3-18. This technique for converting sources and single elements is not ordinarily applied to networks involving more than one passive element in series or in parallel, since the determination of the source equivalent involves the solution of a differential equation.

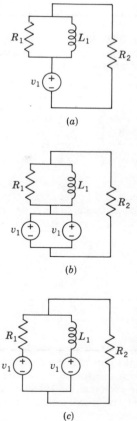

Fig. 3-19. Three equivalent networks illustrating a procedure for a source to be "pushed through a node."

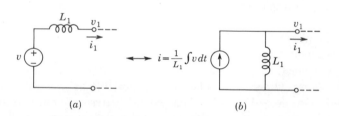

Fig. 3-17. Source transformation for a network with a single inductor.

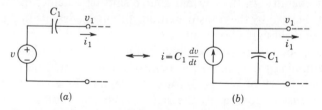

Fig. 3-18. Source transformation involving one capacitor.

In the analysis of networks, we often encounter voltage sources without a series passive element, or current sources without a parallel passive element. If it is desired to transform from one kind of source to the other, it is first necessary to *shift* the source within the network. The technique by which this is accomplished will be explained in terms of the simple example of Fig. 3-19(a). The single voltage source may be considered to be equivalent to two identical sources in parallel as in (b) of the figure. Now the network in (c) is identical to that of (b) since a connection from the positive terminals of the two sources does not affect the network because there would be no current in such a

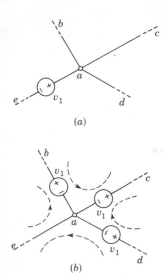

(a)

(b)

Fig. 3-20. Illustrating the procedure by which a voltage source is shifted in a network.

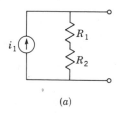

(a)

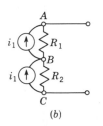

(b)

Fig. 3-21. Two equivalent networks illustrating the manner in which one source is replaced by two such that the Kirchhoff current law is still satisfied at each node.

connection. Thus the network of (c) is seen to be equivalent to that of (a); the source has been "pushed through the node" in obtaining an equivalent network in which the currents throughout the network are unchanged by the transformation. If current sources are now required, the network of Fig. 3-19(c) may be modified by using the equivalences of Figs. 3-16 and 3-17.

The example just given is a special case of a more general form of voltage source shifting illustrated in Fig. 3-20. In traversing the four loops indicated by the dashed lines of (b) of the figure, we observe that Kirchhoff's voltage law gives the same equations for the two networks of (a) and (b). Thus we may "push the voltage source through the node," with a new identical source appearing in every branch connected to the node, without affecting the current distribution in the network. The polarity scheme that must be used for this to be true is shown in Fig. 3-20. Note that the voltage distribution in the network is changed since node a is now at the same voltage as node e, while before shifting the voltage difference of the two nodes was v_1.

Equivalent manipulations are also possible for the current source, following the pattern of duality that has been in evidence since the concept was first introduced in Chapter 2. In the example of Fig. 3-21, the network of (b) is equivalent to that of (a) in applying Kirchhoff's current law at each of the nodes. The current i_1 enters and leaves node B, while the currents at nodes A and C remain the same as in (a) of the figure. Other manipulations are possible as illustrated in the second example of Fig. 3-22. The general pattern of source transformation is seen to be one of maintaining the same currents at all nodes of the network, essentially by adding and then subtracting the same current quantity. In this current source shifting scheme, the voltages are not changed by the transformation, although currents in the active source branches are clearly altered.

The operations we have described may be employed successively in determining the simple equivalent of a complicated network. The resistive network of Fig. 3-23, for example, has three voltage sources and one current source. By the step-by-step reduction shown in the figure, the simplified equivalent networks of (e) or (f) are found.

Source transformations have bearing on the representation of a network by a graph. Two observations may be made: (1) Elements in parallel with voltage sources or in series with current sources can be eliminated from the graph. (2) Since voltage sources may be shifted from one branch to others with the consequent elimination of that branch from the network, in the construction of a graph the voltage source may be shorted out before analysis is started. Similarly, current sources may be open circuited and so eliminated from the graph which represents the network.

We will always follow the practice of *preparing the network* before writing equations from Kirchhoff's laws. This will involve the following practices: (1) If the network is to be analyzed on the node basis with Kirchhoff's current law, we will manipulate the network sources so that only current sources are in the resulting network. (2) If the network is to be analyzed on the loop basis, using Kirchhoff's voltage law, then all sources will be manipulated until equivalent voltage sources are found. (3) If the network is to be analyzed using state variables, then we may prepare the network to have both voltage and current sources, depending upon the variables selected. This policy is adopted in the interest of simplicity, with the understanding that it will suffice for all but the most complex of network problems.

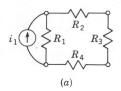

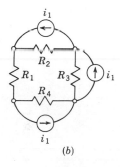

Fig. 3-22. An example illustrating the procedure by which a current source is shifted in a network.

We are now prepared to write sets of equations describing networks. We understand that we may make use of the node method, the loop method, or any other method in writing these equations. Which of the choices shall we make? Our choice will be made in terms of such factors as these: (1) Which method results in the smallest number of variables? (2) What is the objective of the analysis? One voltage? Several currents? (3) Will solution be accomplished through hand calculation or by digital computer? This question may, in turn, determine how concerned we are about the independence of the network equations. If we will use a pencil and pad of paper in obtaining the solution, we are very concerned about keeping the number of variables small because of the number of algebraic operations involved in the solution of simultaneous equations. This number increases very rapidly with the number of variables, a point that will be dramatized by examples when we study the methods of determinant evaluation.

3-4. EXAMPLES OF THE FORMULATION OF NETWORK EQUATIONS

A number of examples will illustrate the formulation of network equations using Kirchhoff's laws and the various rules of this and the previous chapter.

EXAMPLE 1

Figure 3-24(a) shows a series RLC circuit. For this simple circuit, Kirchhoff's voltage law requires that

$$Ri + L\frac{di}{dt} + \frac{1}{C}\int i\, dt = v(t) \qquad (3\text{-}19)$$

at all times. This is an *integrodifferential equation*, which may be changed to a differential equation by differentiation to give

$$L\frac{d^2i}{dt^2} + R\frac{di}{dt} + \frac{1}{C}i = \frac{dv(t)}{dt} \qquad (3\text{-}20)$$

where the derivatives have been arranged in descending order.

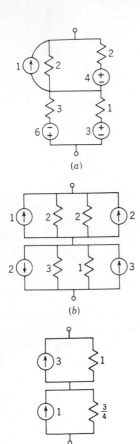

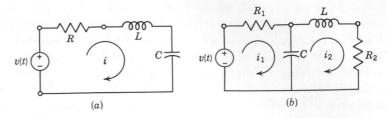

Fig. 3-24. The networks analyzed in Examples 1 and 2.

EXAMPLE 2

Applying Eq. (3-12) to the network of Fig. 3-24(*b*), we see that $b - n + 1 = 5 - 4 + 1 = 2$, a fact that is clear from the "window-pane" rule. With the two loop currents i_1 and i_2 assigned with the positive directions indicated, the equilibrium equations based on Kirchhoff's voltage law are

$$R_1 i_1 + \frac{1}{C} \int (i_1 - i_2)\, dt = v(t) \tag{3-21}$$

$$\frac{1}{C} \int (i_2 - i_1)\, dt + L \frac{di_2}{dt} + R_2 i_2 = 0 \tag{3-22}$$

EXAMPLE 3

A three-node network is shown in Fig. 3-25 with node-to-datum voltages v_1, v_2, and v_3 assigned as indicated. Assuming current out of the node to be positive for each of the nodes in turn gives the three Kirchhoff current equations

$$\frac{1}{R_1} v_1 + C_1 \frac{d}{dt} (v_1 - v_2) = i(t) \tag{3-23}$$

$$C_1 \frac{d}{dt} (v_2 - v_1) + \frac{1}{R_2} v_2 + C_2 \frac{d}{dt} (v_2 - v_3) = 0 \tag{3-24}$$

$$\frac{1}{R_3} v_3 + C_2 \frac{d}{dt} (v_3 - v_2) = 0 \tag{3-25}$$

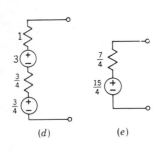

Fig. 3-23. An example of network simplification using successive source transformations.

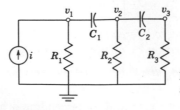

Fig. 3-25. Three-loop network analyzed in Example 3.

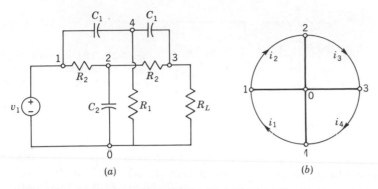

(a) (b)

Fig. 3-26. A twin-T RC network and its graph analyzed in Example 4.

EXAMPLE 4

The network shown in Fig. 3-26(a) is known as a *twin-T network*. This is more complicated than the past three examples, and the construction of a graph will aid in the formulation of the voltage equations. The graph of this network, a choice of a tree, and the choice of chord directions are shown in Fig. 3-26(b). The required four equations are

$$\frac{1}{C_1} \int i_1 \, dt + R_1(i_1 - i_4) = -v_1 \qquad (3\text{-}26)$$

$$R_2 i_2 + \frac{1}{C_2} \int (i_2 - i_3) \, dt = +v_1 \qquad (3\text{-}27)$$

$$\frac{1}{C_2} \int (i_3 - i_2) \, dt + R_2 i_3 + R_L(i_3 - i_4) = 0 \qquad (3\text{-}28)$$

$$\frac{1}{C_1} \int i_4 \, dt + R_1(i_4 - i_1) + R_L(i_4 - i_3) = 0 \qquad (3\text{-}29)$$

Note that in these examples we have used the integral alone as a shorthand notation for the integral with limits; thus

$$\int i_k \, dt \quad \text{represents} \quad \int_{-\infty}^{t} i_k(t) \, dt \qquad (3\text{-}30)$$

EXAMPLE 5

For this and the next example, we will write equations for coupled networks, using the results of Section 2-3. Consider the network shown in Fig. 3-27. This network has two *parts* which are magnetically coupled. For coupled networks, Eq. (3-12) must be modified to the form $b - n + p$ where p is the number of separate parts of the network. Similarly, the number of node equations for

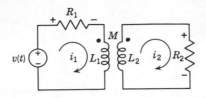

Fig. 3-27. The network of Example 5 containing two parts which are magnetically coupled.

coupled networks is $n - p$ rather than $n - 1$ as previously given. Thus, for this problem, we must write $b - n + p = 5 - 5 + 2 = 2$ loop equations.

To find the polarity of the induced voltages in the network, we apply the methods described in Section 2-3. Thus, in Fig. 3-28(a), current i_1 enters the dotted terminal of winding 1-2 and will induce a voltage in winding 3-4 positive at the dotted terminal, terminal 3. Similarly, i_2 induces a voltage in winding 1-2, with terminal 1—the dotted terminal—positive. In Fig. 3-28(b), the current i_2 has a positive direction reversed from that just considered. This current is positive when it leaves the dotted terminal and hence induces a voltage in winding 1-2 with terminal 2 positive.

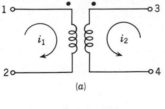

(a)

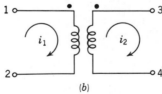

(b)

Fig. 3-28. Networks used to illustrate the rules for determining the polarity of induced voltages.

Applying this rule to the network of Fig. 3-27, the Kirchhoff voltage law is

$$R_1 i_1 + L_1 \frac{di_1}{dt} - M \frac{di_2}{dt} = v(t) \tag{3-31}$$

In the second loop, the equilibrium equation for voltages is

$$L_2 \frac{di_2}{dt} - M \frac{di_1}{dt} + R_2 i_2 = 0 \tag{3-32}$$

EXAMPLE 6

The winding sense of three coils on a flux-conducting material is shown in Fig. 3-29. We are required to write the Kirchhoff voltage equations, taking into account mutual inductance. With the aid of dots, the system of Fig. 3-29 can be replaced by the equivalent circuit

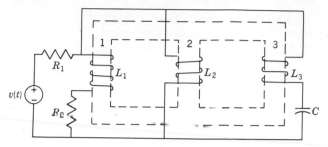

Fig. 3-29. Magnetically coupled network which is analyzed in Example 6.

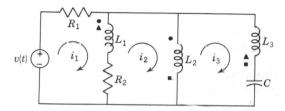

Fig. 3-30. A network representation which is equivalent to that shown in Fig. 3-29.

of Fig. 3-30. If we use a double subscript notation for mutual inductance to indicate the two coils being considered, the Kirchhoff voltage equations are

$$R_1 i_1 + L_1 \frac{d(i_1 - i_2)}{dt} + M_{12} \frac{d(i_2 - i_3)}{dt} - M_{13} \frac{di_3}{dt} + R_2(i_1 - i_2) = v(t)$$

$$(3\text{-}33)$$

$$R_2(i_2 - i_1) + L_1 \frac{d(i_2 - i_1)}{dt} - M_{12} \frac{d(i_2 - i_3)}{dt} + M_{13} \frac{di_3}{dt} + L_2 \frac{d(i_2 - i_3)}{dt}$$

$$+ M_{21} \frac{d}{dt}(i_1 - i_2) + M_{23} \frac{d}{dt} i_3 = 0 \quad (3\text{-}34)$$

$$L_2 \frac{d}{dt}(i_3 - i_2) - M_{23} \frac{di_3}{dt} - M_{21} \frac{d}{dt}(i_1 - i_2) + L_3 \frac{di_3}{dt}$$

$$+ M_{32} \frac{d}{dt}(i_2 - i_3) - M_{31} \frac{d}{dt}(i_1 - i_2) + \frac{1}{C} \int i_3 \, dt = 0 \quad (3\text{-}35)$$

In this particular problem, the equations would have had simpler form if three other loops had been chosen so that each included $v(t)$. (See Prob. 3-25.)

EXAMPLE 7

We next turn to the formulation of equilibrium equations on the node basis. Consider the network shown in Fig. 3-31(a). The voltage

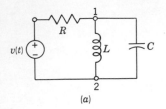

(a)

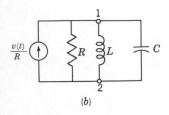

(b)

Fig. 3-31. Network of Example 7.

source may be converted into an equivalent current source by the procedure of Section 3-3, giving the network of Fig. 3-31(*b*). Node 2 is designated the datum node and all branch currents are assigned to be positive if directed out of node 1. By Kirchhoff's current law, the current equation is

$$\frac{1}{R}v_1 + \frac{1}{L}\int v_1\,dt + C\frac{dv_1}{dt} = \frac{v(t)}{R} \tag{3-36}$$

Of course, it is not necessary to make the conversion to the current source before analyzing the network. Since the voltage of the $+$ terminal of the generator is $v(t)$ in Fig. 3-31(*a*), we may write

$$\frac{1}{R}[v_1 - v(t)] + \frac{1}{L}\int v_1\,dt + C\frac{dv_1}{dt} = 0 \tag{3-37}$$

or

$$\frac{1}{R}v_1 + \frac{1}{L}\int v_1\,dt + C\frac{dv_1}{dt} = \frac{v(t)}{R}$$

which is identical with Eq. (3-36). Analysis may be carried out with either the voltage source or the equivalent current source.

EXAMPLE 8

The network shown in Fig. 3-32 is the current source equivalent to the three-loop network shown in Fig. 3-25. Node 3 is the datum node, and the unknown voltages at nodes 1 and 2 are designated v_1 and v_2. At node 1, setting $1/R_1 = G_1$ and $1/R_2 = G_2$,

$$G_1 v_1 + C_1 \frac{dv_1}{dt} + C_2 \frac{d}{dt}(v_1 - v_2) = G_1 v \tag{3-38}$$

and at node 2,

$$C_2 \frac{d}{dt}(v_2 - v_1) + C_3 \frac{dv_2}{dt} + G_2 v_2 = 0 \tag{3-39}$$

In this example, formulation on the node basis has resulted in fewer differential equations than on the loop basis in Example 3. Ordinarily it requires less work in solving two simultaneous differential equations than in solving three. The choice of method of formulation, loop or node, also depends on the objective of analysis. In this example, if the voltage at node 2 is desired, the node method has the advantage over the loop method. But if it is the current flowing in capacitor C_3 that is to be found, we must weigh the relative advantages of the two

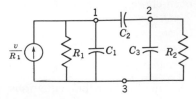

Fig. 3-32. Network which is analyzed in Example 8.

methods. The loop currents can be assigned so that only one loop current flows in C_3, but three simultaneous equations must be solved. Using the node method, we might find the voltage at node 2 first and then determine the current in the capacitor from the equation

$$i_{C_3} = C_3 \frac{dv_2}{dt} \tag{3-40}$$

The second method involves less computation in this particular example.

EXAMPLE 9

The network shown in Fig. 3-33 differs from the networks of other examples in that there is no series resistance with the voltage source. Although this network has three independent loops, there is but one unknown node voltage, that at node 2. From Kirchhoff's current law, we write:

$$C_3 \frac{d}{dt}(v_2 - v_1) + \frac{1}{L} \int (v_2 - v_1)\, dt + Gv_2 + C_2 \frac{dv_2}{dt} = 0 \tag{3-41}$$

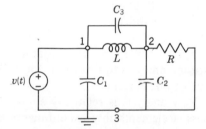

Fig. 3-33. Network of Example 9.

where, as before, $G = 1/R$. Note that C_1 does not appear in the equation. This is because the voltage at node 1 is independent of the capacitor C_1. Capacitor C_1 is an extraneous element. The voltage source must maintain terminal voltage for any load (or it is not an ideal element), and so C_1 may be removed without affecting the network equations.

3-5. LOOP VARIABLE ANALYSIS

Thus far we have progressed from the analysis of very simple networks to more complex network configurations for the loop and node methods. In the next three sections, we will continue the discussion of Section 3-3 for three of the many possible methods for the formulation of equations to describe networks. This statement is illustrated in Fig. 3-34, which is an expansion of a figure presented in Chapter 1. It is intended to illustrate the point that once we have

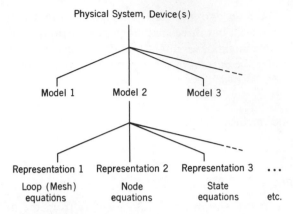

Fig. 3-34. Various representations are available to describe a given model of a physical system.

selected a model of a system of devices, we have a number of alternatives in the representation of that model by a set of network equations. Factors which come into our choice of representation were discussed in Section 3-3, and they include keeping the number of variables small, finding the desired result as directly as possible, and the like. All valid methods are capable of leading to the same end result, the determination of all branch voltages and branch currents in the network. We have noted that it is seldom that we are interested in carrying analysis to this limit, for we are usually interested in one voltage or one current.

Now analysis is relatively simple for networks in which there are passive elements only, excluding mutual inductance and controlled sources. Our approach will be to consider the simple case first, and later outline the modifications needed to treat the more general case.

To begin, let us consider an L-loop network, represented by the graph shown in Fig. 3-35. Consider loop 1. This loop may contain resistance, inductance, and capacitance in any one or all of the

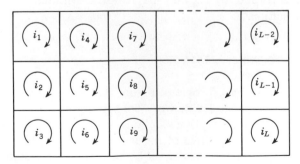

Fig. 3-35. A graph of a network with L independent loop currents identified.

branches that make up the loop. Let

R_{11} be the total resistance in loop 1.
L_{11} be the total inductance in loop 1.
D_{11} be the total elastance of loop 1.

We use elastance instead of capacitance here because elastance terms add directly for a series circuit, while capacitance terms combine as

$$\frac{1}{C_{11}} = \frac{1}{C_1} + \frac{1}{C_2} + \frac{1}{C_3} + \cdots + \frac{1}{C_n} \tag{3-42}$$

There will be voltage drops in loop 1 produced by current flow in loop 2, in loop 3, loop 4—in fact, all loops in the general case. Rather than specialize on loop 1, consider the effect of currents in the jth loop on voltage in loop k, where j and k are any integers from 1 to L. For these two loops, let R_{kj} = the total resistance common to loops k and j; L_{kj} = the total inductance (including mutual) common to loops k and j; D_{kj} = the total elastance common to loops k and j. The voltage drop in loop k produced by current i_j is

$$R_{kj}i_j + L_{kj}\frac{di_j}{dt} + D_{kj}\int i_j\,dt \tag{3-43}$$

At this point, we will adopt a special notation for equations of this form by letting the following equation be the equivalent of Eq. (3-43).

$$\left(R_{kj} + L_{kj}\frac{d}{dt} + D_{kj}\int dt\right)i_j = a_{kj}i_j \tag{3-44}$$

This symbolism implies that the variable i_j is operated upon by multiplication by R_{kj}, multiplication by L_{kj} and differentiation, and finally, multiplication by D_{kj} and integration. All three operations are summarized in the symbol a_{kj}.

The total voltage drop in loop k will be found by successively considering loop k and the currents flowing in every other loop. Mathematically this is done by letting j have all values from 1 to L. This total voltage drop must be equal to the total voltage rise from active sources within loop k, which we write as v_k. Then by Kirchhoff's voltage law, we have

$$\sum_{j=1}^{L} a_{kj}i_j = v_k \tag{3-45}$$

There remains only to repeat this process for all loops, by letting k have all values from 1 to L. Thus the most general form for Kirchhoff's voltage law for an L-loop network is

$$\sum_{j=1}^{L} a_{kj}i_j = v_k, \qquad k = 1, 2, \ldots, L \tag{3-46}$$

The expansion of this concise equation is the following set of equations.

$$a_{11}i_1 + a_{12}i_2 + a_{13}i_3 + \ldots + a_{1L}i_L = v_1$$
$$a_{21}i_1 + a_{22}i_2 + a_{23}i_3 + \ldots + a_{2L}i_L = v_2 \qquad (3\text{-}47)$$
$$a_{L1}i_1 + a_{L2}i_2 + a_{L3}i_3 + \ldots + a_{LL}i_L = v_L$$

It is helpful to arrange these equations given above in the form of a *chart* (or *schedule*) in which the operator coefficients are emphasized. Such a chart is shown below.

Coefficient of

Eq.	Voltage	i_1	i_2	i_3	i_4	i_5	$\ldots$	i_L
1	v_1	a_{11}	a_{12}	a_{13}	a_{14}	a_{15}	$\ldots$	a_{1L}
2	v_2	a_{21}	a_{22}	a_{23}	a_{24}	a_{25}	$\ldots$	a_{2L}
$\ldots$	$\ldots$	$\ldots$	$\ldots$	$\ldots$	$\ldots$	$\ldots$	$\ldots$	$\ldots$
L	v_L	a_{L1}	a_{L2}	a_{L3}	a_{L4}	a_{L5}	$\ldots$	a_{LL}

If the loop currents are all assumed positive in the same path direction, clockwise for example, then all a_{jj} are positive and all $a_{jk}(j \neq k)$ are negative. In actual problems, of course, many of the operator coefficients are zero.

The chart we have just written and Eq. (3-47) can be written compactly as a matrix equation

$$\begin{bmatrix} v_1 \\ v_2 \\ v_3 \\ \cdot \\ \cdot \\ \cdot \\ v_L \end{bmatrix} = \begin{bmatrix} a_{11} & a_{12} & a_{13} & \ldots & a_{1L} \\ a_{21} & a_{22} & a_{23} & \ldots & a_{2L} \\ a_{31} & a_{32} & a_{33} & \ldots & a_{3L} \\ & \cdot & \cdot & \cdot & \\ a_{L1} & a_{L2} & a_{L3} & \ldots & a_{LL} \end{bmatrix} \begin{bmatrix} i_1 \\ i_2 \\ i_3 \\ \cdot \\ \cdot \\ i_L \end{bmatrix} \qquad (3\text{-}48)$$

or simply

$$\mathcal{V} = \mathcal{Q}\mathcal{J} \qquad (3\text{-}49)$$

Here the $\mathcal{V}$ and $\mathcal{J}$ matrices are known as *column* matrices or *vectors* and $\mathcal{Q}$ is a *square* matrix. The matrix multiplication of $\mathcal{Q}$ and $\mathcal{J}$ is carried out in such a way that we reconstruct Eq. (3-47). Thus the first entry in the column of $\mathcal{V}$ is equal to the summation of the product of successive elements in the first row of $\mathcal{Q}$ and the elements of the column of $\mathcal{J}$. By this rule, we see that

$$v_1 = a_{11}i_1 + a_{12}i_2 + a_{13}i_3 + \ldots + a_{1L}i_L \qquad (3\text{-}50)$$

which is Eq. (3-46) for $k = 1$. Note that k and j of a_{kj} in Eq. (3-46) describe the row and column, respectively, of the elements in Eqs. (3-47) and (3-48) as well as for the chart we have constructed. The details of matrix algebra are covered in Appendix B.

EXAMPLE 10

A two-loop network is shown in Fig. 3-36. In this network there are two sources of voltage and no mutual inductance. The Kirchhoff voltage law is

$$\sum_{j=1}^{2} a_{kj}i_j = v_k, \qquad k = 1, 2 \tag{3-51}$$

or in expanded form,

$$a_{11}i_1 + a_{12}i_2 = v_1, \qquad a_{21}i_1 + a_{22}i_2 = v_2 \tag{3-52}$$

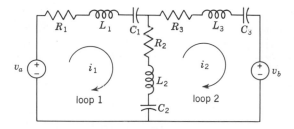

Fig. 3-36. A two-loop network analyzed in Example 10.

The operator coefficients are found by inspection of the network as follows.

$$a_{11} = (R_1 + R_2) + (L_1 + L_2)\frac{d}{dt} + (D_1 + D_2)\int dt \tag{3-53}$$

$$a_{22} = (R_2 + R_3) + (L_2 + L_3)\frac{d}{dt} + (D_2 + D_3)\int dt \tag{3-54}$$

$$a_{12} = a_{21} = -R_2 - L_2\frac{d}{dt} - D_2\int dt \tag{3-55}$$

In Eq. (3-53), a_{11} is found by traversing loop 1 and adding all R's, C's, and D's; a_{22} is similarly found by an addition around loop 2. Equation (3-55) is made up of R, L, and D terms common to loops 1 and 2, the negative sign arising because loop 1 and loop 2 have opposite reference directions. Finally, the voltage terms are seen to be

$$v_1 = v_a, \qquad v_2 = -v_b \tag{3-56}$$

The Kirchhoff equations have an especially simple form for resistive networks when $a_{jk} = R_{jk}$. This will be illustrated by means of an example.

EXAMPLE 11

Consider the network of Fig. 3-36. For this example, let us write the Kirchhoff voltage equations in chart form, where the first row of the chart is equivalent to the equation

$$0 = 4i_1 - i_2 + 0i_3 - i_4 + 0i_5 + 0i_6 + 0i_7 + 0i_8 + 0i_9 \qquad (3\text{-}57)$$

						Coefficient of				
Eq.	Voltage		i_1	i_2	i_3	i_4	i_5	i_6	i_7	i_8	i_9
1	0	=	4	−1	0	−1	0	0	0	0	0
2	1	=	−1	5	−1	0	−1	0	0	0	0
3	0	=	0	−1	4	0	0	−1	0	0	0
4	−1	=	−1	0	0	5	−1	0	−1	0	0
5	0	=	0	−1	0	−1	4	−1	0	−1	0
6	0	=	0	0	−1	0	−1	5	0	0	−1
7	1	=	0	0	0	−1	0	0	4	−1	0
8	0	=	0	0	0	0	−1	0	−1	5	−1
9	0	=	0	0	0	0	0	−1	0	−1	4

Again, this chart may be constructed in a very simple manner. All terms of the principal diagonal of the chart are found as the summation of resistance around each of the nine loops. The terms off the diagonal are all negative, and are all the value of the resistance common to the two loops being considered, identified by the row number (equation number) and the column number (subscript of the current).

From the chart, or from the corresponding matrix of the form of Eq. (3-48), observe: (1) The elements on the principal diagonal are all positive; all others are negative or zero. (2) There is symmetry about the principal diagonal. This symmetry and the sign rule always apply when loops are drawn in the same clockwise or counterclockwise direction. These observations for one example turn out to describe the general case, in the absence of controlled sources.

What about mutual inductance and controlled sources? Mutual inductance will present no problems as shown by the examples of the last section, and the symmetry observations will also hold since $M_{ij} = M_{ji}$. The presence of controlled sources is another matter. Such sources will give rise to terms of the form $v_j = ki_k$ which will appear in the summation around the loop containing v_j but not in the loop defining i_k. Writing the equations will not be a problem, but the rules of symmetry and sign that we have observed will usually not hold (exceptions exist) in the presence of controlled sources. This topic will be explored in greater depth in Chapter 9 in connection with our study of reciprocity.

3-6. NODE VARIABLE ANALYSIS

Consider a network with n nodes and only one part. As discussed in Section 3-3, there are $n - 1$ independent node pairs. Of the many possibilities for node-pair variables, we will select the node-to-datum voltages as our variables exclusively. The form of the voltages for the branch connecting node j to node k with node j positive will be $v_j - v_k$ (from Kirchhoff's voltage law). For each of the $n - 1$ nodes at which the Kirchhoff current law will be formulated, we will assume that currents are directed *out* of the node to be consistent with the voltage sign assignment we have just made. We recall from our earlier discussion that this is an arbitrary choice, and that selecting the other alternative is equivalent to multiplying the resulting equations by -1.

We will follow the practice of converting all voltage sources into equivalent current sources as preparation of the network preceding the writing of the equations. Let us postpone consideration of mutual inductance and controlled sources, and consider a passive network made up of resistors, capacitors, and inductors. Note first that for elements connected as shown in Fig. 3-37, the elements may be replaced by an equivalent system made up as follows: (1) all parallel capacitances replaced by an equivalent capacitance of value $C_{kj} = C_1 + C_2 + \ldots$; (2) an equivalent resistance found by adding conductances as $G_{kj} = 1/R_{kj} = G_1 + G_2 + \ldots$; and (3) an equivalent inductance of value L_{kj}, where $1/L_{kj} = 1/L_1 + 1/L_2 + \ldots$. Applying this network simplification to the elements from node k to all other nodes from $j = 1$ to $j = N$, we have the equation

$$\sum_{j=1}^{N} \left(G_{kj} + C_{kj}\frac{d}{dt} + \frac{1}{L_{kj}}\int dt \right) v_j = i_k, \qquad k = 1, 2, \ldots, N \qquad (3\text{-}58)$$

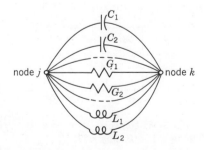

Fig. 3-37. Elements connecting nodes j and k. The three kinds of elements may be combined to give an equivalent parallel *RLC* network between nodes j and k.

which may be written concisely as

$$\sum_{j=1}^{N} b_{kj} v_j = i_k, \qquad k = 1, 2, \ldots, N \qquad (3\text{-}59)$$

by letting b_{kj} summarize the operations

$$\left(G_{kj} + C_{kj} \frac{d}{dt} + \frac{1}{L_{kj}} \int dt \right) = b_{kj} \qquad (3\text{-}60)$$

The expansion of Eq. (3-59) has the same form as the expansion for the loop case, Eq. (3-47), with a's replaced by b's, i's by v's, and v's by i's.

In applying this equation to networks, it is not necessary to simplify the network by combining elements. At node j, the capacitance C_{jj} is the sum of the capacitance *connected* to node j or the capacitance from node j to ground with all other nodes grounded. The value of C_{kj} is the sum of the capacitances *connected between* node j and node k or the capacitance from node j to node k with all other nodes grounded. Similar instructions hold for inverse inductance $1/L$ and for conductance $G = 1/R$. Coefficients can thus be found by inspection by simply noting which elements are "hanging on" or "hanging between" the various nodes.

If the same convention for positive current is maintained in formulating all node equations for a network, the sign of b_{kj} will be positive when $k = j$, and negative when $k \neq j$.

EXAMPLE 12

A network with two independent node pairs is shown in Fig. 3-38. For this network, Kirchhoff's current law is

$$\sum_{j=1}^{2} b_{kj} v_j = i_k, \qquad k = 1, 2 \qquad (3\text{-}61)$$

or

$$b_{11} v_1 + b_{12} v_2 = i_1, \qquad b_{21} v_1 + b_{22} v_2 = i_2 \qquad (3\text{-}62)$$

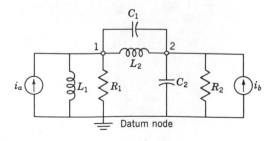

Fig. 3-38. Network with two independent node-pair voltages analyzed in Example 11.

In the form of a matrix equation, we have

$$\begin{bmatrix} i_1 \\ i_2 \end{bmatrix} = \begin{bmatrix} b_{11} & b_{12} \\ b_{21} & b_{22} \end{bmatrix} \begin{bmatrix} v_1 \\ v_2 \end{bmatrix} \tag{3-63}$$

Values for the operator coefficients are summarized in chart form as follows.

Coefficient of

Eq.	Current	v_1	v_2
1	i_a	$G_1 + C_1\dfrac{d}{dt} + \left(\dfrac{1}{L_1} + \dfrac{1}{L_2}\right)\int dt$	$-C_1\dfrac{d}{dt} - \dfrac{1}{L_2}\int dt$
2	i_b	$-C_1\dfrac{d}{dt} - \dfrac{1}{L_2}\int dt$	$+G_2 + (C_1+C_2)\dfrac{d}{dt} + \dfrac{1}{L_2}\int dt$

EXAMPLE 13

Consider the resistive network shown in Fig. 3-39. For this network, the six node variable equations may be routinely written in the following chart form:

Coefficient of

Eq. for node:	Current	v_a	v_b	v_c	v_d	v_e	v_f
a	0	$\frac{5}{2}$	-1	0	0	0	$-\frac{1}{2}$
b	0	-1	2	-1	0	0	0
c	0	0	-1	$\frac{5}{2}$	-1	0	$-\frac{1}{2}$
d	0	0	0	-1	2	-1	0
e	1	0	0	0	-1	$\frac{5}{2}$	$-\frac{1}{2}$
f	0	$-\frac{1}{2}$	0	$-\frac{1}{2}$	0	$-\frac{1}{2}$	$\frac{3}{2}$

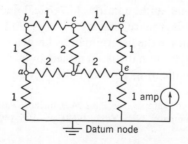

Fig. 3-39. Network of Example 13. Element values are in ohms.

In the form of a matrix equation, we have

$$
\begin{bmatrix} 0 \\ 0 \\ 0 \\ 0 \\ 1 \\ 0 \end{bmatrix} = \begin{bmatrix} \frac{5}{2} & -1 & 0 & 0 & 0 & -\frac{1}{2} \\ -1 & 2 & -1 & 0 & 0 & 0 \\ 0 & -1 & \frac{5}{2} & -1 & 0 & -\frac{1}{2} \\ 0 & 0 & -1 & 2 & -1 & 0 \\ 0 & 0 & 0 & -1 & \frac{5}{2} & -\frac{1}{2} \\ -\frac{1}{2} & 0 & -\frac{1}{2} & 0 & -\frac{1}{2} & \frac{3}{2} \end{bmatrix} \begin{bmatrix} v_a \\ v_b \\ v_c \\ v_d \\ v_e \\ v_f \end{bmatrix} \tag{3-64}
$$

Such equations can be written by inspection, using the "hanging on" and "hanging between" rules and the sign convention for the i_j and G_{kj} entries. Note that all terms of the *principal* diagonal are positive and that symmetry exists with respect to the principal diagonal.

Special problems are encountered in the nodal analysis of networks containing mutual inductance, and a good working rule is to bypass the problem by always analyzing such networks on the loop basis. Should nodal analysis be required, one approach is to replace the coupled coils by an equivalent network without mutual inductance.[3] The presence of controlled sources in the network to be analyzed creates no special problems but generally results in a nonsymmetrical matrix of the form given in Eq. 3-64.

Assuming that we can now write network equations in the two representations, the next problem is to be able to solve the sets of equations, which will require a knowledge of determinants.

3-7. DETERMINANTS: MINORS AND THE GAUSS ELIMINATION METHOD

The array of quantities enclosed by straight-line brackets

$$
\begin{vmatrix} a_{11} & a_{12} & a_{13} & \cdots & a_{1n} \\ a_{21} & a_{22} & a_{23} & \cdots & a_{2n} \\ \cdot & \cdot & \cdot & \cdot & \cdot \\ a_{n1} & a_{n2} & a_{n3} & \cdots & a_{nn} \end{vmatrix} \tag{3-65}
$$

is known as a *determinant of order n*. Quantities in horizontal lines form *rows*, and quantities in vertical lines form *columns*. Such a determinant is square, having n rows and n columns. Each of the n^2 quantities in the determinant is known as an *element*. Element position in the determinant is identified by a double subscript, the first subscript indicating row and the second indicating column (numbered

[3] For example, see Paul M. Chirlian, *Basic Network Theory*, (McGraw-Hill Book Company, New York, 1969), pp. 136–140.

from the upper left-hand corner). Elements along the line extending from a_{11} to a_{nn} form the *principal diagonal* of the determinant.

A determinant has a value which is a function of the values of its elements. In finding this value, we must make use of rules for expansion of the determinant in terms of the elements. Second- and third-order determinants have expansions that are familiar from studies in elementary algebra.

$$\begin{vmatrix} a_{11} & a_{12} \\ a_{21} & a_{22} \end{vmatrix} = a_{11}a_{22} - a_{12}a_{21} \tag{3-66}$$

and

$$\begin{vmatrix} a_{11} & a_{12} & a_{13} \\ a_{21} & a_{22} & a_{23} \\ a_{31} & a_{32} & a_{33} \end{vmatrix} = \begin{array}{l} a_{11}a_{22}a_{33} + a_{12}a_{23}a_{31} + a_{13}a_{21}a_{32} \\ - a_{13}a_{22}a_{31} - a_{23}a_{32}a_{11} - a_{33}a_{21}a_{12} \end{array} \tag{3-67}$$

Expansions for determinants of order higher than third are conveniently made in terms of *minors*.

The *minor* of any element of a determinant a_{jk} is the determinant which remains when the column and row containing a_{jk} are deleted. In terms of the third-order determinant,

$$A = \begin{vmatrix} a_{11} & a_{12} & a_{13} \\ a_{21} & a_{22} & a_{23} \\ a_{31} & a_{32} & a_{33} \end{vmatrix} \tag{3-68}$$

the minor for a_{11}, for example, is

$$M_{11} = \begin{vmatrix} a_{22} & a_{23} \\ a_{32} & a_{33} \end{vmatrix} \tag{3-69}$$

A minor of the element a_{jk} multiplied by $(-1)^{j+k}$ is given the name *cofactor*. The cofactor sign is thus found by raising (-1) to the power found by adding the row and the column, $j + k$, as

$$(\text{cofactor}) = (-1)^{j+k}(\text{minor}) \quad \text{or} \quad \Delta_{jk} = (-1)^{j+k}M_{jk} \tag{3-70}$$

Since, according to this rule, the cofactor signs alternate along any row or column, the proper cofactor sign can be determined by "counting" (plus, minus, plus, etc.) from a positive a_{11} position to any element, proceeding along any combination of horizontal or vertical paths.

Expansion of a determinant in terms of minors (or cofactors) consists of successive reduction of determinant order. A determinant of order n is equal to the sum of the product of the elements of any row or column multiplied by their corresponding $(n - 1)$ order cofactors. Applying this rule to the expansion of the determinant of

Eq. 3-68 along the first column gives

$$A = a_{11}M_{11} - a_{21}M_{21} + a_{31}M_{31}$$

$$= a_{11}\begin{vmatrix} a_{22} & a_{23} \\ a_{32} & a_{33} \end{vmatrix} - a_{21}\begin{vmatrix} a_{12} & a_{13} \\ a_{32} & a_{33} \end{vmatrix} + a_{31}\begin{vmatrix} a_{12} & a_{13} \\ a_{22} & a_{23} \end{vmatrix} \quad (3\text{-}71)$$

There are $2n$ equivalent expansions of the determinant about the n rows or n columns. The minor determinants can, in turn, be expanded by the same rule and the process continued until the value of Δ is given as the sum of $n \times n!$ product factors.

The facts about determinants that we have just reviewed are essential in solving simultaneous equations of the form

$$a_{11}i_1 + a_{12}i_2 + a_{13}i_3 + \ldots + a_{1L}i_L = v_1$$

$$\cdot \ \cdot \ \cdot \ \cdot \ \cdot \ \cdot \ \cdot \ \cdot \ \cdot \ \cdot \ \cdot \ \cdot \ \cdot \ \cdot \ \cdot \quad (3\text{-}72)$$

$$a_{L1}i_1 + a_{L2}i_2 + a_{L3}i_3 + \ldots + a_{LL}i_L = v_L$$

that have resulted from application of Kirchhoff's voltage law (and similar equations from the Kirchhoff current law). The solution to such simultaneous equations is given by *Cramer's rule* as

$$i_1 = \frac{D_1}{\Delta}, \qquad i_2 = \frac{D_2}{\Delta} \ldots i_L = \frac{D_L}{\Delta} \quad (3\text{-}73)$$

where Δ is the *system determinant* given as

$$\Delta = \begin{vmatrix} a_{11} & a_{12} & \cdots & a_{1L} \\ a_{21} & a_{22} & \cdots & a_{2L} \\ \cdot & \cdot & \cdot & \cdot \\ a_{L1} & a_{L2} & \cdots & a_{LL} \end{vmatrix} \quad (3\text{-}74)$$

which must be different from zero for the solutions $i_1, i_2, \ldots, i_n$ to be unique, and D_j is the determinant formed by replacing the jth column of a coefficients by the column $v_1, v_2, \ldots, v_n$.

With Cramer's rule and the method of expansion by minors, simultaneous equations of the form of Eq. (3-47) can be solved. For a third-order equation, the solution for i_1 is

$$i_1 = \frac{D_1}{\Delta} = \frac{v_1\Delta_{11} + v_2\Delta_{21} + v_3\Delta_{31}}{\Delta} \quad (3\text{-}75)$$

or

$$i_1 = \frac{\Delta_{11}}{\Delta}v_1 + \frac{\Delta_{21}}{\Delta}v_2 + \frac{\Delta_{31}}{\Delta}v_3 \quad (3\text{-}76)$$

Similarly,

$$i_2 = +\frac{\Delta_{21}}{\Delta}v_1 + \frac{\Delta_{22}}{\Delta}v_2 + \frac{\Delta_{23}}{\Delta}v_3 \quad (3\text{-}77)$$

and so on. The form of these equations is greatly simplified if all v's except one are zero, corresponding to only one driving voltage source.

EXAMPLE 14

For a certain three-loop network, the following equations are given.

$$5i_1 - 2i_2 - 3i_3 = 10$$
$$-2i_1 + 4i_2 - 1i_3 = 0 \qquad (3\text{-}78)$$
$$-3i_1 - 1i_2 + 6i_3 = 0$$

From Cramer's rule we write the solution for i_1 as

$$i_1 = \frac{D_1}{\Delta} = \frac{10 \begin{vmatrix} 4 & -1 \\ -1 & 6 \end{vmatrix} - 0 \begin{vmatrix} -2 & -3 \\ -1 & 6 \end{vmatrix} + 0 \begin{vmatrix} -2 & -3 \\ 4 & -1 \end{vmatrix}}{\begin{vmatrix} 5 & -2 & -3 \\ -2 & 4 & -1 \\ -3 & -1 & 6 \end{vmatrix}} = \frac{230}{43}$$

$$(3\text{-}79)$$

Similarly,

$$i_2 = \frac{-(+10)\begin{vmatrix} -2 & -1 \\ -3 & 6 \end{vmatrix}}{\Delta} = \frac{150}{43}, \qquad i_3 = \frac{+(10)\begin{vmatrix} -2 & 4 \\ -3 & -1 \end{vmatrix}}{\Delta} = \frac{140}{43}$$

$$(3\text{-}80)$$

When the order of the determinant becomes larger than 4 or 5, the *Gauss elimination method* or its variants offers advantages over expansion by minors, requiring only $n^3/3$ multiplications rather than $n \times n!$. The Gauss elimination method is a systematic way of eliminating variables, which will be introduced by the example of Eqs. (3-78) which we have just solved. Note that both sides of an equation may be multiplied by a constant without changing the equation. If we multiply the first equation in (3-78) by $\frac{2}{5}$ and then add the first and second equations, we have

$$0i_1 + \tfrac{16}{5}i_2 - \tfrac{11}{5}i_3 = 4 \qquad (3\text{-}81)$$

Next we multiply the first equation by $\frac{3}{5}$ and add it to the third equation giving

$$0i_1 - \tfrac{11}{5}i_2 + \tfrac{21}{5}i_3 = 6 \qquad (3\text{-}82)$$

Now if we multiply Eq. (3-81) by $\frac{11}{16}$, we may eliminate i_2 by adding the resulting equation to Eq. (3-82), giving

$$\tfrac{215}{5}i_3 = 140 \qquad (3\text{-}83)$$

The three equations

$$5i_1 - 2i_2 - 3i_3 = 10$$
$$0i_1 + \tfrac{16}{5}i_2 - \tfrac{11}{5}i_3 = 4 \qquad (3\text{-}84)$$
$$0i_1 + 0i_2 + 43i_3 = 140$$

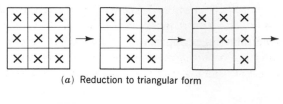

(a) Reduction to triangular form

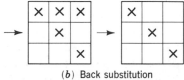

(b) Back substitution

Fig. 3-40. An illustration of the steps involved in the Gauss elimination process. The same pattern holds for larger systems of equations. Spaces without an X have a zero in them.

are in a form characteristic of the Gauss method, a triangular array as shown in Fig. 3-40. The last equation of (3-84) is the solution for i_3, which is $\frac{140}{43}$. That value substituted into the second equation gives the solution for i_2, and both values substituted into the first equation gives the value for i_1, as in Eqs. (3-79) and (3-80).

The basic idea of the Gauss method is the systematic elimination of variables which is called *triangularization*. The generalization of this example gives us a powerful method for arithmetic calculation. Assume that we have a set of n equations in n unknowns. The first equation in this set is

$$a_{11}x_1 + a_{12}x_2 + \ldots + a_{1n}x_n = A_1 \qquad (3\text{-}85)$$

where $a_{11} \neq 0$. If $a_{11} = 0$, then we must shuffle the equations so that the first one satisfies this requirement. Now we multiply this equation, Eq. (3-85), by a suitable factor so that when it is added to the next equation the term containing x_1 is eliminated. This process is repeated for each of the equations remaining so that we have Eq. (3-85) and $n-1$ additional equations in which the multiplier of the x_1 term is 0. One of these equations must have a coefficient of the x_2 term which is nonzero, for otherwise there is no solution to the set of equations. Let this equation be

$$b_{22}x_2 + b_{23}x_3 + \ldots + b_{2n}x_n = B_1 \qquad (3\text{-}86)$$

The basic operation is repeated with this equation: It is multiplied by a suitable factor and then added to each of the $n-2$ remaining equations, one at a time, with a different multiplying factor for each. The result will be $n-2$ equations in which the multiplier of x_2 will be zero. We continue this operation until the last term, which is

$$q_{nn}x_n = Q_1 \qquad (3\text{-}87)$$

and from which x_n is determined. We then work backwards determining each value of x_j until the last step in which x_1 is determined. These steps are known as *back substitution*. Thus the Gauss elimination method may be thought of in two parts: triangularization and back substitution. The algorithm is quite simple once you learn it!

3-8. DUALITY

Several analogous situations will have been noted in the preceding discussions. The statements of the two Kirchhoff laws were almost word for word with voltage substituted for current, independent loop for independent node pair, etc. Likewise, the integrodifferential equations that resulted from the application of the two Kirchhoff laws have been similar in appearance. This repeated similarity is only part of a larger pattern of identical behavior patterns in the roles played by voltage and current in network analysis. This similarity, with all the implications, is termed the principle of *duality*.

Consider the two networks completely different in physical appearance shown in Fig. 3-41. Inspection shows that the first might be analyzed to advantage on the loop basis and the other on the node basis. The resulting equations are

$$L\frac{di}{dt} + Ri + \frac{1}{C}\int i\, dt = v(t) \tag{3-88}$$

$$C'\frac{dv'}{dt} + G'v' + \frac{1}{L'}\int v'\, dt = i'(t) \tag{3-89}$$

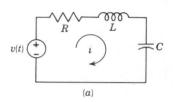

(a)

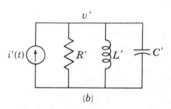

(b)

Fig. 3-41. Networks used to illustrate the concept of dual networks. The networks are duals when $R' = 1/R$, $L' = C$, $C' = L$, and $i'(t) = v(t)$.

These two equations specify identical mathematical operations, the only difference being in letter symbols. The solution of one equation is also the solution of the other. The two networks are *duals*. The roles of current and voltage in the two networks have been interchanged. As a word of caution, one network is not the equivalent of the other in the sense that one can replace the other.

An inspection of the terms of Eqs. (3-88) and (3-89) shows that the following are analogous quantities (with primes omitted).

$$Ri \quad \text{and} \quad Gv$$

$$L\frac{di}{dt} \quad \text{and} \quad C\frac{dv}{dt}$$

$$\frac{1}{C}\int i\, dt \quad \text{and} \quad \frac{1}{L}\int v\, dt$$

Evidently the following pairs are dual quantities.

$$R \quad \text{and} \quad G$$

$$L \quad \text{and} \quad C$$

$$\text{loop current, } i \quad \text{and} \quad v, \text{node-pair voltage}$$

$$\left.\begin{array}{c} q \text{ or} \\ \int i\,dt \end{array}\right\} \quad \text{and} \quad \left\{\begin{array}{c} \psi \text{ or} \\ \int v\,dt \end{array}\right.$$

$$\text{loop} \quad \text{and} \quad \text{node pair}$$

$$\text{short circuit} \quad \text{and} \quad \text{open circuit}$$

A simple graphical construction[4] may be followed in finding the dual of a network.

(1) Inside each loop place a node, giving it a number for convenience. Place an extra node, the datum node, external to the network. Arrange the same numbered nodes on a separate space on the paper for construction of the dual.

(2) Draw lines from node to node through the elements in the original network, traversing only one element at a time. For each element traversed in the original network, connect the dual element—from the listing just given—on the dual network being constructed.

(3) Continue this process until the number of possible paths through single elements is exhausted. (Should you slip and go through a connecting wire which is assumed to be a short circuit, the dual element is an open circuit.)

(4) The network constructed in this manner is the dual network. This construction may be checked by writing the differential equations for the two systems, one on the loop basis and the other on the node basis.

The graphical construction we have just outlined is illustrated in Fig. 3-42. If the graph of a network is planar, the method will always determine the dual network. If the network has a graph which is not planar, the method will always fail.

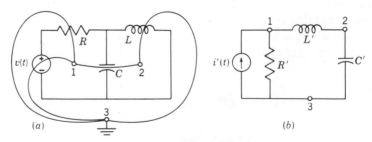

Fig. 3-42. Graphical construction employed in determining the dual for a planar network. The networks are duals when $i'(t) = v(t)$, $R' = 1/R$, $L' = C$, and $C' = L$.

[4]M. F. Gardner and J. L. Barnes, *Transients in Linear Systems* (John Wiley & Sons, Inc., New York, 1942), pp. 46ff.

3-9. STATE VARIABLE ANALYSIS

The third formulation of network equations to be described in this chapter is that using state variables. The state variables usually selected for network analysis are the capacitor voltages and the inductor currents. These replace the loop currents and the node-to-datum voltages in the two methods previously studied. These variables have the same property as those previously studied: Their determination permits all other voltages and currents in the network to be found.

The particular advantage of the state variable formulation is that it is done in a form especially suited for computer solution, either the digital or the analog computer. In addition, this formulation is popular in describing control systems—in fact, systems in general, including the time-variable and nonlinear cases.

We will introduce this method of analysis in terms of the simple network shown in Fig. 3-43. For this series RLC network, the state variables are the capacitor voltage v_C and the inductor current i_L. At node 3, we write from Kirchhoff's current law

$$C\frac{dv_C}{dt} = i_L \tag{3-90}$$

From the Kirchhoff voltage law applied around the only loop,

$$L\frac{di_L}{dt} = v_s - i_L R - v_C \tag{3-91}$$

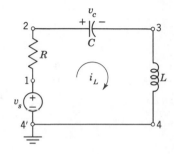

Fig. 3-43. *RLC* network used to introduce state-variable method.

We rearrange these two equations to the form

$$\frac{dv_C}{dt} = 0v_C + \frac{1}{C}i_L$$
$$\frac{di_L}{dt} = \frac{-1}{L}v_C - \frac{R}{L}i_L + v_s \tag{3-92}$$

which are said to be in *state form*. The generalization of these equations is easily constructed using x as the general state variable and y as the general input.

$$\frac{dx_1}{dt} = a_{11}x_1 + a_{12}x_2 + \ldots + a_{1n}x_n + y_1$$

$$\frac{dx_2}{dt} = a_{21}x_1 + a_{22}x_2 + \ldots + a_{2n}x_n + y_2$$

$$\vdots \tag{3-93}$$

$$\frac{dx_n}{dt} = a_{n1}x_1 + a_{n2}x_2 + \ldots + a_{nn}x_n + y_n$$

Given that we wish to write network equations in this form, the question is how this might be done in general with the assurance that the equations so written are independent. As mentioned earlier, the motivation is that the state variable form, Eqs. (3-93), is the most convenient for computer solution.

First, we observe intuitively that if we desire terms like dv_C/dt in Eq. (3-92), then we must write node equations involving capacitors; similarly, the di_L/dt terms suggest loop equations. How might these be written systematically?

The first solution to this problem applied specifically to the network problem was given by T. R. Bashkow in 1957. The strategy he suggested is accomplished in the following steps.

(1) Select a tree containing all capacitors[5] but no inductors.
(2) The state variables are the branch capacitor voltages in this tree and the inductor currents of the chords.
(3) Write a node equation for each capacitor.
(4) Manipulate each equation, if necessary, until it involves only the variables selected in (2) plus the inputs.
(5) Write a loop equation using each inductor as a chord in the tree of (1).
(6) Repeat step (4).
(7) Manipulate the equations as may be necessary (division by constants, for example) until they appear in the standard form of Eq. (3-93).

Let us illustrate these steps by an example.

EXAMPLE 14

Consider the network shown in Fig. 3-44(*a*). Following the steps just given:

(1) The tree indicated by the dark lines meets the requirement that it contain all capacitors, but no inductors.
(2) The state variables are shown on the figure as v_C, i_1, and i_2, with reference directions indicated.
(3) At node c, the Kirchhoff current law gives

$$C\frac{dv_C}{dt} = -i_1 - i_2 \qquad (3\text{-}94)$$

(4) This equation is in appropriate form, except for division by C.

---

[5]If a network contains a loop of capacitors or a node to which only inductors are connected, then the method given must be modified. Consult the advanced textbooks listed in Further Reading.

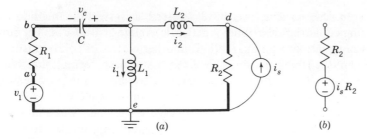

Fig. 3-44. The network for Example 15 with the tree selected shown by heavy lines.

(5) The first loop is formed by the chord containing L_1. The Kirchhoff voltage law gives

$$L_1 \frac{di_1}{dt} = -R_1 i_1 - R_1 i_2 + v_C + v_1 \qquad (3\text{-}95)$$

The second loop is formed by the chord containing L_2. The current source and resistor R_2 are first converted to the form shown in (b) of the figure. Then

$$L_2 \frac{di_2}{dt} = -R_1 i_1 - (R_1 + R_2) i_2 + v_C - R_2 i_s \qquad (3\text{-}96)$$

(6) This equation is written in terms of the state variables of (2) and requires only division by L_2.

(7) In standard form, we have

$$\frac{dv_C}{dt} = 0 v_C - \frac{1}{C} i_1 - \frac{1}{C} i_2$$

$$\frac{di_1}{dt} = \frac{1}{L_1} v_C - \frac{R_1}{L_1} i_1 - \frac{R_1}{L_1} i_2 + \frac{1}{L_1} v_1 \qquad (3\text{-}97)$$

$$\frac{di_2}{dt} = \frac{1}{L_2} v_C - \frac{R_1}{L_1} i_1 - \frac{1}{L_2}(R_1 + R_2) i_2 - \frac{R_2}{L_2} i_s$$

In matrix form, this equation becomes

$$
\begin{bmatrix} \dfrac{dv_C}{dt} \\[2mm] \dfrac{di_1}{dt} \\[2mm] \dfrac{di_2}{dt} \end{bmatrix}
=
\begin{bmatrix} 0 & -\dfrac{1}{C} & -\dfrac{1}{C} \\[2mm] \dfrac{1}{L_1} & -\dfrac{R_1}{L_1} & -\dfrac{R_1}{L_1} \\[2mm] \dfrac{1}{L_2} & -\dfrac{R_1}{L_1} & \dfrac{-(R_1+R_2)}{L_2} \end{bmatrix}
\begin{bmatrix} v_C \\[2mm] i_1 \\[2mm] i_2 \end{bmatrix}
+
\begin{bmatrix} \dfrac{1}{L_1} & \dfrac{-R_2}{L_2} \end{bmatrix}
\begin{bmatrix} v_1 \\[2mm] i_s \end{bmatrix}
\qquad (3\text{-}98)
$$

or more compactly

$$\dot{\mathbf{x}} = \mathbf{A}\mathbf{x} + \mathbf{B}\mathbf{y} \qquad (3\text{-}99)$$

Here $\mathbf{x}$ is the state matrix of *state vector*, $\mathbf{A}$ is a constant matrix known as the *Bashkow A matrix*, $\mathbf{B}$ is a constant vector, and $\mathbf{y}$ is the *vector input*. We will be especially interested in $\mathbf{x}(0)$ which is the *initial state* and will be studied in Chapter 5.

In this section, we have stressed a systematic method for writing equations describing a network in state-space form. The solution of equations like Eq. (3-98) is another matter. If solution is to be accomplished by computer, then the state-space formulation offers advantages, and writing the equations correctly is the only requirement. If the network to be analyzed contains one element or several elements that are nonlinear or time-variable, then the state-space formulation is recommended and, of course, solution by computer methods is the only practical possibility. If solution is to be accomplished using a pencil and pad of paper, then it is ordinarily simpler to use the node or loop formulations.

FURTHER READING

CHIRLIAN, PAUL M. *Basic Network Theory*, McGraw-Hill Book Company, New York, 1969. Chapters 3 and 7. Chapter 7 discusses a graphical interpretation of the solution of state-space equations.

CRUZ, JOSE B., JR., AND M. E. VAN VALKENBURG, *Signals in Linear Circuits*, Houghton Mifflin Company, Boston, Mass., 1974.

DESOER, CHARLES A., AND ERNEST S. KUH, *Basic Circuit Theory*, McGraw-Hill Book Company, New York, 1969. Chapters 11 and 12.

GUILLEMIN, ERNST A., *Introductory Circuit Theory*, John Wiley & Sons, Inc., New York, 1953. A classic!

HAYT, JR., WILLIAM H., AND JACK E. KEMMERLY, *Engineering Circuit Analysis*, 2nd ed., McGraw-Hill Book Company, New York, 1971.

HUANG, THOMAS S., AND RONALD R. PARKER, *Network Theory: An Introductory Course*, Addison-Wesley Publishing Company, Reading, Mass., 1971.

KARNI, SHLOMO, *Intermediate Network Analysis*, Allyn and Bacon, Inc., Boston, 1971. Chapters 10 and 11.

LEON, BENJAMIN J., AND PAUL A. WINTZ, *Basic Linear Networks for Electrical and Electronics Engineers*, Holt, Rinehart & Winston, Inc., New York, 1970.

ROHRER, RONALD A., *Circuit Theory: An Introduction to the State Variable Approach*, McGraw-Hill Book Company, New York, 1970.

WARD, JOHN R., AND ROBERT D. STRUM, *State Variable Analysis*, Prentice-Hall, Inc., Englewood Cliffs, N. J., 1970. This is a programmed instruction manual, with emphasis on the solution of state equations.

DIGITAL COMPUTER EXERCISES

This chapter is rich in topics related to the use of the digital computer. Elementary possibilities include matrix multiplication as described in references from Appendix E-3.2 and the Gauss elimination method from references in Appendix E-4.1. Consider also the analysis of resistive ladder networks as described in references in Appendix E-4.2. For specific suggestions, see Huelsman, reference 7 of Appendix E-10, for the resistive network as related to the solution of simultaneous equations in Chapter 7 and the solution of equations for the *RLC* networks of Chapter 6. More advanced possibilities include the solution of state equations by methods described in references given in Appendix E-4.3 and the use of canned programs for network analysis as given in Appendix E-8.4.

PROBLEMS

3-1. What must be the relationship between C_{eq} and C_1 and C_2 in (a) of the figure of the networks if (a) and (c) are equivalent? Repeat for the network shown in (b).

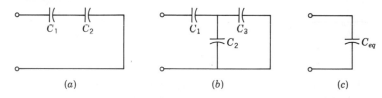

Fig. P3-1.

3-2. What must be the relationship between L_{eq} and L_1, L_2 and M for the networks of (a) and of (b) to be equivalent to that of (c)?

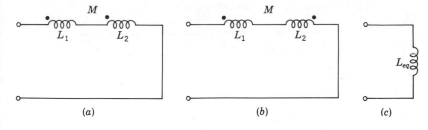

Fig. P3-2.

3-3. Repeat Prob. 3-2 for the three networks shown in the accompanying figure.

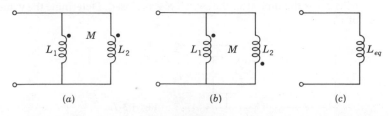

Fig. P3-3.

3-4. The network of inductors shown in the figure is composed of a 1-H inductor on each edge of a cube with the inductors connected to the vertices of the cube as shown. Show that, with respect to vertices a and b, the network is equivalent to that in (b) of the figure when $L_{eq} = \frac{5}{6}$ H. Make use of symmetry in working this problem, rather than writing Kirchhoff laws.

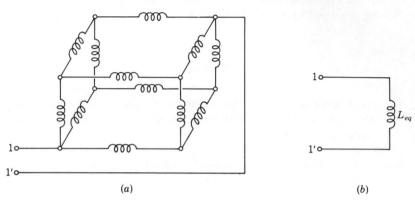

(a)   (b)

Fig. P3-4.

3-5. In the networks of Prob. 3-4, each 1-H inductor is replaced by a 1-H capacitor, and L_{eq} is replaced by C_{eq}. What must be the value of C_{eq} for the two networks to be equivalent?

3-6. This problem may be solved using the two Kirchhoff laws and voltage-current relationships for the elements. At time t_0 after the switch K was closed, it is found that $v_2 = +5$ V. You are required to determine the value of $i_2(t_0)$ and $di_2(t_0)/dt$.

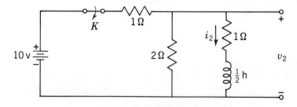

Fig. P3-6.

3-7. This problem is similar to Prob. 3-6. In the network given in the figure, it is given that $v_2(t_0) = 2$ V, and $(dv_2/dt)(t_0) = -10$ V/sec, where t_0 is the time after the switch K was closed. Determine the value of C.

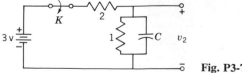

Fig. P3-7.

The series of problems described in the following table all pertain to the network of (g) of the figure with the network in A and B specified in the table. In A, two entries in the column implies a series connection of elements, while in B, two entries implies a parallel connection of elements. In each case, all initial conditions are equal to zero. For the specified waveform for v_2, you are required to determine v_1 in the form of a sketch of the waveform as it might be seen on a cathode ray oscilloscope. Evaluate significant amplitudes, slopes, and so on.

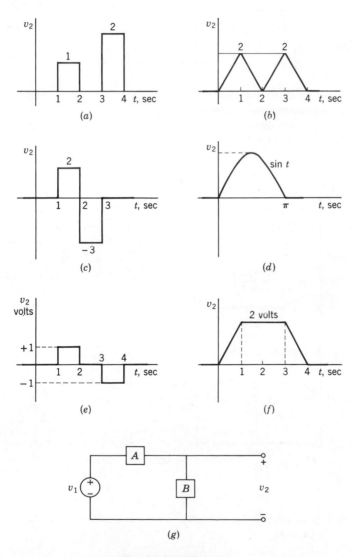

Figs. P3-8 to P3-16.

	Network of A	*Network of B*	*Waveforms of v_2*
3-8.	$R = 2$	$L = \frac{1}{2}$	a, b, c, d, e, f
3-9.	$C = \frac{1}{2}$	$L = 1$	a, b, c, d, e, f
3-10.	$C = \frac{1}{2}, R = 1$	$L = 2$	a, b, c, d, e, f
3-11.	$C = 1, R = \frac{1}{2}$	$L = \frac{1}{2}, R = 1$	a, b, c, d, e, f
3-12.	$R = 2$	$C = 1$	b, d, f
3-13.	$R = 1$	$R = 2, C = 1$	b, d, f
3-14.	$R = 2$	$R = 1, C = 1$	b, d, f
3-15.	$L = \frac{1}{2}$	$R = 1, C = \frac{1}{2}$	b, d, f
3-16.	$L = 1, R = 1$	$R = 1, C = \frac{1}{2}$	b, d, f

3-17. For each of the four networks shown in the figure, determine the number of independent loop currents, and the number of independent node-to-node voltages that may be used in writing equilibrium equations using the Kirchhoff laws.

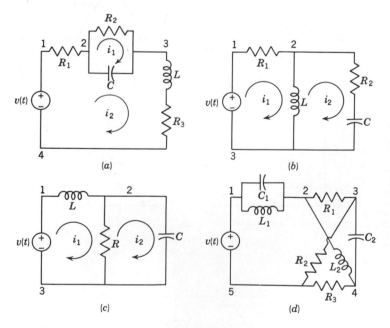

Fig. P3-17.

3-18. Repeat Prob. 3-17 for each of the four networks shown in the figure on page 91.

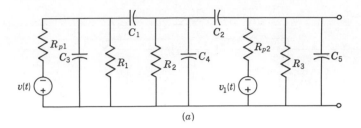

(a)

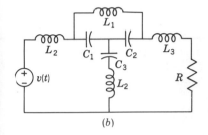

(b)

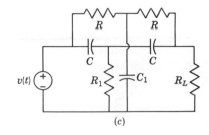

(c)

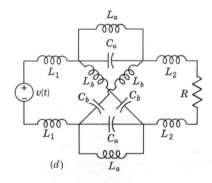

(d)

Fig. P3-18.

3-19. Demonstrate the equivalence of the networks shown in Fig. 3-17 and so establish a rule for converting a voltage source in series with an inductor into an equivalent network containing a current source.

3-20. Demonstrate that the two networks shown in Fig. 3-18 are equivalent.

3-21. Write a set of equations using the Kirchhoff voltage law in terms of appropriate loop-current variables for the four networks of Prob. 3-17.

3-22. Make use of the Kirchhoff voltage law to write equations on the loop basis for the four networks of Prob. 3-18.

3-23. Write a set of equilibrium equations on the loop basis to describe the network in the accompanying figure. Note that the network contains one controlled source. Collect terms in your formulation so that your equations have the general form of Eqs. (3-47).

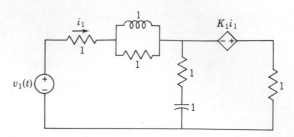

Fig. P-3-23.

3-24. For the coupled network of the figure, write loop equations using the Kirchhoff voltage law. In your formulation, use the three loop currents which are identified.

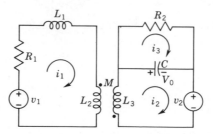

Fig. P3-24.

3-25. The network of the figure is that of Fig. 3-30 but with different loop-current variables chosen. Using the specified currents, write the Kirchhoff voltage law equations for this network.

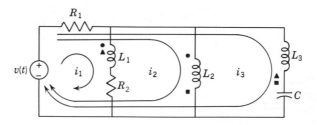

Fig. P3-25.

3-26. A network with magnetic coupling is shown in the figure. For the network, $M_{12} = 0$. Formulate the loop equations for this network using the Kirchhoff voltage law.

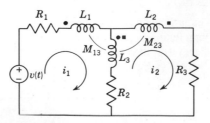

Fig. P3-26.

3-27. Write the loop-basis voltage equations for the magnetically coupled network of Fig. P5- 22 with K closed.

3-28. Write equations using the Kirchhoff current law in terms of node-to-datum voltage variables for the four networks of Prob. 3-17.

3-29. Making use of the Kirchhoff current law, write equations on the node basis for the four networks of Prob. 3-18.

3-30. For the given network, write the node-basis equations using the node-to-datum voltages as variables. Collect terms in your formulation so that the equations have the general form of Eqs. (3-59).

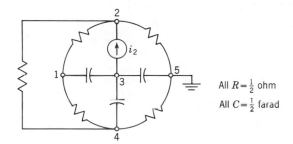

All $R = \frac{1}{2}$ ohm

All $C = \frac{1}{2}$ farad

Fig. P3-30.

3-31. The network in the figure contains one independent voltage source and two controlled sources. Using the Kirchhoff current law, write node-basis equations. Collect terms in the formulation so that the equations have the general form of Eqs. (3-59).

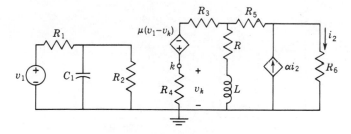

Fig. P3-31.

3-32. The network of the figure is a model suitable for "midband" operation of the "cascode-connected" MOS transistor amplifier. Analyze the

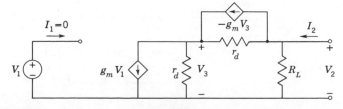

Fig. P3-32.

network on (a) the loop basis, and (b) the node basis. Write the resulting equations in matrix form, but do not solve them.

3-33. In the network of the figure, each branch contains a 1-Ω resistor, and four branches contain a 1-V voltage source. Analyze the network on the loop basis, and organize the resulting equations in the form of a chart as in Example 11. Do not solve the equations.

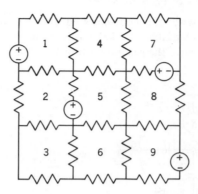

Fig. P3-33.

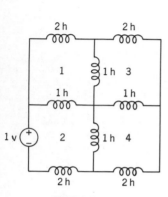

Fig. P3-34

3-34. Repeat Prob. 3-33 for the network of the accompanying figure. In addition, write equations on the node basis, and arrange the equations in the form of the chart of Example 13.

3-35. In the network of the figure, $R = 2\ \Omega$ and $R_1 = 1\ \Omega$. Write equations on (a) the loop basis, and (b) the node basis, and simplify the equations to the form of the chart used in Examples 11 and 13.

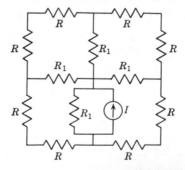

Fig. P3-35.

3-36. For the network shown in the figure, determine the numerical value of the branch current i_1. All sources in the network are time invariant.

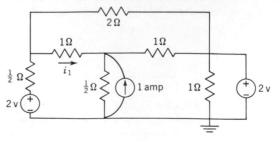

Fig. P3-36.

3-37. In the network of the figure, all sources are time invariant. Determine the numerical value of i_2.

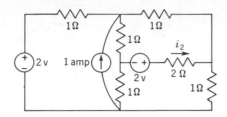

Fig. P3-37.

3-38. In the given network, all sources are time invariant. Determine the branch current in the 2-Ω resistor.

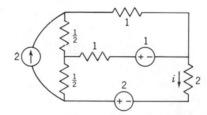

Fig. P3-38.

3-39. In the network of the figure, all voltage sources and current source are time invariant, and all resistors have the value $R = \frac{1}{2}\,\Omega$. Solve for the four node-to-datum voltages.

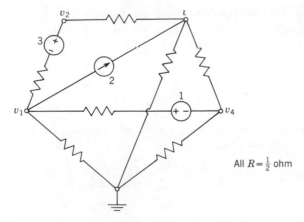

Fig. P3-39.

3-40. In the given network, node d is selected as the datum. For the specified element and source values, determine values for the four node-to-datum voltages.

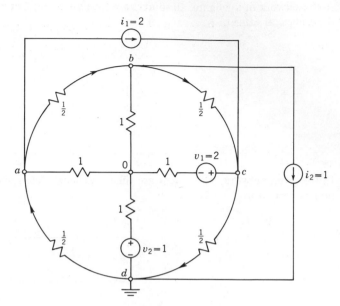

Fig. P3-40.

3-41. Evaluate the determinant:

$$\begin{vmatrix} 2 & -1 & 0 & 0 \\ -1 & 3 & -2 & 0 \\ 0 & -2 & 3 & -1 \\ 0 & 0 & -1 & 2 \end{vmatrix}$$

3-42. Evaluate the determinant:

$$\begin{vmatrix} 1 & -2 & 0 & 3 & 4 \\ -1 & 4 & -1 & 1 & 0 \\ 2 & 0 & 1 & 1 & 3 \\ 4 & -2 & 4 & 2 & -1 \\ 3 & 1 & 3 & -2 & 1 \end{vmatrix}$$

3-43. Solve the following system of equations for i_1, i_2, and i_3, using Cramer's rule.

$$3i_1 - 2i_2 + 0i_3 = 5$$
$$-2i_1 + 9i_2 - 4i_3 = 0$$
$$0i_1 - 4i_2 + 9i_3 = 10$$

3-44. Solve the following system of equations for the three unknowns, i_1, i_2, and i_3 by Cramer's rule.

$$8i_1 - 3i_2 - 5i_3 = 5$$
$$-3i_1 + 7i_2 - 0i_3 = -10$$
$$-5i_1 + 0i_2 + 11i_3 = -10$$

3-45. Solve the equations of Prob. 3-43 using the Gauss elimination method.

3-46. Solve the equations of Prob. 3-44 using the Gauss elimination method.

3-47. Determine i_1, i_2, i_3, and i_4 from the following system of equations.

$$6i_1 - 8i_2 - 10i_3 + 12i_4 = 8$$
$$2i_1 - 4i_2 + 5i_3 + 6i_4 = 33$$
$$-8i_1 + 20i_2 + 14i_3 - 16i_4 = 10$$
$$5i_1 + 7i_2 + 2i_3 - 10i_4 = -15$$

3-48. Consider the equations

$$3x - y - 3z = 1$$
$$x - 3y + z = 1$$
$$4x + 0y - 5z = 1$$

(a) Is (4, 2, 3) a solution? Is $(-1, -1, -1)$ a solution? (b) Can these equations be solved by determinants? Why? (c) What can you conclude regarding the three lines represented by these equations?

3-49. Find duals for the four networks of Prob. 3-17.

3-50. Find the dual networks for the four networks given in Prob. 3-18.

3-51. Find the dual of the network of Prob. 3-31.

3-52. If one exists, find a dual of the network of Prob. 3-40.

3-53. Analyze the network of Prob. 3-17(c) using the state variable formulation.

3-54. Consider the network shown in Prob. 3-23. Analyze this network using appropriate state variables.

3-55. Analyze the network shown in Fig. P3-18(b) using the state variable formulation.

3-56. Analyze the network of Prob. 3-30 using state variables.

3-57. Apply the method of state variables to analyze the network shown in Fig. P3-31.

3-58. The element represented in the network is a *gyrator* which is described by the equations

$$v_1 = R_0 i_2$$
$$v_2 = -R_0 i_1$$

Find the two-element equivalent network shown in (b) of the figure.

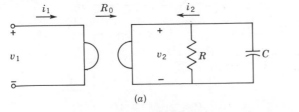 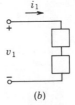

(a) (b)

Fig. P3-58.

3-59. For the gyrator-*RL* network of the figure, write the differential equation relating v_1 to i_1. Find a two-element equivalent network, as in Prob. 3-49, in which neither of the elements is a gyrator.

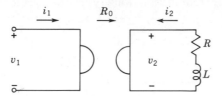

Fig. P3-59.

3-60. In the network of (*a*) of the figure, all self inductance values are 1 H, and mutual inductance values are $\frac{1}{2}$ H. Find L_{eq}, the equivalent inductance, shown in (*b*) of the figure.

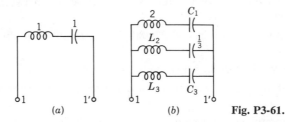

Fig. P3-60.

3-61. It is intended that the two networks of the figure be equivalent with respect to the pair of terminals which are identified. What must be the values for C_1, L_2, and L_3?

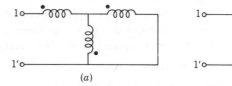

Fig. P3-61.

3-62. It is intended that the two networks of the figure be equivalent with respect to two pairs of terminals, terminal pair 1-1′ and terminal pair 2-2′. For this equivalence to exist, what must be the values for C_1, C_2, and C_3?

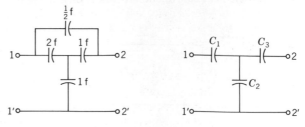

Fig. P3-62.

First-order Differential Equations 4

4-1. GENERAL AND PARTICULAR SOLUTIONS

In this chapter we will study a number of techniques for the solution of the simplest form of linear differential equations with constant coefficients which describe linear networks, that of the first order written

$$a_0 \frac{di(t)}{dt} + a_1 i(t) = v(t) \tag{4-1}$$

which is the form for $n = 1$ of the nth-order equation

$$a_0 \frac{d^n i}{dt^n} + a_1 \frac{d^{n-1} i}{dt^{n-1}} + \dots + a_{n-1} \frac{di}{dt} + a_n i = v(t) \tag{4-2}$$

In these equations, $a_0, a_1, a_2, \dots$ are constants; $i(t)$, the *dependent variable*, is usually a current, a voltage, a charge, or a flux; t, the *independent variable* is time; and $v(t)$ is the *forcing function* representing a linear combination of voltage sources and current sources. The solution of the equation will often be called the *response*; similarly, $v(t)$ is sometimes called the *excitation*.

Assume that we are given a network of passive elements and sources which is initially in a known state with respect to all voltages and currents. At a reference instant of time designated $t = 0$, the system is altered in a manner that can be represented by the opening or closing of one or more switches. The objective of analysis is to obtain equations for current, voltage, charge, etc., in terms of time

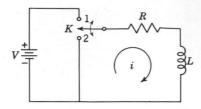

Fig. 4-1. *RL* network with the switch *K* moved from position 1 to position 2 at the reference time, $t = 0$.

measured from the instant that equilibrium was altered by the switching.

In the network shown in Fig. 4-1, the switch *K* is changed from position 1 to position 2 at the reference time $t = 0$.[1] After the switching has taken place, the Kirchhoff voltage equation is

$$L \frac{di}{dt} + Ri = 0 \tag{4-3}$$

This is a homogeneous first-order linear differential equation with constant coefficients. It can be solved if the variables can be separated. This may be accomplished by rearranging Eq. (4-3) in the form

$$\frac{di}{i} = -\frac{R}{L} dt \tag{4-4}$$

With the variables separated, the equation can be integrated to give

$$\ln i = -\frac{R}{L} t + K \tag{4-5}$$

where ln designates that the logarithm is of base $e = 2.718 \ldots$. To simplify the form of this equation, the constant *K* is redefined in terms of the logarithm of another constant as

$$K = \ln k \tag{4-6}$$

Equation (4-5) may then be written

$$\ln i = \ln e^{-Rt/L} + \ln k \tag{4-7}$$

since, by the definition of a logarithm, $\ln e^x = x$, or $\log_{10} 10^x = x$. Also, from logarithms we know that

$$\ln y + \ln z = \ln yz \tag{4-8}$$

so that Eq. (4-7) may be written

$$\ln i = \ln (ke^{-Rt/L}) \tag{4-9}$$

With the equation in this form, the antilogarithm may be taken to give,

$$i = ke^{-Rt/L} \tag{4-10}$$

Redefining the constant *K* as the logarithm of another constant has

---

[1]It is assumed that the switch is a "make-before-break" type and that the transition from position 1 to position 2 does not cause an interruption of the current *i*. The switch is marked *K* (for *knife* switch) to reserve *S* for another role.

simplified the form of the solution. Equation (4-10) is the network response or solution. This solution is free of derivatives and expresses the relationship between the dependent and independent variables. That it is the solution can be verified by substituting Eq. (4-10) into Eq. (4-3).

In the form of Eq. (4-10), the solution is known as the *general solution*. If the constant of integration is evaluated, the solution is a *particular solution*. The general solution applies to any number of situations. A particular solution fits the specifications of a particular problem.

To evaluate the constant k, we must know something new about the problem, such as any pair of values of i and t. In this particular problem, we know that the current after switching has taken place must be just the same as before switching because of the inductor in the circuit. Thus at $t = 0$, we know that the current has the value

$$i(0) = \frac{V}{R} \tag{4-11}$$

This value is known as the *initial condition* of the circuit or the *zero state* of the inductor. Substituting this required condition into Eq. (4-10) gives

$$\frac{V}{R} = ke^0 = k \tag{4-12}$$

The particular solution of this example becomes

$$
\begin{aligned}
i &= \frac{V}{R}\, e^{-Rt/L}, & t \geq 0 \\
&= \frac{V}{R}, & t < 0
\end{aligned}
\tag{4-13}
$$

This equation is plotted for various parameter values in Fig. 4-2. In (*a*) of that figure, V and R are held constant and L is allowed to vary; in (*b*) R is variable for constant V and L. These curves are known

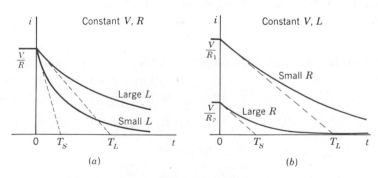

Fig. 4-2. The decaying exponential form of Eq. 4-13: (*a*) for two values of L for R constant, and (*b*) for two values of R for L constant.

as decaying exponentials; the characteristics of the decay are determined by only R and L, the two parameters of the network.

A physical interpretation of this result begins by considering the energy in the network. Prior to the instant of switching, the energy $Li^2/2$ is stored in the inductor, and energy is dissipated by the resistor at the rate Ri^2 as determined in Chapter 1. After the switching, the energy source is removed, and energy stored in the inductor is totally dissipated by the resistor with the passage of time. Since the energy is dissipated at a maximum rate at the initial instant after switching, the current decreases most rapidly at that time. The rate of decrease is controlled by the ratio L/R. Eventually, all energy is dissipated and the current becomes zero.

These statements are illustrated by the curves of Fig. 4-2. In (a), the current decreases less rapidly for large L than for small L; in (b), the current decreases less rapidly for the smaller value of R.

In the network of Fig. 4-3, switching action occurs at $t = 0$ so that the *zero state* of the capacitor is $v_C(0) = V$, the battery voltage. Prior to switching, the current has zero value. At $t = 0$, the current changes abruptly from zero value to $-V/R$. With the switch closed, the Kirchhoff voltage equation is

$$\frac{1}{C} \int_{-\infty}^{t} i(t)\,dt + Ri(t) = \frac{1}{C} \int_{0}^{t} i(t)\,dt + V + Ri(t) = 0, \qquad t > 0$$

(4-14)

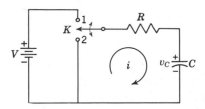

Fig. **4-3.** *RC* network with a capacitor voltage $v_c = V$ at the time the switch is moved from position 1 to position 2.

Differentiating this equation and dividing through by R, we find that

$$\frac{di}{dt} + \frac{1}{RC} i = 0$$

(4-15)

Now this equation has the same form as Eqs. (4-3) and (4-4), and the general solution is the same as Eq. (4-10) with R/L replaced by $1/RC$:

$$i = ke^{-t/RC}$$

(4-16)

Since $i(0) = -V/R$, we see that the particular solution for the network of Fig. 4-3 is

$$i = \frac{-V}{R} e^{-t/RC}, \qquad t \geq 0$$
$$= 0, \qquad\qquad t < 0$$

(4-17)

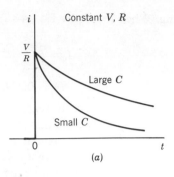

(a)

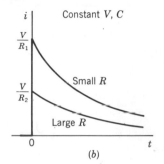

(b)

Fig. 4-4. The current waveform which is a decaying exponential: (a) for two values of C with constant R, and (b) for two values of R with constant C.

The physical explanation of this result parallels that given for the RL series network. The energy stored in the capacitor is $Cv^2/2$. Charge flows through the resistor from one plate of the capacitor to the other and energy is dissipated in the resistor at the rate Ri^2. The reduction in stored energy due to dissipation reduces the voltage of the capacitor and the current, which is $i = v/R$, is reduced, eventually becoming zero. As shown in Fig. 4-4, the rate of decay of the current is controlled by the product RC: the larger the value of RC, the smaller will be the rate of decay.

4-2. TIME CONSTANTS

The responses given by Eqs. (4-13) and (4-17) for $t \geqq 0$ may be written in the general form

$$\frac{i}{I_0} = e^{-t/T} \tag{4-18}$$

where I_0 is the initial value of current at $t = 0$ and T is the *time constant* of the system. The form of Eq. (4-18) is the solution of all homogeneous first-order differential equations, where I_0 and T have different values for different problems. The physical significance attached to the time constant is of great importance in electrical engineering. When $t = T$, by Eq. (4-18),

$$\frac{i(T)}{I_0} = e^{-1} \cong 0.37 \qquad (4\text{-}19)$$

or

$$i(T) = 0.37I_0 \qquad (4\text{-}20)$$

In other words, the current decreases to 37 per cent of its initial value in one time constant. By a similar computation, it can be shown that the current decreases to approximately 2 per cent of its initial value in four time constants. Several other useful values of $e^{-t/T}$ for integer values of t/T are given in the following tabulation:

t/T	i/I_0
0	1.0
1	0.37
2	0.14
3	0.05
4	0.018
5	0.0067

With these values, a plot of i/I_0 as a function of t/T may be made as in Fig. 4-5.

In the next section, we will be concerned with responses to a constant forcing function which may be written in the general form

$$\frac{i}{I_0} = 1 - e^{-t/T} \qquad (4\text{-}21)$$

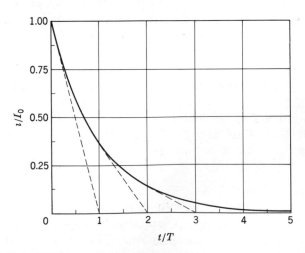

Fig. 4-5. The normalized form of the decaying exponential in which I_o is the initial value of the current, and T is the time constant.

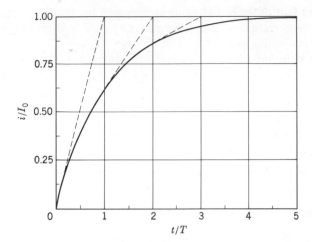

Fig. 4-6. The normalized saturating exponential given to Eq. 4-21 in which I_o is the final value of the current and T is the time constant.

This equation is plotted in Fig. 4-6. Observe that when $t = T$,

$$i(T) = (1 - 0.37)I_0 = 0.63I_0 \qquad (4\text{-}22)$$

meaning that the current reaches 63 per cent of its final value in one time constant. The current increases to approximately 98 per cent of its final value in four time constants.

A useful artifice in visualizing the time constant associated with a given exponential response is contained in the observation that the tangent to the equation $i = I_0 e^{-t/T}$ at $t = 0$ intersects the $i = 0$ line at $t = T$. Thus, if the current decreases at the initial rate, it would be reduced to zero value in one time constant. Similar observations apply to the equation $i = I_0(1 - e^{-t/T})$ which has a tangent at $t = 0$ which intersects the line $i = I_0$ at $t = T$. These two lines are shown on Figs. 4-5 and 4-6.

The time constant is useful in comparing one response with another. It is not possible to compare times at which the response disappears (or reaches its steady state) since, mathematically at least, this requires infinite time. However, the time interval for an exponential function to decrease to 37 per cent of its initial value (or increase to 63 per cent of its final value) is conveniently measured and used as a standard for comparison. As an example, consider a series RC circuit which has a general solution,

$$i = I_0 e^{-t/RC} \qquad (4\text{-}23)$$

The time constant for the circuit is $T = RC$. Suppose that R has the value of 100 Ω and C is 100 pF; then $T = 100 \times 1000 \times 10^{-12} = 0.1$ μsec. However, if $R = 1000\ M\Omega$ and $C = 1\ \mu F$, then $T = 1000$ sec, or 17 min. For one combination of R and C, the current would decrease to 37 per cent of the initial value in the small time of 0.1 μsec; for the other, the current would require about 17 min to decrease to 37 per cent of the initial value.

In experimentally recording a transient, the accuracy of measurement is often of the order of 1 or 2 per cent. For this reason, a transient is sometimes assumed to have disappeared when it reaches 2 per cent of the final value (as accurately as can be determined). Since the time to reach 2 per cent of the final value (or 98 per cent in the case of an increasing exponential) is *four time constants*, it is often assumed that a transient disappears in four time constants. This basis is used sometimes to measure the time constant of a system.

4-3. THE INTEGRATING FACTOR

Consider a nonhomogeneous equation written

$$\frac{di}{dt} + Pi = Q \tag{4-24}$$

where P is a constant and Q may be a function of the independent variable t or a constant. The equation is not altered if every term is multiplied by the same factor. Suppose that we multiply Eq. (4-24) by the quantity e^{Pt}, which is known as an *integrating factor*.[2] There results

$$e^{Pt}\frac{di}{dt} + Pie^{Pt} = Qe^{Pt} \tag{4-25}$$

That this multiplication by a factor "pulled out of the hat" has made possible the solution of Eq. (4-24) can be recognized by recalling the equation for the derivative of a product:

$$d(xy) = x\,dy + y\,dx \tag{4-26}$$

By letting $x = i$ and $y = e^{Pt}$, we have

$$\frac{d}{dt}(ie^{Pt}) = e^{Pt}\frac{di}{dt} + ie^{Pt}P \tag{4-27}$$

which is the left-hand side of Eq. (4-25); thus we have

$$\frac{d}{dt}(ie^{Pt}) = Qe^{Pt} \tag{4-28}$$

This equation may be integrated to give

$$ie^{Pt} = \int Qe^{Pt}\,dt + K \tag{4-29}$$

or[3]

$$i = e^{-Pt}\int Qe^{Pt}\,dt + Ke^{-Pt} \tag{4-30}$$

[2]If P is a function of time, the proper integrating factor is $e^{\int P\,dt}$. See Prob. 4-8.
[3]This equation may also be written

$$i = e^{-Pt}\int_0^t Qe^{Pt}\,dt$$

The form of Eq. (4-30) is used to stress the separation of the solution into two parts.

The first term in Eq. (4-30) is known as the *particular integral*; the second is known as the *complementary function*. Note that the particular integral does not contain the arbitrary constant, and the complementary function does not depend on the forcing function Q.

For any network problem, P will be a positive constant determined by the network parameters, and Q will be either the forcing function or a derivative of the forcing function. In the limit, the complementary function must approach zero, because P is a positive constant; that is,

$$\lim_{t \to \infty} Ke^{-Pt} = 0 \qquad (4\text{-}31)$$

Thus the value of i as time approaches infinity is

$$i(\infty) = \lim_{t \to \infty} i(t) = \lim_{t \to \infty} e^{-Pt} \int Qe^{Pt} \, dt \qquad (4\text{-}32)$$

When the particular integral does not approach zero in the limit, its value as t approaches ∞ is spoken of as the *steady-state value*. For this case, the particular integral must contain no exponential factor, otherwise it would reduce to zero. In electrical engineering, the steady-state values most frequently encountered have the forms

$$i = A \sin(\omega t + \phi) \quad \text{and} \quad i = \text{a constant} \qquad (4\text{-}33)$$

Let the general solution of Eq. (4-30) be written as the sum of the two parts of the solution, letting i_P be the particular integral and i_C be the complementary function; thus

$$i = i_P + i_C \qquad (4\text{-}34)$$

If i_P has either of the forms of Eq. (4-33), it may be written as a steady-state value, designated i_{ss}. A convention has been established for calling the remaining term i_C the *transient* portion of the solution. By this convention, the response is made up of two separate parts:

$$i = i_{ss} + i_t \qquad (4\text{-}35)$$

The steady-state value is regarded as having been established at $t = 0$, and the transient must adjust itself, mathematically, to account for the response at $t = 0$ and all other times. This is an arbitrary division of the solution which nevertheless has utility as a conceptual aid. The division of solution is made purely by convention; the individual electron in a current has no way of knowing whether it is in the transient or the steady-state division of the current.

EXAMPLE 1

To illustrate the transient and steady-state portion of the solution to a problem, consider the network of Fig. 4-1 with the switch moved from position 2 to position 1 at $t = 0$. The Kirchhoff voltage equation is, after division by L,

$$\frac{di}{dt} + \frac{R}{L} i = \frac{V}{L} \tag{4-36}$$

Comparing this equation to Eq. (4-24), we see that

$$P = \frac{R}{L} \quad \text{and} \quad Q = \frac{V}{L} \tag{4-37}$$

The solution to this equation is given as Eq. (4-30) which becomes for this problem,

$$i = e^{-Rt/L} \int \frac{V}{L} e^{Rt/L} \, dt + Ke^{-Rt/L} \tag{4-38}$$

Evaluating the integral, we obtain

$$i = \frac{V}{R} + Ke^{-Rt/L} \tag{4-39}$$

as the general solution. If the current in the network being considered is zero before the switching action, it must be zero immediately afterward because of the inductor. The requirement that $i(0) = 0$ leads to the particular solution

$$i = \frac{V}{R} (1 - e^{-Rt/L}) \tag{4-40}$$

The steady-state and transient divisions of this current are shown in Fig. 4-7 along with their sum or the actual current. The steady-state portion (V/R) is established at $t = 0$, and the transient term is adjusted such that there is zero current at $t = 0$.

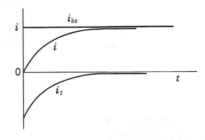

Fig. 4-7. The network response given by Eq. 4-40 which may be regarded as being composed of two components, i_{ss} and i_t.

EXAMPLE 2

As a generalization of Example 1, we observed that for the case in which both Q and P are constants, Eq. (4-30) becomes

$$i = K_1 + K_2 e^{-t/T} \tag{4-41}$$

where K_1 and K_2 are constants, and T is the time constant. Now observe that K_1 is the steady-state value of i, since the second term of Eq. (4-41) vanishes as t approaches ∞. Thus we write

$$K_1 = i(\infty) \tag{4-42}$$

Similarly, when $t = 0$, we see that

$$K_1 + K_2 = i(0) \qquad (4\text{-}43)$$

Combining this equation with Eq. (4-42), we see that

$$K_2 = i(0) - i(\infty) \qquad (4\text{-}44)$$

and Eq. (4-41) may be written

$$i = i(\infty) - [i(\infty) - i(0)]e^{-t/T} \qquad (4\text{-}45)$$

for any network for which $i(\infty)$ is defined.

To illustrate the use of Eq. (4-45), consider the network shown in Fig. 4-8. We first close the switch at $t = 0$. Since the current cannot change instantaneously, we see that $i(0) = V/(R_1 + R_2)$, $i(\infty) = V/R_1$, and $T = L/R_1$. Substituting these values into Eq. (4-45), we find that

$$i = \frac{V}{R_1}\left(1 - \frac{R_2}{R_1 + R_2}e^{-R_1 t/L}\right) \qquad (4\text{-}46)$$

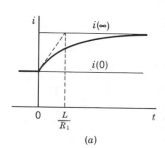

(a)

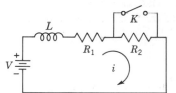

Fig. 4-8. The network is assumed to be in the steady state before the switch K is opened or closed.

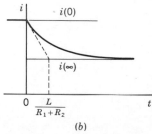

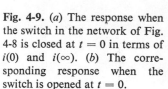

(b)

which is plotted in Fig. 4-9(a). After the steady state is reached with the switch closed, let us solve a new problem, which is to determine the current if the switch is opened at $t = 0$, a new reference time. For this case, the values of $i(0)$ and $i(\infty)$ are exchanged in comparison to the previous case. Thus, we see that $i(0) = V/R_1$, $i(\infty) = V/(R_1 + R_2)$, and $T = L/(R_1 + R_2)$. Substituting these values in Eq. (4-45), we have

$$i = \frac{V}{R_1 + R_2}\left(1 + \frac{R_2}{R_1}e^{-(R_1+R_2)t/L}\right) \qquad (4\text{-}47)$$

which is plotted in Fig. 4-9(b). Observe that the time constants for the two kinds of switching operations are different.

Fig. 4-9. (a) The response when the switch in the network of Fig. 4-8 is closed at $t = 0$ in terms of $i(0)$ and $i(\infty)$. (b) The corresponding response when the switch is opened at $t = 0$.

4-4. MORE COMPLICATED NETWORKS

What other networks may be described by first-order differential equations and so have a response described by one time constant? Thus far, we have considered only simple RC and RL combinations. There are many other networks in this classification, of course, most of which can be described by the following two cases:

(1) Networks containing a single inductor or a single capacitor in combination with any number of resistors.

(2) Networks which can be simplified by using the equivalence conditions studied in the last chapter such that it may be represented by a single equivalent resistor and a single equivalent inductor or capacitor.

For each of these cases, the time constant of the response will be of the form L_{eq}/R_{eq} or $R_{eq} C_{eq}$ where the subscript eq designates an equivalent element. These statements will be illustrated by examples.

EXAMPLE 3

The network shown in Fig. 4-10 contains a single capacitor along with four resistors and so is a network of the first kind, according to our classification. Let the capacitor be initially uncharged. At the reference time $t = 0$, the switch K is closed. For algebraic simplification, element values are assigned: $R_1 = R_2 = R_3 = R_4 = 1 \ \Omega$ and $C = 1$ F. Analyzing on the node basis gives us the equations

$$\frac{dv_1}{dt} + 2v_1 - \frac{dv_2}{dt} = v \tag{4-48}$$

$$-\frac{dv_1}{dt} + \frac{dv_2}{dt} + 2v_2 = v \tag{4-49}$$

The solution of such equations will be studied in Chapter 6. The operations are algebraic involving the operator $D = d/dt$. Manipulating these equations, we solve for v_1 and v_2 in terms of the as yet unspecified v.

$$2\frac{dv_1}{dt} + 2v_1 = \frac{dv}{dt} + v \tag{4-50}$$

and

$$2\frac{dv_2}{dt} + 2v_2 = \frac{dv}{dt} + v \tag{4-51}$$

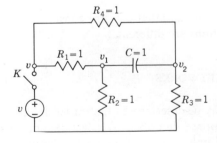

Fig. 4-10. The *RC* network of Example 3 which may be described by one time constant.

We can solve each of these equations once $v(t)$ is specified. But directly from the equations, we observe that both responses have the same time constant, $T = 1$ sec. This is not a coincidence, as we will see in the next chapter, but is always true for the various responses in a given network. For this reason the network of Fig. 4-10 is said to be described by one time constant. The same conclusion can be reached for all networks of our first classification containing a single inductor or capacitor (or a single energy storage element) and any number of resistors.

EXAMPLE 4

The network of Fig. 4-11 contains three capacitors and two resistors, but will be shown to be of our second classification and so describable by a single time constant. The capacitor part of the network is reduced to a single equivalent capacitor by the series and parallel rules developed in Chapter 3. The resistor network is reduced with the aid of a source transformation, from current source to voltage source and then back to current source. The node-to-node voltage of the resulting network, shown in Fig. 4-11(c), is described by the equation

$$C\frac{dv}{dt} + \frac{1}{R}v = \frac{1}{2}i_0 \qquad (4\text{-}52)$$

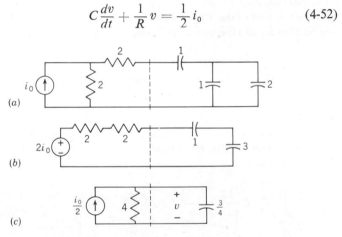

Fig. 4-11. The network of Example 4 which is of a special form that may be reduced to an equivalent *RC* network by replacing combinations of elements by their equivalents. Element values in ohms and farads.

and for the given element values

$$\frac{dv}{dt} + \frac{1}{3}v = \frac{2}{3}i_0 \qquad (4\text{-}53)$$

from which we see that the time constant of the response of the network is $T = 3$ sec.

FURTHER READING

CLOSE, CHARLES M., *The Analysis of Linear Circuits*, Harcourt, Brace & World, New York, 1966. Chapter 2.

Cox, Cyrus W., and William L. Reuter, *Circuits, Signals, and Networks*, The Macmillan Company, New York, 1969. Chapter 4.

Cruz, Jose B., Jr., and M. E. Van Valkenburg, *Signals in Linear Circuits*, Houghton Mifflin Company, Boston, Mass., 1974. Chapter 5.

Huelsman, Lawrence P., *Basic Circuit Theory with Digital Computations*, Prentice-Hall, Inc., Englewood Cliffs, N.J., 1972. Chapter 5.

Leon, Benjamin J., and Paul A. Wintz, *Basic Linear Networks for Electrical and Electronics Engineers*, Holt, Rinehart & Winston, New York, 1970. Chapter 2.

DIGITAL COMPUTER EXERCISES

Exercises relating to the topics of this chapter are concerned with the numerical solution of first-order differential equations in Appendix E-6.1, and the solution of the RLC series circuit in Appendix E-6.2. In particular, see Section 5.2 of Huelsman, reference 7 in Appendix E-10.

PROBLEMS

4-1. In the network of the figure, the switch K is moved from position 1 to position 2 at $t = 0$, a steady-state current having previously been established in the RL circuit. Find the particular solution for the current $i(t)$.

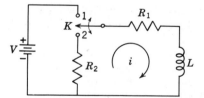

Fig. P4-1.

4-2. The switch K is moved from position a to b at $t = 0$, having been in position a for a long time before $t = 0$. Capacitor C_2 is uncharged at $t = 0$. (a) Find the particular solution for $i(t)$ for $t > 0$. (b) Find the particular solution for $v_2(t)$ for $t > 0$.

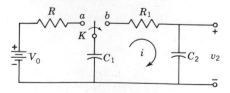

Fig. P4-2.

4-3. In the network given, the initial voltage on C_1 is V_1 and on C_2 is V_2 such that $v_1(0) = V_1$ and $v_2(0) = V_2$. At $t = 0$, the switch K is closed. (a) Find $i(t)$ for all time. (b) Find $v_1(t)$ for $t > 0$. (c) Find $v_2(t)$ for $t > 0$. (d) From your results on (b) and (c), show that $v_1(\infty) = v_2(\infty)$. (e) For the following values of the elements, $R = 1\,\Omega$, $C_1 = 1$ F, $C_2 = \frac{1}{2}$ F, $V_1 = 2$ V, $V_2 = 1$ V, sketch $i(t)$ and $v_2(t)$ and identify the time constant of each.

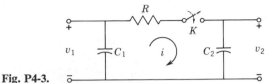

Fig. P4-3.

4-4. In the network of the figure, the switch K is in position a for a long period of time. At $t = 0$, the switch is moved from a to b (by a "make-before-break" mechanism). Find $v_2(t)$ using the numerical values given in the network. Assume that the initial current in the 2-H inductor is zero.

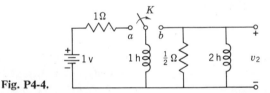

Fig. P4-4.

4-5. The network of the figure reaches a steady state with the switch K open. At $t = 0$, switch K is closed. Find $i(t)$ for the numerical values given, sketch the current waveform, and indicate the value of the time constant.

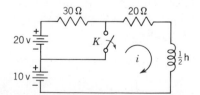

Fig. P4-5.

4-6. The network of Prob. 4-5 reaches a steady state in position 2 and at $t = 0$ the switch is moved to position 1. Find $i(t)$ for the numerical values given for the element, sketch the waveform, and show the value of the time constant.

4-7. In the given network, $v_1 = e^{-t}$ for $t \geq 0$ and is zero for all $t < 0$. If the capacitor is initially uncharged, find $v_2(t)$. Let $R_1 = 10$, $R_2 = 20$, and $C = \frac{1}{20}$ F, and for these values sketch $v_2(t)$ identifying the value of the time constant on the sketch.

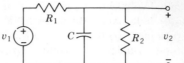

4-8. In the network shown in the figure, switch K is closed at $t = 0$ connecting a source e^{-t} to the RC network. At $t = 0$, it is observed that the capacitor voltage has the value $v_C(0) = 0.5$ V. For the element values given, determine $v_2(t)$.

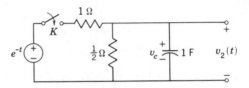

Fig. P4-8.

4-9. In the network shown, $V_0 = 3$ V, $R_1 = 10\ \Omega$, $R_2 = 5\ \Omega$, and $L = \frac{1}{2}$ H. The network attains a steady state, and at $t = 0$ switch K is closed. Find $v_a(t)$ for $t \geq 0$.

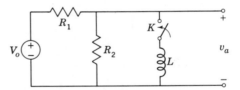

Fig. P4-9.

4-10. The network of the figure consists of a current source of value I_0 (a constant), two resistors, and a capacitor. At $t = 0$, the switch K is opened. For the element values given on the figure, determine $v_2(t)$ for $t \geq 0$.

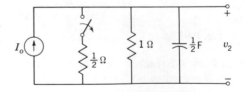

Fig. P4-10.

4-11. We wish to multiply the differential equation

$$\frac{di}{dt} + P(t)i = Q(t)$$

by an "integrating factor" R such that the left-hand side of the equation equals the derivative $d(Ri)/dt$. (a) Show that the required integrating factor is $R = e^{\int P dt}$. (b) Using this integrating factor, find the solution to the differential equation that corresponds to Eq. (4-30).

4-12. In the network shown in the accompanying figure, the switch K is closed at $t = 0$, a steady-state having previously been attained. Solve for the current in the circuit as a function of time.

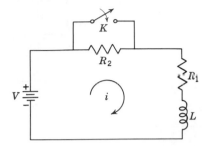

Fig. P4-12.

4-13. In the network shown, the voltage source follows the law $v(t) = Ve^{-\alpha t}$, where α is a constant. The switch is closed at $t = 0$. (a) Solve for the current assuming that $\alpha \neq R/L$. (b) Solve for the current when $\alpha = R/L$.

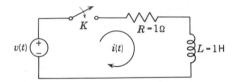

Fig. P4-13.

4-14. In the network shown in Fig. P4-13, $v(t) = 0$ for $t < 0$, and $v(t) = t$ for $t \geq 0$. Show that $i(t) = t - 1 + e^{-t}$ for $t \geq 0$, and sketch this waveform.

4-15. In the network shown, the switch is closed at $t = 0$ connecting a voltage source $v(t) = V \sin \omega t$ to a series RL circuit. For this system, solve for the response $i(t)$.

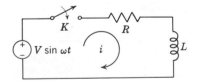

Fig. P4-15.

4-16. Consider the differential equation

$$\frac{di}{dt} + ai = f_k(t)$$

where a is real and positive. Find the general solution of this equation if all $f_k = 0$ for $t < 0$ and for $t \geq 0$ have the following values:

(a) $f_1 = k_1 t$

(b) $f_2 = te^{-2t}$

(c) $f_3 = \sin \omega_0 t$

(d) $f_4 = \cos \omega_0 t$

(e) $f_5 = \sin^2 t$

(f) $f_6 = \cos^2 t$

(g) $f_7 = t \sin 2t$

(h) $f_8 = e^{-t} \sin 2t$

4-17. In the network of the figure, the switch K is open and the network reaches a steady state. At $t = 0$, switch K is closed. Find the current in the inductor for $t > 0$, sketch this current, and identify the time constant.

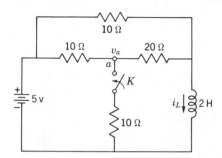

Fig. P4-17.

4-18. Repeat Prob. 4-13, determining the voltage at node a, $v_a(t)$ for $t > 0$.

4-19. The network of the figure is in a steady state with the switch K open. At $t = 0$, the switch is closed. Find the current in the capacitor for $t > 0$, sketch this waveform, and determine the time constant.

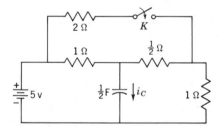

Fig. P4-19.

4-20. In the network shown, the switch K is closed at $t = 0$. The current waveform is observed with a cathode ray oscilloscope. The initial value of the current is measured to be 0.01 amp. The transient appears to disappear in 0.1 sec. Find (a) the value of R, (b) the value of C, and (c) the equation of $i(t)$.

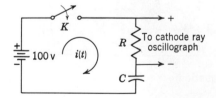

Fig. P4-20.

4-21. The circuit shown in the accompanying figure consists of a resistor and a relay with inductance L. The relay is adjusted so that it is actuated when the current through the coil is 0.008 amp. The switch K is closed at $t = 0$, and it is observed that the relay is actuated when $t = 0.1$ sec. Find: (a) the inductance L of the coil, (b) the equation of $i(t)$ with all terms evaluated.

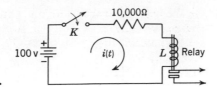

Fig. P4-21.

4-22. A switch is closed at $t = 0$, connecting a battery of voltage V with a series RC circuit. (a) Determine the ratio of energy delivered to the capacitor to the total energy supplied by the source as a function of time. (b) Show that this ratio approaches 0.50 as $t \rightarrow \infty$.

4-23. Consider the exponentially decreasing function $i = Ke^{-t/T}$ where T is the time constant. Let the tangent drawn from the curve at $t = t_1$ intersect the line $i = 0$ at t_2. Show that for any such point, $i(t_1)$, $t_2 - t_1 = T$.

5 Initial Conditions in Networks

5-1. WHY STUDY INITIAL CONDITIONS?

There are many reasons for our study of initial (and final) conditions. The most important at this point is that initial (or final) conditions must be known to evaluate the arbitrary constants that show up in the general solution of differential equations. The bonuses that result from this understanding include: knowledge of the behavior of the elements at the instant of switching, indispensable in understanding nonlinear switching circuits; and knowledge of the initial value of one or more derivatives of a response, helpful in anticipating the form of that response, thus serving as a check on our solution. Finally, the study of initial conditions is a useful artifice in getting to know the elements, individually and in combination, which is essential in the analysis of networks.

Consider our first assertion. In Chapter 4, we found that the general solution of a first-order differential equation contained an unknown designated an arbitrary constant. For differential equations of higher order, the pattern will develop that the number of arbitary constants equals the equation order. If the unknown arbitrary constants are to be evaluated for particular solutions, other things must be known about the network described by the differential equation. We must form a set of simultaneous equations, one of which is the general solution, with additional equations to total the number of unknowns. The additional equations are conveniently given as values of voltage, current, charge, etc., or derivatives of these quantities at the instant network equilibrium is altered by switching action. Conditions existing at this instant are known as *initial conditions* or the *initial state*. Sometimes we may use conditions at $t = \infty$; these are known as *final conditions*.

The student who has previously studied differential equations from a mathematician's point of view may have the impression that initial conditions are somehow always given, preferably with zero value. This, of course, permits concentration on the problem of finding the solution, but it bypasses a problem that can be more difficult than finding the general solution. The mathematician may assume initial conditions, but the engineer must correctly determine them in solving problems.

At this point, we introduce a notation to distinguish two states of the network. At the reference time, $t = 0$, one or more switches operate. We assume that switches act in zero time. To differentiate between the time immediately before and immediately after the operation of a switch, we will use $-$ and $+$ signs. Thus conditions existing just before the switch is operated will be designated as $i(0-)$, $v(0-)$, etc.; conditions after as $i(0+)$, $v(0+)$, etc.

Initial conditions in a network depend on the past history of the network prior to $t = 0-$ and the network structure at $t = 0+$, after switching. Past history will be manifest in the form of capacitor voltages and inductor currents. Details of the history are of no consequence; only the values at the reference instant, $t = 0-$, must be known. After switching, at $t = 0+$, new currents and voltages may appear in the network as the result of the initial capacitor voltages, the initial inductor currents, or because of the nature of the current and voltage sources which are introduced. The evaluation of all voltages and currents and their derivatives at $t = 0+$ constitutes the evaluation of initial conditions.

5-2. INITIAL CONDITIONS IN ELEMENTS

The Resistor. In the ideal resistor, current and voltage are related by Ohm's law, $v = Ri$. If a step input of voltage, shown in Fig. 5-1, is applied to a resistor network, the current will have the same waveform, altered by the scale factor $(1/R)$. The current through a resistor will change instantaneously if the voltage changes instantaneously. Similarly, voltage will change instantaneously if current changes instantaneously.

The Inductor. It was concluded in Section 1-5 that the current cannot change instantaneously in a system of constant inductance.

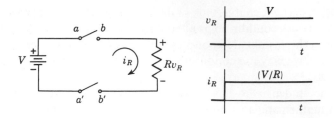

Fig. 5-1. Current-voltage relationship in a resistor: v_r and i_r are always proportional.

Consequently, closing a switch to connect an inductor to a source of energy will not cause current to flow at the initial instant, and the inductor will act as if it were an *open circuit* independent of the voltage at the terminals. If a current of value I_0 flows in the inductor at the instant switching takes place, that current will continue to flow. For the initial instant, the inductor can be thought of as a current source of I_0 amp.

The Capacitor. In Section 1-4, proof was offered that the voltage cannot change instantaneously in a system of fixed capacitance. If an uncharged capacitor is connected to an energy source, a current will flow instantaneously, the capacitor being equivalent to a *short circuit*. This follows because voltage and charge are proportional in a capacitive system, $v = q/C$, so that zero charge corresponds to zero voltage (or a short circuit). With an initial charge in the system, the capacitor is equivalent to a voltage source of value $V_0 = q_0/C$, where q_0 is the initial charge. These conclusions are summarized in Fig. 5-2.

A chart similar to that of Fig. 5-2 may be derived for final conditions in networks in which (1) all excitation is provided by sources of constant output, by sources which reduce to zero value for large t (i.e., e^{-t}, $e^{-t} \sin t$, etc.), or by initial capacitor voltages or inductor

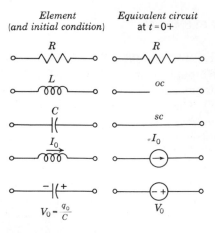

Fig. 5-2. The equivalent form of the elements in terms of the initial condition of the element.

currents, and (2) networks in which the final value for voltages and currents is a constant. This second requirement eliminates LC networks without resistance for which, as we shall see in the next chapter, the final condition of voltages and currents may be one of steady oscillation.

The final-condition equivalent networks are derived from the basic relationships

$$v_L = L \frac{di_L}{dt} \quad \text{and} \quad i_C = C \frac{dv_C}{dt} \tag{5-1}$$

and from the observation that the derivatives have zero value when the quantity in the steady state is a constant. The results are shown in Fig. 5-3.

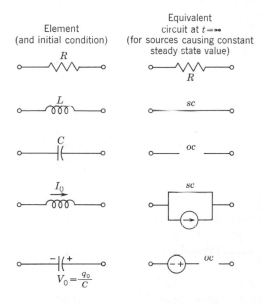

Fig. 5-3. The equivalent form of the elements in terms of the final condition of the element.

Two special cases which represent exceptions to the rules just given require further study. The first special case will be illustrated in terms of the network of Fig. 5-4. Suppose that the capacitors are initially uncharged, and that at $t = 0$ the switch K is closed. Being uncharged, the capacitors act like short circuits at the initial instant, so that the voltage source is shorted by the closing of the switch. Now a short circuit of a voltage source produces infinite current, meaning that we do have an unusual situation on our hands. The charge transferred to the capacitor is

$$q = \int_{0-}^{0+} i(t) \, dt \tag{5-2}$$

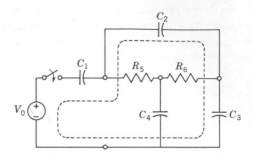

Fig. 5-4. Illustrating the special case in which there is a loop in the network composed of capacitors.

and although the current is infinite, the charge transferred q is finite and of such value that Kirchhoff's voltage law is satisfied,

$$V_0 = v_{C_1}(0+) + v_{C_2}(0+) + v_{C_3}(0+) \qquad (5\text{-}3)$$

or

$$V_0 = \frac{q_1}{C_1} + \frac{q_2}{C_2} + \frac{q_3}{C_3} \qquad (5\text{-}4)$$

where $q_1 = q_2 = q_3$ because the current through the three capacitors is the same. The voltage across each of the capacitors may be found by solving for q from Eq. (5-4) and then determining the voltage

$$v_{C_1}(0+) = \frac{q}{C_1} \qquad (5\text{-}5)$$

Thus our picture of the phenomena occurring with the closing of the switch is that an infinite current flows in the capacitors in the interval from $t = 0-$ to $t = 0+$, depositing sufficient charge on the capacitors so that each has an initial voltage at $t = 0+$ and the Kirchhoff voltage law is satisfied. In Chapter 8, this infinite current will be studied and will there be known as an *impulse* of current.

The same type of analysis may be given if one or more of the capacitors in the loop have an initial voltage prior to the closing of the switch. The final voltage of each capacitor will be independent of the initial value and will be determined as outlined for the uncharged case. We conclude for either case that the voltage across a capacitor *does* change instantaneously, although an infinite current of very short duration is necessary to cause this instantaneous change.

Note that if the capacitor C_1 is replaced by a resistor R_1, the difficulty just described is avoided because the current cannot become infinite. The initial loop current will be limited by R_1 and will have the value $i_{R_1}(0+) = V_a/R_1$ where V_a is the algebraic summation of the battery voltage and the initial capacitor voltages. Thus a voltage will appear across R_1 instantaneously and be of such value that Kirchhoff's voltage law is satisfied.

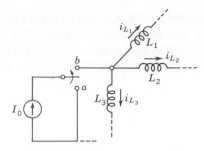

Fig. 5-5. Illustrating the special case in which a node has only inductors connected to it.

The dual situation is illustrated by Fig. 5-5. Prior to the instant the switch is moved from a to b, Kirchhoff's current law requires that

$$i_{L_1}(0-) + i_{L_2}(0-) + i_{L_3}(0-) = 0 \tag{5-6}$$

With the current source connected to the network at $t = 0$, Kirchhoff's current law must still be satisfied, meaning that

$$i_{L_1}(0+) + i_{L_2}(0+) + i_{L_3}(0+) = I_0 \tag{5-7}$$

Clearly the currents must change from $t = 0-$ to $t = 0+$ for both equations to be satisfied. The mechanism by which this is accomplished is that an infinite voltage is generated by the switching action, and this infinite voltage produces a finite flux

$$\psi = \int_{0-}^{0+} v(t)\, dt \tag{5-8}$$

which is just sufficient for Eq. (5-7) to be satisfied. Each current then changes instantaneously from its previous value to a new value determined from equations of the form

$$i_{L_1}(0+) = \frac{\psi}{L_1} \tag{5-9}$$

which is the dual of Eq. (5-5). Just as in the previous case, an *impulse* of voltage will be avoided if there is a resistor connected in parallel with the current source, and without the infinite voltage the instantaneous change in inductor current is avoided.

5-3. GEOMETRICAL INTERPRETATION OF DERIVATIVES

Consider the differential equation that describes an RL circuit connected to a constant voltage source:

$$L\frac{di}{dt} + Ri = V \tag{5-10}$$

This equation may be arranged in the form

$$\frac{di}{dt} = \frac{1}{L}(V - iR) \tag{5-11}$$

to show the relationship that must exist between current and the time derivative of current. If the switch connecting the voltage source to the circuit is closed at $t = 0$, the current in the system at $t = 0$ must be zero. From Eq. (5-11) the initial value of the derivative is

$$\frac{di}{dt}(0+) = \frac{V}{L} \tag{5-12}$$

Now the quantity di/dt is the slope of the required plot of current as a function of time. Equation (5-12) tells us that this slope is positive and has a magnitude V/L. For some small interval of time, this slope must approximate the actual curve found by solving Eq. (5-10). Assume that the current increases linearly at the rate V/L to a new value i_1 at time t_1. A second approximation to the curve of current as a function of time may be made at this point by using Eq. (5-11) as

$$\frac{di}{dt}(t_1) = \frac{1}{L}(V - i_1 R) \tag{5-13}$$

Continuation of this process, illustrated in Fig. 5-6, provides a graphical interpretation of the solution of a differential equation. The smaller the time intervals are chosen, the more closely will the approximate curve approach the actual curve.

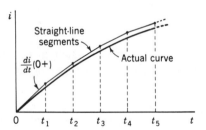

Fig. 5-6. The approximation to $i(t)$ obtained by straight-line segments, the first two being described by Eqs. 5-12 and 5-13.

Just as the first derivative represents slope, so the second derivative represents curvature or the rate of change of the slope with time. From Eq. (5-11), which is general and holds for all time after $t = 0$, we obtain a general expression for the second derivative,

$$\frac{d^2 i}{dt^2} = -\frac{R}{L}\frac{di}{dt} \tag{5-14}$$

Substituting Eq. (5-12) into this equation, we determine the value of the second derivative at $t = 0+$

$$\frac{d^2 i}{dt^2}(0+) = -\frac{VR}{L^2} \tag{5-15}$$

which is negative. Thus the initial curvature of the response, $i(t)$, is negative, meaning that the curve is concave downward, as is observed in Fig. 5-6. Figure 5-7 shows several combinations of initial conditions, with the corresponding interpretation in terms of initial slope and curvature.

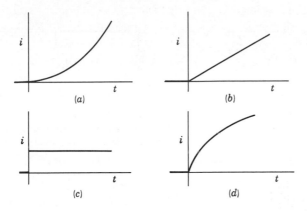

Fig. 5-7. Forms of response illustrating typical initial conditions: (a) $i(0+) = 0$, $di/dt\ (0+) = 0$, and $d^2i/dt^2\ (0+) = K > 0$; (b) $i(0+) = 0$, $di/dt\ (0+) = K > 0$, and $d^2i/dt^2\ (0+) = 0$; (c) $i(0+) = K > 0$, $di/dt\ (0+)$, $d^2i/dt\ (0+)$; (d) $1\ (0+) = 0$, $di/dt\ (0+) = K_1 > 0$, $d^2i/dt^2\ (0+) = K_2 < 0$.

5-4. A PROCEDURE FOR EVALUATING INITIAL CONDITIONS

There is no unique procedure that must be followed in solving for initial conditions. It is much like a game of chess in that the strategy chosen will depend on the particular problem being considered. However, the pattern that will emerge from examples is that we usually solve first for the initial values of the variables—currents, voltages, charges—and then solve for the derivatives. The first step is routine and may be accomplished by drawing an equivalent circuit for $t = 0+$, based on the equivalent element representations given in Fig. 5-2. In solving for initial values of derivatives, the details and order of manipulation will be different for each different network. A successful approach will not be obvious at all, a fact that adds interest and offers a challenge in the solution of initial-value problems.

Initial values of current or voltage may be found directly from a study of the network schematic. For each element in the network, we must determine just what will happen when the switching action takes place. From this analysis, a new schematic of an equivalent network for $t = 0+$ may be constructed according to these rules:

(1) Replace all inductors with open circuits or with current generators having the value of current flowing at $t = 0+$.

(2) Replace all capacitors with short circuits or with a voltage source of value $V_0 = q_0/C$ if there is an initial charge.

(3) Resistors are left in the network without change.

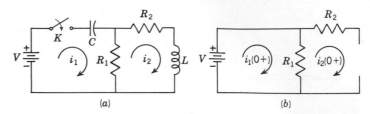

(a)   (b)

Fig. 5-8. Network used in illustrating the determination of initial conditions: (*a*) the two-loop network, and (*b*) the equivalent network at $t = 0+$.

Consider the two-loop network shown in Fig. 5-8(*a*). Suppose that the switch is closed at $t = 0$, no voltage having been applied to the passive network prior to that time. Since there is no initial voltage on the capacitor, it may be replaced by a short circuit; similarly, the inductor may be replaced by an open circuit, there being no initial value of current. The resulting equivalent network is shown as (*b*) in the figure. In this particular case, there is no need to write equations for the resistor network. By inspection the initial values of the currents are $i_1(0+) = V/R_1$ and $i_2(0+) = 0$ because the second loop is open.

The first step in solving initial values of derivatives is to write the integrodifferential equations from Kirchhoff's laws, employing either the loop or node basis which gives the required quantities more directly. In terms of the network of Fig. 5-8(*a*), the Kirchhoff voltage equations are

$$\frac{1}{C}\int i_1\, dt + R_1(i_1 - i_2) = V \qquad (5\text{-}16)$$

$$R_1(i_2 - i_1) + R_2 i_2 + L\frac{di_2}{dt} = 0 \qquad (5\text{-}17)$$

Because these equations hold in general, they hold at $t = 0+$. Now the values of i_1 and i_2 are known at $t = 0+$. Also the term $(1/C)$ $\int_{-\infty}^{0+} i_1\, dt$ has a known value, since this term is the voltage across the capacitor, which is known to be zero because the capacitor acts as a short circuit. (On the node basis, $(1/L)\int_{-\infty}^{0+} v\, dt$ similarly represents current through the inductor at $t = 0+$.)

We observe that Eq. (5-17) contains a derivative term in addition to terms involving only i_1 and i_2, which are known at $t = 0+$. Algebraically solving for (di_2/dt) gives

$$\frac{di_2}{dt} = \frac{1}{L}[R_1 i_1 - (R_1 + R_2)i_2] \qquad \text{(general)} \qquad (5\text{-}18)$$

$$\frac{di_2}{dt}(0+) = \frac{1}{L}\left[R_1\frac{V}{R_1} - (R_1 + R_2)0\right] = \frac{V}{L} \qquad (t = 0+) \qquad (5\text{-}19)$$

The precaution of marking equations as (general) or ($t = 0+$) is suggested as a safeguard against differentiating equations that hold only for $t = 0+$.

Neither Eq. (5-16) nor Eq. (5-17) contains a (di_1/dt) term. However, if Eq. (5-16), which holds in general, is differentiated and manipulated algebraically, there results

$$\frac{i_1}{C} + R_1\frac{di_1}{dt} - R_1\frac{di_2}{dt} = 0 \qquad \text{(general)} \qquad (5\text{-}20)$$

$$\frac{di_1}{dt} = \frac{di_2}{dt} - \frac{i_1}{R_1C} \qquad \text{(general)} \qquad (5\text{-}21)$$

Both di_2/dt and i_1 are known for $t = 0+$, so that (di_1/dt) may be evaluated as

$$\frac{di_1}{dt}(0+) = \frac{V}{L} - \frac{V}{R_1^2 C} \qquad (t = 0+) \qquad (5\text{-}22)$$

Suppose that it is required to evaluate (d^2i_2/dt^2) at $t = 0+$. From a practical point of view, second- and higher-order derivatives are less frequently required than the first derivative in the solution of differential equations. However, the procedure of continued differentiation and algebraic manipulation can be applied in solving for all derivatives. Differentiation of Eq. (5-18) gives

$$\frac{d^2i_2}{dt^2} = \frac{1}{L}\left[R_1\frac{di_1}{dt} - (R_1 + R_2)\frac{di_2}{dt}\right] \qquad \text{(general)} \qquad (5\text{-}23)$$

$$\frac{d^2i_2}{dt^2}(0+) = -V\left(\frac{1}{R_1LC} + \frac{R_2}{L^2}\right) \qquad (t = 0+) \qquad (5\text{-}24)$$

In each case, the initial conditions have been given in terms of constants (network and driving force parameters); solutions to problems should not be given in terms of integral or derivative expressions.

EXAMPLE 1

In the circuit shown in Fig. 5-9, $V = 10$ V, $R = 10\,\Omega$, $L = 1$ H, $C = 10\ \mu$F, and $v_C(0) = 0$. Let it be required to find $i(0+)$, $di/dt(0+)$, and $d^2i/dt^2(0+)$. From the Kirchhoff voltage law,

$$V = L\frac{di}{dt} + Ri + \frac{1}{C}\int i\,dt \qquad \text{(general)} \qquad (5\text{-}25)$$

Analyzing the circuit in terms of equivalent element values for $t = 0$

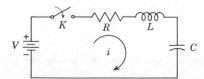

Fig. 5-9. *RLC* network of Example 1.

shows that because of the open circuit,

$$i(0+) = 0 \qquad (t = 0+) \tag{5-26}$$

The last term in Eq. (5-25), $(1/C) \int i \, dt$, represents the voltage across the capacitor, which is zero at $t = 0$. The general expression in Eq. (5-25) becomes the following for $t = 0+$:

$$V = L\frac{di}{dt}(0+) + R0 + 0 \qquad (t = 0+) \tag{5-27}$$

from which

$$\frac{di}{dt}(0+) = \frac{V}{L} = 10\frac{\text{amp}}{\text{sec}} \qquad (t = 0+) \tag{5-28}$$

To find the second derivative, Eq. (5-25) must be differentiated as

$$L\frac{d^2i}{dt^2} + R\frac{di}{dt} + \frac{1}{C}i = 0 \qquad (\text{general}) \tag{5-29}$$

In Eq. (5-29), values for the second and third terms are known at $t = 0+$; thus

$$\frac{d^2i}{dt^2}(0+) = -\frac{R}{L}\frac{di}{dt}(0+) = -100\frac{\text{amp}}{\text{sec}^2} \tag{5-30}$$

EXAMPLE 2

In the network shown in Fig. 5-10, a steady state is reached with the switch K *open*, and at $t = 0$ the switch is *closed*. Let it be required to find the initial value of all three loop currents. We must first find the various currents and voltages in the network at $t = 0-$, before the switch is closed. The current through R_2, R_1, and L will be

$$i_{R_1}(0-) = i_L(0-) = \frac{V}{R_1 + R_2} \qquad (t = 0-) \tag{5-31}$$

The total voltage across the capacitors will be the same as the voltage across R_1; that is,

$$V_{C_1} + V_{C_2} = \frac{R_1}{R_1 + R_2}V \tag{5-32}$$

Since the charges on the capacitors must be equal when connected in series, we have $q_1 = q_2$ or $C_1V_{C_1} = C_2V_{C_2}$. Hence, the voltage across

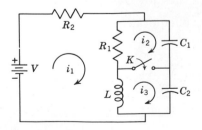

Fig. 5-10. Network of Example 2.

the capacitors will divide such that

$$\frac{V_{C_1}}{V_{C_2}} = \frac{C_2}{C_1} = \frac{D_1}{D_2} \quad \text{(general)} \tag{5-33}$$

where $D_1 = 1/C_1$ and $D_2 = 1/C_2$, so that

$$V_{C_1} = \frac{R_1}{R_1 + R_2}\left[\frac{D_1}{D_1 + D_2}\right]V; \quad V_{C_2} = \frac{R_1}{R_1 + R_2}\left[\frac{D_2}{D_1 + D_2}\right]V \tag{5-34}$$

To find i_1 at $t = 0+$, apply Kirchhoff's voltage law around the outside loop (not drawn on the diagram). Traversing this loop, we write

$$i_1 R_2 = V - V_{C_1} - V_{C_2} = \frac{R_2}{R_1 + R_2}V \tag{5-35}$$

so that

$$i_1(0+) = \frac{V}{R_1 + R_2} \quad (t = 0+) \tag{5-36}$$

Now,

$$i_1(0+) - i_3(0+) = i_L(0+) = \frac{V}{R_1 + R_2} \quad (t = 0+) \tag{5-37}$$

since the current i_L cannot change instantaneously. Comparing the last two equations, we see that

$$i_3(0+) = 0 \tag{5-38}$$

Next, consider the current flowing in the resistor R_1. Since the voltage across the capacitor cannot change instantaneously,

$$i_1(0+) - i_2(0+) = \frac{V_{C_1}}{R_1} \tag{5-39}$$

or

$$i_2(0+) = i_1(0+) - \frac{V_{C_1}}{R_1} \tag{5-40}$$

so that

$$i_2(0+) = \frac{V}{R_1 + R_2} - \frac{D_1}{D_1 + D_2}\frac{V}{R_1 + R_2} \tag{5-41}$$

Finally,

$$i_2(0+) = \frac{V}{R_1 + R_2}\frac{D_2}{D_1 + D_2} \tag{5-42}$$

This completes the solution of the problem.

5-5. INITIAL STATE OF A NETWORK

In our study of the state variable method of analysis in Section 3-9, we made use of the capacitor voltages and inductor currents as the state variables. We have not studied techniques for the solution of the state equations, but in that solution it is found that the initial conditions required are the values of the state variables at $t = 0$ (or at the more general initial time, $t = t_0$). These initial conditions are known as the *initial state*, designated in vector matrix notation as

$$x_0 = x(0) = \begin{bmatrix} v_{C_1}(0) \\ v_{C_2}(0) \\ \cdot \\ \cdot \\ \cdot \\ i_{L_1}(0) \\ \cdot \\ \cdot \\ \cdot \end{bmatrix} \tag{5-43}$$

If all of the initial conditions have zero value, then the network is said to be in the *zero state*, and the solution with these initial conditions is known as the *zero-state response*.

We note that the techniques developed in Section 5-2 are applicable for the determination of the initial state. Further, since capacitor voltages and inductor currents cannot change in zero time (in the absence of impulses), we may designate the time as $0-$, 0, or $0+$, although $t = 0-$ is the most logical choice.

An interesting difference between the state equation solution and the solution of equations written on the loop or node basis is that the state variable formulation does not require a knowledge of the initial value of the derivatives of the variables. However, if this information is required, then it is readily available from the state equations as an example will show. Consider Fig. 5-8 which was used as the first example in Section 5-4. This network is redrawn as Fig. 5-11 to emphasize choice of a tree in the state variable analysis. At node a, we write

$$C\frac{dv_C}{dt} = -i_L - i_a \tag{5-44}$$

In this equation i_a is not a state variable, but it may be replaced by writing the loop equation formed by replacing the chord containing R_1. Then

$$i_a R_1 - V - v_C = 0 \tag{5-45}$$

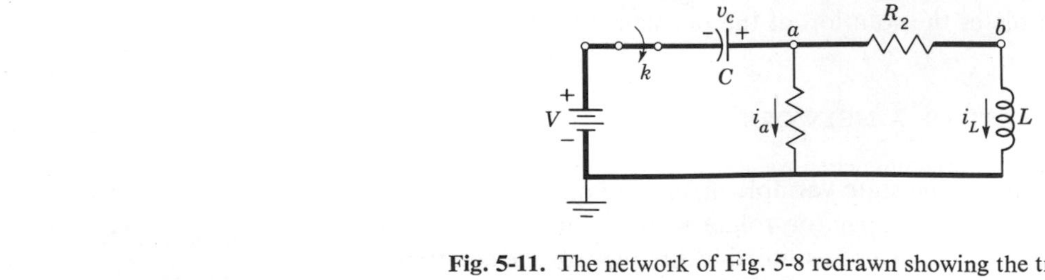

Fig. 5-11. The network of Fig. 5-8 redrawn showing the tree (dark lines) used in state variable analysis.

and Eq. (5-44) becomes

$$\frac{dv_C}{dt} = -\frac{1}{R_1 C} v_C - \frac{1}{C} i_L - \frac{V}{R_1 C} \tag{5-46}$$

The second equation is written around the loop defined by replacing the chord containing L. After division by L, we have

$$\frac{di_L}{dt} = \frac{1}{L} v_C - \frac{R}{L} i_L + \frac{V}{L} \tag{5-47}$$

Now Eqs. (5-46) and (5-47) are general equations applying at all times after $t = 0$ including $t = 0+$. Since the network is in zero state at the time K is closed, we see that

$$\frac{di_L}{dt}(0+) = \frac{V}{L} \quad \text{and} \quad \frac{dv_C}{dt}(0+) = -\frac{V}{R_1 C} \tag{5-48}$$

Should second derivatives of the state variables be required, Eqs. (5-46) and (5-47) can be differentiated and the values given in Eq. (5-48) substituted into the resulting equations.

FURTHER READING

BRENNER, EGON, AND MANSOUR JAVID, *Analysis of Electric Circuits*, 2nd ed., McGraw-Hill Book Company, New York, 1959. Chapter 5 covers topics of this chapter. Section 5-7 describes a method by which the state equations may be constructed from a knowledge of the initial conditions in the network.

CHIRLIAN, PAUL M., *Basic Network Theory*, McGraw-Hill Book Company, New York, 1969. Chapter 7.

CLOSE, CHARLES M., *The Analysis of Linear Circuits*, Harcourt, Brace & World, Inc., New York, 1969. Section 4.2.

HUELSMAN, LAWRENCE P., *Basic Circuit Theory with Digital Computations*, Prentice-Hall, Inc., Englewood Cliffs, N.J., 1972. Chapters 5 and 6.

DIGITAL COMPUTER EXERCISES

The determination of initial conditions in networks as part of the analysis of networks is accomplished with ordinary analysis coupled with a knowledge of the behavior of the network elements at the specific instant of time, $t = 0$. Such analysis is often subtle and not especially suited to computer analysis. It is suggested that time available be used in completing additional exercises suggested at the end of Chapter 3.

PROBLEMS

5-1. In the network of the figure, the switch K is closed at $t = 0$ with the capacitor uncharged. Find values for i, di/dt and d^2i/dt^2 at $t = 0+$, for element values as follows: $V = 100$ V, $R = 1000\ \Omega$, and $C = 1\ \mu$F.

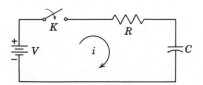

Fig. P5-1.

5-2. In the given network, K is closed at $t = 0$ with zero current in the inductor. Find the values of i, di/dt, and d^2i/dt^2 at $t = 0+$ if $R = 10\ \Omega$, $L = 1$ H, and $V = 100$ V.

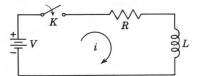

Fig. P5-2.

5-3. In the network of the figure, K is changed from position a to b at $t = 0$. Solve for i, di/dt, and d^2i/dt^2 at $t = 0+$ if $R = 1000\ \Omega$, $L = 1$ H, $C = 0.1\ \mu$F, and $V = 100$ V.

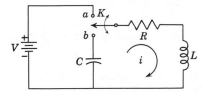

Fig. P5-3.

5-4. For the network and the conditions stated in Prob. 4-3, determine the values of dv_1/dt and dv_2/dt at $t = 0+$.

5-5. For the network described in Prob. 4-7, determine values of d^2v_2/dt^2 and d^3v_2/dt^3 at $t = 0+$.

5-6. The network shown in the accompanying figure is in the steady state with the switch K closed. At $t = 0$, the switch is opened. Determine the voltage across the switch, v_K, and dv_K/dt at $t = 0+$.

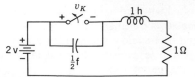

Fig. P5-6.

5-7. In the given network, the switch K is opened at $t = 0$. At $t = 0+$, solve for the values of v, dv/dt, and d^2v/dt^2 if $I = 10$ amp, $R = 1000\,\Omega$, and $C = 1\,\mu\text{F}$.

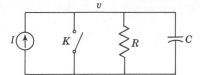

Fig. P5-7.

5-8. The network shown in the figure has the switch K opened at $t = 0$. Solve for v, dv/dt, and d^2v/dt^2 at $t = 0+$ if $I - 1$ amp, $R - 100\,\Omega$, and $L = 1$ H.

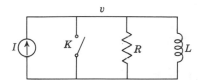

Fig. P5-8.

5-9. In the network shown in the figure, a steady state is reached with the switch K open. At $t = 0$, the switch is closed. For the element values given, determine the value of $v_a(0-)$ and $v_a(0+)$.

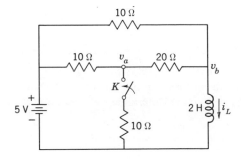

Fig. P5-9.

5-10. In the accompanying figure is shown a network in which a steady state is reached with switch K open. At $t = 0$, the switch is closed.

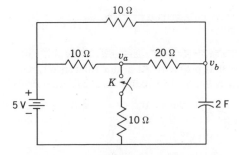

Fig. P5-10.

For the element values given, determine the values of $v_a(0-)$ and $v_a(0+)$.

5-11. In the network of Fig. P5-9, determine $i_L(0+)$ and $i_L(\infty)$ for the conditions stated in Prob. 5-9.

5-12. In the network given in Fig. P5-10, determine $v_b(0+)$ and $v_b(\infty)$ for the conditions stated in Prob. 5-10.

5-13. In the accompanying network, the switch K is closed at $t = 0$ with zero capacitor voltage and zero inductor current. Solve for (a) v_1 and v_2 at $t = 0+$, (b) v_1 and v_2 at $t = \infty$, (c) dv_1/dt and dv_2/dt at $t = 0+$, (d) d^2v_2/dt^2 at $t = 0+$.

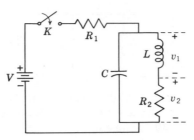

Fig. P5-13.

5-14. The network of Prob. 5-13 reaches a steady state with the switch K closed. At a new reference time, $t = 0$, the switch K is *opened*. Solve for the quantities specified in the four parts of Prob. 5-13.

5-15. The switch K in the network of the figure is closed at $t = 0$ connecting the battery to an unenergized network. (a) Determine i, di/dt, and d^2i/dt^2 at $t = 0+$. (b) Determine v_1, dv_1/dt, and d^2v_1/dt^2 at $t = 0+$.

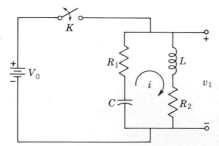

Fig. P5-15.

5-16. The network of Prob. 5-15 reaches a steady state under the conditions specified in that problem. At a new reference time, $t = 0$, the switch K is opened. Solve for the quantities specified in Prob. 5-15 at $t = 0+$.

5-17. In the network shown in the accompanying figure, the switch K is changed from a to b at $t = 0$ (a steady state having been established at position a). Show that at $t = 0+$,

$$i_1 = i_2 = -\frac{V}{R_1 + R_2 + R_3}, \qquad i_3 = 0$$

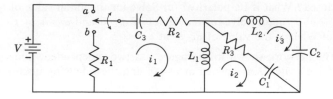

Fig. P5-17.

5-18. In the given network, the capacitor C_1 is charged to voltage V_0 and the switch K is closed at $t = 0$. When $R_1 = 2\,M\Omega$, $V_0 = 1000\,V$, $R_2 = 1\,M\Omega$, $C_1 = 10\,\mu F$, and $C_2 = 20\,\mu F$, solve for $d^2 i_2/dt^2$ at $t = 0+$.

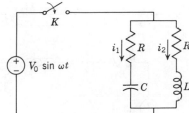

Fig. P5-18.

5-19. In the circuit shown in the figure, the switch K is closed at $t = 0$ connecting a voltage, $V_0 \sin \omega t$, to the parallel RL-RC circuit. Find (a) di_1/dt and (b) di_2/dt at $t = 0+$.

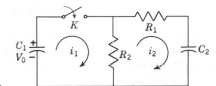

Fig. P5-19.

5-20. In the network shown, a steady state is reached with the switch K open with $V = 100\,V$, $R_1 = 10\,\Omega$, $R_2 = 20\,\Omega$, $R_3 = 20\,\Omega$, $L = 1\,H$, and $C = 1\,\mu F$. At time $t = 0$, the switch is closed. (a) Write the integrodifferential equations for the network after the switch is closed. (b) What is the voltage V_0 across C before the switch is

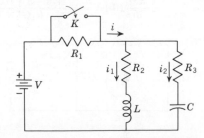

Fig. P5-20.

closed? What is its polarity? (c) Solve for the initial value of i_1 and $i_2(t = 0+)$. (d) Solve for the values of di_1/dt and di_2/dt at $t = 0+$. (e) What is the value of di_1/dt at $t = \infty$?

5-21. The network shown in the figure has two independent node pairs. If the switch K is opened at $t = 0$, find the following quantities at $t = 0+$: (a) v_1, (b) v_2, (c) dv_1/dt, (d) dv_2/dt.

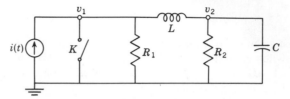

Fig. P5-21.

5-22. In the network shown in the figure, the switch K is closed at the instant $t = 0$, connecting an unenergized system to a voltage source. Let $M_{12} = 0$. Show that if $v(0) = V$, then:

$$\frac{di_1}{dt}(0+)$$
$$= \frac{V(L_2 + L_3 + 2M_{23})}{(L_1 + L_3 + 2M_{13})(L_2 + L_3 + 2M_{23}) - (L_3 + M_{13} + M_{23})^2}$$

$$\frac{di_2}{dt}(0+)$$
$$= \frac{V(L_3 + M_{13} + M_{23})}{(L_1 + L_3 + 2M_{13})(L_2 + L_3 + 2M_{23}) - (L_3 + M_{13} + M_{23})^2}$$

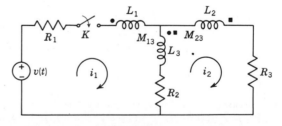

Fig. P5-22.

5-23. For the network of the figure, show that if K is closed at $t = 0$,

$$\frac{d^2 i_1}{dt^2}(0+) = -\frac{1}{R_1}\left\{ \frac{-1}{R_1 C}\left[\frac{v(0)}{R_1 C} - \frac{dv}{dt}(0) \right] - \frac{d^2 v}{dt^2}(0) \right\}$$

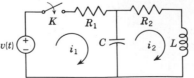

Fig. P5-23.

5-24. The given network consists of two coupled coils and a capacitor. At $t = 0$, the switch K is closed connecting a generator of voltage, $v(t) = V \sin (t/\sqrt{MC})$. Show that

$$v_a(0+) = 0, \qquad \frac{dv_a}{dt}(0+) = (V/L)\sqrt{M/C}, \qquad \text{and} \qquad \frac{d^2 v_a}{dt^2}(0+) = 0$$

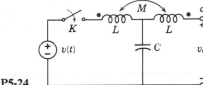

Fig. P5-24.

5-25. In the network of the figure, the switch K is *opened* at $t = 0$ after the network has attained a steady state with the switch closed. (a) Find an expression for the voltage across the switch at $t = 0+$. (b) If the parameters are adjusted such that $i(0+) = 1$ and $di/dt \, (0+) = -1$, what is the value of the derivative of the voltage across the switch, $dv_K/dt \, (0+)$?

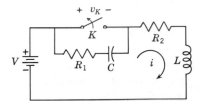

Fig. P5-25.

5-26. In the network shown in the figure, the switch K is closed at $t = 0$ connecting the battery with an unenergized system. (a) Find the voltage v_a at $t = 0+$. (b) Find the voltage across capacitor C_1 at $t = \infty$.

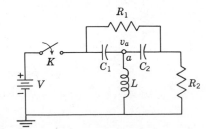

Fig. P5-26.

5-27. In the network of the figure, the switch K is closed at $t = 0$. At $t = 0-$, all capacitor voltages and inductor currents are zero. Three node-to-datum voltages are identified as v_1, v_2, and v_3. (a) Find v_1 and dv_1/dt at $t = 0+$. (b) Find v_2 and dv_2/dt at $t = 0+$. (c) Find v_3 and dv_3/dt at $t = 0+$.

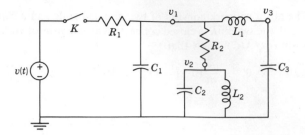

Fig. P5-27.

5-28. In the network of the figure, a steady state is reached, and at $t = 0$, the switch K is opened. (a) Find the voltage across the switch, v_K at $t = 0+$. (b) Find dv_K/dt at $t = 0+$.

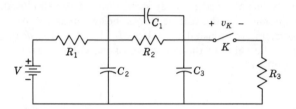

Fig. P5-28.

5-29. In the network of the accompanying figure, a steady state is reached with the switch K closed and with $i = I_0$, a constant. At $t = 0$, switch K is opened. Find: (a) $v_2(0-)$, (b) $v_2(0+)$, and (c) (dv_2/dt) $(0+)$.

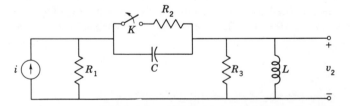

Fig. P5-29.

Differential Equations, Continued 6

The differential equations studied in Chapter 4 were limited to linear equations of the first order with constant coefficients. In this chapter, we will continue our study of differential equations with the same restrictions as to linearity and constant coefficients but of high order. The mathematical procedures given in these two chapters are included under the heading of the *classical* method of solution. As we will see, the classical method affords a better insight into the interpretation of differential equations and the requirements of a solution. Aside from conceptual advantages, the operational method that uses the Laplace transformation is better suited to our purpose. For this reason, topics which are ordinarily covered by using the classical method but are more easily developed with the aid of the Laplace transformation will be reserved for the next chapter.

6-1. SECOND-ORDER EQUATION; INTERNAL EXCITATION

A second-order homogeneous differential equation with constant coefficients may be written in the general form

$$a_0 \frac{d^2i}{dt^2} + a_1 \frac{di}{dt} + a_2 i = 0 \qquad (6\text{-}1)$$

The solution of this differential equation must be of such form that the solution itself, its first derivative, and its second derivative—each multiplied by a constant coefficient—add to zero. To satisfy this requirement, the three terms must be of the same form, differing only in their coefficients. Is there such a function? By whatever method we search, perhaps trying possible functions, the search always leads to the exponential[1]

$$i(t) = ke^{st} \tag{6-2}$$

where k and s are constants which may be real, imaginary, or complex. Substituting the exponential solution into Eq. (6-1) gives

$$a_0 s^2 k e^{st} + a_1 s k e^{st} + a_2 k e^{st} = 0 \tag{6-3}$$

or, since ke^{st} can never be zero for finite t,

$$a_0 s^2 + a_1 s + a_2 = 0 \tag{6-4}$$

as the requirement for ke^{st} to be the solution. This equation is known as the *characteristic* (or *auxiliary*) *equation*. It is satisfied by the two roots given by the quadratic formula

$$s_1, s_2 = -\frac{a_1}{2a_0} \pm \frac{1}{2a_0}\sqrt{a_1^2 - 4a_0 a_2} \tag{6-5}$$

We now have discovered that there are two forms of the exponential solution ke^{st}; they are

$$i_1 = k_1 e^{s_1 t} \quad \text{and} \quad i_2 = k_2 e^{s_2 t} \tag{6-6}$$

Now, if i_1 and i_2 are each solutions of the differential equation of Eq. (6-1), the sum of these solutions,

$$i_3 = i_1 + i_2 \tag{6-7}$$

is also a solution. This may be shown by direct substitution of Eq. (6-7) into Eq. (6-1), giving

$$a_0 \frac{d^2}{dt^2}(i_1 + i_2) + a_1 \frac{d}{dt}(i_1 + i_2) + a_2(i_1 + i_2) = 0 \tag{6-8}$$

$$\left(a_0 \frac{d^2 i_1}{dt^2} + a_1 \frac{di_1}{dt} + a_2 i_1\right) + \left(a_0 \frac{d^2 i_2}{dt^2} + a_1 \frac{di_2}{dt} + a_2 i_2\right) = 0 \tag{6-9}$$

or $0 + 0 = 0$. The general solution of the differential equation is thus

$$i(t) = k_1 e^{s_1 t} + k_2 e^{s_2 t} \tag{6-10}$$

The magnitude of the coefficients in Eq. (6-1) determines the form of the roots of the characteristic equation. In Eq. (6-5), the radical $\pm\sqrt{a_1^2 - 4a_0 a_2}$ may be real, zero, or imaginary, depending on the value of a_1^2 compared with $4a_0 a_2$. The forms of the solutions for these three cases will be given by three simple examples.

---

[1] Taken two at a time, the sine and the cosine or the hyperbolic sine and the hyperbolic cosine satisfy the requirement; however, the exponential solution will be shown to simplify to these forms.

EXAMPLE 1

The differential equation for the current in the circuit of Fig. 6-1 is given by Kirchhoff's law as

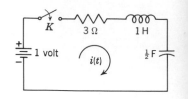

Fig. 6-1. Network for Examples 1 and 3.

$$L\frac{di}{dt} + Ri + \frac{1}{C}\int i\,dt = V \tag{6-11}$$

Differentiating and using numerical values for R, L, and C shown in Fig. 6-1 gives

$$\frac{d^2i}{dt^2} + 3\frac{di}{dt} + 2i = 0 \tag{6-12}$$

The characteristic equation can be found by substituting the trial solution $i = e^{st}$ or by the equivalent of substituting s^2 for (d^2i/dt^2), and s for (di/dt); thus

$$s^2 + 3s + 2 = 0 \tag{6-13}$$

This equation has the roots $s_1 = -1$ and $s_2 = -2$, so that the general solution is

$$i(t) = k_1 e^{-t} + k_2 e^{-2t} \tag{6-14}$$

The arbitrary constants k_1 and k_2 can be evaluated for a specific problem by a knowledge of the initial conditions. If the switch K is closed at $t = 0$, then $i(0+) = 0$, because current cannot change instantaneously in the inductor. In Eq. (6-11), the second and third voltage terms are zero at the instant of switching, $Ri(0+)$ being zero because $i(0+) = 0$ and $(1/C)\int_{-\infty}^{0+} i\,dt$ being zero because it is the initial voltage across the capacitor. Hence

$$\frac{di}{dt}(0+) = \frac{V}{L} = 1 \qquad \text{amp/sec}$$

The two initial conditions, substituted into the general solution, Eq. (6-14), give the equations,

$$k_1 + k_2 = 0, \qquad -k_1 - 2k_2 = 1 \tag{6-15}$$

The solution of these equations is $k_1 = +1$ and $k_2 = -1$; hence the particular solution to Eq. (6-12) is

$$i(t) = e^{-t} - e^{-2t} \tag{6-16}$$

A plot of the separate parts and their combination is shown in Fig. 6-2. As discussed in Chapter 4, the total current may be thought of as being made up of two components which exist from $t = 0$ and combine in such a way as to satisfy the initial conditions.

EXAMPLE 2

The equilibrium equation for the network shown in Fig. 6-3 formulated on the node basis is

$$C\frac{dv}{dt} + Gv + \frac{1}{L}\int v\,dt = I \tag{6-17}$$

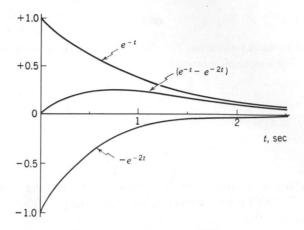

Fig. 6-2. Response described by Eq. 6-16 including the two component parts that sum to give the response.

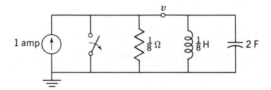

Fig. 6-3. Network for Example 2 in which the switch is opened at $t = 0$.

or, by differentiation,

$$C\frac{d^2v}{dt^2} + G\frac{dv}{dt} + \frac{v}{L} = 0 \tag{6-18}$$

Substituting numerical values into this equation as shown in Fig. 6-3 gives

$$2\frac{d^2v}{dt^2} = 8\frac{dv}{dt} + 8v = 0 \tag{6-19}$$

The corresponding characteristic equation is

$$2s^2 + 8s + 8 = 0 \tag{6-20}$$

which has as roots $s_1 = -2$ and $s_2 = -2$, or repeated roots. Substituting into the general form of the solution, Eq. (6-10), gives

$$v(t) = k_1 e^{-2t} + k_2 e^{-2t} = K e^{-2t} \tag{6-21}$$

where $K = k_1 + k_2$. This is not a complete form of the solution, since the general solution to a second-order differential equation must contain two arbitrary constants. The solution $v = ke^{-2t}$ must be modified in some manner for the condition of repeated roots. If we assume the new solution to be $v = ye^{-2t}$, where y is a factor to be determined, and substitute into Eq. (6-19), we arrive at the requirement that y satisfy the differential equation

$$\frac{d^2 y}{dt^2} = 0 \qquad (6\text{-}22)$$

Two successive integrations of the equation lead to the solution

$$y = k_1 + k_2 t \qquad (6\text{-}23)$$

Thus the solution to our problem with repeated roots becomes

$$v(t) = k_1 e^{-2t} + k_2 t e^{-2t} \qquad (6\text{-}24)$$

To obtain a particular solution for this problem will require knowledge of two initial conditions. From the network of Fig. 6-3, $v(0+)$ must equal zero, since the capacitor acts as a short circuit at the initial instant. In Eq. (6-17), the second and third terms are equal to zero, the former because $v(0+) = 0$ and the latter because there is no current in the inductor at the initial instant. Then, by Eq. (6-17), dv/dt $(0+) = I/C = \frac{1}{2}$ V/sec for this network. Substituting these initial conditions into Eq. (6-24) leads to the result that $k_1 = 0$ and $k_2 = \frac{1}{2}$. The desired particular solution is

$$v(t) = \frac{1}{2} t e^{-2t} \qquad (6\text{-}25)$$

A plot of this solution is shown in Fig. 6-4.

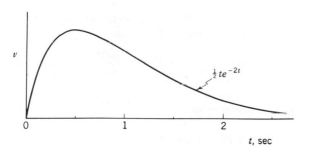

Fig. 6-4. Voltage response of the network of Fig. 6-3 as given by Eq. 6-25.

EXAMPLE 3

For this example, we will use the network of Fig. 6-1 with the following network parameter values: $V = 1$ V, $L = 1$ H, $R = 2\ \Omega$, and $C = \frac{1}{2}$ F. The characteristic equation becomes

$$s^2 + 2s + 2 = 0 \qquad (6\text{-}26)$$

with roots[2]

---

[2]We will use the letter j for the operator $\sqrt{-1}$ to reserve the letter i for current. The letter j in textbooks of electrical engineering is equivalent to $i = \sqrt{-1}$ in textbooks of mathematics and physics.

$$s_1, s_2 = -1 \pm j1 \qquad (6\text{-}27)$$

The general solution, Eq. (6-10), with these values for s, becomes

$$i(t) = k_1 e^{(-1+j1)t} + k_2 e^{(-1-j1)t} = e^{-t}(k_1 e^{jt} + k_2 e^{-jt}) \qquad (6\text{-}28)$$

Our next task is to find equivalent expressions for the two exponential functions, e^{jt} and e^{-jt}, in terms of the sine and cosine functions. This is accomplished by using a relationship which stems from the series expansion of the exponential e^x which is given by the familiar equation

$$e^x = 1 + x + \frac{x^2}{2!} + \frac{x^3}{3!} + \cdots \qquad (6\text{-}29)$$

Here x may be a real variable or it may be complex. For our objectives, we are interested in the case $x = j\omega t$ which, with $\omega = 1$, will reduce to the required exponentials of Eq. (6-28). Substituting $x = j\omega t$ into Eq. (6-29) gives

$$e^{j\omega t} = 1 + j\omega t + \frac{(j\omega t)^2}{2!} + \frac{(j\omega t)^3}{3!} + \cdots \qquad (6\text{-}30)$$

Now, when $j = \sqrt{-1}$, then $j^2 = -1$, $j^3 = -j$, $j^4 = 1$, etc. Using these identities, and collecting together the real and imaginary parts of Eq. (6-30), gives

$$e^{j\omega t} = 1 - \frac{(\omega t)^2}{2!} + \frac{(\omega t)^4}{4!} - \cdots + j\left[\omega t - \frac{(\omega t)^3}{3!} + \cdots\right] \qquad (6\text{-}31)$$

The real and imaginary parts of this series are next identified with other familiar series expansions. These are

$$\cos \omega t = 1 - \frac{(\omega t)^2}{2!} + \frac{(\omega t)^4}{4!} - \cdots \qquad (6\text{-}32)$$

and

$$\sin \omega t = \omega t - \frac{(\omega t)^3}{3!} + \frac{(\omega t)^5}{5!} - \cdots \qquad (6\text{-}33)$$

Then Eq. (6-31) may be written as

$$e^{j\omega t} = \cos \omega t + j \sin \omega t \qquad (6\text{-}34)$$

Observe next that with $-\omega t$ replacing ωt in the last equation,

$$e^{-j\omega t} = \cos(-\omega t) + j \sin(-\omega t) \qquad (6\text{-}35)$$

or

$$e^{-j\omega t} = \cos \omega t - j \sin \omega t \qquad (6\text{-}36)$$

Equations (6-34) and (6-36) may be combined into one compact equation

$$e^{\pm j\omega t} = \cos \omega t \pm j \sin \omega t \qquad (6\text{-}37)$$

This is known as *Euler's identity;* we will make frequent use of this identity in our study.

Now, with $\omega = 1$ in the last equation, we have the result we require:

$$e^{\pm jt} = \cos t \pm j \sin t \qquad (6\text{-}38)$$

which may be used to determine Eq. (6-28) in the equivalent form

$$i(t) = e^{-t}(k_3 \cos t + k_4 \sin t) \qquad (6\text{-}39)$$

where $k_3 = k_1 + k_2$ and $k_4 = j(k_1 - k_2)$. The initial conditions are the same as in Example 1: $i(0+) = 0$ and $di/dt\,(0+) = 1$ amp/sec. Substituting into the solution, we have

$$i(0+) = 0 = e^{-0}(k_3 \cos 0 + k_4 \sin 0) = k_3 \qquad (6\text{-}40)$$

With k_3 equal to zero,

$$\frac{di}{dt}(0+) = k_4(e^{-0} \cos 0 - \sin 0\, e^{-0}) = 1 \qquad (6\text{-}41)$$

whence $k_4 = 1$. The particular solution is

$$i(t) = e^{-t} \sin t \qquad (6\text{-}42)$$

A plot of the two factors in this solution and their product is shown in Fig. 6-5.

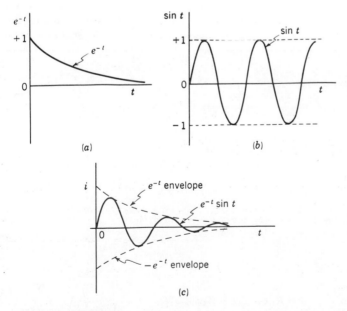

Fig. 6-5. The current response for Example 3. The response is the product of the waveform of (a) and that of (b). The envelope of the response is defined in (c) of the figure.

The results that we have found in terms of these three examples are easily generalized. If the differential equation being considered is

$$\frac{d^2i}{dt^2} + a_1 \frac{di}{dt} + a_2 i = 0 \qquad (6\text{-}43)$$

which we obtain by setting $a_0 = 1$ in Eq. (6-1) (by dividing the equation by the coefficient of the highest-ordered term), then the characteristic equation is

$$s^2 + a_1 s + a_2 = 0 \qquad (6\text{-}44)$$

The roots of this equation are found by the quadratic formula, and are

$$s_1, s_2 = \frac{-a_1}{2} \pm \sqrt{\left(\frac{a_1}{2}\right)^2 - a_2} \qquad (6\text{-}45)$$

The three possible forms of the roots, depending on the value of a_1^2 in comparison with $4a_2$ are (a) real and unequal, (b) real and equal, and (c) conjugate complex. There is also the special case for $a_1 = 0$ in which the roots are conjugate imaginary. These possibilities are summarized in Table 6-1.

Table 6-1. RESPONSE IN TERMS OF COEFFICIENT CONDITIONS IN THE CHARACTERISTIC EQUATION, $s^2 + a_1 s + a_2 = 0$ FOR a_1 AND a_2 REAL AND NONNEGATIVE

Case	Coefficient Condition	Nature of Roots	Descriptive Name	Form of Solution	Graph of Response
1	$a_1^2 > 4a_2$	Negative real and unequal	Overdamped	$i = K_1 e^{s_1 t} + K_2 e^{s_2 t}$	
2	$a_1^2 = 4a_2$	Negative real and equal	Critically damped	$i = K_1 e^{s_1 t} + K_2 t e^{s_1 t}$	
3	$a_1^2 < 4a_2$	Conjugate complex (real part negative)	Underdamped	$i = e^{\sigma_1 t}(K_1 \cos \omega_1 t + K_2 \sin \omega_1 t)$ $s_1, s_2 = \sigma_1 \pm j\omega_1$	
4	$a_1 = 0$ $a_2 \neq 0$	Conjugate imaginary	Oscillatory	$i = K_1 \cos \omega_1 t + K_2 \sin \omega_1 t$ $s_1, s_2 = \pm j\omega_1$	

6-2. HIGHER-ORDER EQUATIONS; INTERNAL EXCITATION

The method of solution discussed for first- and second-order differential equations may be followed in the solution of higher-order equations. For an nth-order differential equation, the characteristic equation will be[3]

$$a_0 s^n + a_1 s^{n-1} + \ldots + a_{n-1} s + a_n = 0 \qquad (6\text{-}46)$$

A fundamental theorem of algebra states that an equation of order n has n roots. These roots can be found by factoring Eq. (6-46).

$$a_0(s - m_1)(s - m_2) \ldots (s - m_n) = 0 \qquad (6\text{-}47)$$

Each root gives rise to a factor of the form $k_1 e^{m_1 t}$ in the solution. The sum of all such factors constitutes the solution of the differential equation. Thus, solution of higher-order homogeneous differential equations is primarily a matter of finding the roots of the characteristic equation.

Fortunately, there is some simplification in finding these roots because the coefficients of Eq. (6-46) are *positive* and *real* coefficients. This follows because these coefficients are made up of the system parameters, R, L, and C. And since R, L, and C are assumed to be positive and real, so the a-coefficients must be positive and real. The coefficients may be zero or negative and real with controlled sources in the network.

There are three possible forms for the roots: (1) real roots, (2) imaginary roots, and (3) complex roots. For the first-order characteristic equation

$$a_0 s + a_1 = 0 \qquad (6\text{-}48)$$

the root is $s = -a_1/a_0$, which is negative and real because a_0 and a_1 are always positive and real. For a second-order characteristic equation

$$a_0 s^2 + a_1 s + a_2 = 0 \qquad (6\text{-}49)$$

we let $a_0 = 1$ (or divide through by a_0 so that this is the case). The roots are given by Eq. (6-45), which is

$$s_1, s_2 = \frac{-a_1}{2} \pm \sqrt{\left(\frac{a_1}{2}\right)^2 - a_2} \qquad (6\text{-}50)$$

With the positive real restrictions on the a-coefficients, these roots may have any of the three possible forms—real, imaginary, or complex. But

[3] In describing algebraic equations, order and degree are used interchangeably in electrical engineering literature. Since our emphasis here is the differential equation of order n, we will use the word *order* in chapters concerned with the time domain. Beginning with Chapter 9, we will use the word *degree* rather than order to conform with usual practice in frequency-domain analysis.

if the roots are complex, they occur in conjugate pairs, since this is the only way complex roots can combine to give positive real coefficients. Thus, for characteristic equation roots to be complex, they must occur in conjugate pairs.

Consider next a third-order characteristic equation. In this case, because of the rule just given for complex roots, at least one root must be real. The other two may be both real or a conjugate pair of complex roots.[4] For a fourth-order characteristic equation, there are more possibilities: four real roots, two real roots and a conjugate complex pair, or two sets of conjugate complex roots. The general pattern is thus established and the following rules may be given:

(1) If the roots are complex, they occur in conjugate pairs.
(2) If the characteristic equation is of odd order, at least one root is real. The remaining roots may be real or occur in conjugate complex pairs.
(3) If the characteristic equation is of even order, the roots may be real or occur in conjugate complex pairs.

Summarizing this discussion, an equation of any order can be factored into its roots, and the roots determine the solution of the homogeneous differential equation as the sum of first-order (or second-order) solutions which have already been considered.

An example will illustrate the method of solution of higher-order homogeneous differential equations. The differential equation

$$\frac{d^5 i}{dt^5} + 6\frac{d^4 i}{dt^4} + 17\frac{d^3 i}{dt^3} + 28\frac{d^2 i}{dt^2} + 24\frac{di}{dt} + 8i = 0 \qquad (6\text{-}51)$$

has a characteristic equation which may be factored as

$$(s + 1)(s + 1)(s + 2)(s^2 + 2s + 4) = 0 \qquad (6\text{-}52)$$

In this equation, there are two repeated real roots, one nonrepeated real root, and one conjugate complex pair. Using the equations already derived for first- and second-order systems, we see that the solution is

$$i = (K_1 + K_2 t)e^{-t} + K_3 e^{-2t} + e^{-t}(K_4 \sin \sqrt{3}\, t + K_5 \cos \sqrt{3}\, t)$$
$$(6\text{-}53)$$

6-3. NETWORKS EXCITED BY EXTERNAL ENERGY SOURCES

In the nonhomogeneous differential equation, the right-hand side of the equation is not zero, being equal to the forcing function or some derivative of the forcing function, $v(t)$. In studying such equations,

---

[4]In this discussion, imaginary roots are considered as a special case of complex roots.

we first observe that the solution to the corresponding homogeneous differential equation is a part of the solution of the nonhomogeneous equation. To illustrate by a simple example, consider the equation

$$\frac{d^2i}{dt^2} + 5\frac{di}{dt} + 6i = v(t) \tag{6-54}$$

This equation has as roots of its characteristic equation, $s_1 = -2$ and $s_2 = -3$. Thus the complete solution for the case $v(t) = 0$ is

$$i_C = k_1 e^{-2t} + k_2 e^{-3t} \tag{6-55}$$

Suppose that some function i_p, which we will presently find, satisfies the nonhomogeneous equation, Eq. (6-54). Then i_P plus i_C given above is *also* a solution, since substituting either $k_1 e^{-2t}$ or $k_2 e^{-3t}$ into Eq. (6-54) would add nothing to the right-hand side of the equation. In other words, part of the solution of a nonhomogeneous differential equation is the solution to the homogeneous differential equation. That part, by analogy to the discussion in Section 4-3, is termed the *complementary function*. The remaining part of the solution—needed to make the operations of the differential equation add to $v(t)$—is the *particular integral*. Thus we write the total solution as the sum of two parts of the solution

$$i = i_P + i_C \tag{6-56}$$

Since we can find i_C for any equation, as discussed in the last section, there remains to be found only the particular integral i_P.

In the analysis of electric circuits, the term $v(t)$ in the differential equation is the driving force or a derivative of the driving force. As a practical matter, driving forces are represented by only a few mathematical forms like V (a constant), $\sin \omega t$, kt, $e^{-\alpha t}$, or products of these terms (or linear combinations to give square waves, pulses, etc.). We do not ordinarily encounter physical generators of such functions as the tangent. Several mathematical methods are available for determining the particular integral. If only driving forces of the practical forms mentioned are considered, the method of *undetermined coefficients* is particularly suited to our use.

Ordinarily, the method of undetermined coefficients is applied by selecting trial functions of all possible forms that might satisfy the differential equation. Each trial function is assigned an undetermined coefficient. The sum of the trial functions is substituted into the differential equation, and a set of linear algebraic equations is formed by equating coefficients of like functions in the equation resulting from this substitution. The undetermined coefficients are thus determined by solution of this set of equations. If any trial function is not a solution, its coefficient will be zero.

It is not necessary to study rules for selecting trial functions for the forms of driving force function $v(t)$ we are considering. The

Table 6-2.

Factor* in $v(t)$	Necessary Choice for the Particular Integral**
1. V (a constant)	A
2. $a_1 t^n$	$B_0 t^n + B_0 t^{n-1} + \ldots + B_{n-1} t + B_n$
3. $a_2 e^{rt}$	$C e^{rt}$
4. $a_3 \cos \omega t$	$D \cos \omega t + E \sin \omega t$
5. $a_4 \sin \omega t$	
6. $a_5 t^n e^{rt} \cos \omega t$	$(F_1 t^n + \ldots + F_{n-1} t + F_n) e^{rt} \cos \omega t$
7. $a_6 t^n e^{rt} \sin \omega t$	$\qquad + (G_1 t^n + \ldots + G_n) e^{rt} \sin \omega t$

*When $v(t)$ consists of a sum of several terms, the appropriate particular integral is the sum of the particular integrals corresponding to these terms individually.

**Whenever a term in any of the trial integrals listed in this column is already a part of the complementary function of the given equation, it is necessary to modify the indicated choice by multiplying it by t before using it. If such a term appears r times in the complementary function, the indicated choice should be multiplied by t^r.
(By permission from *Advanced Engineering Mathematics*, 3rd ed., by C. R. Wylie, Jr., McGraw-Hill Book Company, New York, 1966.)

required form of the trial functions is given in Table 6-2. In using this table, the following procedure is suggested:

(1) Determine the complementary function i_C. Compare each part of the complementary function with the form of $v(t)$. The rules given in Table 6-2 are modified if these two functions have terms of the same mathematical form.

(2) Write the trial form of the particular integral, using Table 6-2. Each different trial solution should be assigned a different letter coefficient, and all similar functions should be combined.

(3) Substitute the trial solution into the differential equation. By equating coefficients of all like terms, form a set of algebraic equations in the undetermined coefficients.

(4) Solve for the undetermined coefficients and so find the particular integral. These coefficients must be in terms of circuit and driving force parameters. There are no arbitrary constants in the particular integral.

Having determined the particular integral, the total solution may be found by adding the complementary function to the particular integral. If a particular solution is required, the arbitrary constants of i_C can be evaluated from a knowledge of the initial conditions. As a precaution, the initial conditions must always be applied to the total solution—never to the complementary function alone unless $i_p = 0$ [when $v(t) = 0$].

EXAMPLE 4

Consider a series RL circuit with the driving force voltage of the form $v(t) = Ve^{-\alpha t}$ for $t \geq 0$, where V and α are constants. By Kirchhoff's voltage law, the differential equation is, after division by L,

$$\frac{di}{dt} + \frac{R}{L}i = \frac{V}{L}e^{-\alpha t} \tag{6-57}$$

The characteristic equation is $s + (R/L) = 0$, so that the complementary function is

$$i_C = ke^{-Rt/L} \tag{6-58}$$

From Table 6-2, the trial solution should be

$$i_p = Ae^{-\alpha t} \tag{6-59}$$

if $\alpha \neq R/L$, where A is the undetermined coefficient. Substituting this trial solution into the differential equation gives

$$-\alpha Ae^{-\alpha t} + \frac{R}{L}Ae^{-\alpha t} = \frac{V}{L}e^{-\alpha t} \tag{6-60}$$

or

$$A = \frac{V}{R - \alpha L}, \qquad \alpha \neq \frac{R}{L} \tag{6-61}$$

The solution is the sum of i_p and i_C, or

$$i = \frac{V}{R - \alpha L}e^{-\alpha t} + Ke^{-Rt/L}, \qquad \alpha \neq \frac{R}{L} \tag{6-62}$$

The arbitrary constant can be evaluated from knowledge of the initial conditions. If $\alpha = R/L$, the form of the trial solution should be

$$i_p = Ate^{-\alpha t} \tag{6-63}$$

Substituting this solution into the differential equation gives

$$A(-\alpha te^{-\alpha t} + e^{-\alpha t}) + \alpha Ate^{-\alpha t} = \frac{V}{L}e^{-\alpha t} \tag{6-64}$$

or

$$A = \frac{V}{L} \tag{6-65}$$

The solution for this case is thus

$$i = \frac{V}{L}te^{-\alpha t} + Ke^{-Rt/L}, \qquad \alpha = \frac{R}{L} \tag{6-66}$$

EXAMPLE 5

As a second example, consider a series RC circuit with a sinusoidal driving force voltage $v(t) = V \sin \omega t$ for $t \geq 0$. The Kirchhoff voltage equation is

$$Ri + \frac{1}{C} \int i \, dt = V \sin \omega t \qquad (6\text{-}67)$$

or, differentiating and dividing by R,

$$\frac{di}{dt} + \frac{1}{RC} i = \frac{\omega V}{R} \cos \omega t \qquad (6\text{-}68)$$

From Table 6-2, the assumed i_P should be the sum of a sine and a cosine term, as

$$i_P = A \cos \omega t + B \sin \omega t \qquad (6\text{-}69)$$

If this assumed solution is substituted into the differential equation and coefficients of like functions are equated, the following system of linear equations results.

$$\frac{A}{RC} + \omega B = \frac{\omega V}{R}, \qquad \frac{B}{RC} - \omega A = 0 \qquad (6\text{-}70)$$

Solving for A and B yields

$$A = \frac{\omega C V}{1 + \omega^2 R^2 C^2}, \qquad B = \frac{\omega^2 R C^2 V}{1 + \omega^2 R^2 C^2} \qquad (6\text{-}71)$$

Substituting these values into the assumed solution, there results, after some simplification,

$$i_P = \frac{V}{R^2 + (1/\omega^2 C^2)} \left(\frac{1}{\omega C} \cos \omega t + R \sin \omega t \right) \qquad (6\text{-}72)$$

This equation can be reduced to a single sinusoid by defining $1/\omega C = K \cos \phi$ and $R = K \sin \phi$, and making use of the trigonometric identity for the cosine of the difference of two angles. Thus

$$K(\cos \phi \cos \omega t + \sin \phi \sin \omega t) = K \cos (\omega t - \phi) \qquad (6\text{-}73)$$

where, since $\sin^2 \phi + \cos^2 \phi = 1$,

$$K^2 = R^2 + \frac{1}{\omega^2 C^2} \qquad (6\text{-}74)$$

and

$$\tan \phi = \omega RC \qquad (6\text{-}75)$$

Finally

$$i_P = \frac{V}{\sqrt{R^2 + (1/\omega^2 C^2)}} \cos (\omega t - \tan^{-1} \omega RC) \qquad (6\text{-}76)$$

To this value of i_P must be added $i_C = K e^{-t/RC}$ for the complete general solution.

EXAMPLE 6

Knowledge of the response of systems with sinusoidal driving force voltages is important in studies of power generation and distribution systems. Consider the circuit equivalent of such a system shown in Fig. 6-6. With K closed, the Kirchhoff voltage equation for this system is

$$L\frac{di}{dt} + Ri = V \sin(\omega t + \theta) \qquad (6\text{-}77)$$

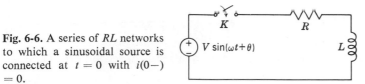

Fig. 6-6. A series of RL networks to which a sinusoidal source is connected at $t = 0$ with $i(0-) = 0$.

The method for finding the particular integral is like that illustrated in the last example. The result is

$$i_P = \frac{V}{\sqrt{R^2 + \omega^2 L^2}} \sin\left(\omega t + \theta - \tan^{-1}\frac{\omega L}{R}\right) \qquad (6\text{-}78)$$

To this result must be added the complementary function which, from Example 4, is

$$i_C = Ke^{-Rt/L} \qquad (6\text{-}79)$$

The total solution thus becomes

$$i = \frac{V}{\sqrt{R^2 + \omega^2 L^2}} \sin\left(\omega t + \theta - \tan^{-1}\frac{\omega L}{R}\right) + Ke^{-Rt/L} \qquad (6\text{-}80)$$

Now, if the switch is closed at $t = 0$, the initial current has zero value because of the inductor, requiring that

$$\frac{V}{\sqrt{R^2 - \omega^2 L^2}} \sin\left(\theta - \tan^{-1}\frac{\omega L}{R}\right) + Ke^0 = 0 \qquad (6\text{-}81)$$

or

$$K = -\frac{V}{\sqrt{R^2 + \omega^2 L^2}} \sin\left(\theta - \tan^{-1}\frac{\omega L}{R}\right) \qquad (6\text{-}82)$$

If the angle θ, which represents the angle of the sinusoid at the time the switch is closed, has the value

$$\theta = \tan^{-1}\frac{\omega L}{R} \qquad (6\text{-}83)$$

the constant K will have zero value, and the transient term i_C will vanish. In other words, if the switch is closed at the proper instant, there will be no transient. The same conclusion can be reached for the RC series network, but not for an RLC network.

6-4. RESPONSE AS RELATED TO THE s-PLANE LOCATION OF ROOTS

In this section we will amplify our previous discussion of the solution of second-order differential equations, particularly in relating the form of the response of the network to the location of the roots of the characteristic equation in the complex s plane. We will do this by studying the response of a specific network, the RLC series network excited by a voltage source, $v(t)$. For this network, the Kirchhoff voltage law gives

$$L\frac{di}{dt} + Ri + \frac{1}{C}\int i\,dt = v(t) \tag{6-84}$$

The corresponding homogeneous equation is of second order and is

$$\frac{d^2i}{dt^2} + \frac{R}{L}\frac{di}{dt} + \frac{1}{LC}i = 0 \tag{6-85}$$

The two roots of the corresponding characteristic equation may be found by the quadratic formula to be

$$s_1, s_2 = -\frac{R}{2L} \pm \sqrt{\left(\frac{R}{2L}\right)^2 - \frac{1}{LC}} \tag{6-86}$$

To convert Eq. (6-85) to a standard form, we define the value of resistance that causes the radical term in the above equation to vanish as the *critical resistance*, R_{cr}. This value is found by solving the equation

$$\left(\frac{R_{cr}}{2L}\right)^2 = \frac{1}{LC} \tag{6-87}$$

or

$$R_{cr} = 2\sqrt{\frac{L}{C}} \tag{6-88}$$

We will next introduce two definitions; we define the quantity

$$\zeta = \frac{R}{R_{cr}} = \frac{R}{2}\sqrt{\frac{C}{L}} \tag{6-89}$$

as the dimensionless *damping ratio*. (ζ is the lower-case Greek letter zeta.) The damping ratio is the ratio of the actual resistance to the critical value of resistance. The other definition is

$$\omega_n = \frac{1}{\sqrt{LC}} \tag{6-90}$$

where ω_n is the *undamped natural frequency* or simply the *natural frequency*. Now the product $2\zeta\omega_n$ has the value

$$2\zeta\omega_n = 2\frac{R}{2}\sqrt{\frac{C}{L}}\frac{1}{\sqrt{LC}} = \frac{R}{L} \tag{6-91}$$

and

$$\omega_n^2 = \frac{1}{LC} \tag{6-92}$$

Substituting these relationships into Eq. (6-85) gives

$$\frac{di^2}{dt^2} + 2\zeta\omega_n \frac{di}{dt} + \omega_n^2 i = 0 \tag{6-93}$$

Note that in these manipulations we have simply substituted two new constants for the two coefficients in Eq. (6-85). As we shall see, these new constants are convenient for interpreting the geometry of the root locations in the s plane, and, in addition, they have significance in understanding the response.

Another quantity often encountered in studies in electrical engineering is the *circuit Q*, which will be further discussed in Chapter 10. The Q of the RLC series circuit we are studying is defined by the equation

$$Q = \frac{\omega_n L}{R} \tag{6-94}$$

In terms of Q and ω_n, Eq. (6-85) becomes

$$\frac{d^2 i}{dt^2} + \frac{\omega_n}{Q}\frac{di}{dt} + \omega_n^2 i = 0 \tag{6-95}$$

meaning, of course, that $Q = 1/2\zeta$.

Let us now turn to the location of the roots of the characteristic equation in terms of Q, ζ, and ω_n. The characteristic equation obtained from Eq. (6-93) is

$$s^2 + 2\zeta\omega_n s + \omega_{n}^2 = 0 \tag{6-96}$$

and the roots of the characteristic equation are

$$s_1, s_2 = -\zeta\omega_n \pm \omega_n \sqrt{\zeta^2 - 1} \tag{6-97}$$

The general solution may now be written

$$i = K_1 e^{[-\zeta\omega_n + \omega_n\sqrt{(\zeta^2-1)}]t} + K_2 e^{[-\zeta\omega_n - \omega_n\sqrt{(\zeta^2-1)}]t} \tag{6-98}$$

Before simplifying this solution, let us examine the behavior of the roots of the characteristic equation as the dimensionless damping ratio ζ varies from zero (corresponding to $R = 0$) to infinity (corresponding to $R = \infty$). There are evidently three different forms for the roots:

Case 1: $\zeta > 1$, the roots are real.
Case 2: $\zeta = 1$, the roots are real and repeated.
Case 3: $\zeta < 1$, the roots are complex and conjugates.

If we follow the form of the roots for a variation of ζ from 0 to ∞, we will recognize a locus of roots in the complex plane. To start with, for

$\zeta = 0,$

$$s_1, s_2 = \pm j\omega_n \qquad (6\text{-}99)$$

that is, the roots are purely imaginary. For $\zeta < 1$, the roots are complex conjugates as

$$s_1, s_2 = -\zeta\omega_n \pm j\omega_n\sqrt{1 - \zeta^2} \qquad (6\text{-}100)$$

In terms of s-plane locations, we see that the real and imaginary parts of $s = \sigma + j\omega$ are

$$\sigma = -\zeta\omega_n \qquad (6\text{-}101)$$

and the imaginary part is

$$\omega = \pm\omega_n\sqrt{1 - \zeta^2} \qquad (6\text{-}102)$$

as shown in Fig. 6-7. Observe next that since

$$\sigma^2 + \omega^2 = \zeta^2\omega_n^2 + \omega_n^2(1 - \zeta^2) = \omega_n^2 \qquad (6\text{-}103)$$

it follows that the locus of the roots in the complex s plane is a *circle* of radius ω_n and that this locus is formed by ζ varying from 0 to 1. This locus is shown in Fig. 6-8.

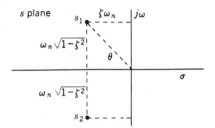

Fig. 6-7. The roots of the second-order characteristic equation are shown plotted in the complex s plane.

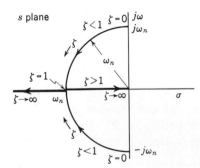

Fig. 6-8. The locus of the roots shown in Fig. 6-7 as ζ varies from 0 to ∞.

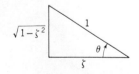

Fig. 6-9. Illustrating a useful right triangle relationship.

Let the angle between the line from the origin to either root and the $-\sigma$ axis be designated θ as in Fig. 6-7. The line to s_1 is of length ω_n as we have just shown, meaning that all lengths of the triangle are proportional to ω_n. A scaled representation of this triangle is shown in Fig. 6-9, with $\omega_n = 1$. From this figure, observe that

$$\theta = \cos^{-1}\zeta \qquad (6\text{-}104)$$

Thus we have shown that the s-plane location of the roots s_1 and s_2 may be specified by ω_n and ζ; ω_n is the radius and $\cos^{-1} \zeta$ is the angle measured with respect to the negative real axis. Alternatively, the angle may be given in terms of Q of Eq. (6-94) as

$$\theta = \cos^{-1} \frac{1}{2Q} \qquad (6\text{-}105)$$

Next, let us trace the path of the roots, the root locus of s_1 and s_2, for a fixed ω_n as ζ varies from 0 to ∞. With $\zeta = 0$, the roots are on the imaginary axis of the s plane. The range of values of ζ, $0 < \zeta < 1$, corresponds to complex values of the roots. When $\zeta = 1$, the roots come together (repeated) and have the real value $-\omega_n$. For all other values of ζ, $1 < \zeta < \infty$, the roots are negative real and given by

$$s_1, s_2 = (-\zeta \pm \sqrt{\zeta^2 - 1})\,\omega_n \qquad (6\text{-}106)$$

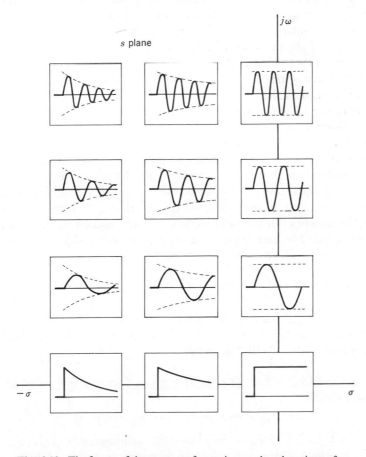

Fig. 6-10. The forms of the response for various s-plane locations of the roots of the characteristic equation in the left half of the s plane. Points on the σ axis correspond to single roots and other points assume that the conjugate is present.

As ζ becomes so large that 1 is small compared to ζ^2, the roots approach $-2\zeta\omega_n$ and 0, and one root approaches infinity as the other approaches the origin.

Now the position of the roots of the characteristic equation in the complex s plane determines the form of the response, as shown by the examples of Section 6-1 and by the summary in Table 6-1. From these we see that roots on the imaginary axis correspond to oscillatory response (zero damping), roots in the complex plane correspond to damped oscillations (underdamped response), and that roots on the negative real axis correspond either to the critically damped case ($\zeta = 1$) or to an overdamped form of response expressible as the difference of two exponential functions. The form of response as a function of s-plane root location is illustrated in Fig. 6-10.

Returning to the RLC circuit from which our discussion started, observe that ω_n is controlled by L and C, but not by R, and that ζ or Q is determined by R. Thus, if R represents a variable resistor, we see that any of the cases we have described for the form of the response can be realized by a simple adjustment of the resistor. If R has a very small value (and zero values are attainable in superconductive materials), then the current in the circuit will oscillate when a switch connecting a battery is closed. This oscillatory form of response with increasing damping will continue to result from the closing of a switch as R is increased until a critical value is reached. This is the critical value of resistance we have described. For larger values of R the current will not oscillate, but will increase and decrease exponentially. We see that ζ and ω_n are not only convenient quantities for determining the s-plane location of roots, but also for describing the form of response in the network. In the next section, we will make these conclusions more explicit by carrying out the solution in terms of ζ, Q, and ω_n.

6-5. GENERAL SOLUTIONS IN TERMS OF ζ, Q, AND ω_n

In this section we will study the cases corresponding to the classification of Table 6-1, and in each case we will find a simplified equation, starting with Eq. (6-98).

Case 1: $\zeta > 1$ or $Q < \frac{1}{2}$. If $e^{-\zeta\omega_n t}$ is factored from Eq. (6-98), there results

$$i = e^{-\zeta\omega_n t}K_1 e^{\omega_n\sqrt{\zeta^2-1}\,t} + K_2 e^{-\omega_n\sqrt{\zeta^2-1}\,t}) \qquad (6\text{-}107)$$

where K_1 and K_2 are arbitrary constants of integration. This equation is sometimes more convenient to evaluate in terms of hyperbolic functions. The *hyperbolic cosine* of x is defined as

$$\cosh x = \tfrac{1}{2}(e^x + e^{-x}) \qquad (6\text{-}108)$$

and the *hyperbolic sine* of x is defined as

$$\sinh x = \tfrac{1}{2}(e^x - e^{-x}) \tag{6-109}$$

An equivalent relationship can be obtained by successively adding or subtracting these two equations; that is,

$$e^x = \sinh x + \cosh x \tag{6-110}$$

and

$$e^{-x} = \cosh x - \sinh x \tag{6-111}$$

These two identities may be used to convert Eq. (6-107) to terms involving hyperbolic functions; thus

$$i = e^{-\zeta \omega_n t}\{K_1[\cosh(\omega_n\sqrt{\zeta^2 - 1}\, t) + \sinh(\omega_n\sqrt{\zeta^2 - 1}\, t)]$$
$$+ K_2[\cosh(\omega_n\sqrt{\zeta^2 - 1}\, t) - \sinh(\omega_n\sqrt{\zeta^2 - 1}\, t)]\} \tag{6-112}$$

or

$$i = e^{-\zeta \omega_n t}[K_3 \cosh(\omega_n\sqrt{\zeta^2 - 1}\, t) + K_4 \sinh(\omega_n\sqrt{\zeta^2 - 1}\, t)] \tag{6-113}$$

where

$$K_3 = K_1 + K_2 \tag{6-114}$$

and

$$K_4 = K_1 - K_2 \tag{6-115}$$

This equation is the equivalent of Eq. (6-107). Each has two arbitrary constants, which are usually evaluated to find a particular solution in terms of the initial conditions.

Case 2: $\zeta = 1$ or $Q = \tfrac{1}{2}$. For this case, we have shown that the two roots become identical. With repeated roots, the solution of the equation is

$$i = (K_1 + K_2 t)e^{-\omega_n t} \tag{6-116}$$

The limit of the quantity $te^{-\omega_n t}$ may be investigated by l'Hospital's rule. If this quantity is written as

$$\frac{t}{e^{+\omega_n t}} \tag{6-117}$$

differentiation of numerator and denominator with respect to t shows that

$$\lim_{t \to \infty} te^{-\omega_n t} = 0 \tag{6-118}$$

Case 3: $\zeta < 1$ or $Q > \tfrac{1}{2}$. For Case 3 the roots become complex, and Eq. (6-98) may be written

$$i = e^{-\zeta \omega_n t}[K_1 e^{j\omega_n\sqrt{1-\zeta^2}\,t} + K_2 e^{-j\omega_n\sqrt{1-\zeta^2}\,t}) \tag{6-119}$$

This equation may be written in terms of sine and cosine quantities by making use of Euler's identity,

$$e^{\pm jx} = \cos x \pm j \sin x \tag{6-120}$$

The solution for Case 3 reduces to

$$i = e^{-\zeta \omega_n t}[K_5 \cos (\omega_n \sqrt{1 - \zeta^2}\, t) + K_6 \sin (\omega_n \sqrt{1 - \zeta_2}\, t)] \qquad (6\text{-}121)$$

where

$$K_5 = K_1 + K_2 \quad \text{and} \quad K_6 = j(K_1 - K_2) \qquad (6\text{-}122)$$

which are, again, arbitrary constants of integration. This equation may be written in different form by defining

$$K_5 = K \sin \phi \qquad (6\text{-}123)$$

$$K_6 = K \cos \phi \qquad (6\text{-}124)$$

Using the trigonometric identity

$$\sin (x + y) = \sin x \cos y + \sin y \cos x \qquad (6\text{-}125)$$

Eq. (6-121) becomes

$$i = K e^{-\zeta \omega_n t} \sin (\omega_n \sqrt{1 - \zeta^2}\, t + \phi) \qquad (6\text{-}126)$$

These algebraic manipulations have resulted in an equation of one sinusoid equivalent to Eq. (6-121), which contains two sinusoids of the same frequency. In the revised form, the two arbitrary constants are K and ϕ, which may be related to K_5 and K_6 by means of Eqs. (6-123) and (6-124). Summing the squares of K_5 and K_6 gives the relationship

$$K = \sqrt{K_5^2 + K_6^2} \qquad (6\text{-}127)$$

Dividing Eq. (6-123) by Eq. (6-124) gives an equation for ϕ in terms of K_5 and K_6:

$$\phi = \tan^{-1} \frac{K_5}{K_6} \qquad (6\text{-}128)$$

From factors of the form $\sin (\omega_n \sqrt{1 - \zeta^2}\, t)$, in the Case 3 solution, we see that

$$\omega = \omega_n \sqrt{1 - \zeta^2} \qquad (6\text{-}129)$$

is the frequency[5] of oscillation of the decaying sinusoid. In the absence of damping ($R = 0$ or $\zeta = 0$), the frequency becomes $\omega = \omega_n$. Now ω_n was originally defined as the undamped natural frequency, and we now see that indeed it is a special case of Eq. (6-129) with $\zeta = 0$. If, in the absence of damping, a system oscillates with frequency ω_n, then the addition of damping will cause the frequency of oscillation to be reduced, the reduction being determined solely by the factor $\sqrt{1 - \zeta^2}$.

To illustrate the application of the general solutions we have found to a particular network with a given set of initial conditions,

---

[5]The use of ω as frequency in radians per second is not to be confused with the similar use of ω as angular velocity in rotating mechanical systems, such as rotating machinery. Note that we also make use of frequency f in hertz, the two frequencies being related by the equation, $\omega = 2\pi f$.

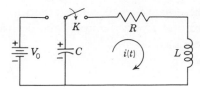

Fig. 6-11. Network for which particular solutions are found from the general solutions of the second-order differential equation.

consider the network of Fig. 6-11. Let the capacitor be charged to a voltage V_0, and at time $t = 0$ let the switch K be closed. The value of the resistance R with respect to the critical resistance R_{cr} will determine whether the system is overdamped, critically damped, or underdamped. Consider these three possibilities in turn.

With $R > R_{cr}$, the system is overdamped. The general solution can be reduced to a particular solution for a given set of initial conditions. For the circuit shown in Fig. 6-11, $i(0+) = 0$ because of the inductance. The term $(1/C) \int_{-\infty}^{0} i \, dt = -V_0$ at $t = 0$ (the initial voltage on the capacitor) such that

$$\frac{di}{dt}(0+) = +\frac{V_0}{L} \tag{6-130}$$

The requirement that $i(0+) = 0$ means that K_3 in Eq. (6-113) has zero value; that is,

$$i = K_4 e^{-\zeta \omega_n t} \sinh \omega_n \sqrt{\zeta^2 - 1} \, t \tag{6-131}$$

The constant K_4 can be evaluated from the second initial condition:

$$\frac{di}{dt} = K_4 [e^{-\zeta \omega_n t} \cosh (\omega_n \sqrt{\zeta^2 - 1} \, t) \cdot \omega_n \sqrt{\zeta^2 - 1}$$
$$+ \sinh (\omega_n \sqrt{\zeta^2 - 1} \, t) e^{-\zeta \omega_n t} (-\zeta \omega_n)] \tag{6-132}$$

The hyperbolic cosine term approaches unity as $t \to 0$, and the hyperbolic sine term approaches zero as $t \to 0$. Hence,

$$\frac{di}{dt}(0+) = K_4 \omega_n \sqrt{\zeta^2 - 1} = \frac{V_0}{L} \tag{6-133}$$

and

$$K_4 = \frac{V_0}{\omega_n L \sqrt{\zeta^2 - 1}} \tag{6-134}$$

The particular solution for the overdamped case thus becomes

$$i = \frac{V_0}{\omega_n L \sqrt{\zeta^2 - 1}} e^{-\zeta \omega_n t} \sinh (\omega_n \sqrt{\zeta^2 - 1} \, t) \tag{6-135}$$

The general shape of the current against time curve for this equation is shown in Fig. 6-12(a).

For the critically damped case, $R = R_{cr}$ and the solution in general is

$$i = (K_1 + K_2 t)e^{-\omega_n t} \tag{6-136}$$

subject to the same initial conditions as the overdamped case. The

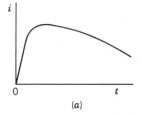

(a)

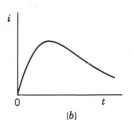

(b)

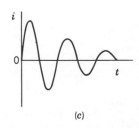

(c)

Fig. 6-12. Network response for the three cases: (a) overdamped, (b) critically damped, and (c) underdamped or oscillatory.

initial current condition implies that $K_1 = 0$, since otherwise this equation does not reduce to zero at $t = 0$. To apply the derivative condition, Eq. (6-136) is differentiated as

$$\frac{di}{dt} = K_2[te^{-\omega_n t}(-\omega_n) + e^{-\omega_n t}1]] \tag{6-137}$$

Hence

$$\frac{di}{dt}(0+) = K_2 = \frac{V_0}{L} \tag{6-138}$$

and the particular solution for the critically damped case is

$$i = \frac{V_0}{L} te^{-\omega_n t} \tag{6-139}$$

This curve is shown in Fig. 6-12(*b*) and has much the same appearance as that of Fig. 6-12(*a*).

For the underdamped, or oscillatory case, $R < R_{cr}$, and the solution is given by Eq. (6-121). The initial condition for the current requires that K_5 be zero, so that the solution can be written

$$i = K_6 e^{-\zeta \omega_n t} \sin (\omega_n \sqrt{1 - \zeta^2} \, t) \tag{6-140}$$

The constant K_6 is evaluated by using the initial condition of the derivative of the current; thus

$$\frac{di}{dt} = K_6 e^{-\zeta \omega_n t}[\omega_n \sqrt{1 - \zeta^2} \cos (\omega_n \sqrt{1 - \zeta^2} \, t)$$
$$- \zeta \omega_n \sin (\omega_n \sqrt{1 - \zeta^2} \, t)] \tag{6-141}$$

such that

$$\frac{di}{dt}(0+) = K_6 \omega_n \sqrt{1 - \zeta^2} = \frac{V_0}{L} \tag{6-142}$$

The particular solution for the oscillatory case is

$$i = \frac{V_0}{\omega_n L \sqrt{1 - \zeta^2}} e^{-\zeta \omega_n t} \sin (\omega_n \sqrt{1 - \zeta^2} \, t) \tag{6-143}$$

The variation of current with time for the oscillatory case is shown in Fig. 6-12(*c*). Since the current is the product of the damping factor and the oscillatory term, the damping factor represents an *envelope* or *boundary curve* for the oscillation. The factor $\zeta \omega_n$ determines how rapidly the oscillations are damped. As R approaches zero, the oscillations become undamped, and *sustained* oscillations result.

The physical meaning of this mathematical result may be interpreted in terms of an interchange of energy between the electric energy storage element (C) and the magnetic energy storage element (L). After the switch is closed, the energy which is stored in the electric field is transferred to the inductor as magnetic energy. When the current begins to decrease, energy is being returned to the electric field from the magnetic field. This interchange continues as long as any energy remains. If the resistance has zero value, the oscillatory current will be sustained indefinitely. However, if there *is* resistance present, the current through the resistor will cause energy to be dissipated, and the total energy will decrease with each cycle. Eventually all the energy will be dissipated and the current will be reduced to zero. If a scheme can be devised to supply the energy that is lost in each cycle, the oscillations can be sustained. This is accomplished in the electronic oscillator to produce audio frequency or radio frequency sinusoidal signals.

FURTHER READING

BALABANIAN, NORMAN, *Fundamentals of Circuit Theory*, Allyn and Bacon, Inc., Boston, 1961. Chapter 3.

CHIRLIAN, PAUL M., *Basic Network Theory*, McGraw-Hill Book Company, New York, 1969. Chapter 4.

CLEMENT, PRESTON R., AND WALTER C. JOHNSON, *Electrical Engineering Science*, McGraw-Hill Book Company, New York, 1960. Chapter 7.

CLOSE, CHARLES M., *The Analysis of Linear Circuits*, Harcourt, Brace & World, Inc., New York, 1966. Chapter 4.

HUELSMAN, LAWRENCE P., *Basic Circuit Theory with Digital Computations*, Prentice-Hall, Inc., Englewood Cliffs, N.J., 1972. Chapter 6.

SKILLING, HUGH H., *Electrical Engineering Circuits*, 2nd ed., John Wiley & Sons, Inc., New York, 1965. Chapter 2.

WYLIE, CLARENCE R., JR., *Advanced Engineering Mathematics*, 3rd ed., McGraw-Hill Book Company, New York, 1966. Chapters 2, 3, and 5.

DIGITAL COMPUTER EXERCISES

References that are useful in designing exercises to go with the topics of this chapter are cited in Appendix E-6.3 and are concerned with the numerical solution of higher-order differential equations. In particular, the suggestions contained in Chapters 5, 6, and 7 of Huelsman, reference 7, Appendix E-10, are recommended.

PROBLEMS

6-1. Show that $i = ke^{-2t}$ and $i = ke^{-t}$ are solutions of the differential equation

$$\frac{d^2i}{dt^2} + 3\frac{di}{dt} + 2i = 0$$

6-2. Show that $i = ke^{-t}$ and $i = kte^{-t}$ are solutions of the differential equation

$$\frac{d^2i}{dt^2} + 2\frac{di}{dt} + i = 0$$

6-3. Find the general solution of each of the following equations:

(a) $\dfrac{d^2i}{dt^2} + 3\dfrac{di}{dt} + 2i = 0$ (e) $\dfrac{d^2i}{dt^2} + \dfrac{di}{dt} + 6i = 0$

(b) $\dfrac{d^2i}{dt^2} + 5\dfrac{di}{dt} + 6i = 0$ (f) $\dfrac{d^2i}{dt^2} + \dfrac{di}{dt} + 2i = 0$

(c) $\dfrac{d^2i}{dt^2} + 7\dfrac{di}{dt} + 12i = 0$ (g) $\dfrac{d^2i}{dt^2} + 2\dfrac{di}{dt} + i = 0$

(d) $\dfrac{d^2i}{dt^2} + 5\dfrac{di}{dt} + 4i = 0$ (h) $\dfrac{d^2i}{dt^2} + 4\dfrac{di}{dt} + 4i = 0$

6-4. Find the general solution of each of the following homogeneous differential equations:

(a) $\dfrac{d^2v}{dt^2} + 2\dfrac{dv}{dt} + 2v = 0$ (d) $2\dfrac{d^2v}{dt^2} + 8\dfrac{dv}{dt} + 16v = 0$

(b) $\dfrac{d^2v}{dt^2} + 2\dfrac{dv}{dt} + 4v = 0$ (e) $\dfrac{d^2v}{dt^2} + 2\dfrac{dv}{dt} + 3v = 0$

(c) $\dfrac{d^2v}{dt^2} + 4\dfrac{dv}{dt} + 2v = 0$ (f) $\dfrac{d^2v}{dt^2} + 3\dfrac{dv}{dt} + 5v = 0$

6-5. Find particular solutions for the differential equations of Prob. 6-3 subject to the initial conditions:

$$i(0+) = 1, \qquad \frac{di}{dt}(0+) = 0$$

6-6. Find particular solutions for the differential equations of Prob. 6-3 subject to the initial conditions:

$$i(0+) = 2, \qquad \frac{di}{dt}(0+) = +1$$

6-7. Find particular solutions to the differential equations of Prob. 6-4 subject to the initial conditions:

$$v(0+) = 1, \qquad \frac{dv}{dt}(0+) = -1$$

6-8. Find particular solutions to the differential equations given in Prob. 6-4, given the initial conditions:

$$v(0+) = 2, \qquad \frac{dv}{dt}(0+) = 1$$

6-9. Solve the differential equation

$$3\frac{d^3i}{dt^3} + 8\frac{d^2i}{dt^2} + 10\frac{di}{dt} + 3i = 0$$

6-10. Solve the differential equation

$$2\frac{d^3i}{dt^3} + 9\frac{d^2i}{dt^2} + 13\frac{di}{dt} + 6i = 0$$

subject to the initial conditions $i(0+) = 0$, $di/dt = 1$ at $t = 0+$, and $d^2i/dt^2 = -1$ at $t = 0+$.

6-11. The response of a network is found to be

$$i = K_1 t e^{-\alpha t}, \qquad t \geq 0$$

where α is real and positive. Find the time at which $i(t)$ attains a maximum value.

6-12. In a certain network, it is found that the current is given by the expression

$$i = K_1 e^{-\alpha_1 t} - K_2 e^{-\alpha_2 t}, \qquad t > 0, \quad \alpha_1 > \alpha_2$$

Show that $i(t)$ reaches a maximum value at time

$$t = \frac{1}{\alpha_1 - \alpha_2} \ln \frac{\alpha_1 K_1}{\alpha_2 K_2}$$

6-13. The graph shows a damped sinusoidal waveform having the general form

$$Ke^{-\sigma t} \sin(\omega t + \phi)$$

From the graph, determine numerical values for K, σ, ω, and ϕ.

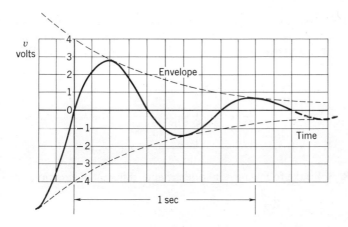

Fig. P6-13.

6-14. Repeat Prob. 6-13 for the waveform of the accompanying figure.

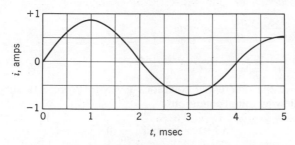

Fig. P6-14.

6-15. In the network of the figure, the switch K is closed and a steady state is reached in the network. At $t = 0$, the switch is opened. Find an expression for the current in the inductor, $i_2(t)$.

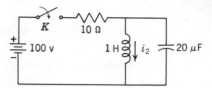

Fig. P6-15.

6-16. The capacitor of the figure has an initial voltage $v_C(0-) = V_1$, and at the same time the current in the inductor is zero. At $t = 0$, the switch K is closed. Determine an expression for the voltage $v_2(t)$.

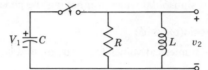

Fig. P6-16.

6-17. The voltage source in the network of the figure is described by the equation, $v_1 = 2\cos 2t$ for $t \geq 0$ and is a short circuit prior to that time. Determine $v_2(t)$. Repeat if $v_1 = K_1 t$ for $t \geq 0$ and $v_1 = 0$ for $t < 0$.

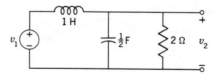

Fig. P6-17.

6-18. Solve the following nonhomogeneous differential equations for $t \geq 0$.

(a) $\dfrac{d^2 i}{dt^2} + 2\dfrac{di}{dt} + i = 1$

(b) $\dfrac{d^2 i}{dt^2} + 3\dfrac{di}{dt} + 2i = 5t$

(c) $\dfrac{d^2 i}{dt^2} + 3\dfrac{di}{dt} + 2i = 10 \sin 10t$

(d) $\dfrac{d^2 q}{dt^2} + 5\dfrac{dq}{dt} + 6q = te^{-t}$

(e) $\dfrac{d^2 v}{dt^2} + 5\dfrac{dv}{dt} + 6v = e^{-2t} + 5e^{-3t}$

6-19. Solve the differential equations given in Prob. 6-18 subject to the following initial conditions:

$$x(0+) = 1 \quad \text{and} \quad \frac{dx}{dt}(0+) = -1$$

where x is the general dependent variable.

6-20. Find the particular solutions to the differential equations of Prob. 6-18 for the following initial conditions:

$$x(0+) = 2 \quad \text{and} \quad \frac{dx}{dt}(0+) = -1$$

where x is the dependent variable in each case.

6-21. Solve the differential equation

$$2\frac{d^3i}{dt^3} + 9\frac{d^2i}{dt^2} + 13\frac{di}{dt} + 6i = K_0 t e^{-t} \sin t$$

which is valid for $t \geq 0$, if $i(0+) = 1$, $di/dt(0+) = -1$, and $d^2i/dt^2(0+) = 0$.

6-22. A special generator has a voltage variation given by the equation $v(t) = t$ V, where t is the time in seconds and $t \geq 0$. This generator is connected to an RL series circuit, where $R = 2\,\Omega$ and $L = 1$ H, at time $t = 0$ by the closing of a switch. Find the equation for the current as a function of time $i(t)$.

6-23. A bolt of lightning having a waveform which is approximated as $v(t) = te^{-t}$ strikes a transmission line having resistance $R - 0.1\,\Omega$ and inductance $L = 0.1$ H (the line-to-line capacitance is assumed negligible). An equivalent network is shown in the accompanying diagram. What is the form of the current as a function of time? (This current will be in amperes per unit volt of the lightning; likewise the time base is normalized.)

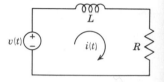

Fig. P6-23.

6-24. In the network of the figure, the switch K is closed at $t = 0$ with the capacitor initially unenergized. For the numerical values given, find $i(t)$.

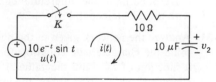

Fig. P6-24.

6-25. In the network shown in the accompanying figure, a steady state is reached with the switch K open. At $t = 0$, the switch is closed. For the element values given, determine the current, $i(t)$ for $t \geq 0$.

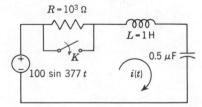

Fig. P6-25.

6-26. In the network shown in Fig. P6-25, a steady state is reached with the switch K open. At $t = 0$, the value of the x resistor R is changed to the critical value, R_{cr} defined by Eq. (6-88). For the element values given, determine the current $i(t)$ for $t \geq 0$.

6-27. Consider the network shown in Fig. P6-24. The capacitor has an initial voltage, $v_C = 10$ V. At $t = 0$, the switch K is closed. Determine $i(t)$ for $t \geq 0$.

6-28. The network of the figure is operating in the steady state with the switch K open. At $t = 0$, the switch is closed. Find an expression for the voltage, $v(t)$ for $t \geq 0$.

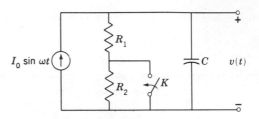

Fig. P6-28.

6-29. Consider a series RLC network which is excited by a voltage source. (a) Determine the characteristic equation corresponding to the differential equation for $i(t)$. (b) Suppose that L and C are fixed in value but that R varies from 0 to ∞. What will be the locus of the roots of the characteristic equation? (c) Plot the roots of the characteristic equation in the s plane if $L = 1$ H, $C = 1$ μF, and R has the following values: 500 Ω, 1000 Ω, 3000 Ω, 5000 Ω.

6-30. Consider the RLC network of Prob. 6-16. Repeat Prob. 6-29, except that in this case the study will concern the characteristic equation corresponding to the differential equation for $v_2(t)$. Compare results with those obtained in Prob. 6-29.

6-31. Analyze the network given in the figure on the loop basis, and determine the characteristic equation for the currents in the network as a function of K_1. Find the value(s) of K_1 for which the roots of the characteristic equation are on the imaginary axis of the s plane. Find the range of values of K_1 for which the roots of the characteristic equation have positive real parts.

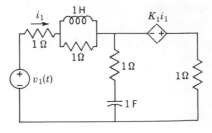

Fig. P6-31.

6-32. Show that Eq. (6-121) can be written in the form

$$i = Ke^{-\zeta\omega_n t}\cos(\omega_n\sqrt{1-\zeta^2}\,t + \phi)$$

Give the values for K and ϕ in terms of K_5 and K_6 of Eq. (6-121).

6-33. A switch is closed at $t = 0$ connecting a battery of voltage V with a series RL circuit. (a) Show that the energy in the resistor as a function of time is

$$w_R = \frac{V^2}{R}\left(t + \frac{2L}{R}e^{-Rt/L} - \frac{L}{2R}e^{-2Rt/L} - \frac{3L}{2R}\right) \text{ joules}$$

(b) Find an expression for the energy in the magnetic field as a function of time. (c) Sketch w_R and w_L as a function of time. Show the steady-state asymptotes, that is, the values that w_R and w_L approach as $t \longrightarrow \infty$. (d) Find the total energy supplied by the voltage source in the steady state.

6-34. In the series RLC circuit shown in the accompanying diagram, the frequency of the driving force voltage is

(1) $\omega = \omega_n$ (the undamped natural frequency)
(2) $\omega = \omega_n\sqrt{1 - \zeta^2}$ (the natural frequency)

These frequencies are applied in two separate experiments. In each experiment we measure (a) the peak value of the transient current when the switch is closed at $t = 0$, and (b) the maximum value of the steady-state current. (a) In which case (that is, which frequency) is the maximum value of the transient greater? (b) In which case (that is, which frequency) is the maximum value of the steady-state current greater?

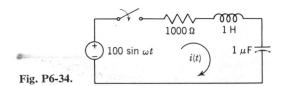

Fig. P6-34.

7 The Laplace Transformation

7-1. INTRODUCTION

The forerunner of the Laplace transformation method of solving differential equations, the *operational calculus*, was invented by the brilliant English engineer Oliver Heaviside (1850–1925). Heaviside was a practical man and his interest was in the practical solution of electric circuit problems rather than careful justification of his methods. He was gifted with an insight into physical problems that enabled him to pick the correct solution from a number of alternatives. This heuristic point of view drew bitter and perpetual criticism from the leading mathematicians of his time. In the years that followed publication of Heaviside's work, the rigor was supplied by such men as Bromwich, Giorgi, Carson, and others. The basis for substantiating the work of Heaviside was found in the writings of Laplace in 1780. As the years have passed, the structural members of the framework of Heaviside's operational calculus have been replaced, piece by piece, by new members derived by the Laplace transformation. This transformation has provided rigorous substantiation of the operational methods; no important errors have been discovered in Heaviside's results.

The Laplace transformation method for solving differential equations offers a number of advantages over the classical methods that were discussed in Chapters 4 and 6. For example:

170

(1) The solution of differential equations is routine and progresses systematically.
(2) The method gives the total solution—the particular integral and the complementary function—in one operation.
(3) Initial conditions are automatically specified in the transformed equations. Further, the initial conditions are incorporated into the problem as one of the first steps rather than as the last step.

What is a transformation? The *logarithm* is an example of a transformation that we have used in the past. Logarithms greatly simplify such operations as multiplication, division, extracting roots, and raising quantities to powers. Suppose that we have two numbers, given to seven-place accuracy, and we are required to find the product, maintaining the accuracy of the given numbers. Rather than just multiplying the two numbers together, we transform these numbers by taking their logarithms. These logarithms are added (or subtracted in the case of division). The resulting sum itself has little meaning. However, if we perform an *inverse transformation*, if we find the antilogarithm, then we have the desired numerical result. The direct division looks more straightforward, but our experience has been that the use of the logarithm often saves time. If the simple problem of multiplying two numbers is not convincing, consider evaluating $(1437)^{0.1328}$ without logarithms!

A flow sheet of the operation of using logarithms to find a product or a quotient is shown in Fig. 7-1. The individual steps are: (1) find the logarithm of each separate number, (2) add or subtract the numbers to obtain the sum of logarithms, and (3) take the antilogarithm to obtain the product or quotient. This is roundabout compared with *direct* multiplication or division, yet we use logarithms to advantage, particularly when a good table of logarithms is available.

The flow sheet idea may be used to illustrate what we will do in using the Laplace transformation to solve a differential equation. The flow sheet for the Laplace transformation is shown in Fig. 7-1(*b*) with a block corresponding to every block of the logarithm flow sheet considered above. The steps will be as follows. (1) Start with an integrodifferential equation and find the corresponding Laplace *transform*. This is a mathematical process, but there are tables of transforms just as there are tables of logarithms (and one is included in this chapter). (2) The transform is manipulated algebraically after the initial conditions are inserted. The result is a *revised transform*. As step (3), we perform an inverse Laplace transformation to give us the solution. In this step, we also can use a table of transforms, just as we use the table of logarithms in the corresponding step for logarithms.

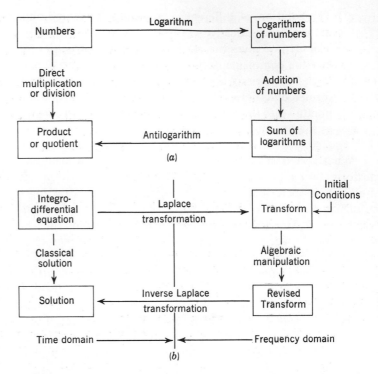

Fig. 7-1. Comparison of logarithms and the Laplace transformation.

The flow sheet reminds us that *there is another way:* the classical solution. It looks more direct (and sometimes it is for simple problems). For complicated problems, an advantage will be found for the Laplace transformation, just as an advantage was found for the use of logarithms.

7-2. THE LAPLACE TRANSFORMATION

To construct a Laplace transform for a given function of time $f(t)$, we first multiply $f(t)$ by e^{-st}, where s is a complex number, $s = \sigma + j\omega$. This product is integrated with respect to time from zero to infinity. The result is the Laplace transform of $f(t)$, which is designated $F(s)$. Denoting the Laplace transformation by the script letter $\mathcal{L}$ (in order to reserve L for inductance), the Laplace transformation is given by the expression

$$\mathcal{L}[f(t)] = F(s) = \int_{0-}^{\infty} f(t)e^{-st}\, dt \tag{7-1}$$

The letter $\mathcal{L}$ can be replaced by the words "the Laplace transform of" in the above expression. The implication of the use of $0-$ as the lower limit of integration in this equation requires discussion. It is conventional in studying electric circuits to subdivide the time $t = 0$ into three parts: $0-$ indicating the time just before the reference time (or $t = -\epsilon$, $\epsilon > 0$, as $\epsilon \rightarrow 0$), 0 indicating the exact reference time, and $0+$ the time just after $t = 0$. When the continuity condition $f(0-) = f(0+)$ is satisfied, the choice of $0-$ or $0+$ is not important. However, if there is an impulse function at $t = 0$, then $t = 0-$ must be used so that the impulse function is included. Rather than make an exception for networks with impulses, we will use $t = 0-$ throughout.

In order that $f(t)$ be transformable, it is sufficient that

$$\int_{0-}^{\infty} |f(t)| e^{-\sigma_1 t} \, dt < \infty \tag{7-2}$$

for a real, positive σ_1. Although Eq. (7-1) may be a rather formidable appearing integral at first glance, the actual evaluation of $F(s)$ for a given $f(t)$ is usually not difficult. Furthermore, once the transform of a function is found, it need not be found again for each new problem, but can be tabulated for future use. The time function $f(t)$ and its transform $F(s)$ are called a *transform pair*. Various workers have compiled extensive tables of transform pairs, so that Eq. (7-1) will seldom be used in solving problems in practice.

Two parts of the last paragraph require further examination. (1) How serious is the limitation imposed by Eq. (7-2) that must be satisfied for $f(t)$ to have a transform? (2) For a given $F(s)$, how do we find the corresponding $f(t)$: Can transform pairs be used in reverse?

The restriction of Eq. (7-2) is satisfied by most $f(t)$ encountered in engineering, since $e^{-\sigma t}$ is a "powerful reducing factor" as a multiplier of $f(t)$. Thus we may show by l'Hospital's rule that

$$\lim_{t \to \infty} t^n e^{-\sigma t} = 0, \qquad \sigma > 0 \tag{7-3}$$

such that the integral of the product, for $n = 1$, is

$$\int_{0-}^{\infty} t e^{-\sigma t} \, dt = \frac{1}{\sigma^2}, \qquad \sigma > 0 \tag{7-4}$$

and the integral for other values of n similarly remains finite for $\sigma \neq 0$.

An example of a function that does not satisfy Eq. (7-2) is e^{at^2} or, in general, e^{at^n}; there is no value of σ for which the integral of Eq. (7-2) for these $f(t)$ remain finite. Now such functions are seldom required to describe the driving function in engineering problems. Furthermore, a generator would produce this function for only a limited range of values of t, and thereafter would saturate at a constant value. The function we have just described which is

$$\begin{aligned} v &= e^{at^2}, & 0 \leq t \leq t_0 \\ &= K, & t > t_0 \end{aligned} \tag{7-5}$$

does, of course, satisfy Eq. (7-2).

EXAMPLE 1

As an example of the evaluation of Eq. (7-1), consider the *unit step function* introduced by Heaviside. This function is described by the equation

$$u(t) = 1, \qquad t \geq 0$$
$$= 0, \qquad t < 0 \qquad (7\text{-}6)$$

as shown in Fig. 7-2. Such a notation is convenient for representing the closing of a switch at $t = 0$; if a battery of voltage V_0 is connected to a network at $t = 0$, then the driving voltage may be represented as $V_0 u(t)$ without the necessity of mentioning the presence of a switch (or showing it in the schematic diagram). For $V_0 = 1$, we have

$$\mathcal{L}[u(t)] = \int_{0-}^{\infty} e^{-st}\, dt = -\frac{1}{s} e^{-st} \Big|_{0-}^{\infty} = \frac{1}{s} \qquad (7\text{-}7)$$

Similarly,

$$\mathcal{L}[V_0 u(t)] = \frac{V_0}{s} \qquad (7\text{-}8)$$

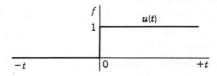

Fig. 7-2. The unit step function.

EXAMPLE 2

As a second example of the calculation of a transform, let $f(t) = e^{at}$, where a is a constant. Substituting into Eq. (7-1), we have

$$\mathcal{L}[e^{at}] = \int_{0-}^{\infty} e^{at} e^{-st}\, dt = \int_{0-}^{\infty} e^{-(s-a)t}\, dt = \frac{1}{s-a}, \qquad \sigma_1 > a \qquad (7\text{-}9)$$

Thus e^{at} and $1/(s-a)$ constitute a transform pair.

These two computations form the beginning of a table of transform pairs as shown below.

TABLE OF TRANSFORM PAIRS

$f(t)$	$F(s)$
$u(t)$	$\dfrac{1}{s}$
e^{at}	$\dfrac{1}{s-a}$

We may now turn to our second question which concerns finding $f(t)$ from $F(s)$. The inverse Laplace transformation is given by the *complex inversion integral*,

$$f(t) = \frac{1}{2\pi j} \int_{\sigma_1 - j\infty}^{\sigma_1 + j\infty} F(s) e^{st}\, ds \qquad (7\text{-}10)$$

which is a contour integral where the path of integration, known as the *Bromwich path*, is along the vertical line $s = \sigma_1$ from $-j\infty$ to $j\infty$, as shown in Fig. 7-3. On the figure is also shown the *abscissa of convergence*, which is the number designated as a in Eq. (7-9). For the proper evaluation of the inversion integral, we require that $\sigma_1 > \sigma_c$ or that the path of integration in Eq. (7-10) be to the right of the abscissa of convergence. In terms of our using the Laplace transformation, what do these results mean? In terms of Eq. (7-9), the fact that a σ_1 may be chosen greater than $\sigma_c = a$ implies that the transform $F(s)$ exists since Eq. (7-2) is satisfied. If we use the inversion integral to compute $f(t)$ by Eq. (7-10), then a proper value of σ_1 could be chosen by knowing σ_c. However, another property of the Laplace transformation makes it unnecessary for us to use the inversion integral in most cases. This property is the *uniqueness* of the Laplace transformation: There cannot be two different functions having the same Laplace transformation, $F(s)$. That being the case, we may use the table of transform pairs to find $f(t)$, provided we can find the necessary form of $F(s)$ in the table. We use the symbol $\mathcal{L}^{-1}$ to indicate the *inverse Laplace transformation*. Then

$$\mathcal{L}^{-1}\{\mathcal{L}[f(t)]\} = \mathcal{L}^{-1}[F(s)] = f(t) \qquad (7\text{-}11)$$

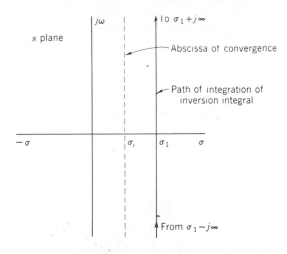

Fig. 7-3. Figure showing the abscissa of convergence which relates to the inverse Laplace transformation.

7-3. SOME BASIC THEOREMS FOR THE LAPLACE TRANSFORMATION

(1) Transforms of linear combinations. If $f_1(t)$ and $f_2(t)$ are two functions of time and a and b are constants, then

$$\mathcal{L}[af_1(t) + bf_2(t)] = aF_1(s) + bF_2(s) \qquad (7\text{-}12)$$

This theorem is established with Eq. (7-1). It follows from the fact that the integral of a sum of terms is equal to the sum of the integrals of the terms; that is,

$$\mathcal{L}[af_1(t) + bf_2(t)] = \int_{0-}^{\infty} [af_1(t) + bf_2(t)]e^{-st}\, dt$$

$$= a \int_{0-}^{\infty} f_1(t)e^{-st}\, dt + b \int_{0-}^{\infty} f_2(t)e^{-st}\, dt \qquad (7\text{-}13)$$

$$= aF_1(s) + bF_2(s)$$

We will make use of this theorem in finding the Laplace transformation of the sum of terms that appear in network equations.

EXAMPLE 3

As an example of the use of this result, let us find the transform of $\cos \omega t$ and $\sin \omega t$. From Euler's identity, Eq. 6-37, which is

$$e^{\pm j\omega t} = \cos \omega t \pm j \sin \omega t \qquad (7\text{-}14)$$

we see that by adding the two equations of Eqs. 7-14, we have

$$\cos \omega t = \frac{e^{j\omega t} + e^{-j\omega t}}{2} \qquad (7\text{-}15)$$

and by subtracting,

$$\sin \omega t = \frac{e^{j\omega t} - e^{-j\omega t}}{2j} \qquad (7\text{-}16)$$

Since the transform of the exponential function is

$$\mathcal{L}[e^{+j\omega t}] = \frac{1}{s - j\omega}, \qquad \sigma_1 > 0 \qquad (7\text{-}17)$$

we see that

$$\mathcal{L}[\cos \omega t] = \frac{1}{2}\left(\frac{1}{s - j\omega} + \frac{1}{s + j\omega}\right) = \frac{s}{s^2 + \omega^2}, \qquad \sigma_1 > 0 \qquad (7\text{-}18)$$

and

$$\mathcal{L}[\sin \omega t] = \frac{1}{2j}\left(\frac{1}{s - j\omega} - \frac{1}{s + j\omega}\right) = \frac{\omega}{s^2 + \omega^2}, \qquad \sigma_1 > 0 \qquad (7\text{-}19)$$

These results may be added to our collection of transform pairs.

(2) *Transforms of derivatives.* From the defining equation for the Laplace transformation, we write

$$\mathcal{L}\left[\frac{d}{dt}f(t)\right] = \int_{0-}^{\infty} \frac{d}{dt}f(t)e^{-st}\,dt \tag{7-20}$$

This equation may be integrated by parts by letting

$$u = e^{-st} \quad \text{and} \quad dv = df(t) \tag{7-21}$$

in the equation

$$\int_a^b u\,dv = uv\Big|_a^b - \int_a^b v\,du \tag{7-22}$$

Then

$$du = -se^{-st}\,dt \quad \text{and} \quad v = f(t) \tag{7-23}$$

so that the transform of a derivative becomes

$$\mathcal{L}\left[\frac{d}{dt}f(t)\right] = e^{-st}f(t)\Big|_{0-}^{\infty} + s\int_{0-}^{\infty} f(t)e^{-st}\,dt = sF(s) - f(0-) \tag{7-24}$$

provided $\lim_{t\to\infty} f(t)e^{-st} = 0$, which follows from l'Hospital's rule, provided that $f(t)$ and its successive derivatives are finite at $t = \infty$ and $\sigma > 0$.

To find the transform of the second derivative, we follow a similar procedure but make use of the result of Eq. (7-24). Since

$$\frac{d^2}{dt^2}f(t) = \frac{d}{dt}\frac{d}{dt}f(t) \tag{7-25}$$

then

$$\mathcal{L}\left[\frac{d^2 f(t)}{dt^2}\right] = s\mathcal{L}\left[\frac{df}{dt}(t)\right] - \frac{df}{dt}(0-)$$

$$= s[sF(s) - f(0-)] - \frac{df}{dt}(0-) \tag{7-26}$$

$$= s^2 F(s) - sf(0-) - \frac{df}{dt}(0-)$$

In this expression, the quantity $df/dt(0-)$ is the derivative of $f(t)$ evaluated at $t = 0-$. The general expression for an nth derivative is

$$\mathcal{L}\frac{d^n f(t)}{dt^n} = s^n F(s) - s^{n-1}f(0-) - s^{n-2}\frac{df}{dt}(0-) - \dots - \frac{d^{n-1}}{dt^{n-1}}f(0-) \tag{7-27}$$

(3) *Transforms of integrals.* The transform for an integral is found by starting from the definition

$$\mathcal{L}\left[\int_{0-}^{t} f(t)\,dt\right] = \int_{0-}^{\infty}\left[\int_{0-}^{t} f(t)\,dt\right]e^{-st}\,dt \tag{7-28}$$

The integration is carried out by parts where we let

$$u = \int_0^t f(t)\, dt, \qquad du = f(t)\, dt \tag{7-29}$$

and

$$dv = e^{-st}\, dt, \qquad v = -\frac{1}{s} e^{-st} \tag{7-30}$$

Hence

$$\mathscr{L}\left[\int_{0-}^t f(t)\, dt\right] = -\frac{e^{-st}}{s} \int_{0-}^t f(t)\, dt \bigg|_{0-}^{\infty} + \frac{1}{s} \int_{0-}^{\infty} f(t) e^{-st}\, dt \tag{7-31}$$

Now the first term vanishes since e^{-st} approaches zero for infinite t, and at the lower limit

$$\int_{0-}^t f(t)\, dt \bigg|_{t=0-} = 0 \tag{7-32}$$

Hence we conclude that

$$\mathscr{L}\left[\int_{-0}^t f(t)\, dt\right] = \frac{F(s)}{s} \tag{7-33}$$

Now the formulation of the Kirchhoff laws for a network often involve an integral with limits from $-\infty$ to t. Such integrals may be divided into two parts

$$\int_{-\infty}^t f(t)\, dt = \int_{-\infty}^{0-} f(t)\, dt + \int_{0-}^t f(t)\, dt \tag{7-34}$$

where the first term on the right of this equation is a constant. When $f(t)$ is current, then this integral is the initial value of charge, $q(0-)$, and when $f(t)$ is voltage, then the integral is flux linkages $\psi(0-) = Li(0-)$. In either case, this term should be included in the equation

Table 7-1. SUMMARY OF RESULTS*

Name	Property	Equation
Definition	$\mathscr{L} f(t) = F(s) = \int_0^{\infty} f(t) e^{-st}\, dt$	(7-1)
Linearity	$\mathscr{L}[af_1(t) + bf_2(t)] = aF_1(s) + bF_2(s)$	(7-12)
Time Differentiation	$\mathscr{L} \dfrac{d}{dt} f(t) = sF(s) - f(0-)$	(7-24)
Time Integration	$\mathscr{L} \int_0^t f(x)\, dx = \dfrac{1}{s} F(s)$	(7-33)

*See Table 8-1 for an extension of this table.

formulation; the transform of a constant $q(0-)$ is, from (Eq. 7-8),

$$\mathcal{L}[q(0-)] = \frac{q(0-)}{s} \tag{7-35}$$

and a similar equation may be written for $\psi(0-)$.

Our results thus far are summarized in Table 7-1.

7-4. EXAMPLES OF THE SOLUTION OF PROBLEMS WITH THE LAPLACE TRANSFORMATION

With the short table of transforms that has been given on page 174 and the three basic theorems that have been derived in the previous section, we are now equipped to solve a network problem (elementary as yet, to be sure) using the Laplace transformation.

EXAMPLE 4

For this example, we will write the Kirchhoff voltage law for a series RC network shown in Fig. 7-4. It will be assumed that the switch K is closed at $t = 0$. This information will be included in the formation of the network equations by writing the voltage expression as $Vu(t)$. Hence

$$\frac{1}{C} \int_{-\infty}^{t} i \, dt + Ri = Vu(t) \tag{7-36}$$

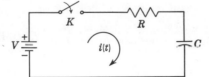

Fig. 7-4. RC series network for Example 4.

This is the integral equation we wish to solve. The transforms of the linear combination of terms is

$$\frac{1}{C}\left[\frac{I(s)}{s} + \frac{q(0-)}{s}\right] + RI(s) = V \cdot \frac{1}{s} \tag{7-37}$$

In terms of the flow chart of Fig. 7-1, we have found the Laplace transformation of the integral equation and there has resulted a transform expression. The required initial conditions are automatically specified and may be inserted as the second step (rather than as the final step as in differential equations solved by classical methods). Now $q(0-)$ is the charge on the capacitor at $t = 0-$. If the capacitor is initially uncharged, $q(0-) = 0$, the last equation reduces to the

form

$$I(s)\left(\frac{1}{Cs} + R\right) = \frac{V}{s} \tag{7-38}$$

The next step, again according to the flow chart, is algebraic manipulation. The objective of this manipulation is to solve for $I(s)$. This is accomplished by multiplying by s and dividing by R to give

$$I(s) = \frac{V/R}{s + (1/RC)} \tag{7-39}$$

which is a "revised transform" expression. The next step on our flow chart is to perform the inverse Laplace transformation and obtain the solution. That is,

$$\mathcal{L}^{-1}[I(s)] = \mathcal{L}^{-1}\left[\frac{V/R}{s + (1/RC)}\right] = i(t) \tag{7-40}$$

Using the second transform pair of our short table, the solution is

$$\begin{aligned} i(t) &= \frac{V}{R} e^{-t/RC}, & t \geq 0 \\ &= 0, & t < 0 \end{aligned} \tag{7-41}$$

This is the complete solution. The arbitrary constant emerges evaluated (and has the magnitude V/R).

EXAMPLE 5

As our second example, consider the RL series circuit shown in Fig. 7-5 with the switch closed at $t = 0$. The differential equation for the circuit is, by Kirchhoff's law,

$$L\frac{di}{dt} + Ri = Vu(t) \tag{7-42}$$

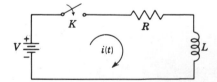

Fig. 7-5. RL series network for Example 5.

The corresponding transform equation is

$$L[sI(s) - i(0-)] + RI(s) = \frac{V}{s} \tag{7-43}$$

The initial condition specified by the last equation is $i(0-)$. Because the inductance is unfluxed, $i(0-) = 0$. Our equation may now be manipulated to solve for $I(s)$; thus

$$I(s) = \frac{V}{L}\frac{1}{s[s + (R/L)]} \tag{7-44}$$

This transform, however, is not in our short table. We need something new (or a larger table). Notice that this term is made up of the product of the term $1/s$ and the term $1/[s + (R/L)]$. We know the inverse Laplace transformation of each of these individual terms. This suggests that the inverse operation could be performed if there were some way to break the transform terms into several parts. As an attempt to perform this operation, let us try the following expansion:

$$\frac{V/L}{s[s + (R/L)]} = \frac{K_0}{s} + \frac{K_1}{s + (R/L)} \tag{7-45}$$

In this equation K_0 and K_1 are unknown coefficients. As the first step, let us simplify the equation by putting all terms over a common denominator. Then

$$\frac{V}{L} = K_0\left(s + \frac{R}{L}\right) + K_1 s \tag{7-46}$$

By equating coefficients of like functions, we obtain a set of linear algebraic equations:

$$K_0 \cdot \frac{R}{L} = \frac{V}{L}, \qquad K_0 + K_1 = 0 \tag{7-47}$$

From these two equations, we find the required values for K_0 and K_1:

$$K_0 = \frac{V}{R} \quad \text{and} \quad K_1 = -\frac{V}{R} \tag{7-48}$$

This algebraic manipulation has permitted Eq. (7-44) to be written

$$I(s) = \frac{V}{L}\frac{1}{s[s + (R/L)]} = \frac{V}{R}\left[\frac{1}{s} - \frac{1}{s + (R/L)}\right] \tag{7-49}$$

We have transform pairs corresponding to each of these expressions. The current as a function of time is found by taking the inverse Laplace transformation of the individual expressions; thus

$$i(t) = \frac{V}{R}\left[\mathcal{L}^{-1}\frac{1}{s} - \mathcal{L}^{-1}\frac{1}{s + (R/L)}\right] \tag{7-50}$$

or

$$i(t) = \frac{V}{R}(1 - e^{-Rt/L}), \qquad t \geq 0 \tag{7-51}$$

This is the final (time-domain) solution. The method we used to expand a transform into the sum of several separate parts is known under the heading of *partial fraction expansion*. It is this subject that we study next.

7-5. PARTIAL FRACTION EXPANSION

The examples in the last section have suggested the general procedure in applying the Laplace transformation to the solution of integrodifferential equations. A differential equation of the general

form

$$a_0 \frac{d^n i}{dt^n} + a_1 \frac{d^{n-1} i}{dt^{n-1}} + \ldots + a_{n-1} \frac{di}{dt} + a_n i = v(t) \qquad (7\text{-}52)$$

becomes, as a result of the Laplace transformation, an algebraic equation which may be solved for the unknown as

$$I(s) = \frac{\mathcal{L}[v(t)] + \text{initial condition terms}}{a_0 s^n + a_1 s^{n-1} + \ldots + a_{n-1} s + a_n} \qquad (7\text{-}53)$$

The general form of this equation is a quotient of polynomials in s. Let the numerator and denominator polynomials be designated $P(s)$ and $Q(s)$, respectively, as

$$I(s) = \frac{P(s)}{Q(s)} \qquad (7\text{-}54)$$

Note that $Q(s) = 0$ is the characteristic equation of Chapter 6. If the transform term $P(s)/Q(s)$ can now be found in a table of transform pairs, the solution $i(t)$ can be written directly. In general, however, the transform expression for $I(s)$ must be broken into simpler terms before any practical transform table can be used.

As the first step in the expansion of the quotient $P(s)/Q(s)$, we check to see that the order[1] of the polynomial P is less than that of Q. If this condition is not fulfilled, divide the numerator by the denominator to obtain an expansion in the form

$$\frac{P(s)}{Q(s)} = B_0 + B_1 s + B_2 s^2 + \ldots + B_{m-n} s^{m-n} + \frac{P_1(s)}{Q(s)} \qquad (7\text{-}55)$$

where m is the order of the numerator and n the order of the denominator. The new function $P_1(s)/Q(s)$ has now been "prepared" and the order rule is satisfied.[2]

EXAMPLE 6

Consider the quotient

$$\frac{P(s)}{Q(s)} = \frac{s^2 + 2s + 2}{s + 1} \qquad (7\text{-}56)$$

By direct division,

$$
\begin{array}{r}
s + 1 \overline{)\, s^2 + 2s + 2} \quad (s + 1 \\
\underline{s^2 + \ \ s } \\
s + 2 \\
\underline{s + 1} \\
1
\end{array}
$$

---

[1] Beginning with Chapter 9, we will use *degree* rather than *order*. See footnote 3 of Chapter 6.

[2] The inverse Laplace transforms of terms like B_0 and $B_1 s$ will be considered in Chap. 9 since they are singularity functions such as the impulse function.

or

$$\frac{s^2 + 2s + 2}{s + 1} = 1 + s + \frac{1}{s + 1} \qquad (7\text{-}57)$$

so that in Eq. (7-55), $B_0 = 1$, $B_1 = 1$, and $P_1(s)/Q(s) = 1/(s+1)$.

Next, we factor the denominator polynomial, $Q(s)$,

$$Q(s) = a_0 s^n + a_1 s^{n-1} + \ldots + a_n = a_0(s - s_1) \ldots (s - s_n) \qquad (7\text{-}58)$$

or, very compactly,

$$Q(s) = a_0 \prod_{j=1}^{n} (s - s_j) \qquad (7\text{-}59)$$

where Π indicates a product of factors, and $s_1, s_2, \ldots, s_n$ are the n roots of the equation $Q(s) = 0$. Now the possible form of these roots was discussed in Chapter 6: (1) real and simple (or distinct) roots, (2) conjugate complex roots, and (3) multiple roots. We will consider these possibilities separately.

(1) If all roots of $Q(s) = 0$ are simple, then the partial fraction expansion is

$$\frac{P_1(s)}{(s - s_1)(s - s_2) \ldots (s - s_n)}$$
$$= \frac{K_1}{s - s_1} + \frac{K_2}{s - s_2} + \ldots + \frac{K_n}{s - s_n} \qquad (7\text{-}60)$$

where the K's are real constants called *residues*.

(2) If a root of $Q(s) = 0$ is of multiplicity r, the partial fraction expansion for the repeated root is

$$\frac{P_1(s)}{(s - s_1)^r} = \frac{K_{11}}{s - s_1} + \frac{K_{12}}{(s - s_1)^2} + \ldots + \frac{K_{1r}}{(s - s_1)^r} \qquad (7\text{-}61)$$

and there will be similar terms for every other repeated root.

(3) An important special rule may be given for two roots which form a complex conjugate pair. For this case, the partial fraction expansion is

$$\frac{P_1(s)}{Q_1(s)(s + \alpha + j\omega)(s + \alpha - j\omega)}$$
$$= \frac{K_1}{(s + \alpha + j\omega)} + \frac{K_1^*}{(s + \alpha - j\omega)} + \ldots \qquad (7\text{-}62)$$

where K_1^* is the complex conjugate of K_1. In other words, when the roots are conjugates, so are the partial fraction expansion coefficients. An expansion of the type shown above is necessary for each pair of complex conjugate roots.

In an expansion of a quotient of polynomials by partial fractions, it may be necessary to use a combination of the three rules given above.

Several examples will illustrate the expansion and the determination of the K's.

EXAMPLE 7

Consider the quotient of polynomials,

$$I(s) = \frac{2s + 3}{s^2 + 3s + 2} \tag{7-63}$$

The first step is to factor the denominator polynomial and then expand by the appropriate rule. For this example, the expansion is

$$\frac{2s + 3}{(s + 1)(s + 2)} = \frac{K_1}{(s + 1)} + \frac{K_2}{(s + 2)} \tag{7-64}$$

since the roots are real and simple. As the first step, multiply the equation by $(s + 1)$ as

$$\frac{(2s + 3)(s + 1)}{(s + 1)(s + 2)} = K_1 \frac{s + 1}{s + 1} + K_2 \frac{s + 1}{s + 2} \tag{7-65}$$

or, canceling common factors,

$$\frac{2s + 3}{s + 2} = K_1 + K_2 \frac{s + 1}{s + 2} \tag{7-66}$$

In this equation, the coefficient K_1 is not multiplied by any function of s. Now s is merely an algebraic factor that can have any value. If $s = -1$, the coefficient of K_2 reduces to zero and we can solve for K_1 as

$$K_1 = \frac{2s + 3}{s + 2}\bigg|_{s=-1} = \frac{-2 + 3}{-1 + 2} = 1 \tag{7-67}$$

To evaluate K_2 and to follow the same pattern, multiply Eq. (7-64) by $(s + 2)$ to obtain

$$\frac{2s + 3}{s + 1} = K_1 \frac{s + 2}{s + 1} + K_2 \tag{7-68}$$

To evaluate K_2, we set $s = -2$ in order to reduce the coefficient of K_1 to zero. Then

$$K_2 = \frac{2s + 3}{s + 1}\bigg|_{s=-2} = \frac{-4 + 3}{-2 + 1} = 1 \tag{7-69}$$

The result of the partial fraction expansion is thus

$$\frac{2s + 3}{s_2 + 3s + 2} = \frac{1}{s + 1} + \frac{1}{s + 2} \tag{7-70}$$

The expansion should always be checked by combining the two terms.

EXAMPLE 8

For this example, consider a quotient of polynomials with repeated denominator roots:

$$\frac{s+2}{(s+1)^2} = \frac{K_{11}}{s+1} + \frac{K_{12}}{(s+1)^2} \tag{7-71}$$

Multiplying by $(s+1)^2$ gives

$$s+2 = (s+1)K_{11} + K_{12} \tag{7-72}$$

and when $s = -1$, K_{12} is readily evaluated as $K_{12} = 1$. If we attempt to follow the same pattern to evaluate K_{11}, trouble develops. That is,

$$\frac{s+2}{s+1} = K_{11} + \frac{K_{12}}{s+1} \tag{7-73}$$

If, in this equation, $s = -1$, one term becomes infinite and K_{11} cannot be evaluated. However, the problem can be resolved if we return to Eq. (7-72) and differentiate with respect to s:

$$1 + 0 = K_{11} + 0 \quad \text{or} \quad K_{11} = 1$$

The constants are now evaluated and the partial fraction expansion is

$$\frac{s+2}{(s+1)^2} = \frac{1}{s+1} + \frac{1}{(s+1)^2} \tag{7-74}$$

Again, this expansion can be checked, in this case by multiplying the first term in the expansion by $(s+1)/(s+1)$.

EXAMPLE 9

This example will illustrate the expansion of a quotient of polynomials where the denominator roots are a complex conjugate pair. Consider the quotient

$$\frac{1}{s^2 + 2s + 5} = \frac{K_1}{(s+1-j2)} + \frac{K_1^*}{(s+1+j2)} \tag{7-75}$$

Multiplying the equation by $s+1-j2$ and then letting $s = -1+j2$ gives $K_1 = +j\frac{1}{4}$; similarly $K_1^* = -j\frac{1}{4}$, and the expansion is

$$\frac{1}{s^2 + 2s + 5} = \frac{-j\frac{1}{4}}{(s+1-j2)} + \frac{j\frac{1}{4}}{(s-1+j2)} \tag{7-76}$$

To use some transform tables, such terms should be revised by completing the square. In this example,

$$(s^2 + 2s + 5) = (s^2 + 2s + 1) + 4 = (s+1)^2 + 2^2$$

so that

$$\frac{1}{s^2 + 2s + 5} = \frac{1}{(s+1)^2 + 2^2} \tag{7-77}$$

In the general form $[(s + a)^2 + b^2]$, a is the real part of the root, and b is the imaginary part.

7-6. HEAVISIDE'S EXPANSION THEOREM

The method of partial fraction expansion illustrated by the last three examples is known as the Heaviside partial fraction expansion method. To generalize the method, again consider the case in which $Q(s)$ has only distinct roots. Let

$$\frac{P_1(s)}{Q(s)} = \frac{K_1}{s - s_1} + \frac{K_2}{s - s_2} + \frac{K_3}{s - s_3} + \ldots + \frac{K_n}{s - s_n} \tag{7-78}$$

Then any of the coefficients $K_1, K_2, K_3, \ldots, K_n$ can be evaluated by multiplying by the denominator of that coefficient and setting s to the value of the root of the denominator. In other words, to find the coefficient K_j,

$$K_j = \left[(s - s_j) \frac{P_1(s)}{Q(s)} \right]_{s=s_j} \tag{7-79}$$

To consider a general case of r-repeated roots, let

$$\frac{P(s)}{Q(s)} = \frac{R(s)}{(s - s_j)^r} = \frac{K_{j1}}{s - s_j} + \frac{K_{j2}}{(s - s_j)^2}$$
$$+ \ldots \frac{K_{jn}}{(s - s_j)^n} + \ldots + \frac{K_{jr}}{(s - s_j)^r} \tag{7-80}$$

where n is any term in the partial fraction expansion and $R(s)$ is defined as

$$R(s) = \frac{P(s)}{Q(s)} (s - s_j)^r \tag{7-81}$$

Multiplying Eq. (7-80) by $(s - s_j)^r$ gives

$$R(s) = K_{j1}(s - s_j)^{r-1} + K_{j2}(s - s_j)^{r-2} + \ldots + K_{jr} \tag{7-82}$$

From this equation, we can visualize the method to be used to evaluate each coefficient. If we let $s = s_j$, all terms in the equation disappear except K_{jr}, which can be evaluated. Next, differentiate the equation once with respect to s. The term K_{jr} will vanish, but $K_{j,r-1}$ will remain without a multiplying function of s. Again, $K_{j,r-1}$ can be evaluated by letting $s = s_j$. To find the general term K_{jn}, differentiate Eq. (7-82) $(r - n)$ times and let $s = s_j$; then

$$K_{jn} = \frac{1}{(r - n)!} \frac{d^{r-n}R(s)}{ds^{r-n}} \bigg|_{s=s_j} \tag{7-83}$$

or

$$K_{jn} = \frac{1}{(r-n)!} \frac{d^{r-n}}{ds^{r-n}} \left[\frac{P(s)}{Q(s)} (s - s_j)^r \right]\Bigg|_{s=s_j} \tag{7-84}$$

EXAMPLE 10

The actual use of this idea is easier than might appear from the complexity of this general equation. For example, consider

$$\frac{2s^2 + 3s + 2}{(s+1)^3} = \frac{K_{11}}{(s+1)} + \frac{K_{12}}{(s+1)^2} + \frac{K_{13}}{(s+1)^3} \tag{7-85}$$

Multiplying the equation by $(s+1)^3$, we have

$$2s^2 + 3s + 2 = K_{11}(s+1)^2 + K_{12}(s+1) + K_{13} \tag{7-86}$$

From this equation,

$$K_{13} = 2s^2 + 3s + 2 \Big|_{s=-1} = 2 - 3 + 2 = 1 \tag{7-87}$$

Next, we differentiate with respect to s to obtain

$$4s + 3 = 2K_{11}(s+1) + K_{12} \tag{7-88}$$

so that

$$K_{12} = 4s + 3 \Big|_{s=-1} = -1 \tag{7-89}$$

Again, we differentiate the last equation to give

$$4 = 2K_{11} \quad \text{or} \quad K_{11} = 2 \tag{7-90}$$

The partial fraction expansion is

$$\frac{2s^2 + 3s + 2}{(s+1)^3} = \frac{2}{s+1} + \frac{-1}{(s+1)^2} + \frac{1}{(s+1)^3} \tag{7-91}$$

EXAMPLE 11

If $Q(s)$ contains both simple and repeated roots, a combination of both rules may be used. As an example, let

$$\frac{P(s)}{Q(s)} = \frac{s+2}{(s+1)^2(s+3)} \tag{7-92}$$

The form of the partial fraction expansion is

$$\frac{s+2}{(s+1)^2(s+3)} = \frac{K_{11}}{s+1} + \frac{K_{12}}{(s+1)^2} + \frac{K_2}{s+3} \tag{7-93}$$

In this expansion, K_2 may be evaluated by Eq. (7-79) and K_{11} and K_{12} may be found from Eq. (7-84); then

$$K_2 = \frac{s+2}{(s+1)^2}\Big|_{s=-3} = -\frac{1}{4} \tag{7-94}$$

Multiplying Eq. (7-93) by $(s+1)^2$, we have

$$\frac{s+2}{s+3} = K_{11}(s+1) + K_{12} + \frac{(s+1)^2}{s+3} K_2 \qquad (7\text{-}95)$$

The constant K_{12} is evaluated directly by letting $s = -1$; thus

$$K_{12} = \frac{s+2}{s+3}\bigg|_{s=-1} = \frac{1}{2} \qquad (7\text{-}96)$$

and K_{11} will be found by differentiating Eq. (7-95) before letting $s = -1$:

$$\frac{(s+3)\cdot 1 - (s+2)\cdot 1}{(s+3)^2} = K_{11} + K_2 \frac{d}{ds}\left[\frac{(s+1)^2}{s+3}\right] \qquad (7\text{-}97)$$

The coefficient of K_2 vanishes when $s = -1$ because an $(s+1)$ term remains common to all terms in the numerator. In the example

$$\frac{d}{ds}\left[\frac{(s+1)^2}{s+3}\right] = \frac{(s+3)2(s+1) - (s+1)^2 \cdot 1}{(s+3)^2} \qquad (7\text{-}98)$$

and this term vanishes when $s = -1$, because each term in the differentiation contains $(s+1)$. This is always the case, since the order of the multiplying factor $(s - s_j)^r$ is higher than the number of times differentiation is required.

By using these methods, all the coefficients of the partial fraction expansion can be found and the transform equation can be written

$$F(s) = \sum_{j=1}^{n} \frac{K_j}{s - s_j} \qquad (7\text{-}99)$$

for simple roots of $Q(s) = 0$ and as

$$F(s) = \sum_{k=1}^{r} \frac{K_{jk}}{(s - s_j)^k} \qquad (7\text{-}100)$$

for a single root s_j repeated r times. The corresponding $f(t)$ may now be found, for the general case, by taking the inverse Laplace transformation of $F(s)$ as

$$f(t) = \mathcal{L}^{-1}\left[\frac{P(s)}{Q(s)}\right] = \sum_{j=1}^{n} (s - s_j)\frac{P(s)}{Q(s)} e^{st}\bigg|_{s=s_j} \qquad (7\text{-}101)$$

as the time-domain solution for simple roots. Likewise, for repeated roots,

$$f(t) = e^{s_j t} \sum_{n=1}^{r} \frac{1}{(r-n)!} \frac{d^{r-n}R(s_j)}{ds^{r-n}} \frac{t^{n-1}}{(n-1)!} \qquad (7\text{-}102)$$

where s_j, in this equation, is *the root* that is repeated r times. By using both equations for the case of both simple and repeated roots, a general solution is obtained in the form originally given as *Heaviside's expansion theorem.*

The method of the Heaviside partial fraction expansion may be used to give a simplified procedure for finding the inverse transform of the terms for a conjugate complex pair of roots. Suppose that these roots have a real part α and imaginary parts $\pm\omega$. The first coefficient is evaluated by the procedure,

$$K_1 = \frac{P(s)}{Q(s)}(s + \alpha - j\omega)\Big|_{s=-\alpha+j\omega} = Re^{j\theta} \qquad (7\text{-}103)$$

and the second as

$$K_1^* = \frac{P(s)}{Q(s)}(s + \alpha + j\omega)\Big|_{s=-\alpha-j\omega} = Re^{-j\theta} \qquad (7\text{-}104)$$

The inverse transformation of these two terms gives

$$f_1(t) = Re^{j\theta}e^{(-\alpha+j\omega)t} + Re^{-j\theta}e^{(-\alpha-j\omega)t} \qquad (7\text{-}105)$$

This equation may be rearranged to the form

$$\begin{aligned} f_1(t) &= 2Re^{-\alpha t}\left[\frac{e^{j(\omega t+\theta)} + e^{-j(\omega t+\theta)}}{2}\right] \\ &= 2Re^{-\alpha t}\cos(\omega t + \theta) \end{aligned} \qquad (7\text{-}106)$$

The factors R and θ are the magnitude and the phase angle of K_1 in Eq. (7-103). K_1 is the residue associated with the root with a positive imaginary part.

7-7. EXAMPLES OF SOLUTION BY THE LAPLACE TRANSFORMATION

EXAMPLE 12

As an example of the total solution, now that the methods of partial fraction expansion have been reviewed, consider the differential equation

$$\frac{d^2i}{dt^2} + 4\frac{di}{dt} + 5i = 5u(t) \qquad (7\text{-}107)$$

The Laplace transformation of this differential equation is

$$\left[s^2I(s) - si(0-) - \frac{di}{dt}(0-)\right] + 4[sI(s) - i(0-)] + 5I(s) = \frac{5}{s} \qquad (7\text{-}108)$$

Notice that the required initial conditions are automatically specified in this equation. We must know, from the physical system, $i(0-)$ and $di/dt(0-)$. Suppose the following values are found:

$$i(0-) = 1 \quad \text{and} \quad \frac{di}{dt}(0-) = 2 \qquad (7\text{-}109)$$

Inserting these initial conditions simplifies the transform equation to

$$I(s)(s^2 + 4s + 5) = \frac{5}{s} + s + 6 \qquad (7\text{-}110)$$

or

$$I(s) = \frac{s^2 + 6s + 5}{s(s^2 + 4s + 5)} \qquad (7\text{-}111)$$

This equation may be expanded by partial fractions as

$$I(s) = \frac{s^2 + 6s + 5}{s(s + 2 + j1)(s + 2 - j1)} = \frac{K_1}{s} + \frac{K_2}{s + 2 - j1} + \frac{K_2^*}{s + 2 + j1} \qquad (7\text{-}112)$$

To evaluate K_2, multiply the equation by s and let $s = 0$. Then

$$K_1 = \frac{s^2 + 6s + 5}{s^2 + 4s + 5}\bigg|_{s=0} = 1 \qquad (7\text{-}113)$$

To evaluate K_2, multiply the equation by $(s + 2 - j1)$ and let $s = -2 + j1$ as

$$K_2 = \frac{s^2 + 6s + 5}{s(s + 2 + j1)}\bigg|_{s=-2+j1} = \frac{-4 + j2}{(-2 + j1)(j2)} = \frac{2}{j2} = -j = e^{-j90°} \qquad (7\text{-}114)$$

The complete partial fraction expansion becomes

$$I(s) = \frac{1}{s} + \frac{-j}{s + 2 - j1} + \frac{j}{s + 2 + j1} \qquad (7\text{-}115)$$

To obtain $i(t)$ from this transform equation, we take the inverse Laplace transformation of the first term and use Eq. (7-106) with $R = 1$ and $\theta = -90°$ for the second and third terms to give the solution

$$i(t) = 1 + 2e^{-2t} \sin t, \qquad t \geq 0 \qquad (7\text{-}116)$$

EXAMPLE 13

For this example, consider a series RLC circuit with the capacitor initially charged to voltage V_0 as indicated in Fig. 7-6. The differential equation for the current $i(t)$ is

$$L\frac{di}{dt} + Ri + \frac{1}{C}\int_{-\infty}^{t} i\,dt = 0 \qquad (7\text{-}117)$$

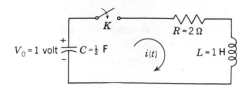

Fig. 7-6. RLC series network for Example 13.

and the corresponding transform equation is

$$L[sI(s) - i(0-)] + RI(s) + \frac{1}{Cs}[I(s) + q(0-)] = 0 \qquad (7\text{-}118)$$

The parameters have been specified as $C = \frac{1}{2}$ F, $R = 2\,\Omega$, and $L = 1$H. The initial current $i(0-) = 0$ because of the inductor, and if C is initially charged to voltage V_0 (with the given polarity),

$$\frac{q(0-)}{Cs} = -\frac{V_0}{s} \qquad (7\text{-}119)$$

or $-1/s$ if $V_0 = 1$ V. The transform equation for $I(s)$ then becomes

$$I(s) = \frac{1}{s^2 + 2s + 2} \qquad (7\text{-}120)$$

or, completing the square,

$$I(s) = \frac{1}{(s + 1)^2 + 1} \qquad (7\text{-}121)$$

Using transform pair 15 of page 192,

$$i(t) = \mathcal{L}^{-1}I(s) = e^{-t} \sin t \cdot u(t) \qquad (7\text{-}122)$$

EXAMPLE 14

In the network shown in Fig. 7-7, the switch is closed at $t = 0$. With the network parameter values shown, the Kirchhoff voltage equations are

$$\frac{di_1}{dt} + 20i_1 - 10i_2 = 100u(t), \qquad (7\text{-}123)$$

$$\frac{di_2}{dt} + 20i_2 - 10i_1 = 0 \qquad (7\text{-}124)$$

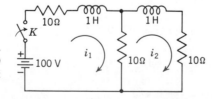

Fig. 7-7. Network for Example 14. The switch is closed at $t = 0$ with zero inductor current at $t = 0$.

If the network is unenergized before the switch is closed, both i_1 and i_2 are initially zero, and the transform equations may be written

$$(s + 20)I_1(s) - 10I_2(s) = \frac{100}{s}, \qquad -10I_1(s) + (s + 20)I_2(s) = 0$$

$$(7\text{-}125)$$

Suppose that we are required to find the current i_2 as a function of time. The transform current $I_2(s)$ may be found from the last two

Table 7-2. TABLE OF TRANSFORMS

$f(t)*$	$F(s)$
1. $u(t)$	$\dfrac{1}{s}$
2. t	$\dfrac{1}{s^2}$
3. $\dfrac{t^{n-1}}{(n-1)!}$, $\quad n = $ integer	$\dfrac{1}{s^n}$
4. e^{at}	$\dfrac{1}{s-a}$
5. te^{at}	$\dfrac{1}{(s-a)^2}$
6. $\dfrac{1}{(n-1)!}\, t^{n-1}e^{at}$	$\dfrac{1}{(s-a)^n}$
7. $\dfrac{1}{a-b}(e^{at} - e^{bt})$	$\dfrac{1}{(s-a)(s-b)}$
8. $\dfrac{e^{-at}}{(b-a)(c-a)}$ $\quad + \dfrac{e^{-bt}}{(a-b)(c-b)} + \dfrac{e^{-ct}}{(a-c)(b-c)}$	$\dfrac{1}{(s+a)(s+b)(s+c)}$
9. $1 - e^{+at}$	$\dfrac{-a}{s(s-a)}$
10. $\dfrac{1}{\omega}\sin \omega t$	$\dfrac{1}{s^2+\omega^2}$
11. $\cos \omega t$	$\dfrac{s}{s^2+\omega^2}$
12. $1 - \cos \omega t$	$\dfrac{\omega^2}{s(s^2+\omega^2)}$
13. $\sin(\omega t + \theta)$	$\dfrac{s \sin \theta + \omega \cos \theta}{s^2+\omega^2}$
14. $\cos(\omega t + \theta)$	$\dfrac{s \cos \theta - \omega \sin \theta}{s^2+\omega^2}$
15. $e^{-\alpha t}\sin \omega t$	$\dfrac{\omega}{(s+\alpha)^2+\omega^2}$
16. $e^{-\alpha t}\cos \omega t$	$\dfrac{s+\alpha}{(s+\alpha)^2+\omega^2}$
17. $\sinh \alpha t$	$\dfrac{\alpha}{s^2-\alpha^2}$
18. $\cosh \alpha t$	$\dfrac{s}{s^2-\alpha^2}$

*All $f(t)$ should be thought of as being multiplied by $u(t)$, i.e., $f(t) = 0$ for $t < 0$.

algebraic equations by determinants as

$$I_2(s) = \frac{\begin{vmatrix} s+20 & 100/s \\ -10 & 0 \end{vmatrix}}{\begin{vmatrix} s+20 & -10 \\ -10 & s+20 \end{vmatrix}} = \frac{1000}{s(s^2 + 40s + 300)} \quad (7\text{-}126)$$

The partial fraction expansion of this equation is

$$\frac{1000}{s(s+10)(s+30)} = \frac{3.33}{s} - \frac{5}{s+10} + \frac{1.67}{s+30} \quad (7\text{-}127)$$

The inverse Laplace transformation gives $i_2(t)$ as

$$i_2(t) = 3.33 - 5e^{-10t} + 1.67e^{-30t}, \quad t \geq 0 \quad (7\text{-}128)$$

which is the required solution.

The properties of Laplace transforms derived in this chapter are summarized in the next chapter in Table 8-1 (together with additional properties derived in Chapter 8). A short table of Laplace transforms is given as Table 7-2; a more extensive table is given in Appendix D.

FURTHER READING

CHIRLIAN, PAUL M., *Basic Network Theory*, McGraw-Hill Book Company, New York, 1969. Chapter 5.

CLOSE, CHARLES M., *The Analysis of Linear Circuits*, Harcourt, Brace & World, Inc., New York, 1966. Chapter 10.

COOPER, GEORGE R., AND CLARE D. MCGILLEM, *Methods of Signal and System Analysis*, Holt, Rinehart & Winston, Inc., New York, 1967. Chapter 6.

DESOER, CHARLES A., AND ERNEST S. KUH, *Basic Circuit Theory*, McGraw-Hill Book Company, New York, 1969. See Chapter 13.

KUO, FRANKLIN F., *Network Analysis and Synthesis*, 2nd ed., John Wiley & Sons, Inc., New York, 1966. Chapter 6.

STRUM, ROBERT D., AND JOHN R. WARD, *Laplace Transform Solution of Differential Equations* (a programmed text), Prentice-Hall, Inc., Englewood Cliffs, N.J., 1968. This is an excellent manual for self study of the topics of this chapter.

For interesting reading in the historical aspects of the subject, the reader should consult the summary titled "The Work of Oliver Heaviside" by Behrend in an appendix to E. J. Berg's *Heaviside's Operational Calculus* (2nd ed.) (McGraw-Hill Book Company, New York, 1936), pp. 173–208. Heaviside's original writings have been reprinted as *Electromagnetic Theory* (Dover Publications, Inc., New York, 1950) and contain an extensive presentation of the method. See also the historical notes in Appendix C of M. F. Gardner and J. L. Barnes, *Transients in Linear Systems* (John Wiley & Sons, Inc., New York, 1942).

DIGITAL COMPUTER EXERCISES

In the study of the Laplace transformation, two operations that are routinely accomplished with the aid of the computer are the determination of the roots of a polynomial and the calculation of residues. References relating to these operations are given in Appendices E-1 and E-5.1. In particular, Chapter 8 of reference 7 by Huelsman in Appendix E-10 which relates to the manipulation of polynomials and the determination of residues is recommended. See also Case Study 23 given by McCracken, reference 12 cited in Appendix E-10.

PROBLEMS

7-1. Verify that the Laplace transform of cos ωt determined from Eq. (7-1) is

$$\mathcal{L}\left[\cos \omega t\right] = \frac{s}{s^2 + \omega^2}$$

In each of the problems that follow, repeat the procedure of Prob. 7-1 for the various transform pairs of the following table.

	$f(t)$ for $t \geq 0$	$F(s)$
7-2.	t^2	$\dfrac{2}{s^3}$
7-3.	$\sinh \alpha t$	$\dfrac{\alpha}{s^2 - \alpha^2}$
7-4.	$\cosh \alpha t$	$\dfrac{s}{s^2 - \alpha^2}$
7-5.	$e^{-\alpha t} \sin \omega t$	$\dfrac{\omega}{(s + \alpha)^2 + \omega^2}$
7-6.	$e^{-\alpha t} \cos \omega t$	$\dfrac{s + \alpha}{(s + \alpha)^2 + \omega^2}$
7-7.	$\sin(\omega t + \theta)$	$\dfrac{s \sin \theta + \omega \cos \theta}{s^2 + \omega^2}$
7-8.	$\cos(\omega t + \theta)$	$\dfrac{s \cos \theta - \omega \sin \theta}{s^2 + \omega^2}$

For the following $f(t)$, determine the other member of the transform pair, $F(s) = \mathcal{L}[f(t)] = P(s)/Q(s)$ where $P(s)$ and $Q(s)$ are polynomials in s.

7-9. $f_1(t) = \sin^2 t$

7-10. $f_2(t) = t \cos at$

7-11. $f_3(t) = [1/(2a)] \sin at$

7-12. $f_4(t) = \cos^2 t$

7-13. $f_5(t) = [1/(2a^3)](\sinh at - \sin at)$

7-14. $f_6(t) = [1/(2a^2)] \sin at \sinh at$

7-15. $f_7(t) = [1/(2a^2)](\cosh at - \cos at)$

7-16. $f_8(t) = [1/(2a)](\sin at + at \cos at)$

7-17. $f_9(t) = t^2 e^{-at} \cos \omega t$

7-18. $f_{10}(t) = (1/t) \sin^2 \omega t$

7-19. $f_{11}(t) = t^{1/2}$

7-20. $f_{12}(t) = e^{-at^2}$

7-21. In the network shown in the figure, C is charged to V_0, and the switch K is closed at $t = 0$. Solve for the current $i(t)$ using the Laplace transformation method.

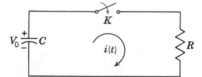

Fig. P7-21.

7-22. In the network shown in the figure, the switch K is moved from position a to position b at $t = 0$, a steady state having previously been established at position a. Solve for the current $i(t)$, using the Laplace transformation method.

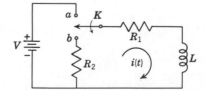

Fig. P7-22.

7-23. In the network shown, C is initially charged to V_0. The switch K is closed at $t = 0$. Solve for the current $i(t)$, using the Laplace transformation method.

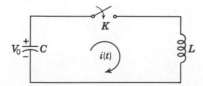

Fig. P7-23.

7-24. In the network shown, the switch K is moved from position a to position b at $t = 0$ (a steady state existing in position a before $t = 0$). Solve for the current $i(t)$, using the Laplace transformation method.

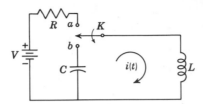

Fig. P7-24.

7-25. Work Prob. 4-2 using the Laplace transformation method.

7-26. Work Prob. 4-3 using the Laplace transformation method of this chapter.

7-27. Work Prob. 4-4 using the Laplace transformation method.

7-28. Work Prob. 4-5 using the Laplace transformation method rather than the classical method of Chapter 4.

7-29. Check the following equations by completing a partial fraction expansion. Determine which two expansions in the set are in error.

(a) $\dfrac{2s}{s^2 - 1} = \dfrac{1}{s + 1} + \dfrac{1}{s - 1}$

(b) $\dfrac{7s + 2}{s^3 + 3s^2 + 2s} = \dfrac{1}{s} + \dfrac{2}{s + 2} + \dfrac{-3}{s + 1}$

(c) $\dfrac{5s + 13}{s^2 + 5s + 6} = \dfrac{2}{s + 3} + \dfrac{3}{s + 2}$

(d) $\dfrac{s^2}{s - 1} = \dfrac{1}{s - 1} + s + 1$

(e) $\dfrac{2(s + 1)}{s^2 + 1} = \dfrac{1 + j1}{s + j1} + \dfrac{1 - j1}{s - j1}$

(f) $\dfrac{s^2 + 4s + 1}{s(s + 1)^2} = \dfrac{1}{s + 1} + \dfrac{1}{(s + 1)^2} + \dfrac{2}{s}$

(g) $\dfrac{3s^3 - s^2 - 3s + 2}{s^2(s - 1)^2} = \dfrac{1}{s} + \dfrac{2}{s^2} + \dfrac{2}{s - 1} + \dfrac{1}{(s - 1)^2}$

(h) $\dfrac{s^3 - 5s^2 + 9s + 9}{s^2(s^2 + 9)} = \dfrac{1}{s} + \dfrac{1}{s^2} + \dfrac{-j}{s + j3} + \dfrac{+j}{s - j3}$

7-30. Expand the following functions as partial fractions:

(a) $F_1(s) = \dfrac{(s + 1)(s + 3)}{s(s + 2)(s + 4)}$

(b) $F_2(s) = \dfrac{(s^2 + 1)(s^2 + 3)}{s(s^2 + 2)(s^2 + 4)}$

(c) $F_3(s) = \dfrac{s}{s^2(s + 1)^2(s + 2)}$

(d) $F_4(s) = \dfrac{1}{s^3 + 2s^2 + 2s + 1}$

(e) $F_5(s) = \dfrac{s + 4}{(s + 1)^2(s + 3)^2}$

(f) $F_6(s) = \dfrac{s^2(s + 3)}{(s + 1)(s + 2)^4}$

7-31. Expand the following functions as partial fractions:

(a) $F_a(s) = \dfrac{s^2 + 7s + 8}{(s + 1)(s + 2)(s + 3)}$

(b) $F_b(s) = \dfrac{s^2 + s + 1}{(s + 1)(s + 2)(s + 3)}$

(c) $F_c(s) = \dfrac{s + 4}{s(s + 1)(s + 3)^2(s + 2)^3}$

7-32. Verify the following inverse Laplace transformations $\mathcal{L}^{-1}F(s) = f(t)$:

(a) $\mathcal{L}^{-1} \dfrac{3s}{(s^2 + 1)(s^2 + 4)} = \cos t - \cos 2t$

(b) $\mathcal{L}^{-1} \dfrac{s + 1}{s^2 + 2s} = \tfrac{1}{2}(1 + e^{-2t})$

(c) $\mathcal{L}^{-1} \dfrac{1}{s(s^2 - 2s + 5)} = \tfrac{1}{5}[1 + \tfrac{1}{2}e^t(-2\cos 2t + \sin 2t)]$

(d) $\mathcal{L}^{-1} \dfrac{1}{(s + 1)(s + 2)^2} = e^{-t} - e^{-2t}(1 + t)$

(e) $\mathcal{L}^{-1} \dfrac{1}{s^3(s^2 - 1)} = -1 - \dfrac{t^2}{2} + \cosh t$

(f) $\mathcal{L}^{-1} \dfrac{s^2 + 2s + 1}{(s + 2)(s^2 + 4)} = \tfrac{1}{8}e^{-2t} + \tfrac{7}{8}\cos 2t + \tfrac{1}{8}\sin 2t$

(g) $\mathcal{L}^{-1} \dfrac{s^2}{(s^2 + 1)^2} = \tfrac{1}{2}t \cos t + \tfrac{1}{2}\sin t$

7-33. Solve the differential equations given in Prob. 6-5 using the Laplace transformation method.

7-34. Solve the differential equations of Prob. 6-6 using the Laplace transformation method.

Solve the following equations using the Laplace transformation method:

7-35. $\dfrac{d^2i}{dt^2} - i = 25 + e^{2t}$

7-36. $\dfrac{d^2i}{dt^2} + 4i = \sin t - \cos 2t$

7-37. $\dfrac{d^2i}{dt^2} + \dfrac{di}{dt} = t^2 + 2t, \qquad i(0-) = 4, \qquad \dfrac{di}{dt}(0-) = -2$

7-38. Solve the differential equation given in Prob. 6-21 using the Laplace transformation method.

7-39. In the series *RLC* circuit shown, the applied voltage is $v(t) = \sin t$ for $t > 0$. For the element values specified, find $i(t)$ if the switch K is closed at $t = 0$.

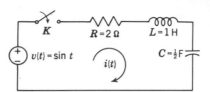

Fig. P7-39.

7-40. At $t = 0$, a switch is closed, connecting a voltage source $v = V \sin \omega t$ to a series *RL* circuit. By the method of the Laplace transformation, show that the current is given by the equation

$$i = \frac{V}{Z}\sin(\omega t - \phi) + \frac{\omega L V}{Z^2}e^{-Rt/L}$$

where

$$Z = \sqrt{R^2 + (\omega L)^2} \quad \text{and} \quad \phi = \tan^{-1}\frac{\omega L}{R}$$

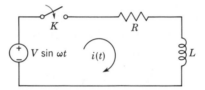

Fig. P7-40.

7-41. Dr. L. A. Woodbury of the University of Utah School of Medicine has made use of an electrical analog in studies of convulsions. In the network shown in the figure, the following quantities are duals: C_1 represents the volume of drug-containing fluid, R_1 is the "resistance" to the passage of the drug from the pool to the blood stream, C_2 represents the volume of the blood stream, and R_2 is equivalent to the body's excretion mechanism (kidney, etc.). The concentration of the drug dose is represented as V_0 and the voltage $v_a(t)$ at node a is analogous to the amount of drug in the blood stream. The analog network has the advantage that the elements may be readily changed and the effects studied (to say nothing of the saving of cats). Find the transform equation for $V_a(s)$ with the coefficient of the highest-order term normalized to unity.

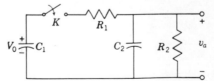

Fig. P7-41.

7-42. This problem is a continuation of Prob. 7-41 concerning Dr. Woodbury's analog. The following constants for the network are selected: $C_1 = 1\ \mu F$, $C_2 = 8\ \mu F$, $R_1 = 9\ M\Omega$, and $R_2 = 5\ M\Omega$. If $V_0 =$

100 V and the switch is closed at $t = 0$, solve for $v_A(t)$, the equivalent of the concentration of drug in the blood stream, as a function of time.

7-43. Find the time t_m when the concentration of drug in the blood stream for Prob. 7-42 is a maximum. (This information is desired so that a second dose may be given at that time to build up the concentration to the point where a convulsion is induced.)

7-44. If a second dose (the voltage equivalent having a magnitude of 100 V) is injected at $t = t_m$ as found in Prob. 7-43, what will be v_A as a function of time, and what will be the maximum v_A obtained? (*Note:* In giving the second dose we will assume that the total voltage is then 100 V plus the voltage on the plates at the time the addition is made.)

7-45. In the network shown, the switch K is closed at $t = 0$ with the network previously unenergized. For the element values shown on the diagram: (a) find $i_1(t)$, (b) find $i_2(t)$.

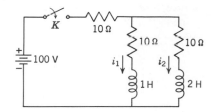

Fig. P7-45.

7-46. With switch K in a position a, the network shown in the figure attains equilibrium. At time $t = 0$, the switch is moved to position b. Find the voltage across R_2 as a function of time.

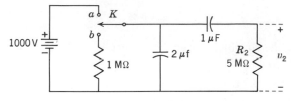

Fig. P7-46.

7-47. (a) Find $i_1(t)$ resulting from closing the switch at $t = 0$ with the circuit previously unenergized. The circuit constants are: $L_1 = 1$ H, $L_2 = 4$ H, $M = 2$ H, $R_1 = R_2 = 1$ Ω, $V = 1$ V. (b) Repeat part (a) in solving for $i_2(t)$.

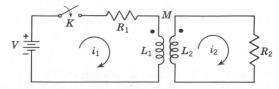

Fig. P7-47.

7-48. In the network given in the figure, the current source is described by $i_1 = 10^{-3}e^{-t}u(t)$ amp. For the element values given, determine $v_2(t)$, assuming that all elements are initially unenergized. Sketch $v_2(t)$ using an enlarged time scale for small values of t.

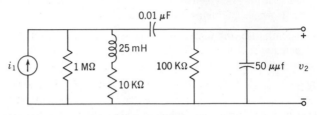

Fig. P7-48.

7-49. The network given in the figure contains a current-controlled voltage source. For the element values given and with $v_1(t) = 5u(t)$, determine $v_a(t)$ if the network is not energized at $t = 0$. Let $K_1 = -3$.

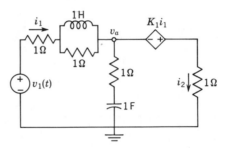

Fig. P7-49.

7-50. Using the conditions given for the network of Prob. 7-49, solve for $i_2(t)$ if $K_1 = +3$.

7-51. (a) Determine $\mathcal{L}bf(t/a)$ in terms of $\mathcal{L}f(t)$, given that a and b are constants.

(b) Determine $\mathcal{L}^{-1}F(s/c)$ in terms of $\mathcal{L}^{-1}F(s)$, given that c is a constant.

7-52. Show that

$$\mathcal{L}[tf(t)] = -\frac{dF(s)}{ds}$$

Using this result, find the Laplace transforms of $t \sin \alpha t$, $t^2 e^{-at}$, and $t \sinh \beta t$. α, a, and β are constants.

7-53. Show that

$$\mathcal{L}\frac{f(t)}{t} \int_s^\infty F(s)\, ds$$

Using this result, find the Laplace transforms of $t^{-1}e^{-t}$ and $t^{-1}(1 - e^{-t})$.

7-54. Show that

$$\mathcal{L}(t^n) = \frac{n!}{s^{n+1}}$$

7-55. Find

$$\mathcal{L}^{-1} \frac{n!}{s(s+1)(s+2)\ldots(s+n)}$$

by partial fraction expansion, and show that the answer may be given in the closed form $(1 - e^{-t})^n$.

7-56. The accompanying figure shows six different signal waveforms. Each figure suggests the general form of the signal, but no numerical values are given. For each waveform, write a transform function, $V(s)$. Be sure that any poles in $V(s)$ required to produce the waveform given are present.

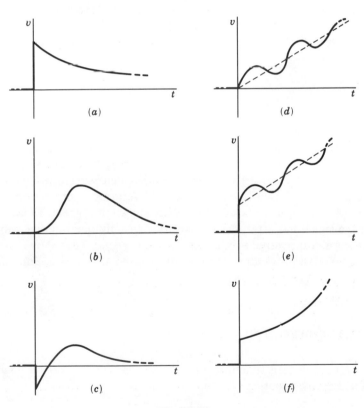

Fig. P7-56.

8 Transforms of Other Signal Waveforms

In past chapters, our study of the response of networks has been with simple excitation or signal waveforms such as the step function or the sine wave. In this chapter we employ transform methods to determine the response of networks to more complicated waveforms by first determining the transforms for these waveforms. Examples of the types of waveforms to be considered are shown in Fig. 8-1. In (a) and (b), the signal has a prescribed variation in only one interval and is zero for all other time; this is a *nonrecurring* signal. That shown in (c) is a square wave which continues for all time after $t = 0$, and is an example of a *recurring* waveform.

8-1. THE SHIFTED UNIT STEP FUNCTION

The unit step function was introduced in Chapter 7 and was defined by the equation

$$u(t) = \begin{cases} 1, & t \geq 0 \\ 0, & t < 0 \end{cases} \tag{8-1}$$

for a function which changes abruptly from zero to a unit value at the time $t = 0$. This expression may be generalized by the definition

$$u(t - a) = \begin{cases} 1, & t \geq a \\ 0, & t < a \end{cases} \tag{8-2}$$

for a step function which changes abruptly at the time $t = +a$. In general, the step function has unit value when the quantity $(t - a)$, the *argument* of the function u, is positive, and has zero value when $(t - a)$ is negative. This definition will apply for any form of the variable. Hence the function $u(t + a)$ is one that changes from zero to unit value at $t = -a$. Similarly, the function $u(a - t)$ is one that changes from unit to zero value (with increasing time) at the time $t = a$. These functions are represented in Fig. 8-2.

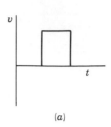

(a)

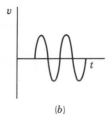

(b)

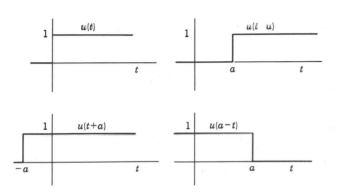

Fig. 8-2. The unit step function shown as $u(t)$, shifted each direction with respect to $t = 0$ and reversed.

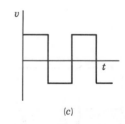

(c)

Fig. 8-1. Examples of the waveforms to be considered: (a) rectangular pulse, (b) a pulse of sine wave, and (c) a square wave.

The Laplace transform of $u(t - a)$ is determined from the defining equation

$$F(s) = \int_{0-}^{\infty} f(t)e^{-st}\, dt \qquad (8\text{-}3)$$

With $f(t) = u(t - a)$, we have,

$$\mathcal{L}u(t - a) = \int_{a}^{\infty} 1e^{-st}\, dt = \frac{-e^{-st}}{s}\bigg|_{a}^{\infty}$$

$$\mathcal{L}u(t - a) = e^{-as}\left(\frac{1}{s}\right) \qquad (8\text{-}4)$$

This equation is made up of the product of two factors: the factor $1/s$ is the transform of the unit step function beginning at the time $t = 0$; the term e^{-as} is a function which effectively *shifts* the transform from one beginning at $t = 0$ to one beginning at $t = a$.

The example given for a unit step function may be generalized for any time function $f(t)$ which is delayed in its beginning to some other time, $t = a$. Such a time-shifted function is represented as

$$f(t - a)u(t - a) \qquad (8\text{-}5)$$

To find the transform of this equation, we write the defining equation in terms of a new variable, t'; that is,

$$F(s) = \int_{0-}^{\infty} f(t')e^{-st'}\, dt' \qquad (8\text{-}6)$$

Let the variable t' be defined as $t' = t - a$ such that the defining equation becomes

$$F(s) = \int_a^\infty f(t-a)e^{-(t-a)s}\, dt \tag{8-7}$$

or

$$= \int_a^\infty f(t-a)e^{-st}(e^{as})\, dt \tag{8-8}$$

The constant factor e^{as} may be removed from within the integral and the lower limit of the integral changed to 0 if $f(t-a)$ is multiplied by $u(t-a)$; thus

$$F(s) = e^{as}\int_{0-}^\infty f(t-a)u(t-a)e^{-st}\, dt \tag{8-9}$$

The integral expression is recognized as the transform of the time function $f(t-a)u(t-a)$, so that

$$\mathcal{L}f(t-a)u(t-a) = e^{-as}\mathcal{L}f(t) \tag{8-10}$$

or, conversely,

$$\mathcal{L}^{-1}e^{-as}F(s) = f(t-a)u(t-a) \tag{8-11}$$

These equations tell us that the transform of any function delayed to begin at the time $t = a$ is e^{-as} times the transform of the function when it begins at $t = 0$. This is a useful result, as we shall see. It is known as the *real translation theorem* or simply the *shifting theorem*.

Before giving an illustration of the application of this theorem, we will show by an example that unit step functions with appropriate shifting may be used as building blocks to represent other signals. Consider the signal formed by the difference of two step functions,

$$v(t) = u(t) - u(t-a) \tag{8-12}$$

These two functions are shown in Fig. 8-3. Their sum is a function that has unit value from $t = 0$ to $t = a$ and zero value for all other time. This function is described as a *pulse* of unit magnitude and duration a. We may generalize our description to the form

$$v(t) = V_0[u(t-a) - u(t-b)] \tag{8-13}$$

which is a pulse of magnitude V_0 of duration $b - a$ occurring at $t = a$.

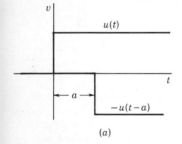

(a)

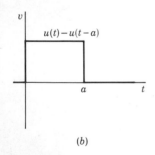

(b)

Fig. 8-3. The linear combination of two unit step functions to describe a pulse of amplitude 1 and duration a.

EXAMPLE 1

At $t = 0$, a pulse of width a is applied to the *RL* network of Fig. 8-4. We are required to determine an expression for the current $i(t)$. The pulse is described by Eq. (8-12) and, from Eq. (8-10), the transform of $v(t)$ is

$$V(s) = \frac{1}{s}(1 - e^{-as}) \tag{8-14}$$

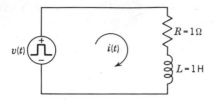

Fig. 8-4. *RL* network with pulse excitation.

Substituting this value of $V(s)$ into the transform equation,

$$L[sI(s) - i(0+)] + RI(s) = \frac{1}{s}(1 - e^{-as}) \qquad (8\text{-}15)$$

Substituting the element values and the initial condition, $i(0+) = 0$, gives

$$I(s) = \frac{(1 - e^{-as})}{s(s + 1)} \qquad (8\text{-}16)$$

This expression may be written as a sum of terms,

$$I(s) = \frac{1}{s(s + 1)} - \frac{e^{-as}}{s(s + 1)} \qquad (8\text{-}17)$$

The first term of this equation is easily expanded by partial fractions to give

$$\frac{1}{s(s + 1)} = \frac{1}{s} - \frac{1}{s + 1} \qquad (8\text{-}18)$$

In terms of this expansion, Eq. (8-17) may be written

$$I(s) = \frac{1}{s} - \frac{1}{s + 1} - \frac{e^{-as}}{s} + \frac{e^{-as}}{s + 1} \qquad (8\text{-}19)$$

The inverse Laplace transformation may be carried out term by term in this equation to give

$$\begin{aligned}
\mathcal{L}^{-1}I(s) &= i(t) \\
&= (1 - e^{-t})u(t) - [1 - e^{-(t-a)}]u(t - a) \qquad (8\text{-}20)
\end{aligned}$$

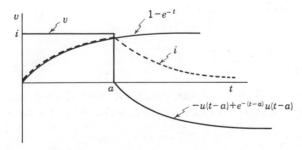

Fig. 8-5. The excitation v, the two parts of the response with their sum indicated by the dashed line.

The third and fourth terms of this expression differ from the first and second only in that they are shifted in time and are opposite in sign. The waveform represented by this equation is plotted as Fig. 8-5.

The result we have obtained in Eq. (8-20) is the same as would be found by using two series-connected voltage sources, one for $u(t)$ and the other for $u(t - a)$, and then summing the two responses to find the total response of Eq. (8-20).

Another theorem of importance is the *scaling theorem* which relates scale changes in the s or frequency domain to the consequent changes in scale in the t or time domain. By a change in scale, we mean that either s or t is multiplied by a positive constant. Given a time-variable function $f(t)$, we change scale by forming a new function, $f(t/t_0)$. Its transform is found as follows. From the defining equation,

$$\mathcal{L}\left[f\left(\frac{t}{t_0}\right)\right] = \int_{0-}^{\infty} f\left(\frac{t}{t_0}\right) e^{-st}\, dt \tag{8-21}$$

or

$$= t_0 \int_{0-}^{\infty} f\left(\frac{t}{t_0}\right) e^{-(t_0 s)t/t_0}\, d\left(\frac{t}{t_0}\right) \tag{8-22}$$

If we let $t/t_0 = \tau$, then the last equation becomes

$$\mathcal{L}\left[f\left(\frac{t}{t_0}\right)\right] = t_0 \int_{0-}^{\infty} f(\tau) e^{-t_0 s\tau}\, d\tau \tag{8-23}$$

Finally, the integral defines $F(t_0 s)$ so that we may write

$$\mathcal{L}\left[f\left(\frac{t}{t_0}\right)\right] = t_0 F(t_0 s) \tag{8-24}$$

The corresponding inverse transform is

$$f\left(\frac{t}{t_0}\right) = t_0 \mathcal{L}^{-1} F(t_0 s) \tag{8-25}$$

EXAMPLE 2

From the transform current

$$I(s) = \frac{1}{s(s + 1)} \tag{8-26}$$

the corresponding $i(t)$ is

$$i(t) = 1 - e^{-t} \tag{8-27}$$

The scaling theorem tells us that the new function

$$i_1(t) = \mathcal{L}^{-1} 2I(2s) = 1 - e^{-t/2} \tag{8-28}$$

is related to $i(t)$ in Eq. (8-27) by a simple change of the time scale. Such time scaling is important in the use of analog computers.

8-2. THE RAMP AND IMPULSE FUNCTIONS

Let us start our discussion in terms of the now familiar unit step function. If this function represents the voltage across an inductor, then the inductor current is

$$i_L = \frac{1}{L} \int_{-\infty}^{t} v_L \, dt \tag{8-29}$$

If $v_L = u(t)$, and $i_L(0) = 0$, $i_L = t/L$ for $t > 0$, and the current will increase linearly with time so long as the constant voltage is applied. This response is known as a *ramp function* (or a *linear ramp*). It is useful in describing various inputs or responses. A wheel turning with constant angular velocity, for example, has an angular displacement which is a ramp function.

Next, let us imagine that the unit step function represents current in an inductor. Then the inductor voltage is given by the equation

$$v_L = L \frac{di_L}{dt} \tag{8-30}$$

The derivative of $i_L = u(t)$ has zero value for all time except the instant stepping occurs, $t = 0$. At that time, the derivative (being the slope) is infinite. This function which has only one nonzero value (and that infinite) is known as an *impulse function*. Representations of the ramp and the impulse function are shown in Fig. 8-6.

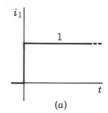

(a)

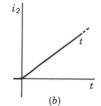

(b)

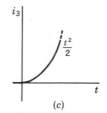

(c)

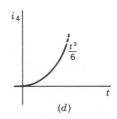

(d)

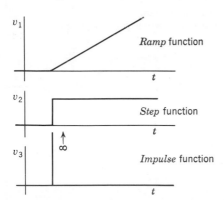

Fig. 8-6. Functions which are related by integration (from bottom to top) and by differentiation (from top to bottom).

Our discussion has been in terms of the inductor, but it is clear that we would obtain the same results on a dual basis for the capacitor. Clearly, for the inductor and capacitor, we will be interested in functions which are related through differentiation and integration. And since the unit step is an important waveform, we will also be concerned with ramps and impulses.

Given the unit step function shown in Fig. 8-7(*a*), we may integrate three times and obtain the three waveforms shown in (*b*),

Fig. 8-7. (*a*) A unit step function and the waveforms resulting from one integration (*b*), two integrations (*c*), and three integrations (*d*).

(c), and (d) of this figure. Starting with the waveform in (d), we may similarly differentiate three times and return to the unit step function. In this sense, each of the four functions of Fig. 8-7 are *unit* functions: The differentiation or integration of one unit function gives another unit function. If we are to extend this concept to the derivative of the unit step function, and beyond, we must define a *unit impulse*. What is a unit impulse?

Consider the modified ramp function of Fig. 8-8. The function is a ramp from $t = b$ to $t = c$ and then has a constant value of unity for all time. The time interval $(c - b)$ is defined as a. The derivative of this modified ramp function is a pulse of width a as shown in the figure. (Conversely, the integral of the pulse function is the linear ramp.) If the ramp function is designated as the variable i, the pulse has a magnitude di/dt, the slope of the ramp. The slope of the ramp is

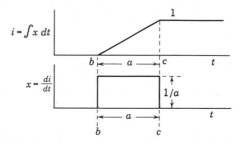

Fig. 8-8. The derivative of the modified ramp function is the pulse. The unit impulse function is defined for a approaching zero.

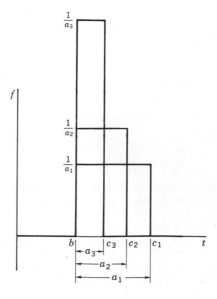

Fig. 8-9. Three steps in the sequence of smaller values of a in Fig. 8-8 with $a_3 < a_2 < a_1$.

the distance 1 divided by the distance a; that is,

$$\frac{di}{dt} = \frac{1}{a} \tag{8-31}$$

Now the area of the pulse is $a \times 1/a = 1$, for any value of a. As a approaches zero, the modified ramp function approaches a unit step function. At the same time, the pulse approaches infinite height and zero width *with the area remaining constant and equal to one*. Three steps in the sequence we have described are shown in Fig. 8-9. In the limit,[1] this function is known as a unit impulse,[2] and is designated as $\delta(t - b)$. Now we see how the unit impulse fits into the pattern of Fig. 8-7. The integral

$$\int_{-\infty}^{t} \delta(t - b)\, dt = 1, \qquad t \geq b$$
$$= 0, \qquad t < b \tag{8-32}$$

which is a unit step function, $u(t - b)$. In the opposite direction, the derivative of the unit step is a unit impulse. Thus

$$\frac{d}{dt}u(t - b) = \delta(t - b) = \infty, \qquad t = b$$
$$= 0, \qquad t \neq b \tag{8-33}$$

The impulse function is important in system analysis, and we will make frequent use of it in the studies to follow.[3]

---

[1]Alternatively, we may regard the impulse function as being defined by the following limits:

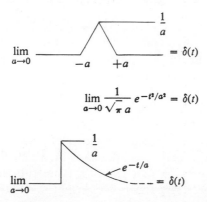

$$\lim_{a \to 0} \frac{1}{\sqrt{\pi}\, a} e^{-t^2/a^2} = \delta(t)$$

[2]In mathematical physics, this function is called a Dirac delta function.

[3]The reader interested in the mathematical properties of impulse functions and a discussion of their justification in terms of Schwartz distributions is referred to S. Seshu and N. Balabanian, *Linear Network Analysis* (John Wiley & Sons, Inc., New York, 1959), pp. 112–119 or, at a more advanced level, to Bernard Friedman, *Principles and Techniques of Applied Mathematics* (John Wiley & Sons, Inc., New York, 1956), pp. 136–144 or A. H. Zemanian, *Distribution Theory and Transform Analysis* (McGraw-Hill Book Company, New York, 1965).

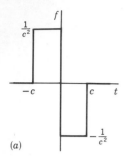

Fig. 8-10. (a) A function whose limit as $c \to 0$ is the unit doublet represented in (b).

The sequence of Fig. 8-7 may be extended as far to the left as we wish by defining other unit functions. Thus,

$$\delta'(t - b) = \frac{d}{dt}\delta(t - b) = +\infty \quad \text{and} \quad -\infty, \qquad t = b$$
$$= 0, \qquad\qquad\qquad t \neq b \qquad (8\text{-}34)$$

is a *unit doublet*, and is the function obtained in the limit $c \to 0$ in Fig. 8-10.

Let us next return to the inductor and capacitor in terms of which we introduced the ramp and the impulse. With the voltage source shown in Fig. 8-11, the current is

$$i_L = \frac{1}{L} \int_{-\infty}^{t} \delta(t - a)\, dt = \frac{1}{L}u(t - a) \qquad (8\text{-}35)$$

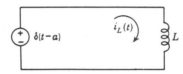

Fig. 8-11. An inductor subjected to a unit impulse of voltage such that the current in the inductor changes instantaneously.

This equation tells us that a unit impulse of voltage applied to an inductor causes a current of $1/L$ amperes to be established immediately. This is unusual behavior for an inductor, but then the impulse of voltage is an unusual driving force. An infinite voltage is required to change the current in an inductor instantaneously. A similar result may be found for capacitor voltage. With an impulse current source shown in Fig. 8-12, the capacitor voltage is

$$v_C = \frac{1}{C}u(t - a) \qquad (8\text{-}36)$$

Thus a unit impulse of current causes $1/C$ volts to appear instantaneously on the plates of the capacitor. Viewed in another way, the integral of current is charge, so that a unit impulse of current corresponds to the delivery to the capacitor of a unit of charge, which results in a voltage change of $1/C$ units.

Our discussion has been in terms of unit functions, but our results are easily extended to the case in which the functions are multiplied by a constant, K. Then $K\delta(t)$ represents an impulse of area K. Its integral gives a step function of height K, and the integral of this step gives a ramp of slope K. A notation often used in representing impulse functions may be introduced in terms of the waveforms of

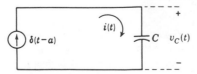

Fig. 8-12. A capacitor subjected to a unit impulse of current such that the capacitor voltage changes instantaneously.

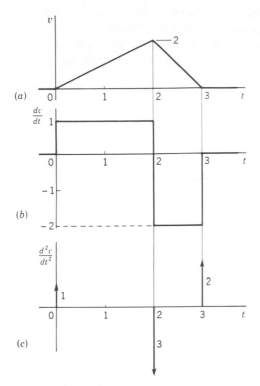

Fig. 8-13. (*a*) A triangular pulse, (*b*) its first derivative, and (*c*) its second derivative. Part (*c*) of the figure illustrates two conventions used to indicate the strength of the impulse: a number placed along the arrow, and a variable length for the arrow.

Fig. 8-13. In (*a*) of this figure is shown a triangular waveform; its derivative is shown in (*b*), and the second derivative in (*c*). The second derivative involves three impulse functions. The impulse functions are represented by arrows directed either up or down, depending on the sign of the constant multiplier. The number at the side of the arrow represents the area of the impulse or, if you like, the *strength* of the impulse. The lengths of the arrows are sometimes drawn proportional to the area, as in Fig. 8-13(*c*). The waveform consisting of a number of impulses is known as an *impulse train*.

Besides being a convenient mathematical function, the impulse is an idealization or a model for an energy source with an output which is very short compared to the response of the network. Thus the response of a network to a short pulse, shorter than any time constant in the natural response of the network, is approximately the impulse response of the network. The impulse is also a model for other familiar events. For example, anyone who has been struck a glancing blow on the head by a fast baseball will not object to the description of this blow as an impulse. And certainly the ball was gone long before there was any response, such as a body dropping to the ground.

The transform of the unit impulse function is obtained in terms of quantities identified in Fig. 8-14 from the equation

$$\delta(t - b) = \lim_{a \to 0} \frac{1}{a} [u(t - b) - u(t - b - a)] \qquad (8\text{-}37)$$

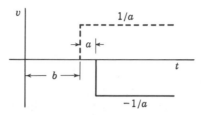

Fig. 8-14. Quantities pertinent to Eq. 8-37 from which the Laplace transform of the impulse is derived.

The Laplace transform of this limit equation is

$$\mathcal{L}\delta(t - b) = \lim_{a \to 0} \frac{e^{-bs} - e^{-(b+a)s}}{as} \qquad (8\text{-}38)$$

This limit may be found by the application of l'Hospital's rule. The result is

$$\mathcal{L}\delta(t - b) = e^{-bs} \qquad (8\text{-}39)$$

where b is the time of appearance of the unit impulse. When $b = 0$ (that is, the impulse occurs at $t = 0$), we have

$$\mathcal{L}\delta(t) = 1 \qquad (8\text{-}40)$$

This result may be obtained in another way by using Eq. (7-24):

$$\mathcal{L}[\delta(t)] = \mathcal{L}\left[\frac{d}{dt} u(t)\right] = s\mathcal{L}[u(t)] = 1 \qquad (8\text{-}41)$$

The transform of the ramp function, $r(t)$, is found from Eq. (7-33):

$$\mathcal{L}[r(t)] = \mathcal{L}\left[\int_0^t u(t)\, dt\right] = \frac{1}{s} \mathcal{L}[u(t)] = \frac{1}{s^2} \qquad (8\text{-}42)$$

From these examples, we see the pattern of the equations of the transforms of the unit functions. All are of the form $1/s^k$ and when one function is found from another by differentiating, the new transform is found from the other by multiplication by $1/s$. The following table illustrates this result.

Function	*Laplace transform*
Unit parabolic function	$1/s^3$
Unit ramp function	$1/s^2$
Unit step function	$1/s$
Unit impulse function	1
Unit doublet function	s
Unit triplet function	s^2

EXAMPLE 3

Suppose that the network of Fig. 8-4 is driven by an impulse source rather than a pulse source. Then $V(s) = 1$, and, with $i(0-) = 0$, the transform voltage equation is

$$LsI(s) + RI(s) = 1 \qquad (8\text{-}43)$$

Substituting values $L = 1$ and $R = 1$ and solving for $I(s)$,

$$I(s) = \frac{1}{s+1} \qquad (8\text{-}44)$$

and

$$i(t) = e^{-t} \qquad (8\text{-}45)$$

This simple example illustrates an important result. The response to an impulse is determined only by the constants of the network. The impulse response thus characterizes the network and may be used for identification purposes.

8-3. WAVEFORM SYNTHESIS

The step function and other functions derived from it are useful as building blocks in constructing other waveforms. In this section, we illustrate this statement with a number of examples, and also determine the transform expressions for these waveforms. Figure 8-15 shows three methods by which we may describe a pulse function in terms of step functions. In Section 8-1, we showed that by taking the difference between two step functions,

$$v(t) = u(t - a) - u(t - b) \qquad (8\text{-}46)$$

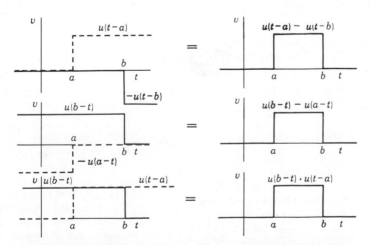

Fig. 8-15. The synthesis of a rectangular pulse from various unit step functions.

a *pulse* is formed of unit amplitude from $t = a$ to $t = b$. The same unit pulse may be formed in terms of the unit step function building blocks as

$$v(t) = u(b - t) - u(a - t) \qquad (8\text{-}47)$$

or

$$v(t) = u(b - t) \cdot u(t - a) \qquad (8\text{-}48)$$

as shown in Fig. 8-15.

Using the same philosophy as in the construction of a pulse function, suppose that we wish to represent a half-cycle of a sine wave, shown in Fig. 8-16(a). In (b) of this figure is shown a sine wave defined only for positive time, and zero for all negative time. This waveform is described by sin $\pi t u(t)$. Using the shifting theorem of Section 8-1, we see that the sine wave of (c) is the same as (b) shifted one unit to the right, and so is sin $\pi(t - 1)u(t - 1)$. Since the sum of these two waveforms gives the required half-cycle of the sine wave, the required representation is

$$v(t) = \sin \pi t \, u(t) + \sin \pi(t - 1) \, u(t - 1) \qquad (8\text{-}49)$$

Using the transform for the sine function of Chapter 7, and the shifting theorem, we see that the transform of a half-cycle of the sine wave is

$$V(s) = \frac{\pi}{s^2 + \pi^2}(1 + e^{-s}) \qquad (8\text{-}50)$$

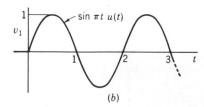

(a)

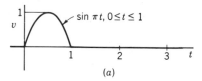

(b)

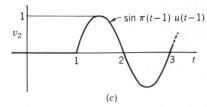

(c)

Fig. 8-16. (a) A half cycle of sine wave or a sinusoidal pulse which is described in terms of the sum of the sine wave of (b) and the shifted wave of (c).

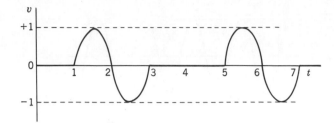

Fig. 8-17. A waveform which is synthesized in terms of sinusoids by Eq. 8-53.

As a more complicated example of signal synthesis, suppose that we wish to represent the waveform of Fig. 8-17. The waveform is sinusoidal from $t = 1$ to $t = 3$ and from $t = 5$ to $t = 7$. It has zero value for all other times from $t = -\infty$ to $t = +\infty$. A sine wave with a period of T is given as $\sin (2\pi/T)t$. In this particular example, $T = 2$ and the time axis is shifted by 1 unit of time for the first wave and by 5 units of time for the second. We will follow a step-by-step procedure in constructing a function to represent this waveform.

(1) The function $\sin \pi(t - 1)$ has the waveform shown in the interval $t = 1$ to $t = 3$, but the waveform also exists for all other time.

(2) Multiplying $\sin \pi(t - 1)$ by $u(t - 1)$ eliminates all waveforms at times *before* $t = 1$. Subtracting from this product a similar product shifted to the time $t = 3$ cancels all times *after* $t = 3$. This product is $u(t - 3) \sin \pi(t - 3)$. Hence the first cycle of sine wave is completely represented by

$$u(t - 1) \sin \pi(t - 1) - u(t - 3) \sin \pi(t - 3) \qquad (8\text{-}51)$$

(3) By the same reasoning, the second cycle of sine wave is represented as

$$u(t - 5) \sin \pi(t - 5) - u(t - 7) \sin \pi(t - 7) \qquad (8\text{-}52)$$

(4) The total waveform is the sum of the two expressions. This follows because each function is defined only for its interval (1 to 3 and 5 to 7, respectively) and is zero for all other time. Hence the waveform of Fig. 8-17 may be represented by the equation

$$v(t) = u(t - 1) \sin \pi(t - 1) - u(t - 3) \sin \pi(t - 3)$$
$$+ u(t - 5) \sin \pi(t - 5) - u(t - 7) \sin \pi(t - 7) \qquad (8\text{-}53)$$

Another way of representing this function is shown with the aid of Fig. 8-18. The product of the two waveforms is, of course, the waveform of Fig. 8-17. We may visualize the train of pulses as *gates*

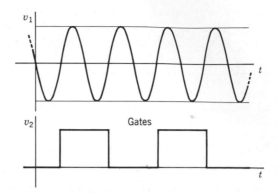

Fig. 8-18. A sine wave shifted one-half cycle and two gate functions. The product of v_1 and v_2 gives v in Fig. 8-17.

which are open when of value 1 and closed when 0. Let us represent a gate which occurs at time t_0 and is of duration T as

$$\sqcap_{t_0,T}(t) = u(t - t_0) - u(t - t_0 - T) \qquad (8\text{-}54)$$

Then the waveform of Fig. 8-17 is simply

$$v(t) = -\sin \pi t[\sqcap_{1,2}(t) + \sqcap_{5,2}(t)] \qquad (8\text{-}55)$$

Similarly, the waveform of Fig. 8-16(a) may be described as

$$v(t) = \sin \pi t \sqcap_{0,1}(t) \qquad (8\text{-}56)$$

The ramp function is also useful in constructing useful waveforms. From Fig. 8-19, we see that a *triangular* waveform of unit height and duration T sec may be constructed from three ramp functions. From the figure, we see that

$$\Lambda_{0,T}(t) = \frac{2}{T} r(t) - \frac{4}{T} r\left(t - \frac{T}{2}\right) + \frac{2}{T} r(t - T) \qquad (8\text{-}57)$$

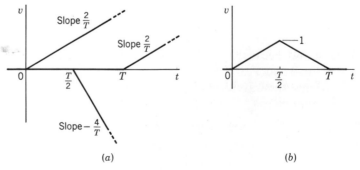

(a)   (b)

Fig. 8-19. The synthesis of an isosceles triangular pulse in terms of ramp functions.

where the Greek letter *lambda* represents the triangular function, and in $\Lambda_{t_0, T}$, t_0 is the time the waveform begins, and T represents its duration. This compact notation allows a sawtooth waveform to be represented by the infinite series,

$$v(t) = \Lambda_{0,T} + \Lambda_{T,T} + \Lambda_{2T,T} + \ldots \tag{8-58}$$

The transform for the triangular waveform described by Eq. (8-57) is found from the sum of transforms

$$V(s) = \frac{2}{Ts^2} - \frac{4}{Ts^2}e^{-Ts/2} + \frac{2}{Ts^2}e^{-Ts} \tag{8-59}$$

$$= \frac{2}{T}\frac{(1 - e^{-Ts/2})^2}{s^2} \tag{8-60}$$

We next consider the problem of determining the Laplace transform of a periodic function of period T. Such a function is the so-called *square wave* shown in Fig. 8-20. This waveform may be represented in terms of an infinite sum of step functions as

$$v(t) = u(t) - 2u(t - a) + 2u(t - 2a)$$
$$- 2u(t - 3a) + \ldots \tag{8-61}$$

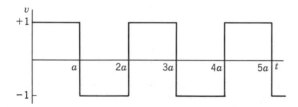

Fig. 8-20. A rectangular wave, also called a square wave.

The Laplace transformation may be applied to this expression term by term to give

$$V(s) = \frac{1}{s} - 2\frac{e^{-as}}{s} + 2\frac{e^{-2as}}{s} - 2\frac{e^{-3as}}{s} + \ldots \tag{8-62}$$

By factoring out common terms, the equation becomes

$$V(s) = \frac{1}{s}[1 - 2e^{-as}(1 - e^{-as} + e^{-2as} - e^{-3as} + \ldots)] \tag{8-63}$$

The infinite series appearing in this equation may be identified by the following expansion from the binomial theorem,

$$\frac{1}{1 + e^{-as}} = 1 - e^{-as} + e^{-2as} - e^{-3as} + \ldots \tag{8-64}$$

such that $V(s)$ becomes

$$V(s) = \frac{1}{s}\left(1 - \frac{2e^{-as}}{1 + e^{-as}}\right) = \frac{1}{s}\left(\frac{1 - e^{-as}}{1 + e^{-as}}\right) \tag{8-65}$$

or, finally,

$$V(s) = \frac{1}{s} \tanh \frac{as}{2} \qquad (8\text{-}66)$$

The procedure outlined in the example may be applied to any periodic function of period T satisfying $f(t + nT) = f(t)$ where n is a positive or negative integer. The transform of any such function is

$$F(s) = \int_0^\infty f(t)e^{-st}\,dt = \int_0^T f(t)e^{-st}\,dt + \int_T^{2T} f(t)e^{-st}\,dt + \dots \qquad (8\text{-}67)$$

By successively shifting each transform term by e^{-nsT}, where n is the number of shifts necessary to make the limits of the integral expression 0 to T, we have

$$F(s) = (1 + e^{-sT} + e^{-2sT} + \dots)\int_0^T e^{-st}f(t)\,dt \qquad (8\text{-}68)$$

Using the binomial theorem to identify the series,

$$F(s) = \frac{1}{1 - e^{-Ts}}\int_0^T e^{-st}f(t)\,dt \qquad (8\text{-}69)$$

Now the integral of this equation represents the transform of $f(t)$ if it existed only from 0 to T. Letting this transform be $F_1(s)$, we have

$$F(s) = \frac{1}{1 - e^{-Ts}}F_1(s) \qquad (8\text{-}70)$$

This equation relates the transform of a periodic function to the transform of the first cycle.

EXAMPLE 4

Let it be required that we find the transform of a pulse train of period T where each pulse is of width a. The transform of this pulse is given by Eq. (8-14). Substituting into Eq. (8-70) gives the final result:

$$F(s) = \frac{1}{s}\frac{1 - e^{-as}}{1 - e^{-Ts}} \qquad (8\text{-}71)$$

Clearly, Eq. (8-70) is a useful result.

We have already made note of the *impulse train* as a waveform obtained by differentiating a waveform represented in terms of step functions. We will have need for a representation of a train of impulses, each of unit area spaced a units of time apart. An appropriate symbol to represent such a train of unit impulses is the Cyrillic letter **Ш**, pronounced *shah*. Then

$$\text{III}_a(t) = \delta(t) + \delta(t - a) + \delta(t - 2a) + \ldots \qquad (8\text{-}72)$$

$$= \sum_{n=0}^{\infty} \delta(t - na) \qquad (8\text{-}73)$$

When $\text{III}_a(t)$ is multiplied by a time function, $f_1(t)$, the product

$$v^*(t) = f_1(t)\text{III}_a(t) \qquad (8\text{-}74)$$

is known as a *discrete sampled function*, indicated by the asterisk superscript, indicating that $v^*(t)$ has a value only at times $t - na$, and there $f_1(na)$ represents the area of the impulse (sometimes called the weight of the function). For example, the derivative of Eq. (8-61), the waveform of Fig. 8-20, is

$$v_1^*(t) = \delta(t) - 2\delta(t - a) + 2\delta(t - 2a) - \ldots \qquad (8\text{-}75)$$

as shown in Fig. 8-21(*b*). The function $f_1(t)$ which multiplies $\text{III}_a(t)$ to represent this discrete sampled function is one for which $f_1(0) = 1$ and $f_1(na) = 2(-1)^n$ for $n \neq 0$. The function

$$v_2^*(t) = t\,\text{III}_a(t) \qquad (8\text{-}76)$$

is the impulse train shown in Fig. 8-21(*c*). Such impulse trains are extensively used in analysis of sampled-data systems, such as those which include a digital computer in the system. They are also used

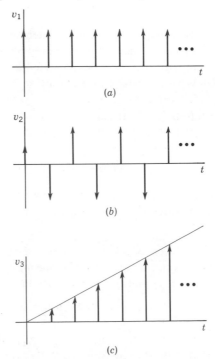

Fig. 8-21. Three different trains of impulses.

with t representing spatial distance in representing sources in the theory of antenna arrays.

The transform of $\text{III}_a(t)$ is easily computed from the series representation of Eq. (8-73), and is

$$\text{III}_a(s) = \mathcal{L}[\text{III}_a(t)] = 1 + e^{-as} + e^{-2as} + \dots \qquad (8\text{-}77)$$

This is the series in Eq. (8-68) and may be written in the closed form

$$\mathcal{L}[\text{III}_a(t)] = \frac{1}{1 - e^{-as}} \qquad (8\text{-}78)$$

This result can be obtained from Eq. (8-70), noting that for the impulse function in the first period, $F_1(s) = 1$. For the weighted impulse train,

$$v_3^*(t) = f_1(t)\text{III}_a(t) = f_1(0) + f_1(a)\delta(t - a) + f_1(2a)\delta(t - 2a) + \dots \qquad (8\text{-}79)$$

the transform becomes

$$\mathcal{L}[f_1(t)\text{III}_a(t)] = f_1(0) + f_1(a)e^{-as}$$
$$+ f_1(2a)e^{-2as} + \dots \qquad (8\text{-}80)$$

As we shall see in Section 8-5, this is a special case of the transform of a product of two time-variable functions.

8-4. THE INITIAL AND FINAL VALUE OF $f(t)$ FROM $F(s)$

It is frequently useful to determine the initial and the final value of a response function $f(t)$ directly from the transform function $F(s)$. We may, for example, be interested in checking $F(s)$ to see if it corresponds to a given initial condition prior to determining $f(t)$ from $F(s)$. This is accomplished by applying two theorems which relate to the initial and final value of $f(t)$.

We first show that if $f(t)$ and its first derivative are Laplace transformable, then

$$f(0-) = \lim_{t \to 0-} f(t) = \lim_{s \to \infty} sF(s) \qquad (8\text{-}81)$$

To prove this statement, we allow s to approach infinity in the equation for the Laplace transform for the derivative of $f(t)$; thus,

$$\lim_{s \to \infty} \int_{0-}^{\infty} \left[\frac{df(t)}{dt}\right] e^{-st}\, dt = \lim_{s \to \infty} [sF(s) - f(0-)] \qquad (8\text{-}82)$$

Since s is not a function of t, we may let $s \to \infty$ before integrating. Under this condition, the integral has zero value so that

$$\lim_{s \to \infty} [sF(s) - f(0-)] = 0 \qquad (8\text{-}83)$$

or

$$f(0-) = \lim_{t \to 0-} f(t) = \lim_{s \to \infty} sF(s) \qquad (8\text{-}84)$$

as required.

EXAMPLE 5

Consider the transform function

$$I_1(s) = \frac{2s+5}{(s+1)(s+2)} \tag{8-85}$$

Applying Eq. (8-84), we have

$$i_1(0-) = \lim_{s\to\infty} \frac{2s^2+5s}{s^2+3s+2} = 2 \tag{8-86}$$

That this is the correct value may be seen from the inverse transform of Eq. (8-85), which is

$$i_1(t) = 3e^{-t} - e^{-2t} \tag{8-87}$$

We next show that if $f(t)$ and its first derivative are Laplace transformable, then

$$\lim_{t\to\infty} f(t) = \lim_{s\to0} sF(s) \tag{8-88}$$

Our starting point is Eq. (8-82) evaluated for $s \to 0$. Since s is not a function of t, we let $s \to 0$ so that $e^{-st} = 1$ and integrate, obtaining

$$\int_{0-}^{\infty} \left[\frac{df(t)}{dt}\right] dt = \lim_{t\to\infty} \int_{0-}^{t} \left[\frac{df(t)}{dt}\right] dt \tag{8-89}$$

$$= \lim_{t\to\infty} [f(t) - f(0-)] \tag{8-90}$$

Equating this result to Eq. (8-82) written for the limit $s \to 0$, we conclude that

$$\lim_{t\to\infty} f(t) = \lim_{s\to0} sF(s) \tag{8-91}$$

as required. This result requires that all roots of the denominator of $sF(s)$ have negative real parts, for otherwise the limit of $f(t)$ with $t \to \infty$ does not exist. For example, sin ∞ does not have a definite value, and e^{∞} is infinite. For this reason, Eq. (8-91) does not apply in the case of sinusoidal excitation.

EXAMPLE 6

For the current

$$i_2(t) = 5u(t) - 3e^{-2t} \tag{8-92}$$

the final value is clearly 5. Let us determine this same result by using the final value theorem. For the given $i_2(t)$,

$$I_2(s) = \frac{2s+10}{s(s+2)} \tag{8-93}$$

and from Eq. (8-91)

$$\lim_{t\to\infty} i_2(t) = \lim_{s\to0} \frac{2s+10}{s+2} = 5 \tag{8-94}$$

8-5. THE CONVOLUTION INTEGRAL

The convolution integral, which dates back at least to Duhamel in 1833,[4] finds applications in many fields, including circuit theory and automatic control. One important application that we will study enables us to evaluate the response of a network to an arbitrary input in terms of the impulse response of the network. Let the two functions $f_1(t)$ and $f_2(t)$ be Laplace transformable and have the transforms $F_1(s)$ and $F_2(s)$. The product of $F_1(s)$ and $F_2(s)$ is the Laplace transform of $f(t)$ which results from the *convolution* of $f_1(t)$ and $f_2(t)$ as given by the equation

$$f(t) = \mathcal{L}^{-1}[F_1(s)F_2(s)] = \int_0^t f_1(\tau)f_2(t - \tau)\, d\tau \qquad (8\text{-}95)$$

$$= \int_0^t f_2(\tau)f_1(t - \tau)\, d\tau \qquad (8\text{-}96)$$

where τ is a dummy variable for t. The integrals in these equations are known as *convolution integrals*. The convolution of $f_1(t)$ and $f_2(t)$ is denoted by the special notation

$$f(t) = f_1(t) * f_2(t) \qquad (8\text{-}97)$$

In terms of this notation, we see that

$$F(s) = \mathcal{L}[f_1(t) * f_2(t)] = \mathcal{L}[f_2(t) * f_1(t)]$$
$$= F_1(s)F_2(s) \qquad (8\text{-}98)$$

Thus the inverse Laplace transform of the product of $F_1(s)$ and $F_2(s)$ is found by *convolving* $f_1(t)$ and $f_2(t)$ by Eq. (8-95) or (8-96).

To derive these equations, we observe that $F(s) = F_1(s)F_2(s)$ may be written as a product of Laplace transform integrals involving the dummy variables x and y as

$$F(s) = \left[\int_0^\infty f_1(y)e^{-sy}\, dy \right]\left[\int_0^\infty f_2(x)e^{-sx}\, dx \right] \qquad (8\text{-}99)$$

Because each of the integrals is constant with respect to the other variable of integration, we may rearrange Eq. (8-99) to have the following form:

$$F(s) = \int_0^\infty \int_0^\infty f_1(y)f_2(x)e^{-s(x+y)}\, dy\, dx \qquad (8\text{-}100)$$

At this point, we introduce new variables t and τ which are related to x and y by $t = x + y$ and $\tau = x$. To rewrite Eq. (8-100) in terms of t and τ, we must investigate the relationship of the product $dy\, dx$ to $d\tau\, dt$, and also determine the new limits of integration. The

[4]For a brief history of the development of the convolution integral, see M. F. Gardner and J. L. Barnes, *Transients in Linear Systems* (John Wiley & Sons, Inc., New York, 1942), pp. 364–365.

differential areas $dy\,dx$ and $d\tau\,dt$ are related by the Jacobian[5] by the following equation,

$$dy\,dx = \left| \frac{\partial x}{\partial t}\frac{\partial y}{\partial \tau} - \frac{\partial x}{\partial \tau}\frac{\partial y}{\partial t} \right| d\tau\,dt \qquad (8\text{-}101)$$

Computing the required partial derivatives, we find that $dy\,dx = d\tau\,dt$, meaning that the differential areas are the same for the two systems of variables.

In the x-y coordinates, the limits of integration extend from zero to infinity for both x and y. Now, $t = x + y = \tau + y$. Since the smallest value of y is zero, it follows that we need consider only $t \geq \tau$. The equation $t = \tau$ defines a straight line in the $t - \tau$ plane at 45° with respect to the τ axis. To integrate the area between the line $t = \tau$ and $\tau = 0$, we see that we integrate from $t = 0$ to $t = \infty$ and then from $\tau = 0$ to $\tau = t$.

We may now rewrite Eq. (8-100) in terms of the variables t and τ as follows:

$$F(s) = \int_0^{\infty} \left[\int_0^t f_1(t - \tau)f_2(\tau)\,d\tau \right] e^{-st}\,dt \qquad (8\text{-}102)$$

Then, from the defining equation of the Laplace transformation, we have the identification,

$$f(t) = \int_0^t f_1(t - \tau)f_2(\tau)\,d\tau = f_1(t) * f_2(t) \qquad (8\text{-}103)$$

which demonstrates the validity of Eq. (8-96). If $f_1(t)$ and $f_2(t)$ are interchanged, the same derivation may be carried out for Eq. (8-95).

While the steps in this derivation may be clear, it is important that we have a feeling for the operations described in the convolution integral. We will show next, by means of examples, that convolution may be interpreted in terms of four steps: (1) folding, (2) translating, (3) multiplying, and (4) integrating. These four steps describe the English equivalent to the German word *Faltung;* the convolution integral is also known as the *Faltung integral.*

EXAMPLE 7

For this example, let $F_1(s) = 1/s$, and $F_2(s) = 1/(s + 1)$, so that $f_1(t) = u(t)$ and $f_2(t) = e^{-t}u(t)$. By convolution, we wish to determine

$$f(t) = f_1(t) * f_2(t) = \int_0^t u(t - \tau)e^{-\tau}\,d\tau \qquad (8\text{-}104)$$

The steps in convolving these two functions are illustrated in Fig. 8-22 where $f_1(t)$ and $f_2(t)$ are shown in (a), and in (b), $f_1(\tau)$ and $f_2(\tau)$, an elementary step once you see what is intended. In (c), we have "folded"

---

[5]A. E. Taylor, *Advanced Calculus* (Ginn & Company, Boston, 1955), p. 428.

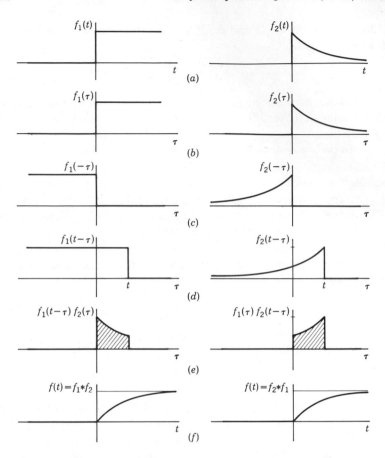

Fig. 8-22. Steps in convolving f_1 with f_2 in (a) to obtain $f = f_1 * f_2$ in step (f) of the figure.

the functions about the line $t = 0$, and in (d) translated for some typical value of t, as described in Section 8-1. In (e) of Fig. 8-22, we have carried out the multiplication indicated within the integral in Eqs. (8-95) and (8-96). The integration of the cross-hatched area gives a point on the curve $f(t)$ for the value of t selected in step (d). Carrying out the multiplication and integration steps for various values of t as shown in Fig. 8-23, we obtain the response $f(t)$ as shown in (d) of that figure. For this simple example, the actual integration of Eq. (8-104) is simple and gives

$$f(t) = \int_0^t e^{-\tau}\, d\tau = 1 - e^{-t} \tag{8-105}$$

which is, of course, the inverse Laplace transform of the product of $F_1(s)$ and $F_2(s)$,

$$f(t) = \mathcal{L}^{-1}\left[\frac{1}{s(s+1)}\right] \tag{8-106}$$

as given by Eq. (8-95).

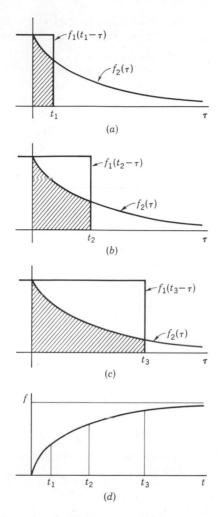

Fig. 8-23. Step (e) in Fig. 8-22 repeated for three values of y. The shaded area represents the value of $f(t)$ which is shown in (d) of the figure for the three values of time.

Consider next the two terminal-pair network shown in Fig. 8-24. We will show in Chapter 10 that if all initial conditions in the network are zero, then the input and output voltage transforms are related by an equation of the form

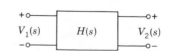

Fig. 8-24. Two-port network described by Eq. 8-107.

$$V_2(s) = H(s)V_1(s) \qquad (8\text{-}107)$$

where $H(s)$ is known as the *transfer function* relating the two voltages: $V_1(s)$, the input voltage transform, and $V_2(s)$, the output voltage transform. Since $V_2(s)$ in Eq. (8-107) is expressed as a product of transforms, we have an obvious application of the convolution integral. Thus,

$$v_2(t) = \mathcal{L}^{-1}[H(s)V_1(s)] = \int_0^t v_1(\tau)h(t-\tau)\,d\tau \qquad (8\text{-}108)$$

$$= \int_0^t v_1(t-\tau)h(\tau)\,d\tau \qquad (8\text{-}109)$$

the two forms corresponding to Eqs. (8-95) and (8-96). Now, if $v_1(t) = \delta(t)$, then the transform of the unit impulse is $V_1(s) = 1$. Under this condition, $V_2(s) = H(s)$ or $v_2(t) = h(t)$ is the *impulse response* of the network as well as being the inverse transform of the transfer function $H(s)$. Just as we will find that the transfer function characterizes the network, so the impulse response is characteristic of the network.

Now Eqs. (8-108) and (8-109) tell us that if $h(t)$, the impulse response, is known, only the input voltage $v_1(t)$ need be specified in order to determine the output voltage through convolution! In other words, any input convolved with the unit impulse response gives the output.

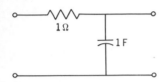

Fig. 8-25. A specific form of the general network of Fig. 8-24 which is considered in Example 8.

EXAMPLE 8

To illustrate the use of Eq. (8-108) or (8-109), consider the network shown in Fig. 8-25. For this network, we see that $V_1(s) = I + (1/s)I$, and $V_2(s) = (1/s)I$ so that

$$\frac{V_2(s)}{V_1(s)} = H(s) = \frac{1}{s+1} \tag{8-110}$$

From the table of transforms, we find that

$$h(t) = e^{-t} \tag{8-111}$$

is the unit impulse response of the network. Then, from the convolution integral, the output for the input voltage $v_1(t) = e^{-2t}$ is found to be

$$v_2(t) = \int_0^t e^{-2(t-\tau)}e^{-\tau}\,d\tau \tag{8-112}$$

or

$$v_2(t) = e^{-2t}(e^t - 1) = e^{-t} - e^{-2t} \tag{8-113}$$

For this particular example, expansion by partial fractions is more direct. For more complicated forms of the input, the convolution integral can be used to advantage.

Suppose that we wish to determine the response of a network whose impulse response is shown in Fig. 8-26(*a*) to the arbitrary input shown in Fig. 8-26(*b*). Clearly, this may be accomplished by convolving the two time functions of the figure. The steps by which this is accomplished are shown in Fig. 8-27 with the same operations shown previously in Fig. 8-22. From this figure, we make two observations: (1) For this arbitrary input, the multiplication and integration steps must be accomplished graphically—or by the equivalent of graphical operations by using a digital or analog computer. (2) The operation of multiplying the input $v_1(\tau)$ by the shifted impulse response $h(t - \tau)$ and then integrating from 0 to t may be thought of as *weighting* all

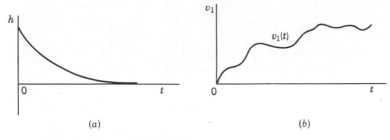

(a)   (b)

Fig. 8-26. The impulse response of the network of Fig. 8-25, and (b) an arbitrary input for which the response will be found by convolution.

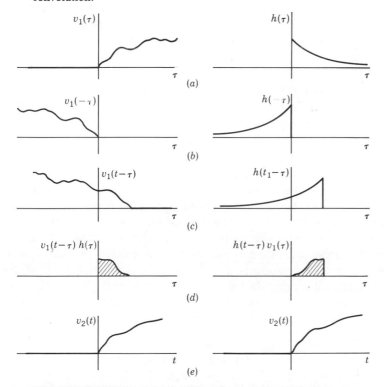

Fig. 8-27. Steps in convolving an arbitrary input v_1 with the impulse response to obtain the output v_2.

past values of the input by the impulse response. This is illustrated by Fig. 8-28 in which the waveforms of Fig. 8-27(a) and (c) are super-imposed. As t increases, $h(t - \tau)$ is shifted to the right, illustrated for increasing times t_1, t_2, and t_3. As this happens, the product of the two functions, shown as the cross-hatched area in Fig. 8-27(d), changes and so the output $v_2(t)$ changes. We may visualize this as $h(t - \tau)$

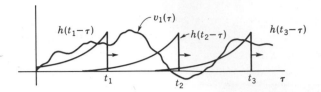

Fig. 8-28. The impulse response as a scanning function.

sliding over or *scanning* $v_1(\tau)$, giving rise to the naming of $h(t - \tau)$ the *scanning function*. As we scan in Fig. 8-28, the output at any time is mainly determined by recent values of the input. "Very old" values of input have very little effect on the present output, although, strictly speaking, we see that the present output is determined by all past history of the input, weighted by the impulse response. This proves to be a useful way of visualizing the form of the response that results from the excitation of a network or a system.

Another way of visualizing the scanning function stems from the observation that, since $F(s) \cdot 1 = F(s)$ and $\mathcal{L}[\delta(t)] = 1$, then from Eq. (8-95),

$$\int_0^t f(\tau)\delta(t - \tau)\,d\tau = f(t) \qquad (8\text{-}114)$$

meaning that the convolution of $f(t)$ with $\delta(t)$, $f(t) * \delta(t)$, gives simply $f(t)$. As the t in $\delta(t - \tau)$ varies from 0 to ∞, we reproduce $f(t)$. Although this is a rather futile operation, it does permit us to identify $\delta(t - \tau)$ as the scanning function. In Fig. 8-29, the impulse at $\tau = t$ has been replaced by a screen with a narrow vertical slit, which is more suggestive of the scanning operation. As the slit scans from 0 to ∞, it reproduces $f(t)$. Since the integral of Eq. (8-114) has a nonzero value only when $\tau = t$, we can replace the limits with $-\infty$ and ∞ without loss of generality, and the scanning operation may be extended over the new limits to reproduce $f(t)$ for all time. In other words,

$$\int_{-\infty}^{\infty} f(\tau)\,\delta(t - \tau)\,d\tau = f(t) \qquad (8\text{-}115)$$

Some of the results of this section are summarized in Table 8-1.

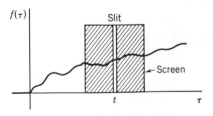

Fig. 8-29. The impulse function, represented by a slit, used as the scanning function.

Table 8-1. SUMMARY OF TRANSFORM OPERATIONS

Name of Operation	$f(t)$	$F(s)$
Definition	$f(t)$	$\mathcal{L}[f(t)] = F(s) = \int_{0-}^{\infty} f(t)e^{-st}\, dt$
Linear Operations		
Addition	$f_1(t) \pm f_2(t)$	$F_1(s) \pm F_2(s)$
Multiplication by a constant	$Kf(t)$	$KF(s)$
Both operations	$a_1 f_1(t) \pm a_2 f_2(t)$	$a_1 F_1(s) \pm a_2 F_2(s)$
Differentiation		
First derivative	$\dfrac{d}{dt} f(t)$	$sF(s) - f(0-)$
Second derivative	$\dfrac{d^2}{dt^2} f(t)$	$s^2 F(s) - sf(0-) - \dfrac{df}{dt}(0-)$
Third derivative	$\dfrac{d^3}{dt^3} f(t)$	$s^3 F(s) - s^2 f(0-) - s\dfrac{df}{dt}(0-) - \dfrac{d^2 f}{dt^2}(0-)$
Integration		
First integral	$\int_0^t f(t)\, dt$	$\dfrac{F(s)}{s}$
First integral	$\int_{-\infty}^t f(t)\, dt = f^{(-1)}(t)$	$\dfrac{F(s)}{s} + \dfrac{f^{(-1)}(0-)}{s}$
Second integral	$\int_{-\infty}^t \int_{-\infty}^t f(t)dt\, dt = f^{(-2)}(t)$	$\dfrac{F(s)}{s^2} + \dfrac{f^{(-1)}(0-)}{s^2} + \dfrac{f^{(-2)}(0-)}{s}$
Shifting	$f(t-a)u(t-a)$	$e^{-as} F(s)$
	$e^{at} f(t)$	$F(s-a)$
Scaling		
Time scaling	$f(at)$	$\dfrac{1}{a} F\left(\dfrac{s}{a}\right)$
Magnitude scaling	$a f(t)$	$aF(s)$
Periodic functions	$f(t) = f(t + nT)$	$\dfrac{1}{1 - e^{-Ts}} F_1(s)$
	n is an integer	where $F_1(s) = \int_0^T f(t)e^{-st}\, dt$
Convolution	$\int_0^t f_1(\tau) f_2(t-\tau)\, d\tau$	$F_1(s)F_2(s)$
	$\int_0^t f_1(t-\tau) f_2(\tau)\, d\tau$	$F_1(s)F_2(s)$
Multiplication by t	$t f(t)$	$\dfrac{-dF(s)}{ds}$
	$t^n f(t)$	$(-1)^n \dfrac{d^n}{ds^n} F(s)$
Multiplication by $e^{-\alpha t}$	$e^{-\alpha t} f(t)$	$F(s + \alpha)$
Initial value	$\lim_{t \to 0} f(t) =$	$= \lim_{s \to \infty} sF(s)$
Final value	$\lim_{t \to \infty} f(t) =$	$= \lim_{s \to 0} sF(s)$ providing poles of $sF(s)$ in left half-plane

8-6. CONVOLUTION AS A SUMMATION

In the evaluation of the convolution integral by means of a digital computer, and in many engineering applications, it is necessary to replace the integral with a finite summation. The steps by which this is accomplished are illustrated for the equation

$$v_2(t) = \int_0^t v_1(\tau)h(t - \tau)\, d\tau \qquad (8\text{-}116)$$

We first replace the infinitesimal $d\tau$ with a time interval of finite width T, and let $nT = \tau$. Then we write Eq. (8-116) as the summation

$$v_2(t) = T \sum_{n=0}^K v_1(nT)h(t - nT) \qquad (8\text{-}117)$$

for

$$KT \le t < (K + 1)T \qquad (8\text{-}118)$$

where K is one of the values of the integer n. In expanded form, Eq. (8-117) is

$$v_2(t) = T[v_1(0)h(t) + v_1(T)h(t - T)$$
$$+ v_1(2T)h(t - 2T) + \ldots] \qquad (8\text{-}119)$$

This result may be interpreted with the aid of Fig. 8-30 which shows the waveforms corresponding to the individual terms in this equation. Each term is the impulse response of magnitude of $v_1(nT)$, which is later multiplied by T, shifted an appropriate distance along the t axis. The summation of the terms in Eq. (8-119) approximates the function $v_2(t)$. The accuracy of the approximation will depend on the size of T, of course; the smaller T is, the better will be the approximation. In the limit, T becomes $d\tau$ and the result is exact.

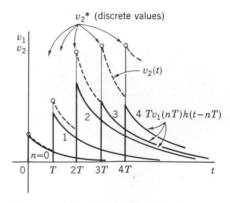

Fig. 8-30. Discrete values of the response v_2 found by considering convolution as a summation. The $v_2(t)$ shown as a dashed line is found from Eq. (8-119).

If values of $v_2(t)$ are computed only at the chosen points of time, KT, then Eq. (8-117) becomes

$$v_2^*(t) = v^2(KT) = \sum_{n=0}^{K} v_1(nT)h(K-n)T \qquad (8\text{-}120)$$

where the asterisk indicates a discrete sampled function, as in Eq. (8-74). By this statement we mean that Eq. (8-120) is not a continuous function, but is represented by real numbers for each integer value of K. If we use Eq. (8-117), we may compute v_2 for all values of t, as shown in Fig. 8-30, with a different equation for each interval as given by Eq. (8-118).

If we use an alternative form of the convolution integral corresponding to Eq. (8-116), namely,

$$v_2(t) = \int_0^t h(\tau)v_1(t-\tau)\,d\tau \qquad (8\text{-}121)$$

then the form of the approximation equation is

$$v_2(t) = T \sum_{n=0}^{K} h(nT)v_1(t-nT) \qquad (8\text{-}122)$$

The steps in the determination of $v_2(t)$ are shown in the representation of this equation shown in Fig. 8-31. The basic operations are time delay, multiplication by a constant, and summation. These steps may be accomplished by using a tapped transmission line to obtain the various values of time delay, an amplifier of adjustable gain set to the values of $Th(nT)$, and a summing amplifier such as is used in analog computers to perform the summation.

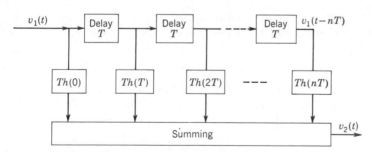

Fig. 8-31. A flow chart illustrating the steps involved in determining a response using Eq. (8-122).

EXAMPLE 9

To illustrate the use of the equations of this section in computing $v_2(t)$, consider the two waveforms of Fig. 8-32, that in (a) for $h(t)$ and in (b) for $v_1(t)$. We will use the values from these waveforms to compute $v_2(5)$ by means of Eq. (8-120). From the figures, we prepare the

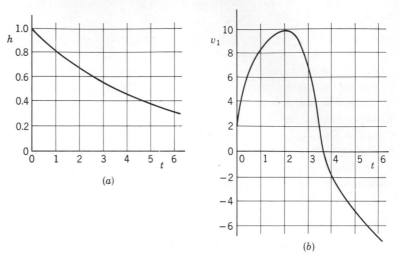

Fig. 8-32. (*a*) The impulse response and (*b*) the input for which the output is to be determined by convolution.

following chart, observing that for this example $T = 1$.

	n						
	0	1	2	3	4	5	6
$h(nT)$	1.00	.819	.670	.549	.449	.368	.301
$v_1(nT)$	2.0	8.3	9.8	6.8	−1.8	−4.8	−7.2
$v_1[(5 - n)T]$	−4.8	−1.8	6.8	9.8	8.3	2.0	0
$v_1[(5 - n)T]h(nT)$	−4.8	−1.47	4.56	5.38	3.73	.74	0

The first two lines are taken from the graphs of Fig. 8-32, and the third is obtained by copying line 2 backwards in the range of n from 0 to 5. The fourth line is the product of line 1 and line 3. Adding together the numbers of line 4, we obtain 8.14; thus,

$$v_2(5) = 8.14 \tag{8-123}$$

A similar computation for each value of n will give $v_2^*(t)$. In this example, the sixth column was included to show that, since $v_1[(5 - n)T]$ is zero for n greater than 5, the series ends at $n = 5$ and the number of computations involved is small for small n. This particular approximation is poor because the interval T is too large for the rate at which $v_1(t)$ is changing. For further study and refinement of this example, see Problems 8-77 and 8-78.

FURTHER READING

ASELTINE, JOHN A., *Transform Method in Linear System Analysis*, McGraw-Hill Book Company, New York, 1958. An excellent treatment is given in Chapter 8.

CHIRLIAN, PAUL M., *Basic Network Theory*, McGraw-Hill Book Company, New York, 1969. Chapter 5.

COOPER, GEORGE R., AND CLARE D. MCGILLEM, *Methods of Signal and System Analysis*, Holt, Rinehart & Winston, Inc., New York, 1967. Chapters 6 and 7.

HAYT, WILLIAM H., JR., AND JACK E. KEMMERLY, *Engineering Circuit Analysis*, 2nd ed., McGraw-Hill Book Company, New York, 1971. Chapter 19.

KUO, FRANKLIN F., *Network Analysis and Synthesis*, John Wiley & Sons, Inc., New York, 1962. Chapter 2.

MANNING, L.A., *Electric Circuits*, McGraw-Hill Book Company, New York, 1966. Chapters 11, 17, 18, and 19.

STRUM, ROBERT D., AND JOHN R. WARD, *Laplace Transform Solution of Differential Equations* (a programmed text), Prentice-Hall, Inc., Englewood Cliffs, N.J., 1968.

DIGITAL COMPUTER EXERCISES

Since the topics of this chapter are a continuation of those of Chapter 7, the exercises suggested at the end of Chapter 7 are also appreciated for this chapter. In addition, the numerical integration of functions by means of the computer as applied to convolution is of interest and exercises may be found in the references of Appendix E-2.4. See, for example, reference 7 by Huelsman cited in Appendix E-10, Chapters 2 and 3, and reference 12 by McCracken in Appendix E-10, especially Case Study 22.

PROBLEMS

8-1. Write an equation for the nonrecurring waveform of the figure as a linear combination of step functions.

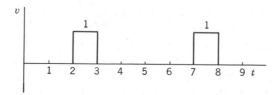

Fig. P8-1.

8-2. Write the equation for the waveform shown in the figure in terms of linear combination of step functions.

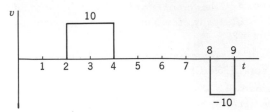

Fig. P8-2.

8-3. The waveform shown in the figure is sinusoidal in the interval from $t = 0$ to $t = 1$, and is an isosceles triangle in the interval from $t = 2$ to $t = 3$. For all other t, v is zero. Write an equation for (vt) using step, ramp, and the sine functions.

Fig. P8-3.

8-4. The waveform shown in the figure is nonrecurring. Write an equation for $v(t)$ in terms of steps and ramps and related functions as needed.

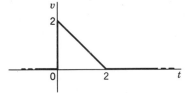

Fig. P8-4.

8-5. The waveform shown in the figure occurs only once. Write an equation for $v(t)$ in terms of steps and related functions as needed.

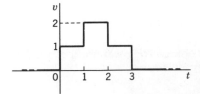

Fig. P8-5.

8-6. The waveform shown in the figure is nonrecurring. Write an equation for this waveform, $v(t)$.

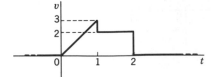

Fig. P8-6.

8-7. The accompanying figure shows a waveform made up of straight-line segments. For this waveform, write an equation for $v(t)$ in terms of steps, ramps, and related waveforms as needed.

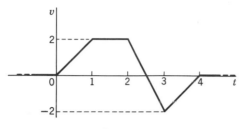

Fig. P8-7.

8-8. Repeat Prob. 8-7 for the waveform shown in the accompanying figure.

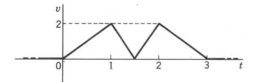

Fig. P8-8.

8-9. The parabolic pulse shown in the figure occurs only once. Write an equation for $v(t)$ using functions multiplied by the appropriate delayed step functions.

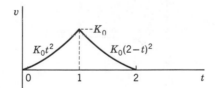

Fig. P8-9.

8-10. In the waveform of the figure, v jumps to value b at $t = 0$ and then decreases exponentially to value a at $t = c$ and then drops to zero value. The cycle is then repeated with negative magnitudes. Write an expression for this waveform using step functions.

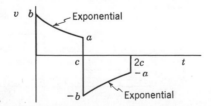

Fig. P8-10.

8-11. A pulse of voltage of 10 V magnitude and 5 μsec duration is applied to the *RL* network shown in the figure. (a) If $R = 2\,\Omega$, and $L = 10$ μH, find $i(t)$ and plot this waveform. (b) Repeat part (a) if $R = 2\,\Omega$ and $L = 2\,\mu$H. (c) Repeat part (a) if $R = 2\,\Omega$ and $L = 5\,\mu$H.

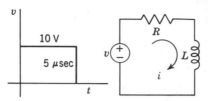

Fig. P8-11.

8-12. A voltage pulse of 10 V magnitude and 5 μsec duration is applied to the *RC* network shown in the figure. Find the current and plot the current waveform if: (a) $R = 100\,\Omega$ and $C = 0.05\,\mu$F. (b) $R = 100$ Ω and $C = 0.02\,\mu$F.

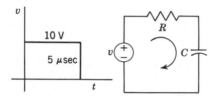

Fig. P8-12.

8-13. Repeat Prob. 8-12 solving for the voltage across the capacitor, $v_C(t)$.

8-14. The waveform shown in the accompanying figure is known as a *truncated ramp* which is used to represent a unit step function having nonzero rise time. Let this waveform replace the rectangular pulse used in Prob. 8-11. Determine and plot $i(t)$ if $t_0 = 1\,\mu$sec.

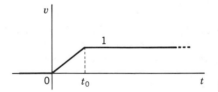

Fig. P8-14.

8-15. Using the waveform of Prob. 8-14 for $v(t)$, solve Prob. 8-12.

8-16. Using the waveform of Prob. 8-14 for $v(t)$, solve Prob. 8-13.

8-17. The waveform shown in the figure is known as a *staircase*. (a) Assuming that the staircase shown is not repeated, write an equation for it

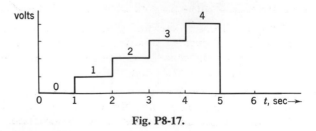

Fig. P8-17.

in terms of unit step functions. (b) If this voltage is applied to an RL series circuit with $R = 1\,\Omega$ and $L = 1$ H, find the current $i(t)$ and sketch its waveform using the same coordinates as the staircase.

8-18. Determine the transform $V(s)$ for the $v(t)$ given in Prob. 8-1.

8-19. A ramp function which has been shifted upward is described by the equation

$$f_1 = K(t + t_0)u(t), \qquad t_0 > 0$$

(a) Sketch $f_1(t)$. (b) Determine $F_1(s)$.

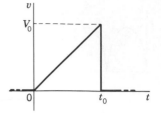
Fig. P8-22.

8-20. A ramp function which has been shifted downward is given by

$$f_2 = K(t - t_0)u(t), \qquad t_0 > 0$$

(a) Sketch $f_2(t)$. (b) Determine $F_2(s)$.

8-21. Find the transform of the waveform given in Prob. 8-3.

8-22. Determine the transform $V(s)$ for $v(t)$ as given in the accompanying figure.

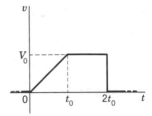
Fig. P8-23.

8-23. The waveform shown in the figure is a terminated truncated ramp. For this $v(t)$, determine the transform $V(s)$.

8-24. The waveform shown in the figure consists of a single triangular pulse. For this $v(t)$, determine the corresponding transform $V(s)$.

8-25. Determine the transform $V(s)$ for $v(t)$ as described in the accompanying figure. Compare your result with that found in Prob. 8-22.

8-26. Determine the transform for the staircase waveform given in Prob. 8-17.

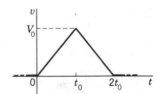
Fig. P8-24.

8-27. Show that the transform of the square wave is

$$F(s) = \frac{1}{s(1 + e^{-as})}$$

8-28. The waveform shown in the figure is that of a full-wave rectified voltage. The equation for the waveform is $\sin t$ from 0 to π, $-\sin t$ from π to 2π, etc. Show that the transform of this function is

$$F(s) = \frac{1}{s^2 + 1}\coth\frac{\pi s}{2}$$

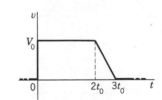
Fig. P8-25.

8-29. The waveform shown is a sweep voltage used to deflect the beam in a cathode ray oscilloscope. Show that the transform of this function is

$$F(s) = \frac{1}{as^2} - \frac{e^{-as}}{s(1 - e^{-as})}$$

8-30. Find the transform of the voltage waveform shown in the figure.

8-31. The staircase waveform of Prob. 8-17 ends at $t = 5$. Consider instead a staircase which extends to infinity and at $t = nt_0$ jumps to the value $n + 1$, being a superposition of unit step functions. Determine the Laplace transform of this waveform.

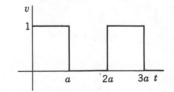
Fig. P8-27.

8-32. A geometrically-increasing staircase is one like that described in Prob. 8-31, but with an amplitude which increases at $t = nt_0$ to the value $(n + 1)^2$. Find the Laplace transform of such a waveform.

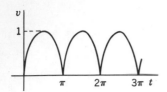

Fig. P8-28.

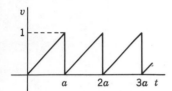

Fig. P8-29.

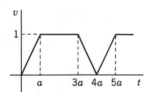

Fig. P8-30.

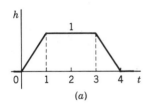
(a)

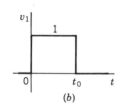
(b)

Fig. P8-36.

8-33. The waveform shown in the figure consists of a train of pulses of magnitude which decays exponentially. For the numerical values given in the figure, find the Laplace transform of the waveform $F(s)$.

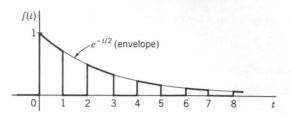

Fig. P8-33.

8-34. Find $\mathcal{L}^{-1}[F_1(s)F_2(s)]$ by using convolution for the following functions:
(a) $F_1 = 1/(s - a)$, $F_2 = 1/(s - a)$
(b) $F_1 = 1/(s + 1)$, $F_2 = s/(s + 2)$
(c) $F_1 = 1/(s^2 + 1)$, $F_2 = 1/(s^2 + 1)$
(d) $F_1 = s/(s + 1)$, $F_2 = 1/(s^2 + 1)$
(e) $F_1 = 1/(s + a)$, $F_2 = 1/(s + b)(s + c)$

8-35. Find $\mathcal{L}^{-1}[F_1(s)F_2(s)]$ by the convolution integral for the following functions:
(a) $F_1 = 1/s$, $F_2 = 1/(s + 1)$
(b) $F_1 = 1/s(s + 1)$, $F_2 = 1/(s + 2)$
(c) $F_1 = 1/s^2$, $F_2 = s/(s^2 + 4)$
(d) $F_1 = s/(s + 1)$, $F_2 = 1/[(s + 1)^2 + 1]$

8-36. Consider a network whose impulse response can be approximated by the straight lines shown in the accompanying figure. Assume that all initial conditions in the network are zero at $t = 0-$. By convolution, determine the response to the network of the input voltage shown in part (b) of the figure if (a) $t_0 = 3$, (b) $t_0 = 4$, and (c) $t_0 = 1$. For each case, sketch $v_2(t)$ giving numerical values for slopes, important magnitudes, and so on.

8-37. Consider the impulse response shown in (a) of the figure for Prob. 8-36. However, for this problem, the input to the network is shown in the accompanying figure. By convolution, determine the response to the network when the input has (a) $\tau = 1$, and (b) $\tau = 3$.

8-38. Consider the impulse response shown as (a) of the figure for Prob. 8-36. Convolve that impulse response with the input function shown in the accompanying figure.

8-39. For this problem, the impulse response is that shown in the figure for Prob. 8-36, shown as (a), and the input is that shown in the accompanying figure. Convolve $h(t)$ and $v_1(t)$ to find the response $v_2(t)$.

8-40. Consider the impulse response shown as (a) of the figure for Prob. 8-36. Convolve the impulse response with the input function shown in the accompanying figure to obtain the network response.

Consider a network having a specified impulse response $h(t)$ and also a specified input $v_1(t)$. Assume that all initial conditions in the network are zero at $t = 0-$. Using graphical convolution, you are required to find the response function $v_2(t)$. The following chart tabulates various combinations of the figures showing functions made up of straight-line segments.

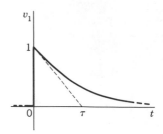

Fig. P8-37.

	$h(t)$ *waveform*	$v_1(t)$ (given in Figure:)
8-41.	(a)	P8-36(b) with $t_0 = 1$
8-42.	(a)	P8-37 with $\tau = 1$
8-43.	(a)	P8-38
8-44.	(a)	P8-39
8-45.	(a)	P8-40
8-46.	(b)	P8-36(b) with $t_0 = 1$
8-47.	(b)	P8-37 with $\tau = 1$
8-48.	(b)	P8-38
8-49.	(b)	P8-39
8-50.	(b)	P8-40
8-51.	(c)	P8-36(b) with $t_0 = 1$
8-52.	(c)	P8-37 with $\tau = 1$
8-53.	(c)	P8-38
8-54.	(c)	P8-39
8-55.	(c)	P8-40
8-56.	(d)	P8-36(b) with $t_0 = 1$
8-57.	(d)	P8-37 with $\tau = 1$
8-58.	(d)	P8-38
8-59.	(d)	P8-39
8-60.	(d)	P8-40
8-61.	(e)	P8-36(b) with $t_0 = 1$
8-62.	(e)	P8-37 with $\tau = 1$
8-63.	(e)	P8-38
8-64.	(e)	P8-39
8-65.	(e)	P8-40

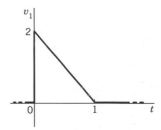

Fig. P8-38.

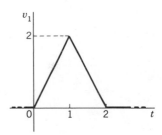

Fig. P8-39.

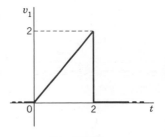

Fig. P8-40.

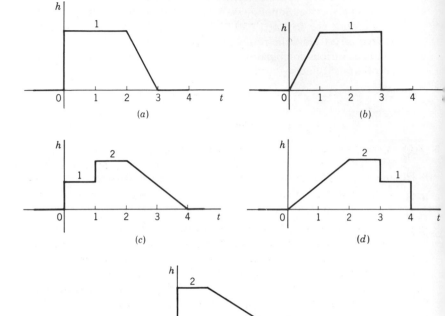

Figs. P8-41 to P8-65.

8-66. Convolve the function

$$h(t) = \sqrt{t}, \quad 0 \leq t \leq 1$$
$$0, \quad \text{all other } t$$

with the function

$$v_1(t) = 1, \quad 0 \leq t \leq 2$$
$$0, \quad \text{all other } t$$

to give $v_2(t) = h(t) * v_1(t)$. Show method.

8-67. The impulse response of a network is given by the equation $h(t) = H_0 e^{-t/\tau}$. The input to this network has a waveform which is a continuously differentiable pulse made up of straight lines and parabolic segments as shown in the accompanying figure. The pulse takes 2 μsec to rise, is perfectly flat for 3 μsec, and then falls in such a way that the waveform is symmetrical. It is known that the time constant τ is in the range half to twice the pulse length of the input.

(a) Sketch the output of the network as accurately as you can, making some assumption as to the value of τ.

(b) Describe a method by which the time constant τ can be determined from the recorded output for the specified input.

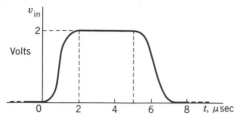

Fig. P8-67.

8-68. Tests on a certain network showed that the current output was $I(t)$ $= -2e^{-t} + 4e^{-3t}$ when a unit voltage was suddenly applied to the input terminals at $t = 0$. What voltage must be applied to give an output current of $i(t) = 2e^{-t}$ if the network remains in the same form as for the previous test?

8-69. A network has a transfer function $H_1(s)$ and the impulse response of the system is approximated by the waveform shown in the figure for Prob. 8-36, part (b) with h replacing v_1 and $t_0 - 1$. Three of these networks are connected together in such a way that the overall transfer function of the new system is $H_3(s) = [H_1(s)]^3$. Using convolution, find the impulse response of the new system, $h_3(t)$.

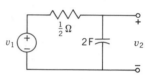

Fig. P8-70.

8-70. Consider the RC network of the figure with v_1 taken as the waveform of Prob. 8-4 with $K_0 = 1$. Make use of the convolution integral to determine $v_2(t)$.

8-71. Repeat Prob. 8-70 for the network of the accompanying figure, but with the same waveform for v_1.

8-72. Using the convolution integral, prove the translation (shifting) theorem of Eq. (8-11).

8-73. Convolve $f(t)$ with $u(t)$ to demonstrate that

$$\mathcal{L}\int_0^t f(t)\, dt = \frac{1}{s} F(s)$$

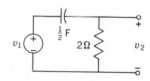

Fig. P8-71.

8-74. Demonstrate the validity of the entries in Table 8-1 described as "multiply by t" and "multiply by $e^{-\alpha t}$."

8-75. Example 9 of Section 8-6 gives the first step in the determination of the response of a network making use of approximate convolution. Carry out the summation for enough values of K so that the response $v_2(t)$ can be plotted.

8-76. Rework Example 9 of Section 8-6 and its continuation as described in Prob. 8-75 using an interval which is half of the interval used in Section 8-6.

9 Impedance Functions and Network Theorems

In this chapter, the operational method we have studied, by means of the Laplace transformation, is used to introduce the concepts of impedance and admittance. We also introduce a number of network theorems in terms of the operational representation of the sources and elements of a network.

9-1. THE CONCEPT OF COMPLEX FREQUENCY

The solution of the differential equations for networks has given rise to time-domain functions of the form

$$K_n e^{s_n t} \tag{9-1}$$

where s_n is a complex number, a root of the characteristic equation, expressed as

$$s_n = \sigma_n + j\omega_n \tag{9-2}$$

Here ω_n, the imaginary part of s_n, is interpreted as *radian frequency* and it appears in time-domain equations in the forms

$$\sin \omega_n t \quad \text{or} \quad \cos \omega_n t \tag{9-3}$$

Radian frequency has the dimensions of radians per second and may be expressed in terms of frequency f_n, in hertz, or in terms of the period T, in seconds, by the equation

$$\omega_n = 2\pi f_n = \frac{2\pi}{T} \tag{9-4}$$

242

By Eq. (9-2) we see that σ_n and ω_n must be identical in dimensions. The dimension of ω_n is $(time)^{-1}$, since the radian is a dimensionless quantity (being *length* of arc per *length* of radius). The dimension of σ_n must be "something" per unit time. Since σ_n appears as an exponential factor,

$$I = I_0 e^{\sigma_n t} \tag{9-5}$$

such that

$$\sigma_n = \frac{1}{t} \ln \frac{I}{I_0} \tag{9-6}$$

it is evident that the "something" per unit time should be a nondimensional logarithmic unit. The usual unit for the natural logarithm is the *neper*,[1] making the dimension for σ the *neper per second*.

The complex sum

$$s_n = \sigma_n + j\omega_n \tag{9-7}$$

is defined as the *complex frequency*. The imaginary part of the complex frequency is the *radian frequency* (or real frequency), and the real part of complex frequency is *neper frequency*[2] (rather than the misleading term "imaginary frequency"). The physical interpretation of complex frequency appearing in the exponential $e^{s_n t}$ will be studied by considering a number of special cases for the value of s_n.

(1) Let $s_n = \sigma_n + j0$ and let σ_n have positive, zero, and negative values. The exponential function of Eq. (9-1) becomes $K_n e^{\sigma_n t}$, an exponential function which increases exponentially for $\sigma_n > 0$ and decreases (or decays) exponentially for $\sigma_n < 0$. When $\sigma_n = 0$, so that $s_n = 0 + j0$, the term becomes

$$K_n e^{s_n t} = K_n e^{0t} = K_n \tag{9-8}$$

a time-invariant quantity which in terms of current and voltage is described as "direct current." The time variation for the three possibilities for real s_n are shown in Fig. 9-1.

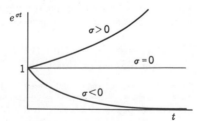

Fig. 9-1. Plots of e^{at} for positive, zero, and negative a. Here a is neper frequency.

[1] The word originates from *Neperus*, the Latin form of the name of Napier, the sixteenth century mathematician.

[2] Early users of the terms *radian frequency* and *neper frequency* were H. A. Wheeler and W. H. Huggins. The term *complex frequency* was used by early contributors to circuit theory, such as Heaviside (about 1900), Kennelly (1915), and Vannevar Bush (1917).

(2) Let $s_n = 0 \pm j\omega_n$ (radian frequency only). In this case, the exponential factor becomes

$$K_n e^{\pm j\omega_n t} = K_n(\cos \omega_n t \pm j \sin \omega_n t) \qquad (9\text{-}9)$$

by Euler's identity. The exponential $e^{\pm j\omega_n t}$ is usually interpreted in terms of the physical model (with no actual physical significance) of a unit rotating *phasor*,[3] the direction of rotation being determined by the sign. A positive sign, $e^{+j\omega_n t}$ implies counterclockwise (or positive) rotation, while a negative sign, $e^{-j\omega_n t}$ implies clockwise (or negative) rotation. For positive rotation, the real part of $e^{j\omega_n t}$ (or the projection on the real axis) varies as the cosine of $\omega_n t$, while the imaginary part (or projection on the imaginary axis) varies as the sine of $\omega_n t$. This concept is illustrated by Fig. 9-2. The variation of the exponential function with time is sinusoidal and corresponds to the case of the sinusoidal steady state.

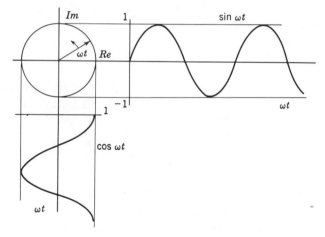

Fig. 9-2. A unit rotating phasor with its imaginary projection which is the sine and its real projection, the cosine.

(3) Let $s_n = \sigma_n + j\omega_n$ (this is the general case and the frequency is complex). For this case,

$$K_n e^{s_n t} = K_n e^{(\sigma_n + j\omega_n)t} = K_n e^{\sigma_n t} e^{j\omega_n t} \qquad (9\text{-}10)$$

This expression shows that such a term has a time variation which is the product of the result for $s_n = \sigma_n$ and for $s_n = \pm j\omega_n$. One term is represented by the rotating phasor model,

---

[3] The student may be more familiar with *vector* than with *phasor*. The phasor will be discussed at greater length in Chapter 11. For the present, a phasor is a complex plane representation characterized by magnitude and phase with respect to a reference angle.

the other term by an exponentially increasing or decreasing function. This result can be thought of as a rotating phasor with a magnitude which changes with time. Such phasors are illustrated in Fig. 9-3. The real and imaginary projections of this phasor are

$$\text{Re}(e^{s_n t}) = e^{\sigma_n t} \cos \omega_n t \tag{9-11}$$

and

$$\text{Im}(e^{s_n t}) = e^{\sigma_n t} \sin \omega_n t \tag{9-12}$$

These projections for both positive and negative σ_n are shown in Fig. 9-4. For $\sigma_n < 0$, the waveform is known as a *damped sinusoid;* for $\sigma_n > 0$, the oscillations are increasing exponentially.

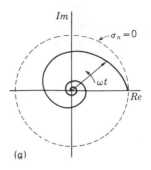

(a)

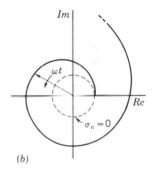

(b)

Fig. 9-3. (*a*) Exponentially decreasing rotating phasor, and (*b*) exponentially increasing rotating phasor.

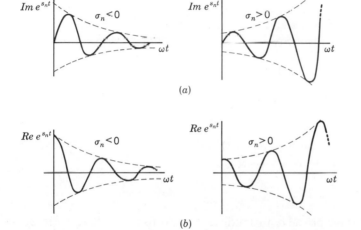

(a)

(b)

Fig. 9-4. The imaginary and real projections of the rotating phasors of Fig. 9-3. Each waveform is described by the complex frequency $s - a + j_w$.

From this discussion, we see that there is nothing really new in the concept of complex frequency. The imaginary part of complex frequency, the radian (or real) frequency, corresponds to oscillations. The real part of complex frequency, neper frequency, corresponds to exponential decay or exponential increase (depending on sign) or to no variation for zero neper frequency. We have talked about such exponential functions before in terms of the time constant. Since the role of the two "kinds" of frequency is the same, even though the consequences are different, we unify the two concepts under one name—complex frequency.

We should constantly guard against semantic difficulties in the use of the word "imaginary" as one part of a complex quantity. The imaginary part of a quantity is *not* physically imaginary (that is invisible or ghostlike) in the sense that it is not physically real. We have borrowed the words "real" and "imaginary" from the mathematicians as designations of two distinct parts of a quantity or function (which we often reinterpret in terms of magnitude and phase). The mathematician's "imaginary" carries no connotation about the physical universe about us!

9-2. TRANSFORM IMPEDANCE AND TRANSFORM CIRCUITS

We next determine network representations for each of the elements in terms of *transform impedance* (or *admittance*) and *initial condition sources*.

Resistance. The time-domain expression relating voltage and current for the resistor is given by Ohm's law in the forms

$$v_R(t) = Ri_R(t) \quad \text{or} \quad i_R(t) = Gv_R(t), \qquad G = \frac{1}{R} \qquad (9\text{-}13)$$

The corresponding transform equations are

$$V_R(s) = RI_R(s) \quad \text{or} \quad I_R(s) = GV_R(s) \qquad (9\text{-}14)$$

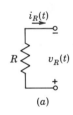

Now the quotient of $V_R(s)$ and $I_R(s)$ is defined as the transform impedance of the resistor; hence,

$$\frac{V_R(s)}{I_R(s)} = Z_R(s) = R \qquad (9\text{-}15)$$

The reciprocal of this ratio is the transform admittance for the resistor, which is

$$\frac{I_R(s)}{V_R(s)} = Y_R(s) = G \qquad (9\text{-}16)$$

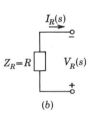

This result tells us that the resistor is frequency insensitive, even to complex frequency. As we shall see next, this is not the case for the other elements.

The familiar network representation of the voltage and current for the resistor is shown in Fig. 9-5(*a*). The corresponding transform representation is that of Fig. 9-5(*b*) in which reference directions for transform voltage and transform current are shown.

Fig. 9-5. The resistor and its transform impedance representation.

Inductance. The time-domain relationship between voltage and current in an inductor is expressed by the following equations.

$$v_L(t) = L\frac{di_L(t)}{dt}$$

and

$$i_L(t) = \frac{1}{L} \int_{-\infty}^{t} v_L(t)\, dt \qquad (9\text{-}17)$$

The equivalent transform equation for the voltage expression is

$$V_L(s) = L[sI_L(s) - i_L(0-)] \qquad (9\text{-}18a)$$

Regrouping the terms, we have

$$L_s I_L(s) = V_L(s) + Li_L(0-) \qquad (9\text{-}18b)$$

In this expression, $V_L(s)$ is the transform of the applied voltage, and $Li_L(0-)$ is the transform voltage resulting from the initial current in the inductor. Designating the transform voltage, which is the sum of the applied voltage and initial-current voltage, as $V_1(s)$, we see that the transform impedance for the inductor becomes

$$\frac{V_1(s)}{I_L(s)} = Z_L(s) = Ls \qquad (9\text{-}19)$$

The transform network representation of the inductor with an initial current is found from Eq. (9-18b), and is shown in Fig. 9-6(a). Observe that $V_1(s)$ is the transform voltage of the inductor of transform impedance Ls.

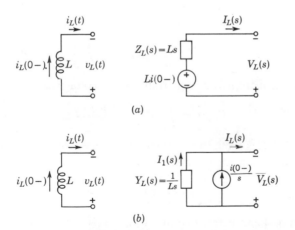

(a)

(b)

Fig. 9-6. The inductor with initial current and its transform representation in terms of (a) impedance, and (b) admittance.

The transform equation for the current expression of Eq. (9-17) is

$$I_L(s) = \frac{1}{L}\left[\frac{V_L(s)}{s} + \frac{v_L^{-1}(0-)}{s}\right] \qquad (9\text{-}20)$$

The initial-value integral $v_L^{-1}(0+)$ can be evaluated in terms of flux linkages Li as

$$v_L^{-1}(0-) = \int_{-\infty}^{t} v_L(t)\, dt \bigg|_{t=0-} = Li_L(0-) \qquad (9\text{-}21)$$

The equation for $I_L(s)$ may be rewritten

$$I_L(s) = \frac{1}{L}\frac{V_L(s)}{s} + \frac{i_L(0-)}{s} \tag{9-22}$$

or

$$\frac{1}{Ls}V_L(s) = I_L(s) - \frac{i_L(0-)}{s} \tag{9-23}$$

In this equation, $i_L(0-)/s$ is an equivalent transform current source resulting from the initial current in the inductor. Designating the transform current in $Y_L(s)$ as $I_1(s) = I_L(s) - i_L(0-)/s$, the transform admittance becomes

$$\frac{I_1(s)}{V_L(s)} = Y_L(s) = \frac{1}{Ls} \tag{9-24}$$

The equivalent transform diagram thus contains an admittance of value $1/Ls$ and an equivalent transform current source defined in Eq. (9-23). This equivalent schematic for the time domain diagram is shown in Fig. 9-6(b). We note that

$$Z_L(s) = \frac{1}{Y_L(s)} = Ls \tag{9-25}$$

Capacitance. The time-domain relationship between voltage and current for a capacitor is given as

$$i_C(t) = C\frac{dv_C(t)}{dt} \quad \text{and} \quad v_C(t) = \frac{1}{C}\int_{-\infty}^{t} i_C(t)\, dt \tag{9-26}$$

The equivalent transform equation for the voltage expression is

$$V_C(s) = \frac{1}{C}\left[\frac{I_C(s)}{s} + \frac{q(0-)}{s}\right] \tag{9-27}$$

where $q(0-)/C$ is the initial voltage of the capacitor which, due to the reference direction for voltage is $-V_0$. This equation may be written

$$\frac{1}{Cs}I_C(s) = V_C(s) + \frac{V_0}{s} \tag{9-28}$$

Designating the transform voltage of $Z_C(s)$ as $V_1(s) = V_C(s) + (V_0/s)$, the ratio of the transform voltage to the transform current is

$$\frac{V_1(s)}{I_C(s)} = Z_C(s) = \frac{1}{Cs} \tag{9-29}$$

The capacitor with an initial charge thus has an equivalent transform diagram with an impedance $1/Cs$ in series with a voltage source having a transform $-v_C(0-)/s$. The schematic of this combination is shown in Fig. 9-7(a).

The transform equation for the current expression of Eq. (9-26) is

$$I_C(s) = C[sV_C(s) - v_C(0-)] \tag{9-30}$$

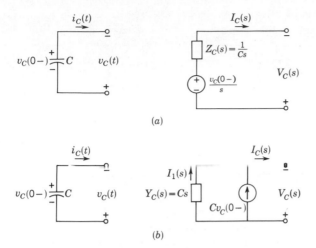

Fig. 9-7. The capacitor with initial voltage and its transform representation in terms of (a) impedance and (b) admittance.

or

$$C_s V_C(s) = I_C(s) - CV_0 \qquad (9\text{-}31)$$

Designating the transform current in $Y_C(s)$ as $I_1(s) = I_C(s) - CV_0$, the ratio of transform current to transform voltage becomes

$$\frac{I_1(s)}{V_C(s)} = Y_C(s) = Cs \qquad (9\text{-}32)$$

The capacitor with an initial charge has an equivalent transform schematic representation of an admittance of value Cs in parallel with a transform current source of value CV_0. This schematic is shown in Fig. 9-7(b). For the capacitor,

$$Z_C(s) = \frac{1}{Y_C(s)} = \frac{1}{Cs} \qquad (9\text{-}33)$$

The pattern of representing network by transform networks which we have illustrated for the three passive elements may be extended to more complicated networks. For example, the controlled source which was discussed in Chapter 2 is described by the equation

$$v_2(t) = \mu v_1(t) \qquad (9\text{-}34)$$

The corresponding transform equation is

$$V_2(s) = \mu V_1(s) \qquad (9\text{-}35)$$

The two networks corresponding to these two equations, Eqs. (9-34) and (9-35), are shown in Fig. 9-8.

Another example is provided by the transformer, first studied in Chapter 2, with reference polarity marks as shown in Fig. 9-9(a). The voltages and currents of the transformer are related by the Kirchhoff

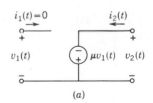

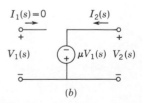

Fig. 9-8. (a) A voltage-controlled voltage source and (b) its transform representation.

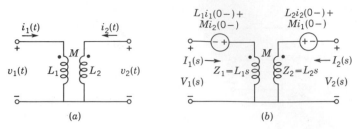

(a)   (b)

Fig. 9-9. (a) A transformer and (b) its transform representation with initial condition voltage generators.

voltage law equations

$$v_1(t) = L_1 \frac{di_1(t)}{dt} = M \frac{di_2(t)}{dt} \tag{9-36}$$

$$v_2(t) = M \frac{di_1(t)}{dt} + L_2 \frac{di_2(t)}{dt} \tag{9-37}$$

The corresponding transform equations are

$$V_1(s) = L_1 s I_1(s) - L_1 i_1(0-) + M s I_2(s) - M i_2(0-) \tag{9-38}$$

$$V_2(s) = M s I_1(s) - M i_1(0-) + L_2 s I_2(s) - L_2 i_2(0-) \tag{9-39}$$

and the transform network representation is that of Fig. 9-9(b) in which the initial conditions in the network are represented by two transform voltage sources.

From the examples of this section, we may state a number of conclusions:

(1) For single elements, the *transform impedance* is defined as the ratio of the transform of the element voltage to the transform of the element current, $V(s)/I(s)$, for zero initial current in an inductor and zero initial voltage in a capacitor.

(2) The reciprocal ratio, with the same initial condition restrictions, is the *transform admittance*.

(3) In transform networks, initial conditions are represented by transform voltage or current sources.

We next extend this discussion from the single element case to more complicated networks.

9-3. SERIES AND PARALLEL COMBINATIONS OF ELEMENTS

Consider the series combination of elements shown in Fig. 9-10(a). Assume that the initial current in all inductors is zero and that the initial voltage of all capacitors is also zero. By Kirchhoff's voltage law, the summation of the voltages of the elements is equal to the

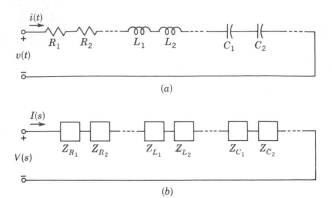

(a)

(b)

Fig. 9-10. (a) Series connected resistors, inductors, and capacitors and (b) corresponding transform representation.

applied voltage, $v(t)$. The transform equation stating this result is

$$V(s) = V_{R_1}(s) + \ldots + V_{L_1}(s) + \ldots + V_{C_1}(s) + \ldots \quad (9\text{-}40)$$

Dividing this equation by $I(s)$ and recognizing that the ratio of the voltage of each element divided by the current for that element is impedance, we have

$$Z(s) = Z_{R_1}(s) + \ldots + Z_{L_1}(s) + \ldots + Z_{C_1}(s) + \ldots \quad (9\text{-}41)$$

or

$$Z(s) = \sum_{k=1}^{n} Z_k(s) \quad (9\text{-}42)$$

for a series combination of elements, where n is the total number of elements in series of all kinds.

The transform impedance network of Fig. 9-10(b) represents Eqs. (9-41) or (9-42) and confirms for the series network our intuitive expectation that each element in a network may be replaced by its transform impedance in determining the transform network.

Of course it should be recognized that in performing such a summation, the elements are *not* being combined. Rather, only a characteristic feature of the element (its impedance) is being summed and added to a characteristic of another element.

Consider next the parallel combination of elements shown in Fig. 9-11(a). As before, we will assume that all initial conditions are zero. For the parallel network, the voltage $v(t)$ is common for all elements. From Kirchhoff's current law, the sum of the currents in the elements is equal to the total current supplied to the network; that is,

$$i(t) = i_{G_1}(t) + \ldots + i_{L_1}(t) + \ldots + i_{C_1}(t) + \ldots \quad (9\text{-}43)$$

and the corresponding transform equation is

$$I(s) = I_{G_1}(s) + \ldots + I_{L_1}(s) + \ldots + I_{C_1}(s) + \ldots \quad (9\text{-}44)$$

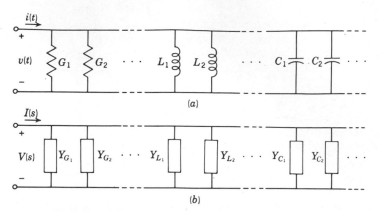

Fig. 9-11. (*a*) A parallel connection of resistors, inductors, and capacitors and (*b*) the corresponding admittance representation.

If this equation is divided by $V(s)$ and it is recognized that the ratio of the current transform to the voltage transform is transform admittance, there results

$$Y(s) = Y_{G_1}(s) + \ldots + Y_{L_1}(s) + \ldots + Y_{C_1}(s) + \ldots \quad (9\text{-}45)$$

or

$$Y(s) = \sum_{k=1}^{n} Y_k(s) \quad (9\text{-}46)$$

for a parallel combination of elements, where n is the total number of all kinds of elements in parallel.

For a series-parallel network, rules for the combination of impedance and of admittance can be used successively to reduce a network to a single equivalent impedance or admittance. This procedure will be illustrated with a number of examples.

EXAMPLE 1

In the series circuit shown in Fig. 9-12, the switch K is held in position a until such a time that a current I_0 flows in the inductor and the capacitor is charged to voltage V_0. At that instant, the switch is thrown to position b, connecting the circuit to a voltage source $v(t)$. The problem is to find $I(s)$ and so $i(t)$. An equivalent circuit diagram

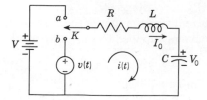

Fig. 9-12. *RLC* network considered in Example 1.

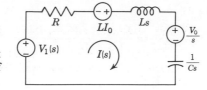

Fig. 9-13. The transform network derived from the network of Fig. 9-12.

marked with transform impedances is shown in Fig. 9-13. In this revised form, the current $I(s)$, a transform current, may be found by Ohm's law for transform networks. The current $I(s)$ is given as the total transform voltage in the network divided by the total transform impedance. Then

$$I(s) = \frac{V(s)}{Z(s)} = \frac{V_1(s) + LI_0 - V_0/s}{R + Ls + 1/Cs} = \frac{sV_1(s) + LI_0 s - V_0}{Ls^2 + Rs + 1/C} \quad (9\text{-}47)$$

This transform equation can be expanded by partial fractions to find the corresponding $i(t)$ by the inverse Laplace transformation. This solution has been found *without* writing the differential equation of the system, and automatically incorporates the required initial conditions.

EXAMPLE 2

The dual of the network of Example 1 is shown in Fig. 9-14. In this network, the switch K_1 is opened at an instant when the inductor current is I_0 and the capacitor is charged to V_0. At the same instant, $t = 0$, the switch K_2 is closed. It is required to find the transform of the node voltage $V(s)$ so that $v(t)$ can be determined. From the equivalent admittance diagram shown in Fig. 9-15, the transform voltage $V(s)$ is

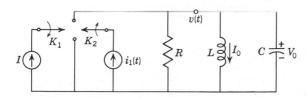

Fig. 9-14. Network of Example 2.

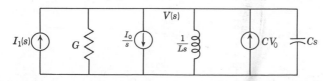

Fig. 9-15. Transform network derived from the network of Fig. 9-14.

found as

$$V(s) = \frac{I(s)}{Y(s)} = \frac{I_1(s) + CV_0 - I_0/s}{Cs + G + 1/Ls} = \frac{sI_1(s) + CV_0s - I_0}{Cs^2 + Gs + 1/L} \qquad (9\text{-}48)$$

This transform is the dual of the transform of Eq. (9-47) (and could therefore have been written by inspection). The corresponding time-domain voltage, $v(t)$ can be found from the inverse Laplace transformation after the above transform has been expanded by partial fractions.

EXAMPLE 3

In this example, we make use of the laws for the series combination of impedance and the parallel combination of admittance to determine current. In the network shown in Fig. 9-16, it will be assumed that the network is initially relaxed (no current, no charge) and that the switch was closed at $t = 0$. We are required to find the current in the generator $i(t)$ by finding the transform of this current $I(s)$. The impedance of the branch containing the 1-Ω resistor and 2-H inductor is, from Fig. 9-17,

$$Z(s) = 1 + 2s \qquad (9\text{-}49)$$

This impedance is in parallel with the impedance $2/s$ of the capacitor. The admittances may be added directly. Thus,

$$Y_{ab}(s) = Y_C + Y_{RL} = \frac{s}{2} + \frac{1}{2s + 1} = \frac{2s^2 + s + 2}{2(2s + 1)} \qquad (9\text{-}50)$$

The impedance from a to b is the reciprocal of the admittance; thus,

$$Z_{ab}(s) = \frac{1}{Y_{ab}(s)} = \frac{2(2s + 1)}{2s^2 + s + 2} \qquad (9\text{-}51)$$

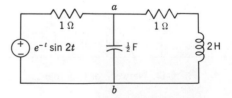

Fig. 9-16. Network of Example 3.

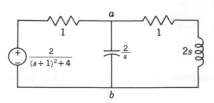

Fig. 9-17. Transform network relating to Example 3.

The total impedance is now found by adding to $Z_{ab}(s)$ the impedance of the 1-Ω resistor. Then the total impedance is

$$Z_{\text{total}}(s) = 1 + \frac{2(2s+1)}{2s^2+s+2} = \frac{2s^2+5s+4}{2s^2+s+2} \qquad (9\text{-}52)$$

This total impedance in series with the transform of the voltage source is shown in Fig. 9-18. The current may now be found; thus,

$$I(s) = \frac{V(s)}{Z(s)} = \frac{2(2s^2+s+2)}{[(s+1)^2+4](2s^2+5s+4)} \qquad (9\text{-}53)$$

Fig. 9-18. Simplified transform network obtained from Fig. 9-17 by combining impedances and admittances.

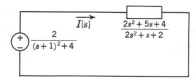

If the inductor has a current flowing through it at $t = 0$ or if the capacitor is charged at $t = 0$, the problem is somewhat more complicated, since several voltage sources are involved.

9-4. SUPERPOSITION AND RECIPROCITY

We next generalize the results we obtained in the last section for the series-parallel networks to the general network. This problem was formulated in Chapter 3 in terms of integrodifferential equations using the a- and b-coefficients where, for example, the coefficient a_{jk} was the operator

$$a_{jk} = R_{jk} + L_{jk}\frac{d}{dt} + D_{jk}\int dt \qquad (9\text{-}54)$$

In the Laplace transform equivalents of Eq. (3-47), the a-coefficients become $Z(s)$ plus initial-condition sources, and the b-coefficients similarly become admittance terms $Y(s)$ and initial-condition sources. Thus the transform form of Eq. (3-47) may be written compactly in the matrix equation

$$[Z][I] = [V] + [V_0] = [V'] \qquad (9\text{-}55)$$

where the impedance matrix is

$$[Z] = \begin{bmatrix} Z_{11}(s) & Z_{12}(s) & \cdots & Z_{1L}(s) \\ Z_{21}(s) & Z_{22}(s) & \cdots & Z_{2L}(s) \\ \cdot & \cdot & \cdots & \cdot \\ Z_{L1}(s) & Z_{L2}(s) & \cdots & Z_{LL}(s) \end{bmatrix} \qquad (9\text{-}56)$$

and $[I]$ and $[V']$ are column matrices in which the entries are $I_1, I_2, \ldots, I_L$ and $V_1 + V_{01}, V_2 + V_{02}, \ldots, V_L + V_{0L}$. The node-basis equations are similarly

$$[Y][V] = [I] + [I_0] = [I']$$
(9-57)

in which the various matrices have a dual interpretation to those provided for Eq. (9-55). The solution to Eq. (9-55), or Eq. (9-57), may be found by Cramer's rule (or by matrix inversion). If we let Δ_z be the determinant of $[Z]$, then for $L = 3$, the transform current which is equivalent to Eq. (3-73) is

$$I_2(s) = \frac{1}{\Delta_z}[\Delta_{12}V'_1(s) + \Delta_{22}V'_2(s) + \Delta_{32}V'_3(s)]$$
(9-58)

which, though a special case, has a representative form like any other solution. Let us note that the voltage transform $V'_j(s) = V_j(s) + V_{0j}(s)$ is a summation of voltages, $V_j(s)$ being the transform of all sources in loop j, and $V_{0j}(s)$ being the transform voltage representation of all initial conditions in loop j. This summation, of course, must be made by taking into account the polarity of the various sources with respect to the reference direction of the loop. We have assumed that any current sources in the network were transformed into equivalent voltage sources before the equations were formulated.

From Eq. (9-58), we observe that we may consider each of the transform sources one at a time and then add the partial responses so found to determine $I_2(s)$. This is, in essence, the *superposition principle*.

Two aspects of superposition are important in network analysis. A given response in a network resulting from a number of independent sources (including initial-condition sources) may be computed by summing the response to each individual source with all other sources made inoperative (reduced to zero voltage or zero current). This statement describes the *additivity* property of linear networks. Furthermore, if all sources are multiplied by a constant, the response is multiplied by the same constant. This statement describes the property of *homogeneity* in linear networks. Both properties are evident from Eq. (9-58). Thus, superposition is the combined properties of additivity and homogeneity of linear networks.

We next consider two networks to illustrate the principle of superposition, especially in networks with initial conditions.

EXAMPLE 4

In the network of Fig. 9-19, the source is $v_1(t)$ and the response is considered to be the current $i(t)$. The initial capacitor voltage is $v_C(0)$. For this simple example, we see that the transform of the current is

$$I(s) = \frac{V_1(s) - v_C(0)/s}{R + 1/Cs}$$
(9-59)

The numerator of this equation illustrates our statement that the

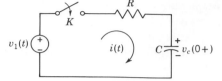

Fig. 9-19. Network of Example 4 used to illustrate the superposition property.

response is due to the superposition of the responses due to each of the sources in the network of Fig. 9-19. Let $V_1(s) = V_0/s$ and $v_C(0) = V_0$. This is an interesting case for which the responses due to the two sources are equal and opposite so that the superimposed response is $i(t) = 0$. If, instead, $V_1(s) = 2V_0/s$ and $v_C(0) = V_0$, then the superimposed responses add to

$$i(t) = \frac{V_0}{R} e^{-t/RC} \qquad (9\text{-}60)$$

With respect to the homogeneity aspect of superposition, observe that if both $V_1(s)$ and $v_C(0)$ are multiplied by K, then $i(t)$ is also multiplied by K.

EXAMPLE 5

Consider the network of Fig. 9-20 for which we impose the further requirement that the initial currents in the two inductors be identical and also that the initial charges on the two capacitors be identical. For this network, the response is $v_2(t)$ and the excitation is due to the current source $i_1(t)$ and initial conditions in the inductors and capacitors. In the application of the superposition principle to this network, the responses due to initial conditions will always cancel, and the response will depend only on the excitation from $i_1(t)$. This is an example of a network in which the initial conditions are *not observable* at the terminals 2-2'. Of course, this conclusion does not apply to, say, the current in the inductors.

In network analysis, we sometimes find it useful to make statements about networks for which the initial conditions (currents in inductors, voltages of capacitors) are all zero. Such networks are described as *initially relaxed*, or in the *zero state*. For such networks, the response depends only on the applied excitation. Example 5 shows

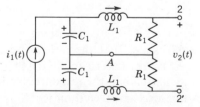

Fig. 9-20. Network of Example 5 in which identical initial currents in the two inductors marked L_1 are not observable at port 2.

that a zero state is sufficient for this condition to exist, but it is not necessary!

Another important property of linear networks is found from a generalization of Eq. (9-58); the current in loop k is

$$I_k = \frac{\Delta_{1k}}{\Delta} V_1 + \frac{\Delta_{2k}}{\Delta} V_2 + \ldots + \frac{\Delta_{LK}}{\Delta} V_L \qquad (9\text{-}61)$$

If all V are zero except V_j, this equation simplifies to the form

$$I_k = \frac{\Delta_{jk}}{\Delta} V_j \qquad (9\text{-}62)$$

Similarly, if V_k is the only source in the network, the current in loop j is

$$I_j = \frac{\Delta_{kj}}{\Delta} V_k \qquad (9\text{-}63)$$

Networks for which these equations apply are shown in Fig. 9-21 in which loops j and k are identified.

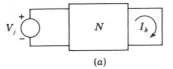

(a)

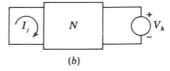

(b)

Fig. 9-21. Networks from which the principle of reciprocity is described.

In terms of these two networks, we make the following observations. In network (a) of the figure, V_j produces current I_k. If V_j is moved to loop k so that $V_k = V_j$, as in (b) of the figure, what will be the value of I_j under this condition? If $[Z]$ is symmetric, the cofactors of Eqs. (9-62) and (9-63) are equal,

$$\Delta_{jk} = \Delta_{kj} \qquad (9\text{-}64)$$

and we see that I_j in Fig. 9-21(b) will be equal to I_k in Fig. 9-21(a). In summary,

$$\frac{I_k}{V_j} = \frac{I_j}{V_k} \qquad (9\text{-}65)$$

The *principle of reciprocity*, which is a word statement equivalent to Eq. (9-65), states that the ratio of response transform to excitation transform is invariant to an interchange of the position in the network of the excitation and the response. Networks for which this condition holds are said to be *reciprocal*.

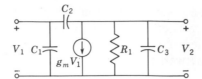

Fig. 9-22. A nonreciprocal network if $g_m \neq 0$.

Under what conditions will the reciprocity principle apply to a network? How might we distinguish a reciprocal network? We have assumed that there is only one source of excitation in the network. For this to be the case, it is necessary that there be no initial-condition transform sources, meaning that the network must be initially relaxed or in the zero state. Equation (9-61) is written on the basis of a network with linear elements only. For $[Z]$ to be symmetric, it is necessary that we admit only R, L, C and transformers as elements. We must exclude dependent (or controlled) sources, even if they are linear, for $[Z]$ to be symmetric. Thus the network shown in Fig. 9-22, which is a model for a transistor, is *nonreciprocal*.

9-5. THÉVENIN'S THEOREM AND NORTON'S THEOREM

When interest is focused on one part of a network under analysis, the remainder of the network may be replaced to advantage by a simple equivalent network determined by using Thévenin's theorem or its dual, Norton's theorem.[4] Thévenin's theorem is especially useful in such applications as determining the load for an electronic circuit which will result in maximum average power delivery to that load.

The two parts of the network of interest are distinguished as network A and network B in Fig. 9-23. Network A is to be replaced by an equivalent network under the condition that the current i and voltage v identified in the figure remain invariant when the replacement

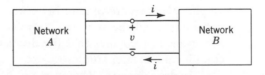

Fig. 9-23. Two networks in terms of which Thévenin's theorem is described.

[4]This theorem was first proposed in 1883 by the French telegraph engineer, Léon Thévenin (1857–1926), in the French scientific journal *Annales Télégraphiques*. An interesting account of the development of the theorem is given by Charles Suchet in *Electrical Engineering*, vol. 68 (October, 1949), 843–844. An earlier statement in the form of this theorem is credited to Helmholtz, and the theorem is sometimes called the Helmholtz-Thévenin theorem. Regarding the earlier work of Helmholtz, see the letters by H. F. Mayer and E. T. Gross in "Léon Charles Thévenin," *Electrical Engineering*, vol. 69 (February, 1950), 186–187. The dual of the theorem is due to E. L. Norton (1898–   ) of the Bell Telephone Laboratories.

is made. Network B is called the *load* which may be a single resistor or a network of greater complexity, perhaps one representing an antenna or a diode. We will make certain assumptions concerning the two networks, and these are summarized in Table 9-1.

Table 9-1.

Network	*Characteristics*
A	Linear elements Voltage and current sources may be independent or dependent (controlled) Initial conditions on passive elements No magnetic or controlled-source coupling to network B
B	Any kind of elements: linear, nonlinear, time-varying Any sources may be present Initial conditions on passive elements No magnetic or controlled-source coupling to network A

We assume that network A contains linear elements which may have initial conditions. It also contains sources, which may be independent or controlled. We shall assume that there is no magnetic coupling to network B, nor does a controlled source in A couple to B. The most significant difference in the assumptions for network B is that the elements need not be linear; they may be nonlinear or time-varying or both. There may be sources present in network B, again of the independent or controlled types, but there is no magnetic or controlled-source coupling to network A. Note that if nonlinear or time-varying elements are present in network B, techniques which we have not studied will be required to solve the network equations. It is important to recognize that the discussion does extend to these two cases for use in later studies.

In describing the Thévenin and Norton equivalent networks, we will have need for a network which is labelled network C in Fig. 9-24. Network C is derived from network A as follows:

(1) All initial conditions are set equal to zero. For all capacitors, $v_C = 0$ and for all inductors $i_L = 0$.

(2) All independent sources are turned off; that is, $v = 0$ (a short circuit) for voltage sources and $i = 0$ (an open circuit) for current sources.

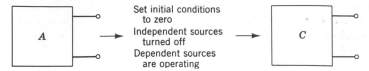

Fig. 9-24. Network A becomes Network C under the conditions specified in the figure. Independent sources are turned off by making $v = 0$ for voltage sources and $i = 0$ for current sources.

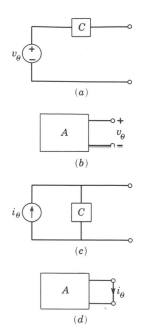

Fig. 9-25. (*a*) The general Thévenin equivalent network, (*b*) and the condition under which v_θ is determined. (*c*) The general Norton equivalent network, (*d*) and the condition under which i_θ is determined.

(3) Controlled sources continue to operate, as distinguished from the independent sources of statement (2).

(4) Under these conditions, we sometimes measure the driving-point impedance or admittance at the input terminals and designate these as Z_θ and Y_θ.

The *Thévenin equivalent network* is that shown in Fig. 9-25(*a*). The voltage source v_θ is the voltage at the open terminals of network A, with network B removed, as shown in Fig. 9-25(*b*). The voltage source v_θ is connected in series with network C.

The *Norton equivalent network* is the dual of the Thévenin equivalent network as is shown in Fig. 9-25(*c*). The current source i_θ has a value equal to the current in the shorted terminals of network A as illustrated in (*d*) of Fig. 9-25. The current source is connected in parallel with network C to form the Norton equivalent network.

Let us consider three different situations to illustrate the determination of the Thévenin equivalent network. In some sense, the situations will represent classes of problems to which the Thévenin theorem will be applied which differ in complexity and in objective.

Case 1. Consider the case in which both networks A and B contain only resistors and sources, both independent and controlled. In this case, we will be concerned with algebraic equations, not differential equations.

The network shown in Fig. 9-26(*a*) is such a network containing only one independent source. We first removed R_4 with the objective of finding the Thévenin equivalent for the remainder of the network. The open-circuit voltage is v_2, and this may be found by solving the following set of equations

$$(G_1 + G_2)v_1 - G_2 v_2 = i_s \qquad (9\text{-}66)$$

$$-G_2 v_1 + (G_2 + G_3)v_2 = 0 \qquad (9\text{-}67)$$

Substituting the given numerical values into these equations and solving them, we find that $v_2 = 3i_s$. The Thévenin equivalent resistance is found by open circuiting the current source; then

$$R_\theta = \frac{(R_1 + R_2)R_3}{R_1 + R_2 + R_3} = \frac{2}{5}\,\text{ohm} \qquad (9\text{-}68)$$

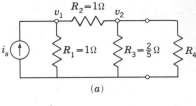

(a)

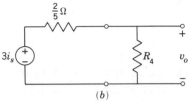

(b)

Fig. 9-26. The network used to illustrate Case 1.

Thus the Thévenin equivalent network has been determined and is shown in Fig. 9-26(b). From this network, we may determine the voltage across R_4, $v_o(t)$, as

$$v_o(t) = \frac{3i_s R_4}{2/5 + R_4} V \tag{9-69}$$

Consider next another network meeting the assumptions for this Case 1 but containing a voltage-controlled voltage source. With the load R_L removed, the open-circuit voltage v_θ may be determined. Since there is no current in R_1 and R_2, the open-circuit voltage is

$$v_\theta = v_a = \beta(v_1 - v_a) \tag{9-70}$$

Solving for v_a which is then equated to v_θ, we have

$$V_\theta = \frac{\beta}{1 + \beta} v_1 \tag{9-71}$$

which is the required voltage. To find the form of network C or, actually, a simplified version of it, we first let $v_1 = 0$. The result is then the network shown in Fig. 9-27. To simplify this further, we have connected a voltage source v to the output terminals, and we will solve for the current i produced by this source. For the one-loop network,

$$v = iR_2 + iR_1 + \beta(0 - v_a) \tag{9-72}$$

To eliminate v_a from this equation, we note that

$$v_a = iR_2 - \beta v_a \tag{9-73}$$

Hence

$$v = i\left(R_1 + R_2 - \frac{\beta}{1 + \beta} R_2\right)$$

$$= i\left[R_1 + R_2\left(1 - \frac{\beta}{1 + \beta}\right)\right] \tag{9-74}$$

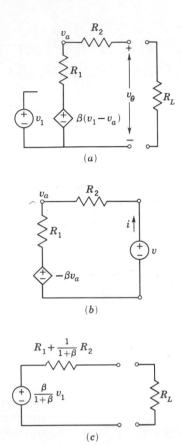

(a)

(b)

Fig. 9-27. A network containing a controlled source which is used as the second example for Case 1.

(c)

Now the ratio v/i is the resistance of the network, R_θ, which may represent network C. Hence

$$R_\theta = R_1 + \left(\frac{1}{1+\beta}\right) R_2 \tag{9-75}$$

is the equivalent resistance of network C. These results are shown in Fig. 9-27(c) which shows the Thévenin equivalent network.

Case 2. Figure 9-28(a) shows an *RLC* network with a voltage source $v(t) = V_0 e^{-t} u(t)$ and zero initial conditions for the passive elements. Let us first compute the equivalent voltage source v_θ which is identified on the network. This voltage is that across the resistor since with the terminals open there is no current in the capacitor. This is a simple *RL* network for which the current may be found by techniques discussed in Chapters 6 or 7 and v_R found by multiplying i by R. The result is

$$v_\theta(t) = V_0 \frac{R}{L-R} (e^{-Rt/L} - e^{-t}) \tag{9-76}$$

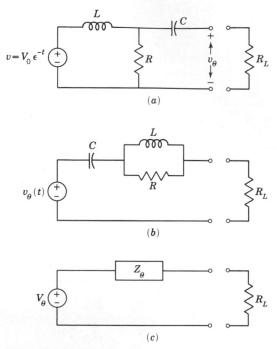

Fig. 9-28. An *RLC* network used to illustrate Case 2.

for $t \geq 0$. To find network C, we short the voltage source giving us a network with C in series with the parallel combination of L and R. This result is shown in Fig. 9-28(*b*) with v_θ defined by Eq. (9-76). The result does not appear to be simpler than the original network, but it is equivalent with respect to the voltage and current of R_L.

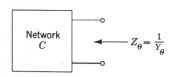

Fig. 9-29. Network *C* may be characterized by the impedance Z_θ which is the reciprocal of Y_θ.

Case 3. For this case, we will study the network used for Case 2, but now use transform quantities for voltage, current and impedance. For the transform network, the impedance $Z_\theta = 1/Y_\theta$ may be used to represent network C. This use is similar to that of Case 1 in which a network was represented by its simplified equivalent, a single resistor. Here we represent the network by its transform impedance. This usage is illustrated by Fig. 9-29. To compute the voltage $V_\theta(s)$, we recognize that V_R in Fig. 9-28(*a*) may be found by considering L and R in series as a voltage divider; thus

$$V_\theta(s) = \frac{R}{Ls + R} V_1 = \frac{V_0 R/L}{[s + (R/L)](s + 1)} \tag{9-77}$$

is the transform voltage of the network represented in Fig. 9-28(*c*). To find Z_θ, we compute the transform impedance with source V_1

replaced by a short circuit. Thus

$$Z_\theta = \frac{1}{Cs} + \frac{1}{(1/R) + (1/Ls)}$$

$$= R\frac{s^2 + (1/RC)s + (1/LC)}{s[s + (R/L)]} \qquad (9\text{-}78)$$

and the transform Thévenin equivalent network has been determined.

Having illustrated the use of Thévenin's theorem, there remains the task of demonstrating its validity and describing the reasons for the restrictions under which the theorem may be used. The proof will be carried out for a transform network, the most general of the three cases we have considered. The steps in the derivation of the theorem are carried out in terms of the descriptions shown in Fig. 9-30. In (a) of this figure, we see the network of Fig. 9-23 repeated with the current, whose transform is $I(s)$, identified. In (b), a voltage source of transform

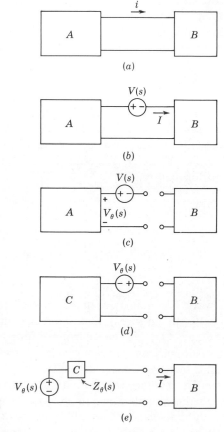

Fig. 9-30. Illustrating steps in the derivation of Thévenin's theorem.

$V(s)$ is inserted where $V(s)$ is chosen so that it will produce current $I_1(s)$ such that $I_1(s) = -I(s)$ and the net current from network A to network B is zero. With zero current, the two networks can be broken apart as shown in Fig. 9-30(c) without affecting conditions in network A of (b). Now the voltage at the broken terminals will be zero because the current is zero; network B is acting as a short and so can be shorted without affecting conditions in network A. Let the voltage at the terminals of network A, with these terminals open as in (c) of the figure, have the transform value $V_\theta(s)$. By Kirchhoff's voltage law, $V_\theta(s) - V(s) = 0$, meaning that $V_\theta(s) = V(s)$. In other words, $V(s)$ must have the value of the voltage transform at the open-circuit terminals of the network A for the zero current condition of (c) in Fig. 9-30 to apply.

We next show that the network of (d) in Fig. 9-30 is equivalent to that shown in (a). We do this by observing that two changes have been made in going from (c) to (d). First, all *independent* sources in network A are reduced to zero by opening current sources and shorting voltage sources. Dependent sources are not changed, however. With this accomplished, network A is transformed to network C, having the properties summarized in Table 9-1. Next, the polarity of the voltage source $V_\theta(s)$ is reversed from that given in (c) such that now $I(s) = +I_1(s)$. When network B is connected in Fig. 9-30(d), the current will be $I(s)$, as it was in the network of (a). Thus, as far as operations in network B are concerned, the network of (d) is equivalent to that of (a). In (e) of the figure, the network is rearranged with the voltage source in its traditional position, and the impedance of network C is identified as $Z_\theta(s)$. In terms of this network, we observe that the transform current is

$$I(s) = \frac{V_\theta(s)}{Z_\theta(s) + Z_B(s)} \tag{9-79}$$

The impedance of network C, which we have identified as $Z_\theta(s)$, requires additional comment. If network A contains only independent sources, then $Z_\theta(s)$ is the impedance of the passive network, network C, computed at the open terminals. However, in the case in which network A contains dependent sources (as is often the case in models of solid-state devices), then the dependent sources are not reduced to zero, and $Z_\theta(s)$ must be determined for an active network. This presents no computational problems as is illustrated by an example at the end of this section.

The source transformation studied in Section 3-3, and summarized for single elements in Figs. 3-16, 3-17, and 3-18, generalized for the impedance $Z_\theta(s)$ of a network, is used next to convert the voltage source representation to a current source representation. This is accomplished as shown in Fig. 9-31 to give a current source of value

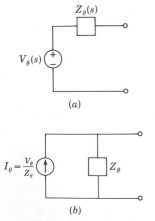

Fig. 9-31. General form of the (a) Thévenin's theorem network, and (b) Norton's theorem network.

$I_\theta(s) = V_\theta(s)/Z_\theta(s)$ in parallel with the impedance $Z_\theta(s)$. This is the Norton equivalent network. With this network replacing that shown in Fig. 9-30(e), we observe that, with $Y_\theta(s) = 1/Z_\theta(s)$, and $Y_B(s) = 1/Z_B(s)$, the voltage of network B is

$$V_B(s) = \frac{I_\theta(s)}{Y_\theta(s) + Y_B(s)} \tag{9-80}$$

Observe that the $I_\theta(s)$ is the transform current between the two terminals of network A when the terminals are shorted together.

Our results may be summarized in the following statement. Let A and B be two networks connected by two conductors, but not magnetically or controlled-source coupled. Network A may be replaced by either of two equivalent networks for computations with respect to network B. The *Thévenin equivalent network* consists of a voltage source in series with a network of impedance $Z_\theta(s)$. The voltage source $V_\theta(s)$ is the transform of the voltage at the open terminals of network A. $Z_\theta(s)$ is the transform impedance at the two terminals of A with all independent sources reduced to zero. The *Norton equivalent network* consists of a current source $I_\theta(s)$ in parallel with a network of impedance $Z_\theta(s)$. The impedance is the same as that found for the Thévenin equivalent, and $I_\theta(s)$ is the transform of the current between the two terminals of A when these terminals are short-circuited.

We conclude our discussion of the Thévenin and Norton theorems with three additional examples.

EXAMPLE 6

The network of Fig. 9-32 is in the zero state until $t = 0$ when the switch is closed. It is required to find the current $i_1(t)$ in the resistor R_3. Figure 9-33 shows the same network in terms of transform impedances, with the Thévenin equivalent network to be found as shown. From the network, we see the transform voltage at the open terminals is

$$V_\theta(s) = \frac{10(100/s)}{10 + s + 10} = \frac{1000}{s(s + 20)} \tag{9-81}$$

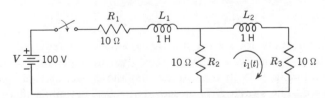

Fig. 9-32. Network for Example 6.

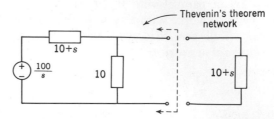

Fig. 9-33. Transform network derived from the network of Fig. 9-26. The dashed line and arrows indicate the portion of the network to which Thévenin's theorem is to be applied.

The impedance of the network with the voltage source short-circuited is

$$Z_\theta(s) = \frac{10(s + 10)}{s + 20} \tag{9-82}$$

The current transform, by Eq. (9-75), is

$$I_1(s) = \frac{V_\theta(s)}{Z_\theta(s) + Z_B(s)} = \frac{1000/s(s + 20)}{10(s + 10)/(s + 20) + (s + 10)} \tag{9-83}$$

which simplifies to

$$I_1(s) = \frac{1000}{s(s^2 + 40s + 300)} \tag{9-84}$$

This equation may be expanded by partial fractions as

$$\frac{1000}{s(s^2 + 40s + 300)} = \frac{K_1}{s} + \frac{K_2}{(s + 10)} + \frac{K_3}{(s + 30)} \tag{9-85}$$

With K_1, K_2, and K_3 evaluated, the current transform becomes

$$I_1(s) = \frac{3.33}{s} + \frac{-5}{s + 10} + \frac{1.67}{s + 30} \tag{9-86}$$

The time-domain current $i_1(t)$ is found by the inverse Laplace transformation as

$$i_1(t) = 3.33 - 5e^{-10t} + 1.67e^{-30t} \tag{9-87}$$

As a check, we note that this equation reduces to the correct values for initial and final conditions.

EXAMPLE 7

In the network shown in Fig. 9-34, we are required to find the current in the resistor R_2. The equivalent impedance schematic is shown in Fig. 9-35. It is assumed that the capacitor C_2 is initially uncharged. The switch K is closed at $t = 0$. Thévenin's theorem is

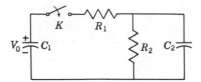

Fig. 9-34. Network for Example 7.

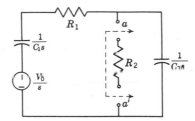

Fig. 9-35. Transform network derived from the network of Fig. 9-34. Thévenin's theorem is applied to the network excluding R_2.

applied at terminals $a\text{-}a'$, and the equivalent impedance and equivalent voltage at these terminals will be found. The equivalent impedance is

$$Z_\theta(s) = \frac{(R_1 + 1/C_1 s)1/C_2 s}{R_1 + 1/C_1 s + 1/C_2 s} \tag{9-88}$$

and

$$V_\theta(s) = \frac{(V_0/s)(1/C_2 s)}{R_1 + 1/C_1 s + 1/C_2 s} \tag{9-89}$$

The current through R_2 is

$$I_2(s) = \frac{V_\theta(s)}{Z_\theta(s) + R_2}$$

$$= \frac{V_0/C_2}{R_1 R_2 s^2 + (R_1/C_2 + R_2/C_1 + R_2/C_2)s + 1/C_1 C_2} \tag{9-90}$$

Suppose that the following values are given for the network: $C_1 = 8\ \mu\text{F}$, $C_2 = 8\ \mu\text{F}$, $R_1 = 9\ \text{M}\Omega$, $R_2 = 5\ \text{M}\Omega$, and $V_0 = 75\ \text{V}$. With these parameter values, Eq. 9-90 reduces to

$$I_2(s) = \frac{0.208 \times 10^{-6}}{(s + 0.045)(s + 0.0077)} \tag{9-91}$$

This equation is expanded by partial fractions to give

$$I_2(s) = 5.55 \times 10^{-6} \left(\frac{1}{s + 0.0077} - \frac{1}{s + 0.045} \right) \tag{9-92}$$

The inverse Laplace transformation is

$$i_2(t) = 5.55 \times 10^{-6}(e^{-0.0077t} - e^{-0.045t}) \tag{9-93}$$

which is the required current. If the current in any *other* branch is required, it is necessary to start over and find a new Thévenin equivalent network.

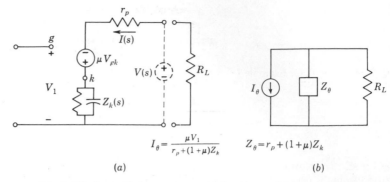

$$I_\theta = \frac{\mu V_1}{r_p + (1+\mu)Z_k} \qquad Z_\theta = r_p + (1+\mu)Z_k$$

(a)   (b)

Fig. 9-36. Controlled source network for Example 6 for which the Norton's equivalent network is found.

EXAMPLE 8

The network of Fig. 9-36(a) contains a controlled source which depends on the voltage

$$V_{gk}(s) = V_1(s) - V_k(s) \tag{9-94}$$

To find the Norton equivalent network, we first find the impedance $Z_\theta(s)$ at the open terminals. We do this by inserting a voltage source $V(s)$ at these terminals and then determining the current $I(s)$ under the condition that independent sources be reduced to zero, meaning that $V_1(s) = 0$. Then

$$V(s) = I(s)r_p + I(s)\mu Z_k(s) + I(s)Z_k(s) \tag{9-95}$$

with IZ_k substituted for V_k in Eq. (9-90). Then the required impedance is

$$Z_\theta(s) = \frac{V(s)}{I(s)} = r_p + (1 + \mu)Z_k(s) \tag{9-96}$$

The current in the shorted terminals is

$$I_\theta(s) = \frac{\mu[V_1(s) - I_\theta(s)Z_k(s)]}{Z_k(s) + r_p} \tag{9-97}$$

or

$$I_\theta(s) = \frac{\mu V_1(s)}{r_p + (1 + \mu)Z_k(s)} \tag{9-98}$$

From these equations, the Norton equivalent network is constructed as shown in Fig. 9-36(b).

Thévenin's theorem should be regarded as a useful artifice that provides a conceptual advantage in visualizing the operation of networks. It is not, in general, a means for simplifying the amount of computation needed although it may sometimes accomplish this.

FURTHER READING

CHIRLIAN, PAUL M., *Basic Network Theory*, McGraw-Hill Book Company, New York, 1969. Chapter 5.

DESOER, CHARLES A., AND ERNEST S. KUH, *Basic Circuit Theory*, McGraw-Hill Book Company, New York, 1969. Chapters 16 and 17.

KUO, FRANKLIN F., *Network Analysis and Synthesis*, 2nd ed., John Wiley & Sons, Inc., New York, 1966. Chapter 7.

DIGITAL COMPUTER EXERCISES

The topics of this chapter are not directly related to the use of the digital computer, since new concepts and theorems are stressed. Use the time available for computer exercises in completing more of those suggested at the end of Chapter 3.

PROBLEMS

9-1. In the network of (a) of the accompanying figure, $v_1 = V_0 e^{-2t} \cos t\, u(t)$, and for the network of (b), $i_1 = I_0 e^{-t} \sin 3t\, u(t)$. The impedance of the passive network N is found to be

$$Z(s) = \frac{(s+2)(s+3)}{(s+1)(s+4)}$$

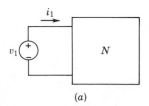

(a)

(a) With N connected to the voltage source as in (a) of the figure, what will be the complex frequencies in the current $i_1(t)$?

(b) With N connected to the current source as in (b) of the figure, what will be the complex frequencies in the voltage $v_1(t)$?

9-2. Repeat Prob. 9-1 if

$$Z(s) = \frac{2s^4 + 3s^3 + 5s^2 + 5s + 1}{(s^2 + 1)(2s^2 + 2s + 4)}$$

Solve part (b) only.

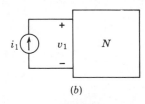

(b)

Fig. P9-1.

9-3. Consider the two series circuits shown in the accompanying figure. Given that $v_1(t) = \sin 10^3 t$, $v_2(t) = e^{-1000t}$ for $t > 0$, and $C = 1\ \mu F$.

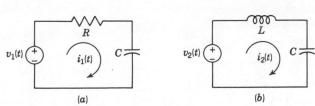

(a)   (b)

Fig. P9-3.

(a) Show that it is possible to have $i_1(t) = i_2(t)$ for all $t > 0$. (b) Determine the required values of R and L for (a) to hold. (c) Discuss the physical meaning of this problem in terms of the complex frequencies of the two series circuits.

9-4. In the network of the figure, the switch is opened at $t = 0$, a steady state having previously been established. With the switch open, draw the transform network for analysis on the loop basis, representing all elements and all initial conditions.

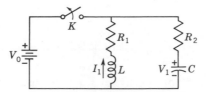

Fig. P9-4.

9-5. This problem is similar to Prob. 9-4, except that the transform network required should be prepared for analysis on the (a) loop basis, and (b) node basis. In this network, initial currents and voltages are a consequence of active elements removed at $t = 0$.

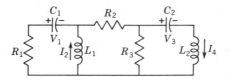

Fig. P9-5.

9-6. In the network of the figure, the switch K is closed at $t = 0$ and at $t = 0-$ the indicated voltages are on the two capacitors. Repeat Prob. 9-4 for this network.

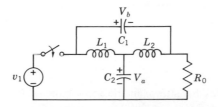

Fig. P9-6.

9-7. Determine the transform impedances for the two networks shown in the accompanying figure.

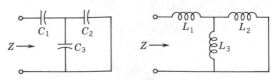

Fig. P9-7.

9-8. For the *RC* network shown in the figure, find the transform impedance, $Z(s)$, in the form of a quotient of polynomials, $p(s)/q(s)$. Factor
$p(s)$ and $q(s)$ so that $Z(s)$ may be written in the form of the impedance
of Prob. 9-1.

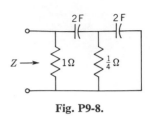

Fig. P9-8.

9-9. Repeat Prob. 9-8 for the *LC* network of the accompanying figure.

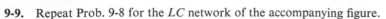

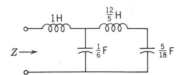

Fig. P9-9.

9-10. Repeat Prob. 9-8 for the *RC* network shown in the accompanying
figure.

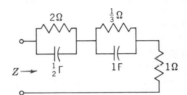

Fig. P9-10.

9-11. Repeat Prob. 9-8 for the *RLC* network of the figure, except that in
this case determine $Y(s)$ rather than $Z(s)$.

9-12. Two black boxes with two terminals each are externally identical. It is
known that one box contains the network shown as (*a*) and the other
contains the network shown as (*b*) with $R = \sqrt{L/C}$. (a) Show that
the input impedance, $Z_{in}(s) = V_{in}(s)/I_{in}(s) = R$ for both networks.[5]
(b) Investigate the possibility of distinguishing the purely resistive
network. Any external measurements may be made, initial and final
conditions may be examined, etc.

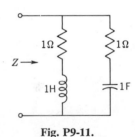

Fig. P9-11.

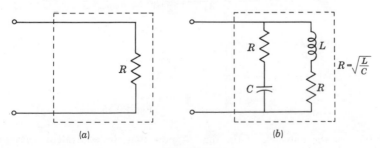

Fig. P9-12.

9-13. Repeat Prob. 9-12 by comparing the network shown in the accompanying figure[6] to that given in (*a*) of the figure for Prob. 9-12.

[5]Slepian, J., letter in *Elec. Engrg.*, **68**, 377; April, 1949.
[6]Macklem, F. S., "Dr. Slepian's black box problem," *Proc. IEEE*, **51**, 1269;
September, 1963.

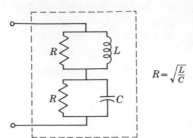

$$R = \sqrt{\frac{L}{C}}$$

Fig. P9-13.

9-14. The network shown in Fig. P9-4 is operated with switch K closed until a steady-state condition is reached. Then at $t = 0$ the switch K is opened. Starting with the transform network found in Prob. 9-4, determine the voltage across the switch, $v_k(t)$, for $t \geq 0$.

9-15. If the capacitors are uncharged and the inductor current zero at $t = 0-$, in the given network, show that the transform of the generator current is

$$I_1(s) = \frac{10(s^2 + s + 1)}{(s^2 + 1)(s^2 + 2s + 2)}$$

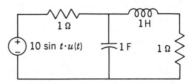

Fig. P9-15.

9-16. Repeat Prob. 9-15 for the network given to show that the generator current is given by the transform

$$I_1(s) = \frac{s(s + 2)(5s + 6)}{(s^2 + 4s + 13)(10s^2 + 18s + 4)}$$

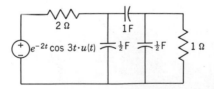

Fig. P9-16.

9-17. For the network of the figure, show that the equivalent Thévenin network is represented by

$$V_\theta = \frac{V_1}{2}(1 + a + b - ab)$$

and

$$Z_\theta = \frac{3 - b}{2}$$

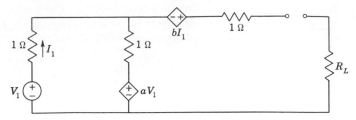

Fig. P9-17.

9-18. The accompanying network consists of resistors and controlled sources in addition to the independent voltage source v_s. For this network, find the Thévenin equivalent network by determining an expression for the voltage v_θ and the Thévenin equivalent resistance.

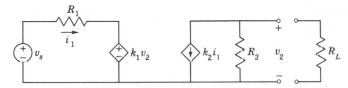

Fig. P9-18.

9-19. The network of the figure contains three resistors and one controlled current source in addition to independent sources. For this network, determine the Thévenin equivalent network at terminals 1-1'.

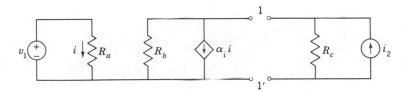

Fig. P9-19.

9-20. The network shown is a simple representation of a transistor. For this network, determine the Thévenin equivalent network for the load R_L.

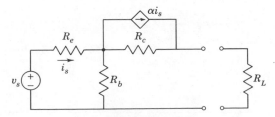

Fig. P9-20.

9-21. The network in the figure contains a resistor and a capacitor in addition to various sources. With respect to the load consisting of R_L in series with L, determine the Thévenin equivalent network.

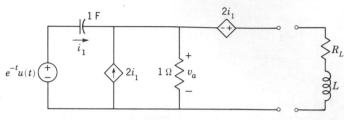

Fig. P9-21.

9-22. Using the network of Prob. 9-18, determine the Norton equivalent network.

9-23. For the network used in Prob. 9-19, determine the Norton equivalent network.

9-24. Determine the Norton equivalent network for the network given in Prob. 9-20.

9-25. Determine the Norton equivalent network for the system described in Prob. 9-21.

9-26. In the given network, the switch is in position a until a steady state is reached. At $t = 0$, the switch is moved to position b. Under that condition, determine the transform of the voltage across the 0.5-F capacitor using (a) Thévenin's theorem, and (b) Norton's theorem.

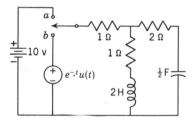

Fig. P9-26.

9-27. In the network of the figure, the switch K is closed at $t = 0$, a steady state having previously existed. Find the current in the resistor R_3 using (a) Thévenin's theorem, and (b) Norton's theorem.

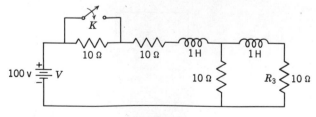

Fig. P9-27.

9-28. The network shown in the figure is a low-pass filter. The input voltage $v_1(t)$ is a unit step function, and the input and load resistors have the value $R = \sqrt{L/C}$. By using Thévenin's theorem, show that the transform of the output voltage is

$$V_2(s) = \frac{4}{(LC)^{3/2}} \left[\frac{1}{s(s^3 + 4\sqrt{1/LC}\, s^2 + 8s/LC + 8/(LC)^{3/2}} \right]$$

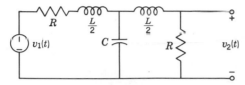

Fig. P9-28.

9-29. In the network shown in the accompanying sketch, the elements are chosen such that $L = CR_1^2$ and $R_1 = R_2$. If $v_1(t)$ is a voltage pulse of 1-V amplitude and T-sec duration, show that $v_2(t)$ is also a pulse, and find its amplitude and time duration.

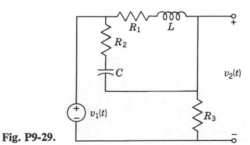

Fig. P9-29.

9-30. Using either Thévenin's or Norton's theorem, determine an equivalent network for the terminals a-b in the figure for zero initial conditions.

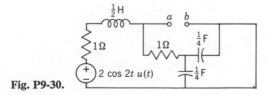

Fig. P9-30.

9-31. The network given contains a controlled source. For the element values given, with $v_1(t) = u(t)$, and for zero initial conditions: (a) determine the equivalent Thévenin network at a-a'. (b) Determine the equivalent Thévenin network at b-b'.

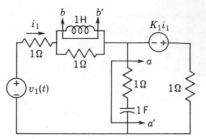

Fig. P9-31.

9-32. For the given network, determine the equivalent Thévenin network to compute the transform of the current in R_L.

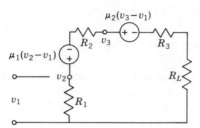

Fig. P9-32.

9-33. Assuming zero initial voltage on the capacitor, determine the equivalent Norton network for the resistor R_K.

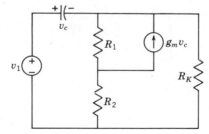

Fig. P9-33.

Network Functions; Poles and Zeros 10

In this chapter, the concept of transform impedance and transform admittance which was introduced in the last chapter is studied and extended. Furthermore, a function relating currents or voltages at different parts of the network, called a *transfer function*, is found to be mathematically similar to the transform impedance function. These functions are called *network functions*.

10-1. TERMINAL PAIRS OR PORTS

Consider an arbitrary network made up entirely of passive elements. To indicate the general nature of the network, let it be represented by the symbol of a rectangle (or a box). If a conductor is fastened to any node in the network and brought out of the box for access, the end of this conductor is designated as a *terminal*.[1] Terminals are required for connecting driving forces to the network, for connecting some other network (say, a load), or for making measurements. The minimum number of terminals that are useful is *two*. Also, the terminals are associated in pairs, one pair for a driving force, another pair for the load, etc. Two associated terminals are given the name *terminal pair*, or *port*, suggesting a port of entry into the network.

---

[1]Terminals are sometimes called *poles*, especially in the German literature. This results in a double usage of the word pole, however, as we shall see later in this chapter.

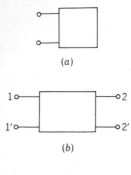

(a)

(b)

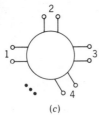

(c)

Fig. 10-1. (a) One-port network, (b) two-port network, and (c) a representation for the *n*-port network.

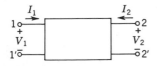

Fig. 10-2. The two-port network with reference directions for the port voltages and currents indicated.

In Figure 10-1(*a*) is shown a representation of a one-port network. The pair of terminals is customarily connected to an energy source which is the driving force for the network, so that the pair of terminals is known as the *driving point* of the network. Figure 10-1(*b*) shows a two-port network. The port designated 1-1′ is assumed to be connected to the driving force (or the *input*), and port 2-2′ is connected to a load (as an *output*). In (*c*) of Fig. 10-1 is shown a representation of an *n*-port network for the general case; we will also be concerned with *n*-terminal networks in some cases, which are different from *n*-port networks. Our emphasis now will be on one-port and two-port networks.

10-2. NETWORK FUNCTIONS FOR THE ONE-PORT AND TWO-PORT

The transform impedance at a port has been defined as the ratio of the voltage transform to current transform for a network in the zero state (no initial conditions) with no internal voltage or current sources except controlled sources. Thus we write

$$Z(s) = \frac{V(s)}{I(s)} \tag{10-1}$$

Similarly, the transform admittance is defined as the ratio

$$Y(s) = \frac{I(s)}{V(s)} = \frac{1}{Z(s)} \tag{10-2}$$

The voltage transform and current transform that define transform impedance and transform admittance must relate to the same port, 1-1′ or 2-2′ in Fig. 10-2. The impedance or admittance found at a given port is called a *driving-point impedance* (*or admittance*).

Because of the similarity of impedance and admittance (and to avoid writing "impedance and admittance"), the two quantities are assigned one name, *immittance* (a combination of *im*pedance and ad*mittance*). An immittance is thus an impedance *or* an admittance.

The transfer function is used to describe networks which have at least two ports, and these functions are computed under the same assumptions that are listed for driving-point functions. In general, the transfer function relates the transform of a quantity at one port to the transform of another quantity at another port. Thus transfer functions which relate voltages and currents have the following possible forms:

(1) The ratio of one voltage to another voltage, or the voltage transfer ratio.

(2) The ratio of one current to another current, or the current transfer ratio.

(3) The ratio of one current to another voltage or one voltage to another current.

It is conventional, although not universal, to define transfer functions as the ratio of an output quantity to an input quantity. In terms of the two-port network of Fig. 10-2, the output quantities are $V_2(s)$ and $I_2(s)$ and the input quantities are $V_1(s)$ and $I_1(s)$. Using this scheme, there are only four transfer functions for the two-port network, and these are tabulated in Table 10-1, together with the assigned designation for each. The manner in which network functions may be found will be illustrated for a number of simple networks.

Table 10-1. TRANSFER FUNCTIONS FOR THE TWO-PORT

	Numerator	
Denominator	$V_2(s)$	$I_2(s)$
$V_1(s)$	$G_{12}(s)$	$Y_{12}(s)$
$I_1(s)$	$Z_{12}(s)$	$\alpha_{12}(s)$

EXAMPLE 1

Figure 10-3 shows an *RLC* series *n*-port network with transform impedances marked for each element. The driving-point impedance is

$$Z(s) = R + Ls + \frac{1}{Cs} = \frac{LCs^2 + RCs + 1}{Cs} \tag{10-3}$$

or

$$Z(s) = L\frac{s^2 + (R/L)s + 1/LC}{s} \tag{10-4}$$

The numerator polynomial for this driving-point impedance is of second degree,[2] while the denominator polynomial is of first degree.

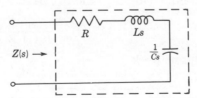

Fig. 10-3. Network for Example 1.

---

[2]Throughout the remainder of the book, we will use the word *degree* in place of *order* as used in earlier chapters. See footnote 3 of Chapter 6.

EXAMPLE 2

Figure 10-4 shows a more complicated network consisting of a series RL network shunted by a capacitor. The driving-point impedance is

$$Z(s) = \frac{1}{Cs + 1/(R + Ls)} = \frac{1}{C}\frac{s + R/L}{s^2 + (R/L)s + 1/LC} \quad (10\text{-}5)$$

In this driving-point impedance function, the numerator is of first degree and the denominator is of second degree. The driving-point admittance function $Y(s)$ for this network is the reciprocal of Eq. (10-5).

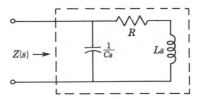

Fig. 10-4. One-port network for Example 2.

EXAMPLE 3

The two-port network shown in Fig. 10-5 has $V_1(s)$ as the input voltage and $V_2(s)$ as the output voltage transform. This network acts as a voltage divider. With no current in the output terminals, the voltage equations are

$$RI(s) + \frac{1}{Cs}I(s) = V_1(s) \quad (10\text{-}6)$$

$$\frac{1}{Cs}I(s) = V_2(s) \quad (10\text{-}7)$$

The ratio of these equations is

$$G_{12}(s) = \frac{V_2(s)}{V_1(s)} = \frac{(1/Cs)I(s)}{(R + 1/Cs)I(s)} = \frac{1/RC}{s + 1/RC} \quad (10\text{-}8)$$

From Eq. (10-6), the ratio of $I(s)$ to $V_1(s)$ is found to be

$$Y_{11}(s) = \frac{I(s)}{V_1(s)}\frac{1}{R}\frac{s}{s + 1/RC} \quad (10\text{-}9)$$

for this network.

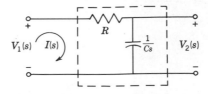

Fig. 10-5. Two-port network for Example 3.

EXAMPLE 4

The two-port network shown in Fig. 10-6 is similar to that of Example 1 except that the resistor has been replaced by an inductor. It is not necessary to write Kirchhoff's equations as above to find the transfer function, since this network is essentially a voltage divider. By this we mean that since from Kirchhoff's voltage law $V_1 = V_L + V_C = LsI(s) + (1/Cs)I(s)$ and from the network we identity $V_2 = V_C = (1/Cs)I(s)$, the voltage ratio is determined as

$$G_{12}(s) = \frac{V_2(s)}{V_1(s)} = \frac{V_C}{V_L + V_C} = \frac{1/Cs}{Ls + 1/Cs} = \frac{1}{LCs^2 + 1} \quad (10\text{-}10)$$

or

$$G_{12}(s) = \frac{1/LC}{s^2 + 1/LC} \quad (10\text{-}11)$$

The numerator polynomial is of zero degree, and the denominator polynomial is of second degree.

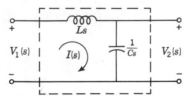

Fig. 10-6. Two-port network for Example 4.

EXAMPLE 5

The same voltage-divider network concept can be used with more than one current loop in the network by network reduction. Figure 10-7 shows such a network. The transform impedances R_1 and $1/Cs$ can be combined into an equivalent impedance having the value

$$Z_{eq}(s) = \frac{1}{Cs + 1/R_1} = \frac{R_1}{R_1 Cs + 1} \quad (10\text{-}12)$$

Then the transfer function becomes

$$G_{12}(s) \frac{V_2(s)}{V_1(s)} = \frac{R_2}{R_2 + Z_{eq}(s)} \quad (10\text{-}13)$$

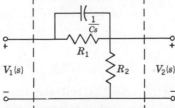

Fig. 10-7. Two-port network for Example 5.

or

$$G_{12}(s) = \frac{R_2 R_1 Cs + R_2}{R_2 R_1 Cs + R_1 + R_2} \tag{10-14}$$

which may be reduced to

$$G_{12}(s) = \frac{s + 1/R_1 C}{s + (R_1 + R_2)/R_1 R_2 C} \tag{10-15}$$

In this transfer function, the degree of the numerator and degree of the denominator are the same. This particular network finds application in automatic control where it is known as a "lead" network.

EXAMPLE 6

For the network of Fig. 10-8, which is driven by a current source, we are required to compute $\alpha_{12}(s)$ and $Z_{12}(s)$. Observe that this is a current-divider network, in the same sense that other examples have involved voltage dividing. Thus,

$$I_1(s) = I_a(s) + I_2(s) = V_1(s)[Y_a(s) + Y_2(s)] \tag{10-16}$$

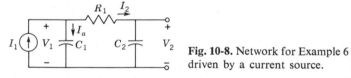

Fig. 10-8. Network for Example 6 driven by a current source.

and since $I_2(s) = Y_2(s)V_1(s)$, we have

$$I_2(s) = \frac{Y_2(s)}{Y_a(s) + Y_2(s)} I_1(s) \tag{10-17}$$

so that

$$\alpha_{12}(s) = \frac{I_2(s)}{I_1(s)} = \frac{Y_2(s)}{Y_a(s) + Y_2(s)} \tag{10-18}$$

Since $Y_2(s) = (s/R_1)/(s + 1/R_1 C_2)$, and $Y_a(s) = C_1 s$, we have

$$\alpha_{12}(s) = \frac{1}{R_1 C_1} \frac{1}{s + (C_1 + C_2)/R_1 C_1 C_2} \tag{10-19}$$

Now, $V_2(s) = (1/C_2 s)I_2(s)$ in the network of Fig. 10-8, so that

$$Z_{12}(s) = \frac{V_2(s)}{I_1(s)} = \frac{1}{R_1 C_1 C_2} \frac{1}{s[s + (C_1 + C_2)/R_1 C_1 C_2]} \tag{10-20}$$

If the elements of the network have the values $C_1 = 1$ F, $R_1 = 1\,\Omega$, and $C_2 = 2$ F, then the transfer functions become

$$\alpha_{12}(s) = \frac{1}{s + 1.5} \quad \text{and} \quad Z_{12}(s) = \frac{0.5}{s(s + 1.5)} \tag{10-21}$$

We observe that all the network functions which we have computed have been quotients of polynomials in s having the general

form

$$\frac{p(s)}{q(s)} = \frac{a_0 s^n + a_1 s^{n-1} + \ldots + a_{n-1} s + a_n}{b_0 s^m + b_1 s^{m-1} + \ldots + b_{m-1} s + b_m} \tag{10-22}$$

which is a *rational* function of s (n and m are integers). In the equation, n is the degree of the numerator polynomial, and m is the degree of the denominator polynomial. This will be the case in general as we see from Eq. (9-62) of the last chapter which is

$$\frac{I_k(s)}{V_j(s)} = \frac{\Delta_{jk}}{\Delta} \tag{10-23}$$

This equation was derived for the conditions we have assumed for the two-port network, or for the one-port network with $j = k = 1$. In this equation, Δ is the determinant of the impedance matrix $[Z]$, and Δ_{jk} is a cofactor of Δ with row j and column k deleted. Now a typical element in Δ, and therefore in Δ_{jk}, is

$$Z_{jk} = R_{jk} + L_{jk}s + \frac{1}{C_{jk}s} \tag{10-24}$$

The evaluation of Δ and Δ_{jk} will involve products, sums, and quotients of terms like that of Eq. (10-24), and will therefore always result in a rational quotient of polynomials of the form of Eq. (10-22). This statement applies to all network functions. The voltage-ratio and current-ratio functions are obtained from Eq. (10-23) by noting that $I_k = Y_k V_k$ and $V_j = I_j Z_j$ so that these network functions may be written as the quotient of Eq. (10-23) multiplied by either Y_k or Z_j. Having established this result, we next consider the polynomials $p(s)$ and $q(s)$ that represent any network function.

10-3. THE CALCULATION OF NETWORK FUNCTIONS

(1) Ladder networks. We first show that simple procedures may be followed in computing the immittance functions for one special class of network structures, the ladder. The ladder network is shown in Fig. 10-9. If each immittance represents one element, the network is known as a *simple* ladder. Otherwise, the ladder network may contain *arms* that are arbitrarily complicated, as shown by the sample of Fig.

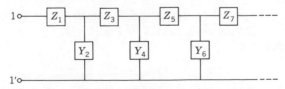

Fig. 10-9. A general ladder network which is described as a simple ladder Z or Y describes only one element.

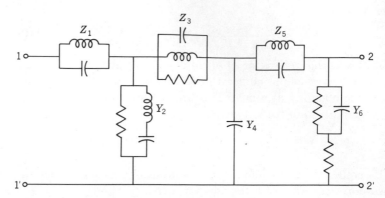

Fig. 10-10. A non-simple ladder network.

10-10. We follow the practice of characterizing series arms by their impedances and shunt arms by their admittances for reasons that will soon be evident.

We first consider the computation of driving-point immittances for the ladder network. If we are finding an open-circuit or short-circuit parameter, we assume that the appropriate port is prepared by being either open or shorted. We begin our computations at the port *other* than the one for which the driving-point immittance is being found. Thus in Fig. 10-10 (with only six arms being considered), we begin with Y_6. It is first inverted and combined with Z_5. Next, this sum is inverted and combined with Y_4. This pattern is continued until the process terminates. The impedance will then be

$$Z = Z_1 + \cfrac{1}{Y_2 + \cfrac{1}{Z_3 + \cfrac{1}{Y_4 + \cfrac{1}{Z_5 + \cfrac{1}{Y_6}}}}} \tag{10-25}$$

This equation is known as a *continued fraction*. It may be simplified, or collapsed, to determine Z for a given ladder network.

EXAMPLE 7

We are required to determine the open-circuit driving-point impedance for the network shown in Fig. 10-11. By Eq. 10-25, the required impedance is

$$Z_{11} = s + \cfrac{1}{s + \cfrac{1}{s + \cfrac{1}{s}}} \tag{10-26}$$

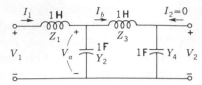

Fig. 10-11. Ladder network of Example 5.

This equation is reduced by starting at the last term and combining terms step by step, giving

$$Z_{11} = \frac{s^4 + 3s^2 + 1}{s^3 + 2s} \tag{10-27}$$

A different method is used to find the transfer functions for the ladder network. It requires that we start at the output port and then work toward the input port, successively applying the Kirchhoff current law and the Kirchhoff voltage law. The equations that result are of particularly simple form in that the first may be substituted into the second, the resulting equation may then be substituted into the third, and this process continued until we find one equation which relates output to input. At no point in this pattern of substitution will simultaneous equations be encountered.

EXAMPLE 8

Again consider the network of Fig. 10-11 and let it be required to find V_2/I_1 and V_2/V_1. Beginning at the right end, we write the following equations:

$$I_b = Y_4 V_2 = sV_2 \tag{10-28}$$

$$V_a = V_2 + I_b Z_3 = (s^2 + 1)V_2 \tag{10-29}$$

$$I_1 = I_b + Y_2 V_a = [s + s(s^2 + 1)]V_2 \tag{10-30}$$

$$V_1 = V_a + Z_1 I_1 = [(s^2 + 1) + s(s^3 + 2s)]V_2 \tag{10-31}$$

From Eq. (10-30), we see that

$$\frac{V_2}{I_1} = \frac{1}{s^3 + 2s} \tag{10-32}$$

and, from Eq. (10-31),

$$\frac{V_2}{V_1} = \frac{1}{s^4 + 3s^2 + 1} \tag{10-33}$$

This pattern of equations may be written for any ladder network, and so any transfer function may be found by routine algebraic operations. It is sometimes easier to make calculations of the type described for a number of specific frequencies of interest than it is to derive a general equation for the transfer function. In such calculations, it is conventional to assume 1 volt or 1 ampere output and then to compute the numerical value of the required input voltage or current.

Since we are interested in a ratio of an output quantity to an input quantity, there is nothing lost by assuming this convenient numerical value for the output. For this reason the method we have described is sometimes called the *unit output method* of computing transfer functions. It applies, of course, only to ladder networks.

(2) *General networks.* The networks used for previous examples of this chapter have been ladder networks. This is an important structure, being the type encountered most frequently in electronic applications, for example. It is important that we observe that there are other network structures to which the techniques described are not applicable, and it is also helpful to review the procedure which should be followed in these cases.

Figure 10-12 shows several nonladder networks. The descriptive names by which they are known include (*a*) the bridged-*T*, (*b*) the parallel-*T* or twin-*T*, and (*c*) the lattice network. From any of these structures, we can see the reason that the techniques for ladders do not apply. Observe that the current at the output is related to a number of node voltages, instead of simply one node voltage, meaning that the equations to be written would be simultaneous equations to be solved in the standard method of node or loop analysis.

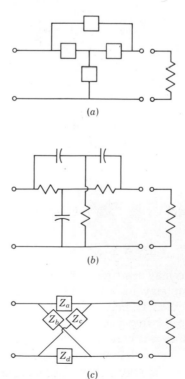

Fig. 10-12. Common forms of non-ladder networks: (*a*) the bridged-*T*, (*b*) parallel-*T*, and (*c*) the lattice.

For nonladder networks, it is necessary to express network functions as a quotient of determinants formulated on the node or loop basis. We have written equations of this form a number of times in previous chapters, such as in Eq. (9-62). We may summarize previous results in the equation

$$Y_{jk} = \frac{\Delta_{kj}}{\Delta} \qquad (10\text{-}34)$$

where Δ is the loop-basis system determinant, and Y is replaced by y, if the output port is shorted, and in

$$Z_{jk} = \frac{\Delta'_{kj}}{\Delta'} \qquad (10\text{-}35)$$

where Δ' is the node-basis system determinant (with a common ground connection between port 1 and port 2) and Z is replaced by z when the output port is open. The two remaining equations for voltage and current ratios are:

$$G_{12} = \frac{V_2}{V_1} = \frac{\Delta'_{12}}{\Delta'_{11}} \qquad (10\text{-}36)$$

and

$$\alpha_{12} = \frac{I_2}{I_1} = \frac{\Delta_{12}}{\Delta_{11}} \qquad (10\text{-}37)$$

EXAMPLE 9

For the bridged-T network of Fig. 10-13, the loop-basis system determinant is

$$\Delta = \begin{vmatrix} \dfrac{1}{s} + 1 & 1 & -\dfrac{1}{s} \\[2mm] 1 & \dfrac{1}{s} + 2 & \dfrac{1}{s} \\[2mm] -\dfrac{1}{s} & \dfrac{1}{s} & \dfrac{2}{s} + 1 \end{vmatrix} \qquad (10\text{-}38)$$

Expanding the determinant and the appropriate cofactors, we find that the input admittance is

$$Y_{11} = \frac{\Delta_{11}}{\Delta} = \frac{2s^2 + 5s + 1}{s^2 + 5s + 2} \qquad (10\text{-}39a)$$

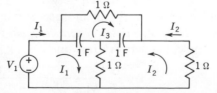

Fig. 10-13. Bridged-T networks analyzed in Example 7.

and the transfer admittance is

$$Y_{12} = \frac{\Delta_{21}}{\Delta} = -\frac{(s^2 + 2s + 1)}{s^2 + 5s + 2} \qquad (10\text{-}39b)$$

Note that when the output, port 2, of the network of Fig. 10-13 is shorted, the network reduces to ladder form.

10-4. POLES AND ZEROS OF NETWORK FUNCTIONS

We have shown that all network functions have the form of a quotient of polynomials in s,

$$N(s) = \frac{p(s)}{q(s)} = \frac{a_0 s^n + a_1 s^{n-1} + \ldots + a_{n-1}s + a_n}{b_0 s^m + b_1 s^{m-1} + \ldots + b_{m-1}s + b_m} \qquad (10\text{-}40)$$

where the a and b coefficients are real and positive for networks of passive elements and no controlled sources. Now the equation $p(s) = 0$ has n roots, and $q(s) = 0$ similarly has m roots. Both $p(s)$ and $q(s)$ may be written as a product of linear factors involving these roots

$$N(s) = H \frac{(s - z_1)(s - z_2)\ldots(s - z_n)}{(s - p_1)(s - p_2)\ldots(s - p_m)} \qquad (10\text{-}41)$$

where $H = a_0/b_0$ is a constant known as the *scale factor*, and $z_1, z_2, \ldots, z_n, p_1, p_2, \ldots, p_m$ are complex frequencies. When the variable s has the values $z_1, z_2, \ldots, z_n$, the network function vanishes; such complex frequencies are known as the *zeros* of the network function. When s has values $p_1, p_2, \ldots, p_m$, the network function becomes infinite; such complex frequencies are the *poles* of the network function. In Eq. (10-41), factors $(s - z_j)$ are known as *zero factors;* $(s - p_j)$ are *pole factors*. Poles and zeros are useful in describing network functions. We see from Eqs. (10-40) and (10-41) that a network function is completely specified by its poles, zeros, and the scale factor.

When r poles or zeros in Eq. (10-41) have the same value, the pole or zero is said to be *of multiplicity r*, although we will often use terms like double, triple, etc., to describe multiplicity. If the pole or zero is not repeated, it is said to be *simple* or *distinct*. Poles or zeros at $s = \infty$ are also assigned a degree: When $n > m$ in Eq. (10-40), the pole at infinity is of degree or multiplicity $n - m$; when $n < m$, the zero at infinity is of degree $m - n$.

If, for any rational network function, poles and zeros at zero and infinity are taken into account in addition to finite poles and zeros, *the total number of zeros is equal to the total number of poles*. For example, the network function

$$N(s) = \frac{s^2(s + 3)}{(s + 1)(s + 2 + j1)(s + 2 - j1)} \qquad (10\text{-}42)$$

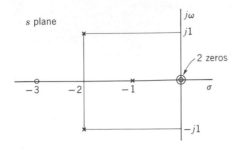

Fig. 10-14. The poles and zeros of $N(s)$ of Eq. 10-42.

has a double zero at $s = 0$ (the origin), a zero at -3, and poles at -1, $-2 + j1$, and $-2 - j1$. Had the factor $(s + 1)$ been squared, then $N(s)$ would have had a double pole at -1, and a zero at infinity. The poles and zeros of $N(s)$ in Eq. (10-42) are plotted in the s plane in Fig. 10-14. It is conventional to use the symbol $\bigcirc$ to designate the location of a zero and the symbol $\times$ for the location of a pole.

Poles and zeros are *critical frequencies*. At poles the network function becomes infinite, while at zeros the network function becomes zero. At other complex frequencies, the network function has a finite, nonzero value. A three-dimensional representation of the magnitude of the transfer function as a function of complex frequency is shown in Fig. 10-15 for one quadrant of the s plane. The portion of the com-

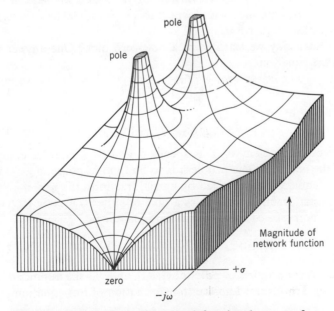

Fig. 10-15. The magnitude of a network function shown as a function of complex frequency with two poles and one zero.

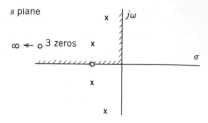

Fig. 10-16. The portion of the s plane shown in Fig. 10-15 is identified.

plex plane represented in Fig. 10-15 is shown in Fig. 10-16. This particular network function has four finite poles, one finite zero, and a third-order zero at infinity.

The pole represents a frequency at which the network function "blows up." The zero represents a frequency at which the opposite behavior takes place: The network function becomes nothing at all. Either "blowing up" or "becoming nothing" sounds like rather drastic behavior for the network function. We may wonder if it would not be wise to completely avoid poles and zeros, to select network functions without poles or zeros. Such is not the case at all. Poles and zeros are the lifeblood of a function; without poles and zeros the function reduces to a dull, drab, grubby constant—a function which does not change under any conditions. Without poles and zeros, the three-dimensional representation of the network function becomes a tedious expanse of mathematical desert—absolutely flat. But add a few poles and a few zeros and we have a land of spectacular peaks (elevation: ∞) and beautiful springs (elevation: 0), the s land whose coordinates (latitude and longitude with respect to $s = 0$ rather than Greenwich) are complex frequencies.

What may we learn from a pole-zero plot? One answer stems from the equation

$$\frac{V_{out}(s)}{V_{in}(s)} = G_{12}(s) \tag{10-43}$$

which may be written

$$V_{out}(s) = G_{12}(s)V_{in}(s) \tag{10-44}$$

In the usual problem, $v_{in}(t)$ is specified, and $G_{12}(s)$ can be computed from the network. The problem is to find the response, $v_{out}(t)$. When the last equation is expanded by partial fractions, the denominator of each partial fraction term gives a pole of either $G_{12}(s)$ or $V_{in}(s)$; that is, with no repeated roots in the denominator of $V_{out}(s)$

$$G_{12}(s)V_{in}(s) = \sum_{j=1}^{u} \frac{K_j}{s - p_j} + \sum_{k=1}^{v} \frac{K_k}{s - p_k} \tag{10-45}$$

where u is the number of poles of $G_{12}(s)$, and v is the number of poles of $V_{in}(s)$. The inverse Laplace transformation of this equation is

$$v_{out}(t) = \mathcal{L}^{-1}[G_{12}(s)V_{in}(s)] = \sum_{j=1}^{u} K_j e^{p_j t} + \sum_{k=1}^{v} K_k e^{p_k t} \tag{10-46}$$

Thus the frequencies s_j are the natural complex frequencies corresponding to *free oscillations*. The frequencies s_k are the driving-force complex frequencies corresponding to *forced oscillations*. The poles, therefore, determine the waveform of the time variation of the response, the output voltage. The zeros determine the magnitude of each part of the response, since they determine the magnitude of K_j and K_k in the partial fraction expansion, as we shall see.

In terms of driving-point immittances, poles and zeros have easily visualized meanings. Since $Z(s) = V(s)/I(s)$, a pole of $Z(s)$ implies zero current for a finite voltage, which means an *open circuit*. A zero of $Z(s)$, on the other hand, means no voltage for a finite current, or a *short circuit*. Thus a one-terminal-pair network is an open circuit for pole frequencies and a short circuit for zero frequencies. This can be visualized easily in terms of single-element networks. For a capacitor, the driving-point impedance is $Z(s) = 1/Cs$. This network function has a pole at $s = 0$ and a zero at $s = \infty$. It behaves as an open circuit at the pole frequency ($s = 0$) and as a short circuit at infinite frequency. Likewise, for an inductor, the driving-point impedance $Z(s) = Ls$ (zero at $s = 0$, pole at $s = \infty$) and this element behaves as a short circuit at zero frequency and as an open circuit at infinite frequency.

10-5. RESTRICTIONS ON POLE AND ZERO LOCATIONS FOR DRIVING-POINT FUNCTIONS

We have already observed that the polynomials $p(s)$ and $q(s)$ in the network function $N(s) = p(s)/q(s)$ are polynomials of the form

$$p(s) = a_0 s^n + a_1 s^{n-1} + \ldots + a_n \qquad (10\text{-}47)$$

in which the coefficients are real and positive. Thus we see that when s is real, $p(s)$ will be real; such a function is known as a *real function*. If $p(s)$ in Eq. (10-47) is a real function, and if one of the zeros of $p(s)$ is complex, then its conjugate must also be a zero of $p(s)$; otherwise some of the coefficients of $p(s)$ would be complex. Thus the product

$$p(s) = (s + a + jb)(s + a - jb) = (s + a)^2 + b^2 \qquad (10\text{-}48)$$

or

$$p(s) = s^2 + 2as + (a^2 + b^2) \qquad (10\text{-}49)$$

has real coefficients when a and b are real. For the second coefficient to be positive, it is also necessary that a be positive. Now the product of other complex factors which are not conjugate, $(s + a + jb)(s + c + jd)$, $a \neq c$, $b \neq d$, do not have real coefficients only, and so are excluded as factors in network functions.

The second property of $N(s)$ comes from the assumption that networks composed only of passive elements are *stable* in the sense that excitation due to an initial condition in the element results in a bounded output, by which we mean an output that never becomes infinite no matter how large time becomes.

Consider a one-port network. If the excitation is a voltage source, the response will be the current at the input. If, however, the excitation is a current source, the response is the voltage at the input. Since the network must be stable for either kind of input, the conclusions we reach for poles will evidently apply to zeros. Suppose that the denominator of the driving-point immittance contains the factor $(s - s_a)$ where $s_a = \sigma_a + j\omega_a$ is complex. This factor will result in a term in the time-domain response of the form

$$K_a e^{s_a t} = K_a e^{\sigma_a t} e^{j\omega_a t} \tag{10-50}$$

and if combined with the term resulting from s_a^*, we will have

$$K_0 e^{\sigma_a t} \sin(\omega_a t + \theta_a) \tag{10-51}$$

as we learned in Chapter 6. For the response to remain bounded, it is necessary that $\sigma_a \leq 0$. Since our arguments apply to both poles and zeros, we see that poles and zeros of driving-point immittances have *negative (or zero) real parts only*. In terms of pole and zero locations in the s plane, all poles and zeros must be in the *left half* of the s plane (LHP) and can never occur in the *right half-plane* (RHP). Poles and zeros can be on the *boundary* (the imaginary axis) subject to the limitation that such poles and zeros be *simple*. The reason for this restriction is that such poles of order two or greater give rise to time-domain terms which increase as t^{n-1} where n is the order of the pole, and such responses are not bounded as required. For example, consider the transform pair

$$\mathcal{L}^{-1}\left[\frac{s}{(s^2 + \omega_a^2)^2}\right] = \frac{t}{2\omega_a} \sin \omega_a t \tag{10-52}$$

corresponding to the pole-zero configuration of Fig. 10-17 with double poles at $+j\omega_a$. The time-domain response is a sinusoid of linearly

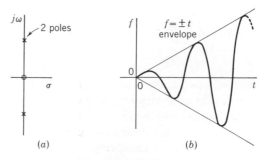

(a)　　　　　　　　　　(b)

Fig. 10-17. The inverse Laplace transform of the function having the poles and zeros of (a) is shown in (b).

increasing amplitude as shown in Fig. 10-17(*b*), which is an unbounded response.

Multiple poles and zeros are permitted at other locations in the left half of the *s* plane, since such poles give rise to terms of the form $t^m e^{-\sigma t}$, having the required zero limit since

$$\lim_{t \to \infty} t^m e^{-\sigma t} = 0 \tag{10-53}$$

for finite *m* by l'Hospital's rule.

In summary, we see that $p(s)$ and $q(s)$ are polynomials which are products of the following factors: K, $(s + a)$ where a may be zero, $s^2 + bs + c$, and $s^2 + d$, where K, a, b, c, and d are all real and positive. When the product of these factors is formed, there is no mechanism for a negative sign to be introduced, meaning that all the coefficients in $p(s)$ and $q(s)$ are real and positive as required. The same reasoning may be applied to show that no coefficient may be zero (no term in the polynomial may be missing) unless all the even or all the odd terms are missing! The crucial observation in making this conclusion is that there is no way for negative signs to be introduced, and without negative terms to cancel with positive terms, there is no way for a coefficient to be zero. Thus, the polynomial $p(s)$ of Eq. (10-47) and so also $q(s)$ have all their coefficients real and positive and, in addition, *nonzero* except for two special cases. If $p(s)$ is composed of factors like $s^2 + d$ only, then $p(s)$ is an even polynomial and the coefficients of all odd terms are zero. If $p(s)$ has a simple zero at the origin, the terms $s^2 + d_j$ are multiplied by s, and an odd polynomial results, meaning that the coefficients of all even terms are zero.

Another interesting property of driving-point functions may be found by considering the behavior of the one-port network at very high and at very low frequencies. The immittance of the inductor and the capacitor change with frequency. At very high frequencies, the immittances of these two elements, $Z_L = Ls$ and $Y_C = Cs$, will dominate the network immittance function in the sense that either Z_L or Y_C will be very large compared to any other element immittance. If L is not present in the network whose impedance we are computing, or if C is not present for the admittance computation (or if these elements are shorted by other elements), then either the resistor will be the dominant element, or the equivalent network will be a short circuit, meaning that $Z_C = 1/Cs$ or $Y_L = 1/Ls$ is the form of the immittance as s approaches infinite value. The equivalent representation of the one port at high frequencies will therefore involve at most one kind of element, and will be one of the three possibilities shown in Fig. 10-18, including $R_{eq} = 0$ or ∞. The same conclusions with respect to single-element representation apply for the low-frequency equivalent of the one port as may be seen from the dual equations to those given, $Y_L = 1/Ls$ and $Z_C = 1/Cs$.

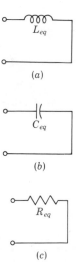

Fig. 10-18. Possible forms of the high-frequency equivalent network for a one-port network.

What must be true of the driving-point immittance functions in order that the equivalent networks of Fig. 10-18 apply at $s = \infty$ and $s = 0$? From Eq. (10-40), where $N(s)$ is a driving-point function, we see that

$$\lim_{s \to \infty} N(s) = \lim_{s \to \infty} \frac{a_0 s^n}{b_0 s^m} = \lim_{s \to \infty} \frac{a_0}{b_0} s^{n-m} \qquad (10\text{-}54)$$

So that $N(s)$ in this limit will have one of the three required forms—a constant times s, 1, or $1/s$—it is necessary that

$$|n - m| \leq 1 \qquad (10\text{-}55)$$

meaning that $n - m$ may have the values of -1, 0, or 1 and no other values. In other words, the degrees of the numerator and denominator polynomials for driving-point functions *may differ by one at most*.

Similarly, at low frequencies as s approaches zero, the terms in $p(s)$ or $q(s)$ of high degree are relatively small and so may be neglected such that $N(s)$ is approximately

$$N(s) \cong \frac{\cdots + a_{n-1}s + a_n}{\cdots + b_{m-1}s + b_m} \qquad (10\text{-}56)$$

In order that $N(s)$ reduce to one of the three permitted forms—a constant times s, 1, or $1/s$—we see that there are only three possibilities. Assume that $a_{n-1} \neq 0$ and $b_{m-1} \neq 0$. If $N(s)$ is impedance, then $a_n = 0$, $b_m \neq 0$ represents an inductor, while $a_n \neq 0$, $b_m = 0$ represents a capacitor. If $N(s)$ is admittance, then the opposite conditions are required to represent the inductor and capacitor. Finally, $a_n \neq 0$ and $b_m \neq 0$ means that $N(0)$ is a constant which may represent either the impedance or admittance of a resistor. In summary, the terms of lowest degree in the numerator and denominator of $N(s)$ *may differ in degree at most by one*. A number of examples are now presented to illustrate the conclusions we have reached.

EXAMPLE 10

Someone asserts that the impedance for a passive one-port network is

$$Z(s) = \frac{4s^4 + s^2 - 3s + 1}{s^3 + 2s^2 + 2s + 40} \qquad (10\text{-}57)$$

This function is not suitable in representing the impedance of a one-port, indicating that an error was made in its determination. In the numerator, one coefficient is missing and one is negative, indicating the presence in $Z(s)$ of zeros in the right half-plane. That this is indeed the case is seen from the factored form of the equation, $4(s^2 + s + 1)$ $(s - 0.5)^2$. Although not so evident, the denominator is also unsuited for a driving-point function since its factored form, $(s + 4)(s^2 - 2s + 10)$, indicates two poles of $Z(s)$ in the right half-plane!

EXAMPLE 11

Another student has determined the impedance of a one-port network to be

$$Z(s) = \frac{15(s^3 + 2s^2 + 3s + 2)}{s^4 + 6s^3 + 8s^2} \tag{10-58}$$

Again, an error in computation is indicated because this function is not suited to represent the impedance of a one-port network. The term of lowest degree in the numerator is of degree 0 while that in the denominator is 2—a difference in excess of that permitted. From another point of view, the denominator factors to the form $s^2(s + 2)(s + 4)$, indicating a double pole at the origin, which is not permitted.

EXAMPLE 12

The impedance function

$$Z(s) = \frac{s^2 + s + 2}{2s^2 + s + 1} \tag{10-59}$$

meets all the requirements we have found for a driving-point function, and so *may* be the impedance of a one-port. We cannot guarantee that $Z(s)$ was found without error, but it does pass all of the necessary conditions we have tabulated. To find conditions which are both necessary and sufficient would require that we understand how to find a network from Eq. (10-59), a topic which is included in the study of *network synthesis.*[3]

Table 10-2. NECESSARY CONDITIONS FOR DRIVING-POINT FUNCTIONS [WITH COMMON FACTORS IN $p(s)$ AND $q(s)$ CANCELLED]

---

1. The coefficients in the polynomials $p(s)$ and $q(s)$ of $N = p/q$ must be real and positive.
2. Poles and zeros must be conjugate if imaginary or complex.
3. (a) The real part of all poles and zeros must be negative or zero; in addition:
   (b) If the real part is zero, then that pole or zero must be simple.
4. The polynomials $p(s)$ and $q(s)$ may not have missing terms between those of highest and lowest degree, unless all even or all odd terms are missing.
5. The degree of $p(s)$ and $q(s)$ may differ by either zero or one only.
6. The terms of lowest degree in $p(s)$ and $q(s)$ may differ in degree by one at most.

---

[3]See, for example, the author's *Introduction to Modern Network Synthesis* (John Wiley & Sons, Inc., New York, 1960), Chapters 3 and 4; a network realization for Eq. (10-59) is given in Fig. 7-17, p. 178. The necessary and sufficient condition for $Z(s)$ to have a network realization is Re $Z(s) \geq 0$ for Re $s \geq 0$ [in addition to the requirement that $Z(s)$ be a real function].

The conclusions of this section, which are useful in checking the determination of driving-point functions, are summarized in Table 10-2.

10-6. RESTRICTIONS ON POLE AND ZERO LOCATIONS FOR TRANSFER FUNCTIONS

Some of the important properties of transfer functions Z_{12}, Y_{12}, G_{12}, and α_{12} may be found by using the same approach as that given for driving-point functions. Consider a two-port network, and let 1-1' be the input and 2-2' the output, as in Fig. 10-2. At port 1-1', we connect a voltage source or current source; and at port 2-2', we connect a load (resistor) and record the output voltage or current. If the source generates a step of voltage or current, the output response will remain bounded for all time, if the network is passive. Then we may conclude that the restriction on the real part of the poles found for the one-port network applies to the poles of transfer functions describing the two-port network. It will not, however, apply to the zeros. For the one-port network, our conclusions concerning the zeros came from the fact that $Y(s) = 1/Z(s)$. For the two-port network, no such relationship exists; $Z_{12} \neq 1/Y_{12}$ and $G_{12} \neq 1/\alpha_{12}$ in general. In summary, the poles of the transfer function must be in the left half-plane, including the boundary (if simple), but the zeros are not so restricted and may be in the right half-plane, provided they are conjugate if complex. Network functions with left half-plane zeros are classified as *minimum phase;* those with zeros in the right half-plane are *nonminimum phase.*

The rules for the difference in degree of $p(s)$ and $q(s)$ for the transfer function $N(s) = p(s)/q(s)$ are now illustrated by means of a number of simple network structures. Figure 10-19 shows two network structures known as the T and the π. We will use the network of (a) to illustrate properties of G_{12} and Z_{12}, and that of (b) for α_{12} and Y_{12}. For open output terminals in (a) and shorted output terminals in (b), Z_3 may be shorted and Y_3 opened without affecting G_{12} for (a) or α_{12} for (b). Then for (a) we have for V_1 as the input

$$G_{12}(s) = \frac{V_2(s)}{V_1(s)} = \frac{Z_2(s)}{Z_1(s) + Z_2(s)} \tag{10-60}$$

while with I_1 as the input for (b),

$$\alpha_{12}(s) = \frac{I_2(s)}{I_1(s)} = \frac{-Y_2(s)}{Y_1(s) + Y_2(s)} \tag{10-61}$$

Since these two equations are duals, the conclusions found for one will apply to the other. For this reason, we will restrict our discussion to G_{12} in Eq. (10-60).

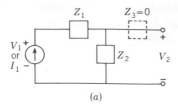

(a)

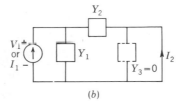

Fig. 10-19. (a) A T network, one arm of which has zero impedance, and (b) a π network, one arm of which has zero admittance.

(b)

In our discussion for the one-port, we asserted that as s approached infinity one kind of element in the network was dominant. The same conclusions will apply to the two-port network, of course, and so we may determine properties for the networks by considering one kind of element for the subnetworks Z_1 and Z_2 in Fig. 10-19(a). Since only three kinds of elements are to be taken two at a time, it is easy to exhaust all the possibilities. The result may be summarized in two statements: (1) If the two elements are of the same kind, G_{12} becomes a constant (the network serving only as a frequency-invariant voltage divider). Since we are considering only the high-frequency equivalent of more general subnetworks Z_1 and Z_2, this case evidently corresponds to the degree of $p(s)$ being equal to the degree of $q(s)$. (2) For other combinations, the degree of the numerator equals, but never exceeds, the degree of the denominator. The two cases are summarized by one statement that applies to voltage-to-voltage and current-to-current transfer functions: The *maximum* degree of the numerator is the degree of the denominator.

To study Z_{12} and Y_{12}, we connect I_1 to the network of Fig. 10-19(a) and V_1 for the network of Fig. 10-19(b). Again exhausting all the possibilities in a search for a network that gives the maximum degree different for the transfer functions, we arrive at the two networks shown in Fig. 10-20. For the two networks shown, we see that

$$Z_{12} = \frac{V_2}{I_1} = Ls \quad \text{and} \quad Y_{12} = \frac{I_2}{V_1} = Cs \qquad (10\text{-}62)$$

with the first result for (a) and the second for (b). Since these equations must represent Z_{12} and Y_{12} as s approaches infinity, the rule covering Z_{12} and Y_{12} must be: The *maximum* degree of the numerator is the degree of the denominator *plus one*. Thus we see that two rules are necessary for transfer functions.

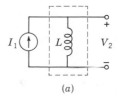

(a)

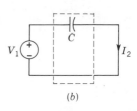

(b)

Fig. 10-20. Special cases of two-port networks which illustrate the degree-difference rules for transfer functions.

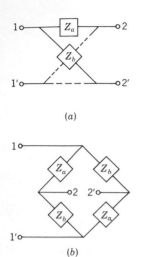

(a)

(b)

Fig. 10-21. Two representations for the symmetrical lattice described by Eq. (10-65). The dashed lines in (a) of the figure indicate that the two series arms have the same impedance Z_a and also that the two shunt arms have the same impedance Z_b.

What may be the *minimum* degree of $p(s)$ compared to $q(s)$? Again, an example is used to illustrate a general result. For the network of Fig. 10-21 (a symmetrical lattice), observe that

$$V_2 = V_1\left(\frac{Z_b}{Z_a + Z_b} - \frac{Z_a}{Z_a + Z_b}\right) \tag{10-63}$$

Since from port 1-1' the network consists of two identical branches connected in parallel, we see that

$$V_1 = I_1[\tfrac{1}{2}(Z_a + Z_b)] \tag{10-64}$$

Substituting this value of V_1 into Eq. (10-63), we obtain

$$Z_{12} = \frac{V_2}{I_1} = \frac{1}{2}(Z_b - Z_a) \tag{10-65}$$

If we let $Z_b = p_b/q_b$ and $Z_a = p_a/q_a$, we see that

$$Z_{12} = \frac{1}{2}\frac{(p_b q_a - p_a q_b)}{q_a q_b} \tag{10-66}$$

This result is useful because it relates a transfer function to the difference of driving-point functions.

What is the lowest possible degree of the numerator? Clearly, the answer is *zero* corresponding to term-by-term cancellation except for the constant term. The Z_{12} of Eq. (10-21) determined for the network of Example 6 was an example of this case, $Z_{12} = (\tfrac{1}{2})/s(s + 1.5)$. Equation (10-66) is useful in examining the conclusions we reached from stability considerations. In the numerator polynomial, any term may be zero or negative, depending on the pattern of cancellation in Eq. (10-66). The only requirement is that the polynomials have zeros which occur in conjugate pairs if complex, whether in

Table 10-3. NECESSARY CONDITIONS FOR TRANSFER FUNCTIONS [WITH COMMON FACTORS IN $p(s)$ AND $q(s)$ CANCELLED]

1. The coefficients in the polynomials $p(s)$ and $q(s)$ of $N = p/q$ must be real, and those for $q(s)$ must be positive.
2. Poles and zeros must be conjugate if imaginary or complex.
3. (a) The real part of poles must be negative or zero; in addition:
   (b) If the real part is zero, then that pole must be simple. This includes the origin.
4. The polynomials $q(s)$ may not have any missing terms between that of highest and lowest degree, unless all even or all odd terms are missing.
5. The polynomial $p(s)$ may have terms missing between the terms of lowest and highest degree, and some of the coefficients may be negative.
6. The degree of $p(s)$ may be as small as zero independent of the degree of $q(s)$.
7. (a) For G_{12} and α_{12}: The maximum degree of $p(s)$ is the degree of $q(s)$.
   (b) For Z_{12} and Y_{12}: The maximum degree of $p(s)$ is the degree of $q(s)$ plus one.

the left or the right half-plane. While this result was developed for Z_{12}, it holds for all the other transfer functions we have considered. The result of this section are summarized in Table 10-3.

10-7. TIME-DOMAIN BEHAVIOR FROM THE POLE AND ZERO PLOT

In this section, we will show that the time-domain response of a system can be determined from the s-plane plot of the poles and zeros of a function and those of the transform of the network sources. Suppose that the transform of a current $I(s)$ is found, and the poles and zeros are determined as

$$I(s) = Y(s)V(s) = \frac{p(s)}{q(s)} \tag{10-67}$$

where

$$\frac{p(s)}{q(s)} = H\frac{(s - s_1)(s - s_2)\ldots(s - s_n)}{(s - s_a)(s - s_b)\ldots(s - s_m)} \tag{10-68}$$

It was shown in Section 10-5 that the poles of this function determine the time-domain behavior of $i(t)$. It was suggested that the zeros together with the poles determine the magnitude of each of the terms of $i(t)$. In this section, we amplify these concepts by showing how $i(t)$ can be determined from a knowledge of the poles, the zeros, and the scale factor H.

In terms of the damping ratio ζ and the undamped natural frequency ω_n, as discussed in Chapter 6, the poles and zeros of the last equation will have the following forms.

$$s_1, s_2 = -\zeta\omega_n \pm j\omega_n\sqrt{1 - \zeta^2}, \quad \zeta < 1 \tag{10-69}$$

$$s_1, s_2 = -\zeta\omega_n \pm \omega_n\sqrt{\zeta^2 - 1}, \quad \zeta > 1 \tag{10-70}$$

$$s_1, s_2 = -\omega_n, \quad \zeta = 1 \tag{10-71}$$

$$s_1, s_2 = \pm j\omega_n, \quad \zeta = 0 \tag{10-72}$$

It was also shown in Chapter 6 that contours of constant ω_n are circles in the s plane, that contours of constant damping ratio are straight lines through the origin, and that contours of constant damping ($\zeta\omega_n$) are straight lines parallel to the $j\omega$ axis of the s plane. Further, lines parallel to the σ axis of the s plane are lines of constant frequency, $\omega_n\sqrt{1 - \zeta^2}$. These facts are summarized in Fig. 10-22.

The location of the poles in the s plane can be interpreted in terms of the general time-domain response in terms of ζ and ω_n.

$$i(t) = K_1 e^{(-\zeta\omega_n + \omega_n\sqrt{\zeta^2 - 1})t} + K_2 e^{(-\zeta\omega_n - \omega_n\sqrt{\zeta^2 - 1})t} \tag{10-73}$$

To illustrate the meaning of the contours of Fig. 10-22, consider the array of poles shown in Fig. 10-23 (zeros have been omitted for clarity).

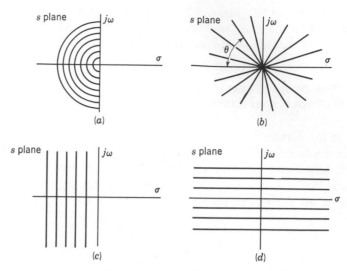

Fig. 10-22. Contours of constant parameter values in the s plane: (a) contours of constant ω_n, (b) lines of constant damping ratio at the angle $\theta = \cos^{-1}\zeta$ with respect to the negative real axis, (c) straight lines parallel to the imaginary axis are contours of constant damping, $\sigma = \zeta\omega_n$, and (d) straight lines parallel to the real axis represent contours of constant frequency of oscillation, $\omega = \omega_n\sqrt{1 - \zeta^2}$.

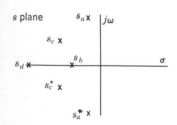

Fig. 10-23. An array of poles in the s plane.

The pair of poles s_a and s_a^* and the pair s_c and s_c^* correspond to oscillatory expressions in the time domain. The frequency of oscillation corresponding to s_a and s_a^* is higher than that of s_c and s_c^*, just as the damping (or rate of decreasing amplitude) is less for s_a and s_a^* than for s_c and s_c^*. The natural frequency ω_n of the two pole pairs is approximately the same, since they are on about the same radius from the origin. The difference in actual frequency of oscillation is due to a lower damping ratio for s_a and s_a^*.

The poles s_b and s_d are quite different from the conjugate pairs just considered. They correspond to the overdamped case, and have an exponential decay form in the time domain. The damping is greater for s_d than for s_b. From another point of view, the time constant for the pole s_b is greater than that for s_d. Typical time-domain response corresponding to each pole is shown in Fig. 10-24 for an arbitrary amplitude for each factor. The total response corresponding to these poles is found by *adding* each of the individual factors as

$$i(t) = K_a e^{s_a t} + K_a^* e^{s_a^* t} + K_b e^{s_b t} + K_c e^{s_c t} + K_c^* e^{s_c^* t} + K_d e^{s_d t} \quad (10\text{-}74)$$

As usual, the terms corresponding to conjugate pairs will combine to give damped sinusoidal expressions. A more detailed figure showing the variation of the response for a variety of pole locations is shown in Fig. 10-25.

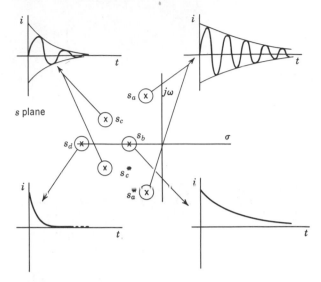

Fig. 10-24. The form of the responses identified with each of the pole positions shown in Fig. 10-23. The responses are shown with arbitrary amplitudes.

There remains the problem of determining the multiplying constant (or magnitude) for each of the terms (or modes). The starting point is Eq. (10-68). To find the time-domain response corresponding to this transform equation, we expand by partial fractions. Hence

$$I(s) = \frac{K_a}{s - s_a} + \frac{K_b}{s - s_b} + \cdots + \frac{K_r}{s - s_r} + \cdots + \frac{K_m}{s - s_m} \quad (10\text{-}75)$$

Any of the coefficients (residues) can be found by the Heaviside method as

$$K_r = H \frac{(s - s_1)(s - s_2)\ldots(s - s_n)}{(s - s_a)\ldots(s - s_r)\ldots(s - s_m)}(s - s_r)\Big|_{s = +s_r} \quad (10\text{-}76)$$

Substituting s_r for s in Eq. (10-76) gives the following value for K_r.

$$K_r = H \frac{(s_r - s_1)(s_r - s_2)\ldots(s_r - s_n)}{(s_r - s_a)(s_r - s_b)\ldots(s_r - s_m)} \quad (10\text{-}77)$$

This equation is composed of factors of the general form $(s_r - s_n)$, where both s_r and s_n are known complex numbers. The difference of two complex numbers is another complex number, which may be written in polar form as

$$(s_r - s_n) = M_{nr}e^{j\phi_{nr}} \quad (10\text{-}78)$$

where M_{nr} is the magnitude of the phasor $(s_r - s_n)$, and ϕ_{nr} is the phase angle of the same phasor. The difference of the two complex quantities s_r and s_n is illustrated in Fig. 10-26 (other poles and zeros

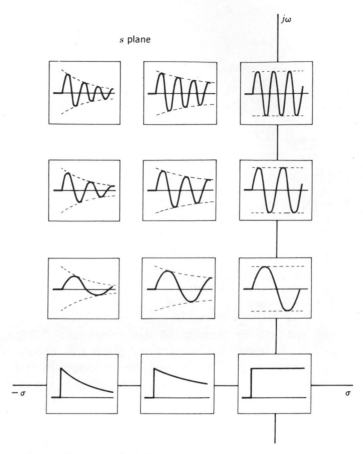

Fig. 10-25. A more detailed exploration of the forms of response for the various s-plane locations.

are omitted again for clarity). The term $(s_r - s_n)$ is interpreted as a phasor directed *from s_n to s_r*. The magnitude M_{nr} is the *distance* from s_n to s_r; the phase angle ϕ_{nr} is the angle of the line from s_n to s_r, measured with respect to the $\phi = 0$ line. The magnitude and phase of the factor $(s_r - s_n)$ are thus easily measured, and so all terms of this general type in Eq. (10-77) are easily found. In terms of M and ϕ for each factor in Eq. (10-77), the value of K_r is seen to become

$$K_r = H \frac{M_{1r} M_{2r} M_{3r} \ldots M_{nr}}{M_{ar} M_{br} M_{cr} \ldots M_{mr}} e^{j(\phi_{1r} + \phi_{2r} + \ldots - \phi_{ar} - \phi_{br} \ldots)} \qquad (10\text{-}79)$$

This equation gives the magnitude and phase of K_r. By performing the operations indicated by this equation, the constant K_r can be evaluated. Determining the quantities in Eq. (10-79) is readily accomplished by a

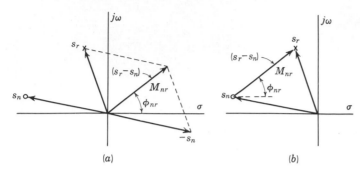

(a)   (b)

Fig. 10-26 The representation of the complex quantity $(s_r - s_n)$ in the complex s plane: (a) the construction of $(s_r - s_n)$, and (b) representation as a phasor directed from s_n to s_r.

graphical procedure which may be outlined as:

(1) Plot the poles and zeros of $I(s) = p(s)/q(s)$ *to scale* on the complex s plane.
(2) Measure (or compute) the distance *from* each of the other finite poles and zeros *to* a given pole s_r.
(3) Measure (or compute) the angle *from* each of the other finite poles and zeros *to* a given pole s_r.
(4) Substitute these quantities into Eq. (10-79) and so evaluate K_r.

EXAMPLE 13

An example will illustrate this procedure. Suppose $I(s)$ has poles $s = -1$ and -3 and a zero at the origin, and H is given as 5. The current transform has the form

$$I(s) = \frac{5s}{(s + 1)(s + 3)} \tag{10-80}$$

This function is easily expanded by partial fractions, but can also be evaluated as outlined above. Referring to Fig. 10-27, it is seen that

$$M_{01}e^{j\phi_{01}} = 1e^{j180°} \tag{10-81}$$

$$M_{31}e^{j\phi_{31}} = 2e^{j0°} \tag{10-82}$$

Fig. 10-27. Poles and zero used in Example 13.

Hence

$$K_1 = H\frac{M_{01}e^{j\phi_{01}}}{M_{31}e^{j\phi_{31}}} = 5 \times \tfrac{1}{2}e^{+j180°} = -2.5 \qquad (10\text{-}83)$$

Similarly,

$$K_3 = H\frac{M_{03}e^{j\phi_{03}}}{M_{13}e^{j\phi_{13}}} = 5 \times \frac{3e^{j180°}}{2e^{j180°}} = 7.5 \qquad (10\text{-}84)$$

Since the poles determine the frequency (in this case neper frequency), we write for the general solution,

$$i(t) = K_1 e^{-t} + K_3 e^{-3t} \qquad (10\text{-}85)$$

and since K_1 and K_3 have been evaluated from a knowledge of the pole and zero locations, we have as a particular solution,

$$i(t) = -2.5e^{-t} + 7.5e^{-3t} \qquad (10\text{-}86)$$

From this discussion and with the aid of Eq. (10-77), the influence of a zero on the time-domain response can be visualized. Consider one pole, say, s_r, in Fig. 10-26. If all other poles and zeros in the s plane remain fixed in position and the zero s_n is moved, the proximity of a zero to a pole is seen by Eq. (10-79) to *reduce* the magnitude of the K coefficient associated with the complex frequency of the pole s_r. Again, from Eq. (10-79), proximity of a pole to s_r is seen to have the opposite effect—since pole magnitudes appear in the denominator—and proximity of another pole to s_r *increases* the magnitude of the coefficient K_r. When the zero s_n is moved so close to s_r that they *coincide*, the pole and zero *cancel* and reduce the value of the particular K_r to zero.

The magnitude of the K coefficient corresponding to a particular pole is thus determined by the proximity of both poles and zeros. If, in the design of a network, the position of the poles and zeros can be selected, they should be selected according to the following pattern:

(1) Select pole locations to give the required time behavior. Do this in terms of complex frequencies.
(2) Fix the position of the zeros in the complex plane to adjust the magnitudes of the various K coefficients.

It should be noted that the graphical interpretation of the position of poles and zeros was discussed for the case of nonrepeated (or simple) poles. In the case of multiple poles, it is suggested that expansion by partial fractions be followed rather than seeking a modification of the procedures that have been discussed to fit the new case.

10-8. STABILITY OF ACTIVE NETWORKS

Thus far in this chapter, we have assumed that the networks we have studied were composed of passive elements only, and our discussion has been based on the assertion that *passive networks are stable* in the sense that the poles of the network functions describing them are excluded from the right half of the s plane. That *active networks*—for example, one containing controlled (or dependent) sources—*are not necessarily stable* will now be shown by a simple example.

The network of Fig. 10-28 contains one controlled source whose voltage is related to V_2 by a positive constant A. Routine analysis of the network gives the transfer function

$$\frac{V_2}{I_1} = \frac{1}{C}\frac{s}{s^2 + [(1/R_1C) + (1 - A)/(R_2C)]s + 1/LC} \qquad (10\text{-}87)$$

If we select specific values for the elements by letting $R_1 = \frac{1}{2}$, $R_2 = 1$, $L = \frac{1}{2}$, and $C = 1$, then Eq. (10-87) becomes

$$\frac{V_2}{I_1} = \frac{s}{s^2 + (3 - A)s + 2} \qquad (10\text{-}88)$$

From this result, we see that the pole locations are determined by the constant A, and also that the poles will migrate in the s plane as A increases from 0 to a very large value. The locus of the roots of the equation $s^2 + (3 - A)s + 2 = 0$ is shown in Fig. 10-29; such plots are widely used in studying systems and are known as *root locus plots.*[4] From this plot we see that the poles are located on the negative real axis for $A = 0$. As A increases, the poles move toward each other, meeting at $s = -\sqrt{2}$. Thereafter, the locus is a circle for the range of values, $3 - 2\sqrt{2} \le A \le 3 + 2\sqrt{2}$. When $A = 3$, the poles are located on the imaginary axis. For $A > 3 + 2\sqrt{2}$, the poles are again on the real axis, but remain in the right half of the s plane—one moving toward zero and the other toward infinity. Clearly, the poles pass the critical boundary which is the imaginary axis for a range of values of A, and the network output is thus capable of being stable, oscillatory, or unstable.

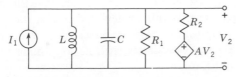

Fig. 10-28. Network with a controlled source which is unstable for a range of values of A.

[4]Root locus plots and rules for their construction were first described by W. R. Evans.

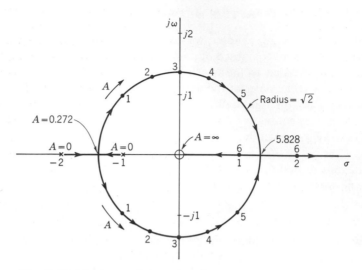

Fig. 10-29. The locus of the poles of the transfer function as A varies from 0 to ∞.

We state that an active network (or any system in general, for that matter) is *stable* if the transfer function relating output to input has poles which are confined to the left half-plane and the imaginary axis, and *strictly stable* if the poles are in the left half-plane only. Thus an active network is stable if it oscillates with constant magnitude, corresponding to the case of poles on the imaginary axis. For an active network to be strictly stable, the possibility of oscillation with constant magnitude is excluded, meaning that poles on the imaginary axis are excluded. In terms of this discussion, and also the various forms of response shown in Fig. 10-25, we see that an equivalent requirement for a stable system is that a *bounded input must give rise to a bounded output*. Thus, in a stable network, an input which is in the form of a step will not cause an output having terms like t, t^2, e^t, $t \sin \omega t$, etc. This is a conceptually convenient definition of a stable network response.

For the network of Fig. 10-28, we see that the network has a response which is strictly stable when $A < 3$, and stable when $A \leq 3$. When $A = 3$, the output voltage oscillates at the frequency $\omega = \sqrt{2}$. For $A > 3$, the output oscillates with an amplitude which increases with time without limit.

Since stability is determined by pole location, as we have just seen in terms of an example, we may state the necessary conditions for stability in terms of a requirement on the denominator polynomial of the transfer function relating output to input. Let this polynomial be

$$p(s) = a_0 s^n + a_1 s^{n-1} + \ldots + a_{n-1} s + a_n \qquad (10\text{-}89)$$

In terms of $p(s)$, we may state the requirement for stability in terms of a simple question: Does $p(s) = 0$ have roots with positive (or zero) real parts? Once this question is answered, the stability of the response of the network is determined. An obvious answer to the question just posed is provided if the roots of the equation $p(s) = 0$ are determined by using a digital computer. However, we are often not so much interested in numerical values of the roots as we are to a simple *yes* or *no* answer to the question: Is the network stable?

Our problem, then, is the following: Given a polynomial having real coefficients for the reasons discussed in Section 10-5, how many of the roots of the equation $p(s) - 0$ have positive real parts? For a stable network, the answer must be none. One of the first to investigate this problem was James Clerk Maxwell in 1868. The topic has since been extensively studied, and there are many rules, algorithms, and criteria available for our use. However, before outlining one of the most useful of these, let us consider some of the properties of the roots of $p(s) = 0$ which can be deduced directly from the coefficients of $p(s)$.

The example

$$p(s) = (s + 4)(s^2 - 2s + 10) = s^3 + 2s^2 + 2s + 40 \quad (10\text{-}90)$$

shows that it is not sufficient that the coefficients of the polynomial all be positive to insure that $p(s) = 0$ have roots with negative real parts. If Eq. (10-89) is factored, it is written as the product

$$a_0(s - s_1)(s - s_2) \dots (s - s_n) = 0 \quad (10\text{-}91)$$

When these factors are multiplied together, we obtain

$$
\begin{aligned}
p(s) = a_0 s^n &- a_0(s_1 + s_2 + \dots + s_n)s^{n-1} \\
&+ a_0(s_1 s_2 + s_2 s_3 + \dots)s^{n-2} \\
&- a_0(s_1 s_2 s_3 + s_4 s_2 s_3 + \dots)s^{n-3} \\
&+ a_0(-1)^n s_1 s_2 s_3 \dots s_n = 0
\end{aligned}
\quad (10\text{-}92)
$$

Equating these coefficients to those of Eq. (10-89), we have (assuming that $a_0 \neq 0$)

$$\frac{a_1}{a_0} = -\text{sum of the roots} \quad (10\text{-}93)$$

$$\frac{a_2}{a_0} = \begin{array}{l}\text{sum of products of roots taken} \\ \text{two at a time}\end{array} \quad (10\text{-}94)$$

$$\frac{a_3}{a_0} = \begin{array}{l}-\text{sum of products of roots taken} \\ \text{three at a time}\end{array} \quad (10\text{-}95)$$

and so on, and finally that

$$\frac{(-1)^n a_n}{a_0} = \text{product of the roots} \quad (10\text{-}96)$$

If the roots all have negative real parts, then we see from these equations that *it is necessary for all the coefficients to have the same sign.* The counterexample of Eq. (10-90) has shown that this necessary condition is not sufficient! We also see that *no coefficient can have zero value*, for this would require cancellation that is possible only if there are roots with positive real parts.

So much for necessary conditions, which are useful for preliminary checking. We next turn to an enumeration of conditions on the coefficients which are both necessary and sufficient to establish stability.

Early contributions to the theory of stability of physical systems were made by Routh in England in 1877, Lyapunov in Russia in 1892, and by Hurwitz in Germany in 1895. More recent additions have been made by Lienard and Chipart in France in 1914, and by Nyquist of the United States in 1932.[5] So much of the work of each of these men was done independently of the others that some confusion has resulted in attributing criteria to the proper authors. The criterion we shall describe next is widely known as the Routh criterion. To acknowledge that it was also developed independently by Hurwitz, engineers now commonly designate it as the *Routh-Hurwitz criterion*. The reader of the literature should not be surprised, however, to find a Routh-Hurwitz criterion described in terms of the evaluation of a number of determinants.

Let us separate $p(s)$ of Eq. (10-89) into two parts, the even part $m(s)$ and the odd part $n(s)$. Note that this is an appropriate choice since M has even symmetry with respect to a vertical line down its center while N has odd symmetry with respect to the same line. The even (odd) part of $p(s)$ is the summation of terms $a_k s^k$ where k is even (odd). We next form a new function

$$\Psi = \frac{m(s)}{n(s)} \quad \text{or} \quad \Psi = \frac{n(s)}{m(s)} \tag{10-97}$$

such that the numerator polynomial is of higher degree than the denominator polynomial. The criterion we shall study is concerned with the coefficients chosen from this $\Psi(s)$. Before describing the criterion, let us consider what happens when $p(s)$ contains a factor of the form $(s^2 + a^2)$ corresponding to conjugate imaginary roots. Then

$$p(s) = p_1(s)(s^2 + a^2) \tag{10-98}$$

If we let $p_1(s) = m_1(s) + n_1(s)$, then the even and odd parts of $p(s)$ will be $m_1(s^2 + a^2)$ and $n_1(s^2 + a^2)$ so that when we form the function specified by Eq. (10-97), we have

$$\Psi(s) = \frac{m_1(s^2 + a^2)}{n_1(s^2 + a^2)} \tag{10-99}$$

---

[5]For a detailed discussion of the Nyquist criterion, see Norman, Balabanian, Theodore A. Bickart, and Sundarem Seshu, *Electrical Network Theory* (John Wiley & Sons, Inc., New York, 1969), pp. 677–689.

assuming that m_1 is of higher degree than n_1. If $\Psi(s)$ is written in this form, we will immediately cancel the common factor in numerator and denominator. Unfortunately, it is not always obvious that there is such a common factor. To illustrate this point, let

$$p(s) = (s + 1)(s + 2)(s^2 + 3)$$
$$= s^4 + 3s^3 + 5s^2 + 9s + 6 \qquad (10\text{-}100)$$

so that

$$\Psi(s) = \frac{s^4 + 5s^2 + 6}{3s^2 + 9s}. \qquad (10\text{-}101)$$

in which the common factor is not immediately apparent. However, if we should factor numerator and denominator, then

$$\Psi = \frac{(s^2 + 2)(s^2 + 3)}{3s(s^2 + 3)} = \frac{s^2 + 2}{3s} \qquad (10\text{-}102)$$

This common factor cancellation relates to one of the special cases of the criterion which we will study soon.

In stating the steps that are to be followed, we first construct the Routh-Hurwitz *array* (which is also known as a *schedule* or *scheme*) which consists of the coefficients of the Ψ function of Eq. (10-97). We separate $p(s)$ into its even and odd parts and form two rows of coefficients. Thus from the given $p(s)$, we select the coefficients for the two rows as follows:

Row 1

$$a_0 s^n + a_1 s^{n-1} + a_2 s^{n-2} + a_3 s^{n-3} + a_4 s^{n-4} + a_5 s^{n-5} + a_6 s^{n-6} + \dots$$

Row 2

$$(10\text{-}103)$$

and written as

$$
\begin{array}{cccccc}
a_0 & a_2 & a_4 & a_6 & a_8 & \dots \\
a_1 & a_3 & a_5 & a_7 & a_9 & \dots
\end{array} \qquad (10\text{-}104)
$$

As the next step, complete the Routh-Hurwitz array of numbers. For $n = 6$, this is

$$
\begin{array}{c|cccc}
s^6 & a_0 & a_2 & a_4 & a_6 \\
s^5 & a_1 & a_3 & a_5 & \\
s^4 & b_1 & b_3 & b_5 & \\
s^3 & c_1 & c_3 & & \\
s^2 & d_1 & d_3 & & \\
s^1 & e_1 & & & \\
s^0 & f_1 & & &
\end{array} \qquad (10\text{-}105)
$$

consisting of $n + 1$ rows, the first two being given in Eq. (10-104). The b, c, d, e, f entries are determined by the algorithm illustrated in Fig. 10-30. The algorithm is applied as follows in determining an element

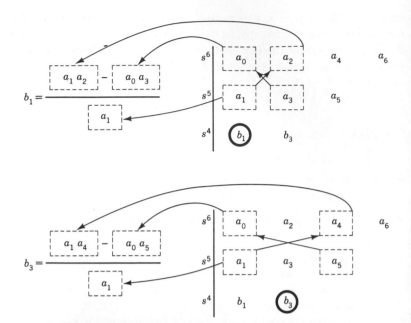

Fig. 10-30. The figure illustrates the pattern of steps involved in determining the elements of the Routh array. For further discussion, see Prob. 10-33.

in row k and column j. The value of this element is determined from four other elements. These elements are in the two rows directly above the element to be determined, rows $k - 1$ and $k - 2$. Further, they are in column 1 of the array and in the column to the immediate right of the element, column $j + 1$. These elements form a determinant-like structure. The elements joined by a line with positive slope have a positive sign; elements joined with a line with negative slope have a negative sign (just the opposite of the rule for determinants). We subtract these two products and divide this difference by the pivotal element of column 1 and row $k - 1$. This process is continued until $n + 1$ rows are formed.[6] To apply the criterion, we focus our attention on the first column of the array. The column to the right of the line in Eq. (10-105) is an index which is useful for accounting purposes.

The Routh-Hurwitz theorem states that the number of changes in sign of the first column (as we scan from top to bottom) *is equal to the number of roots of $p(s) = 0$ with positive real parts.* The Routh-Hurwitz criterion states that the network (or system) described by a network function, for which $p(s)$ is the denominator polynomial, is stable if and only if there are no changes of sign in the first column of the array. This requirement is both necessary and sufficient for stability.

[6]For further study of this algorithm, see Prob. 10-33.

In applying this criterion, it is necessary to distinguish three cases: (1) No element in the first column is zero. (2) There is a zero in the first column, but all other elements of the row containing a zero in the first column are nonzero. (3) There are zeros in the first column and all other elements of the row containing the zero in the first column are also zero. We now illustrate these three cases by means of examples.

EXAMPLE 14 (Case 1)

Consider the identity

$$(s + 1)(s + 2)(s + 3)(s + 4)$$
$$= s^4 + 10s^3 + 35s^2 + 50s + 24 \tag{10-106}$$

The Routh array is formed to give the following:

$$
\begin{array}{c|ccc}
s^4 & 1 & 35 & 24 \\
s^3 & 10 & 50 & \\
s^2 & 30 & 24 & \\
s^1 & 42 & & \\
s^0 & 24 & &
\end{array}
$$

From the first column, it is seen that there are no roots with positive real parts (agreeing with the known roots).

EXAMPLE 15 (Case 1)

As a second example, consider Eq. (10-90),

$$s^3 + 2s^2 + 2s + 40 = 0 \tag{10-107}$$

which is known to have two roots with positive real parts.

$$
\begin{array}{c|cc}
s^3 & 1 & 2 \\
s^2 & 2 & 40 \\
s^1 & -18 & \\
s^0 & 40 &
\end{array}
$$

There are two changes of sign (2 to -18 and -18 to 40) as required.

EXAMPLE 16 (Case 1)

Consider the equation

$$a_0 s^3 + a_1 s^2 + a_2 s + a_3 = 0 \tag{10-108}$$

The Routh-Hurwitz array for this equation is

$$
\begin{array}{c|cc}
s^3 & a_0 & a_2 \\
s^2 & a_1 & a_3 \\
s^1 & \dfrac{a_1 a_2 - a_0 a_3}{a_1} & \\
s^0 & a_3 &
\end{array}
$$

From the array we conclude that it is necessary and sufficient that all coefficients in the equation be positive and, in addition, that $a_1 a_2 > a_0 a_3$ in order that there be no roots with positive real parts.

EXAMPLE 17 (Case 1)

In forming the Routh-Hurwitz array for the equation

$$p(s) = s^4 + s^3 + 2s^2 + 2s + 3 = 0 \qquad (10\text{-}109)$$

the element of the first column, third row is found to be zero, but no other element of this row is zero. To circumvent this difficulty,[7] we replace 0 with the small quantity ϵ, so that the array becomes

$$
\begin{array}{c|ccc}
s^4 & 1 & 2 & 3 \\
s^3 & 1 & 2 & \\
s^2 & \epsilon & 3 & \\
s^1 & 2 - \dfrac{3}{\epsilon} & & \\
s_0 & 3 & &
\end{array}
$$

For small $\epsilon > 0$, the fourth element of the first column is negative, indicating two changes of sign so that two roots have positive real parts. For small $\epsilon < 0$, the third element is negative, leading to the same conclusion.

EXAMPLE 18 (Case 1)

To illustrate the case in which an entire row of the Routh-Hurwitz array vanishes (all elements of the row have zero value), consider a network which is described by the denominator polynomial

$$p(s) = s^3 + 2s^2 + 2s + A = 0 \qquad (10\text{-}110)$$

where A is adjustable, as was the case for the network of Fig. 10-28. The Routh-Hurwitz array is formed as follows:

$$
\begin{array}{c|cc}
s^3 & 1 & 2 \\
s^2 & 2 & A \\
s^1 & \dfrac{4 - A}{2} & \\
s^0 & A &
\end{array}
\qquad (10\text{-}111)
$$

From this array, we see that the system is stable for $A < 4$, but unstable for $A > 4$. Let us examine the case $A = 4$ in more detail.

[7]F. R. Gantmacher, *Applications of the Theory of Matrices* (Interscience Publishers, Inc., New York, 1959), p. 215ff.

For this choice, the Routh-Hurwitz array is

$$\begin{array}{c|cc} s^3 & 1 & 2 \\ s^2 & 2 & 4 \\ s^1 & 0 & \\ s^0 & ? & \end{array} \qquad (10\text{-}112)$$

In this case, an entire row (consisting of one element) has vanished, and it is not possible to complete the array.

The difficulty in this case is related to our earlier study of Eq. (10-98), the presence of a factor of the type $(s^2 + a^2)$ in $p(s)$. This factor causes a premature termination of the formation of the Routh-Hurwitz array. To determine this common factor, we proceed as follows. Let the unknown common factor, an even polynomial, be $p_2(s)$. The index of the vanishing row is s^{v-1}, meaning that the previous row of index s^v has nonzero elements $h_1, h_2, h_3, \ldots$. Then $p_2(s)$ is given by

$$p_2(s) = h_1 s^v + h_2 s^{v-2} + \ldots \qquad (10\text{-}113)$$

Knowing this even polynomial, we may remove it from $p(s)$ and then reapply the Routh-Hurwitz criterion to $p_1(s)$. From array (10-112), we see that

$$p_2(s) = 2(s^2 + 2) \qquad (10\text{-}114)$$

so that by inspection

$$p_1(s) = \tfrac{1}{2}(s + 2) \qquad (10\text{-}115)$$

Observe that the value $A = 4$ caused the roots of $p(s)$ to be located on the imaginary axis, the boundary between stability and instability indicated by our analysis of array (10-111).

EXAMPLE 19 (Case 3)

As a second example of Case 3, consider the polynomial

$$p(s) = (s^2 + 2)(s^2 + 3)(s + 2)(s + 3) \qquad (10\text{-}116)$$

or

$$p(s) = s^6 + 5s^5 + 11s^4 + 25s^3 + 36s^2 + 30s + 36 \qquad (10\text{-}117)$$

The Routh-Hurwitz array is

$$\begin{array}{c|cccc} s^6 & 1 & 11 & 36 & 36 \\ s^5 & 5 & 25 & 30 & \\ s^4 & 6 & 30 & 36 & \\ s^3 & 0 & 0 & 0 & \end{array} \qquad (10\text{-}118)$$

It is clearer here than in array (10-112) that an entire row has vanished! We form the polynomial $p_2(s)$ as indicated by Eq. (10-113) to give

$$p_2(s) = 6(s^4 + 5s^2 + 6) \qquad (10\text{-}119)$$

which is the even polynomial contained in Eq. (10-116). The stability of $p_1(s) = s^2 + 5s + 6$ is evident by inspection.

Case 3 reveals the presence of an even polynomial or an odd polynomial (which is s times an even polynomial). The even polynomial may be a product of factors like $s^2 + a^2 = (s + ja)(s - ja)$ representing zeros on the imaginary axis. There is one other possibility that must be considered before a final conclusion may be reached in Case 3 situations. Consider the zeros shown in Fig. 10-31 which are known as a *quad* of zeros. Their factor product is

$$(s + a + jb)(s + a - jb)(s - a + jb)(s - a - jb)$$
$$= [(s + a)^2 + b^2][(s - a)^2 + b^2] \qquad (10\text{-}120)$$
$$= s^4 + 2(b^2 - a^2)s^2 + (a^2 + b^2)^2$$

By inspection, all coefficients are positive and nonzero if $b > a$. In terms of the s-plane geometry of Fig. 10-31, this situation applies for the quad if $\theta > \pi/4$.

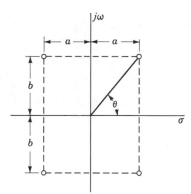

Fig. 10-31. The four zeros, symmetrically located with respect to the s-plane axes, form a *quad* of zeros.

EXAMPLE 20 (Case 3)

The polynomial

$$p(s) = 2s^6 + s^5 + 13s^4 + 6s^3 + 56s^2 + 25s + 25 \qquad (10\text{-}121)$$

has the Routh-Hurwitz array

$$
\begin{array}{c|cccc}
s^6 & 2 & 13 & 56 & 25 \\
s^5 & 1 & 6 & 25 & \\
s^4 & 1 & 6 & 25 & \\
s^3 & 0 & 0 & 0 &
\end{array} \qquad (10\text{-}122)
$$

and again an entire row has vanished. The polynomial $p_2(s)$ is

$$s^4 + 6s^2 + 25 = \prod (s \pm 1) \pm j2 \qquad (10\text{-}123)$$

which is a quad, indicating that $p(s)$ has two zeros in the right half-plane from the quad. Dividing Eq. (10-123) into Eq. (10-121) gives the factor $2s^2 + s + 1$ which may be analyzed by the quadratic formula.

FURTHER READING

DESOER, CHARLES A., AND ERNEST S. KUH, *Basic Circuit Theory*, McGraw-Hill Book Company, New York, 1969. Chapter 15.

KARNI, SHLOMO, *Intermediate Network Analysis*, Allyn and Bacon, Inc., Boston, 1971. Chapter 6.

LATHI, B.P., *Signals, Systems, and Communication*, John Wiley & Sons, Inc., New York, 1965. Chapter 7.

MELSA, JAMES L., AND DONALD G. SCHULTZ, *Linear Control Systems*, McGraw-Hill Book Company, New York, 1969. Chapter 6.

PERKINS, WILLIAM R., AND JOSÉ B. CRUZ, JR., *Engineering of Dynamic Systems*, John Wiley & Sons, Inc., New York, 1969. Chapter 8.

DIGITAL COMPUTER EXERCISES

Two topics of this chapter which lend themselves to computer solution are the determination of the roots of a polynomial and the determination of the locus of roots. The sections of Appendix E devoted to these topics are E-1 and E-9.5. In particular, see Huelsman, reference 7, Appendix E-10, and his discussions of root-locus plots in Section 10.3, and Mc-Cracken, reference 12, Case Studies 21 and 23.

PROBLEMS

10-1. For the network shown in the accompanying figure, determine $Z_{12} = V_2(s)/I_1(s)$.

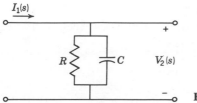

Fig. P10-1.

10-2. Consider the RC two-port network shown in the accompanying figure. For this network show that

$$G_{12} = \left[\frac{s^2 + (R_1C_1 + R_2C_2)s/R_1R_2C_1C_2 + 1/R_1R_2C_1C_2}{s^2 + (R_1C_1 + R_1C_2 + R_2C_2)s/R_1R_2C_1C_2 + 1/R_1R_2C_1C_2} \right]$$

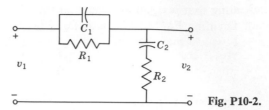

Fig. P10-2.

10-3. (a) For the given network, show that with port 2 open, the input impedance at port 1 is 1 Ω. (b) Find the voltage-ratio transfer function; G_{12} for the two-port network.

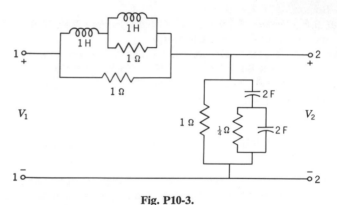

Fig. P10-3.

10-4. For the resistive two-port network of the figure, determine the numerical value for (a) G_{12}, (b) Z_{12}, (c) Y_{12}, and (d) α_{12}.

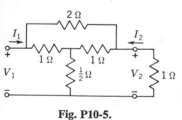

Fig. P10-5.

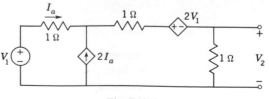

Fig. P10-4.

10-5. The resistive bridged-T, two-port network shown in the figure is to be analyzed to determine (a) G_{12}, (b) Z_{12}, (c) Y_{12}, and (d) α_{12}.

10-6. The given network contains resistors and controlled sources. For this network, compute $G_{12} = V_2/V_1$.

Fig. P10-6.

10-7. For the network of the accompanying figure and the element values specified, determine $\alpha_{12} = I_2/I_1$.

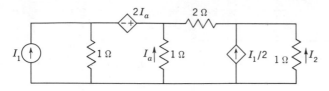

Fig. P10-7.

10-8. For the *RC* two-port network shown in the figure, show that

$$G_{12} = \left[\frac{1/R_1 R_2 C_1 C_2}{s^2 + (R_1 C_1 + R_1 C_2 + R_2 C_2)s/R_1 R_2 C_1 C_2 + 1/R_1 R_2 C_1 C_2} \right]$$

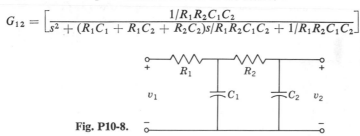

Fig. P10-8.

10-9. For the given network, show that

$$Y_{12} = \frac{K(s+1)}{(s+2)(s+4)}$$

and determine the value and sign of K.

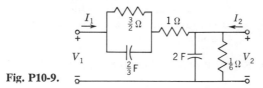

Fig. P10-9.

10-10. For the network shown in the figure, show that the voltage-ratio transfer function is

$$G_{12} = \frac{(s^2+1)^2}{5s^4 + 5s^2 + 1}$$

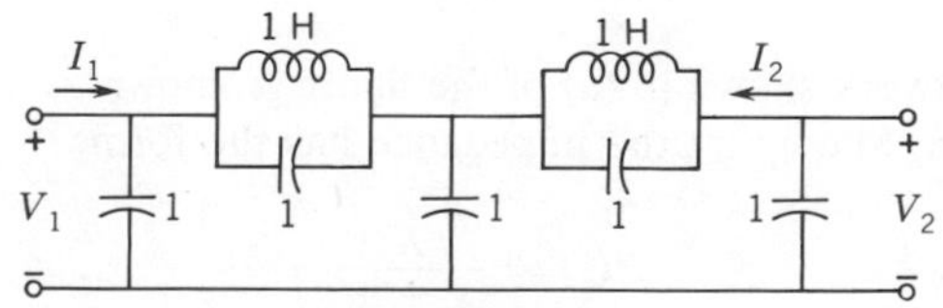

Fig. P10-10.

10-11. For each of the networks shown in the accompanying figure, connect a voltage source V_1 to port 1 and designate polarity references for V_2 at port 2. For each network, determine $G_{12} = V_2/V_1$.

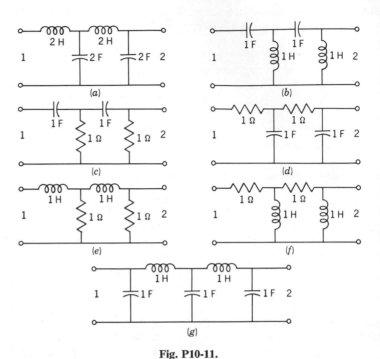

Fig. P10-11.

10-12. For the network given in Fig. P10-11(*a*), terminate port 2 in a 1-Ω resistor and connect a voltage source at port 1. Let I_1 be the current in the voltage source and I_2 be the current in the 1-Ω load. Assign reference directions for each. For this network, compute $G_{12} = V_2/V_1$ and $\alpha_{12} = I_2/I_2$.

10-13. Repeat Prob. 10-12 for the network of Fig. P10-11(*b*).

10-14. Repeat Prob. 10-12 for the network of Fig. P10-11(*g*).

10-15. For the network of Fig. P10-11(*g*), connect a current source I_1 at port 1 and a 1-Ω resistor at port 2. Assign reference directions for all voltages and currents. For this network, compute $Z_{12} = V_2/I_1$ and $\alpha_{12} = I_2/I_1$.

10-16. The network shown in (*a*) of the figure is known as a *shunt peaking* network. Show that the impedance has the form

$$Z(s) = \frac{K(s - z_1)}{(s - p_1)(s - p_2)}$$

and determine z_1, p_1, and p_2 in terms of R, L, and C. If the poles and zeros of $Z(s)$ have the locations shown in (*b*) of the figure with $Z(j0) = 1$, find the values for R, L, and C.

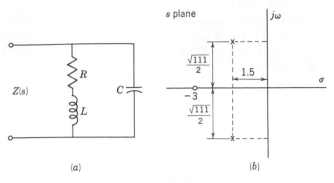

(a)   (b)

Fig. P10-16.

10-17. A system has a transfer function with a pole at $s = -3$ and a zero which may be adjusted in position at $s = -a$. The response of this system to a step input has a term of the form $K_1 e^{-3t}$. Plot the value of K_1 as a function of a for values of a between 0 and 5. This may be done by the graphical procedure of Section 10-7.

10-18. A system has a transfer function with poles at $s = -1 \pm j1$ and a zero which may be adjusted in position at $s = -a$. The response of this system to a step input has a term of the form $K_2 e^{-t} \sin(t + \phi)$. Plot the value of K_2 as a function of a for values of a between 0 and 5. This may be done graphically.

10-19. A system has a transfer function with poles at $s = -1 \pm j1$ and at $s = -3$, and a zero which may be adjusted in position at $s = -a$. One term of the response of this system to a step input is of the form $K_3 e^{-t} \sin(t + \phi)$. Plot the value of K_3 as a function of a for values of a between 0 and 5.

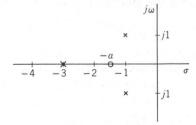

Fig. P10-19.

10-20. Apply the Routh-Hurwitz criterion to the following equations and determine: (a) the number of roots with positive real parts, (b) the number of roots with zero real parts, and (c) the number of roots with negative real parts.
(a) $4s^3 + 7s^2 + 7s + 2 = 0$
(b) $s^3 + 3s^2 + 4s + 1 = 0$
(c) $5s^3 + s^2 + 6s + 2 = 0$
(d) $s^5 + 2s^4 + 2s^3 + 4s^2 + 11s + 10 = 0$

10-21. Given the equation

$$s^3 + 5s^2 + Ks + 1 = 0$$

(a) For what range of values of K will the roots of the equation have negative real parts? (b) Determine the value of K such that the real part vanishes.

10-22. Repeat Prob. 10-20 for the equations:
(a) $5s^4 + 6s^3 + 4s^2 + 2s + 3 = 0$
(b) $s^4 + 3s^3 + 2s^2 + s + 1 = 0$
(c) $2s^4 + 3s^3 + 6s^2 + 7s + 2 = 0$
(d) $3s^6 + s^5 + 19s^4 + 6s^3 + 81s^2 + 25s + 25 = 0$

10-23. Repeat the tests of Prob. 10-20 for the following equations:
(a) $720s^5 + 144s^4 + 214s^3 + 38s^2 + 10s + 1 = 0$
(b) $25s^5 + 105s^4 + 120s^3 + 120s^2 + 20s + 1 = 0$
(c) $s^5 + 5.5s^4 + 14.5s^3 + 8s^2 - 19s - 10 = 0$
(d) $s^5 - s^4 - 2s^3 + 2s^2 - 8s + 8 = 0$
(e) $s^6 + 1 = 0$

10-24. For the following polynomials, (1) determine the number of zeros in the right half of the s plane, (2) determine the number of zeros on the imaginary axis of the s plane. Show method.
(a) $2s^6 + 2s^5 + 3s^4 + 2s^3 + 4s^2 + 3s + 2 = p_1(s)$
(b) $s^6 + 2s^5 + 6s^4 + 10s^3 + 11s^2 + 12s + 6 = p_2(s)$
(c) $2s^6 + 2s^5 + 4s^4 + 3s^3 + 5s^2 + 4s + 1 = p_3(s)$

10-25. For the following polynomial, determine the number of zeros in the right half of the s plane, the left half of the s plane, and on the imaginary axis (the boundary) of the s plane:
(a) $p_1(s) = 2s^7 + 2s^6 + 15s^5 + 17s^4 + 44s^3 + 36s^2 + 24s + 9$
(b) $s^6 + 3s^5 + 4s^4 + 6s^3 + 13s^2 + 27s + 18 = p_2(s)$
(c) $s^8 + 3s^7 + 5s^6 + 9s^5 + 17s^4 + 33s^3 + 31s^2 + 27s + 18 = p_3(s).$

10-26. Consider the equation

$$a_0s^4 + a_1s^3 + a_2s^2 + a_3s + a_4 = 0$$

Use the Routh-Hurwitz criterion to determine a set of conditions necessary in order that all roots of the equation have negative real parts. Assume that all coefficients in the equation are positive.

10-27. For the network of the figure, let $R_1 = R_2 = 1\,\Omega$, $C_1 = 1$ F and $C_2 = 2$ F. For what values of k will the network be stable? In other words, for what values of k will the roots of the characteristic equation have real parts in the left half of the s plane?

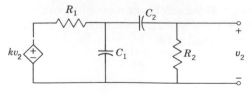

Fig. P10-27.

10-28. For the network of Prob. 10-27, let $k = 2$, $C_1 = 1$ F and $R_2 = 1\,\Omega$. Determine the relationship that must exist between R_1 and C_2 for the system to oscillate, that is, for the roots of the characteristic equation to be conjugate and have zero real parts.

10-29. The amplifier-network shown in the accompanying figure is to be analyzed. (a) What must be the relationship between R_1, R_2, and K for the system to be stable (real parts of the roots of the characteristic equation are zero or negative)? (b) For the system to oscillate without damping, what must be the relationship between R_1, R_2, and K? What will be the frequency of oscillation? Assume that the amplifier has infinite input impedance and zero output impedance.

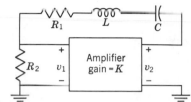

Fig. P10-29.

10-30. The network of the accompanying figure represents a *phase-shift oscillator*. (a) Show that the condition necessary for oscillation is $g_m R_L \geq 29$. (b) Show that the frequency of oscillation when $g_m R_L = 29$ is $\omega_0 = 1/\sqrt{6}\,RC$.

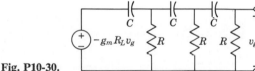

Fig. P10-30.

10-31. Show that with $Z_a Z_b = R_0^2$ in the bridged-T network of the accompanying figure,

$$\frac{V_2}{V_1} = \frac{1}{1 + Z_a/R_0}$$

and the input impedance at port 1 is $Z_{in} = R_0$.

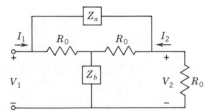

Fig. P10-31.

10-32. An active network is described by the characteristic equation

$$s^2 + (3 + 6K_1)s + 6K_2 = 0$$

It is required that the network be stable and that no component of its response decay more rapidly than $K_1 e^{-3t}$. Show that these conditions are satisfied if $K_2 > 0$, $|K_1| < \frac{1}{2}$, and $K_2 > 3K_1$. Crosshatch the area of permitted values of K_1 and K_2 in the K_1-K_2 plane.

10-33. Values for the elements of the Routh array can also be expressed in terms of second-order determinants multiplied by -1. Thus the formulas shown in Fig. 10-30 become

$$b_1 = \frac{-1}{a_1}\begin{vmatrix} a_0 & a_2 \\ a_1 & a_3 \end{vmatrix}, \qquad b_3 = \frac{-1}{a_1}\begin{vmatrix} a_0 & a_4 \\ a_1 & a_5 \end{vmatrix}$$

$$c_1 = \frac{-1}{b_1}\begin{vmatrix} a_1 & a_3 \\ b_1 & b_3 \end{vmatrix}, \qquad c_3 = \frac{-1}{b_1}\begin{vmatrix} a_1 & a_5 \\ b_1 & b_5 \end{vmatrix}$$

Using the indexing scheme suggested on page 312, give a general formula for the elements of the Routh array.

Two-Port Parameters 11

We next turn our attention to a special class of network functions that is useful in describing the two-port network (and may be generalized to describe an n-port network). As we shall see, these network functions are like those introduced in the last chapter, but with additional restrictions imposed such as the requirement that one of the ports be open or be shorted.

11-1. RELATIONSHIP OF TWO-PORT VARIABLES

In the two-port network of Fig. 11-1, we see four variables identified—two voltages and two currents. There are other voltages and currents that might be identified inside the box, of course. The box enclosing the network has the function of indicating that other voltages and currents are not available for measurement or are not important in a particular problem. We assume that the variables are transform quantities and use V_1 and I_1 as variables at the input, port 1, and V_2 and I_2 as the variables at the output, port 2. Now only two of the four variables are independent, and the specification of any two of them determines the remaining two. For example, if V_1 and V_2 are specified, then I_1 and I_2 are determined. The dependence of two of the four variables on the other two is described in a number of ways, depending on which of the variables are chosen to be the independent variables. In this chapter, we study the six combinations that are given in Table 11-1. As we shall see, the names of the parameters are chosen

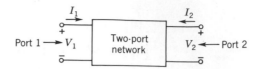

Fig. 11-1. A two-port network with standard reference directions for the voltages and currents indicated.

Table 11-1. TWO-PORT PARAMETERS

Name	Function		Equation
	Express	*In terms of*	
Open-circuit impedance	V_1, V_2	I_1, I_2	$V_1 = z_{11}I_1 + z_{12}I_2$ $V_2 = z_{21}I_1 + z_{22}I_2$
Short-circuit admittance	I_1, I_2	V_1, V_2	$I_1 = y_{11}V_1 + y_{12}V_2$ $I_2 = y_{21}V_1 + y_{22}V_2$
Transmission	V_1, I_1	V_2, I_2	$V_1 = AV_2 - BI_2$ $I_1 = CV_2 - DI_2$
Inverse transmission	V_2, I_2	V_1, I_1	$V_2 = A'V_1 - B'I_1$ $I_2 = C'V_1 - D'I_1$
Hybrid	V_1, I_2	I_1, V_2	$V_1 = h_{11}I_1 + h_{12}V_2$ $I_2 = h_{21}I_1 + h_{22}V_2$
Inverse hybrid	I_1, V_2	V_1, I_2	$I_1 = g_{11}V_1 + g_{12}I_2$ $V_2 = g_{21}V_1 + g_{22}I_2$

to indicate dimensions (impedance, admittance), lack of consistent dimensions (hybrid), or the principal application of the parameter (transmission).

11-2. SHORT-CIRCUIT ADMITTANCE PARAMETERS

The network of Fig. 11-1 is a special case for the general network studied in Chapter 9. We assume that the network does not contain dependent (controlled) sources unless otherwise specified. Then, in Eq. (9-61), all V's are zero except V_1 and V_2. For $k = 1$ and $k = 2$, we write

$$I_1 = \frac{\Delta_{11}}{\Delta} V_1 + \frac{\Delta_{21}}{\Delta} V_2 \tag{11-1}$$

and

$$I_2 = \frac{\Delta_{12}}{\Delta} V_1 + \frac{\Delta_{22}}{\Delta} V_2 \tag{11-2}$$

The quotients of cofactors are dimensionally admittance, and if we introduce the notation $y_{jk} = \Delta_{kj}/\Delta$, they become

$$I_1 = y_{11}V_1 + y_{12}V_2 \qquad (11\text{-}3)$$

and

$$I_2 = y_{21}V_1 + y_{22}V_2 \qquad (11\text{-}4)$$

Observe that if either V_1 or V_2 is zero, the four parameters may be defined in terms of a voltage and current; thus,

$$y_{11} = \left.\frac{I_1}{V_1}\right|_{V_2=0} \qquad (11\text{-}5)$$

$$y_{21} = \left.\frac{I_2}{V_1}\right|_{V_2=0} \qquad (11\text{-}6)$$

$$y_{12} = \left.\frac{I_1}{V_2}\right|_{V_1=0} \qquad (11\text{-}7)$$

and

$$y_{22} = \left.\frac{I_2}{V_2}\right|_{V_1=0} \qquad (11\text{-}8)$$

The condition $V_1 = 0$ or $V_2 = 0$ is accomplished by shorting port 1 or port 2. Thus computations or measurements for the four parameters are made by using one of the connections shown in Fig. 11-2(a) and (c). Since a short-circuit condition is specified for each of the functions in Eqs. (11-5) through (11-8), the parameters are known as *short-circuit*

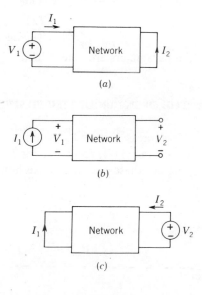

(a)

(b)

(c)

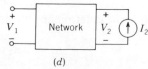

(d)

Fig. 11-2. Four network connections in which (a) and (c) relate to the definitions for the short-circuit admittance functions, and (b) and (d) to the open-circuit impedance functions.

admittance parameters. If the network being studied is *reciprocal*, as discussed in Chapter 9, then by Eq. (9-65) we have

$$\frac{I_1}{V_2}\bigg|_{V_1=0} = \frac{I_2}{V_1}\bigg|_{V_2=0} \tag{11-9}$$

or

$$y_{12} = y_{21} \tag{11-10}$$

and we see that three parameters are sufficient to specify the relationship between I_1, I_2 and V_1, V_2.

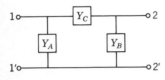

Fig. 11-3. π network used in Example 1.

EXAMPLE 1

Consider the π-network of Fig. 11-3 in which Y_A, Y_B, and Y_C are the admittances of subnetworks. Following the pattern of connections shown in Fig. 11-2, we have from Eqs. (11-5) and (11-7),

$$y_{11} = Y_A + Y_C \tag{11-11}$$

and

$$y_{22} = Y_B + Y_C \tag{11-12}$$

which are two driving-point admittance functions. To find $y_{12} = y_{21}$, we observe from Fig. 11-2(c) that, with port 1 shorted, we have $I_1 = -Y_C V_2$ (the minus sign to account for the reference direction assigned to I_1). Thus from Eq. (11-7),

$$y_{12} = \frac{I_1}{V_2}\bigg|_{V_1=0} = -Y_C \tag{11-13}$$

and the parameters are determined.

11-3. THE OPEN-CIRCUIT IMPEDANCE PARAMETERS

To express V_1 and V_2 in terms of I_1 and I_2, we may begin with Eqs. (11-3) and (11-4) and solve for V_1 and V_2 by using determinants. Carrying out these operations, we obtain

$$V_1 = \frac{y_{22}}{\Delta_y} I_1 + \frac{-y_{12}}{\Delta_y} I_2 \tag{11-14}$$

and

$$V_2 = \frac{-y_{21}}{\Delta_y} I_1 + \frac{y_{11}}{\Delta_y} I_2 \tag{11-15}$$

where

$$\Delta_y = y_{11}y_{22} - y_{12}y_{21} \tag{11-16}$$

which is the determinant of the admittance matrix,

$$\Delta_y = \det[y] = \det\begin{bmatrix} y_{11} & y_{12} \\ y_{21} & y_{22} \end{bmatrix} \tag{11-17}$$

The quantities multiplying I_1 and I_2 in Eqs. (11-14) and (11-15) are dimensionally impedance, and are assigned a single symbol, as in Table 11-1, such that

$$V_1 = z_{11}I_1 + z_{12}I_2 \tag{11-18}$$

and

$$V_2 = z_{21}I_1 + z_{22}I_2 \tag{11-19}$$

These parameters may be interpreted in terms of a single voltage and current by letting either $I_1 = 0$ or $I_2 = 0$. Thus we obtain a set of equations like those given in Eqs. (11-5) through (11-8). They are

$$z_{11} = \frac{V_1}{I_1}\bigg|_{I_2=0} \tag{11-20}$$

$$z_{21} = \frac{V_2}{I_1}\bigg|_{I_2=0} \tag{11-21}$$

$$z_{12} = \frac{V_1}{I_2}\bigg|_{I_1=0} \tag{11-22}$$

and

$$z_{22} = \frac{V_2}{I_2}\bigg|_{I_1=0} \tag{11-23}$$

The condition $I_1 = 0$ or $I_2 = 0$ implies an open circuit at port 1 or port 2, accounting for the designation of the parameters as *open-circuit impedance parameters*. Observe that $y_{12} = y_{21}$ implies

$$z_{12} = z_{21} \tag{11-24}$$

for reciprocal networks. Note that we might have started with Eqs. (11-18) and (11-19) in finding Eq. (11-3) and (11-4). Solving Eqs. (11-18) and (11-19) leads to

$$I_1 = \frac{z_{22}}{\Delta_z}V_1 + \frac{-z_{12}}{\Delta_z}V_2 \tag{11-25}$$

and

$$I_2 = \frac{-z_{21}}{\Delta_z}V_1 + \frac{z_{11}}{\Delta_z}V_2 \tag{11-26}$$

where

$$\Delta_z = z_{11}z_{22} - z_{12}z_{21} \tag{11-27}$$

which is the determinant of the impedance matrix

$$\Delta_z = \det[z] = \det\begin{bmatrix} z_{11} & z_{12} \\ z_{21} & z_{22} \end{bmatrix} \tag{11-28}$$

Comparing terms in Eqs. (11-15), (11-18) and (11-19), we find that

$$z_{11}y_{11} = z_{22}y_{22} \tag{11-29}$$

and also that

$$\Delta_z\Delta_y = 1 \tag{11-30}$$

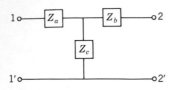

Fig. 11-4. *T* network for Example 2.

EXAMPLE 2

Let it be required to find the open-circuit impedance parameters for the *T*-network of Fig. 11-4. As in Example 1, the two driving-point functions and the one transfer function are found to be

$$z_{11} = Z_a + Z_c \tag{11-31}$$
$$z_{22} = Z_b + Z_c \tag{11-32}$$

and

$$z_{12} = z_{21} = Z_c \tag{11-33}$$

EXAMPLE 3

The network of Fig. 11-5 is a model for a common-base-connected transistor under specified operating conditions. Using Kirchhoff's voltage law, we have

$$V_1 = (R_1 + R_3)I_1 + R_3 I_2 \tag{11-34}$$
$$V_2 = (\alpha R_2 + R_3)I_1 + (R_2 + R_3)I_2 \tag{11-35}$$

and from these equations we see that the open-circuit impedance matrix is

$$[z] = \begin{bmatrix} z_{11} & z_{12} \\ z_{21} & z_{22} \end{bmatrix} = \begin{bmatrix} R_1 + R_3 & R_3 \\ \alpha R_2 + R_3 & R_2 + R_3 \end{bmatrix} \tag{11-36}$$

Since this network contains a controlled source, the network is not reciprocal. This is manifest by $z_{12} \neq z_{21}$.

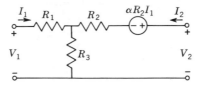

Fig. 11-5. Network which is a model for a common-base-connected transistor in Example 3.

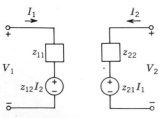

Fig. 11-6. The two-generator equivalent of the general two-port network in terms of the open-circuit impedance functions.

The results of the three examples given thus far may be consolidated by alternate interpretations of the sets of equations in terms of equivalent networks. Applying Kirchhoff's voltage law to the network of Fig. 11-6, we obtain Eqs. (11-18) and (11-19). Evidently this network is equivalent to the general two-port network with respect to the two available ports. This representation of the two-port network is known as the *two-generator equivalent*. To obtain a *one-generator equivalent*, the term $z_{12}I_1$ is added to and subtracted from Eq. (11-19), giving the second equation in the set

$$V_1 = z_{11}I_1 + z_{12}I_2 \tag{11-37}$$

and

$$V_2 = z_{12}I_1 + z_{22}I_2 + (z_{21} - z_{12})I_1 \tag{11-38}$$

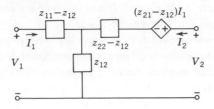

Fig. 11-7. A one-generator equivalent to the general two-port network in terms of the z functions.

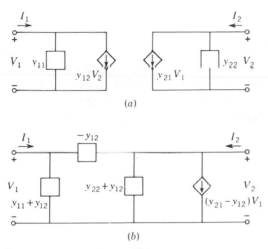

(a)

(b)

Fig. 11-8. (a) Two-generator and (b) one-generator equivalents of the general two-port network in terms of the short-circuit admittance functions.

Again, the application of the Kirchhoff voltage law to the network of Fig. 11-7 gives these equations, indicating that it is another equivalent to the general two-port network for any calculations or measurements at either of the two identified ports. For reciprocal networks, $z_{12} = z_{21}$ and the last term in Eq. (11-38) vanishes, giving a simple T-network. The same results may be found by solving Eqs. (11-31) through (11-33) for Z_a, Z_b, and Z_c.

A dual analysis to the one just presented for the short-circuit admittance equations leads to the one-generator and two-generator equivalent two-port networks shown in Fig. 11-8.

11-4. TRANSMISSION PARAMETERS

The transmission parameters serve to relate the voltage and current at one port to voltage and current at the other port. In equation form,

$$V_1 = AV_2 - BI_2 \qquad (11\text{-}39)$$

$$I_1 = CV_2 - DI_2 \qquad (11\text{-}40)$$

where A, B, C, and D are the transmission parameters. These parameters are known by a variety of other names including the *chain parameters* and, of course, the *ABCD parameters*. Their first use was in the analysis of power transmission lines where they are also known as the *general circuit parameters*. The negative sign for the second term in Eqs. (11-39) and (11-40) arises from two different conventions in assigning a positive direction to I_2. In power transmission problems, it is conventional for the current to be assigned a reference direction which is opposite to that given in Fig. 11-1. Thus the minus signs in Eqs. (11-39) and (11-40) are for I_2, and *not* for B and D!

We next provide interpretations for A, B, C, and D in terms of ratios of transform quantities made under open-circuit or short-circuit conditions. From Eqs. (11-39) and (11-40), we make the following identifications, arranged in reciprocal form to conform to other transfer functions we have used in Chapter 10:

$$\frac{1}{A} = \frac{V_2}{V_1}\bigg|_{I_2=0} \tag{11-41}$$

$$-\frac{1}{B} = \frac{I_2}{V_1}\bigg|_{V_2=0} \tag{11-42}$$

$$\frac{1}{C} = \frac{V_2}{I_1}\bigg|_{I_2=0} \tag{11-43}$$

and

$$-\frac{1}{D} = \frac{I_2}{I_1}\bigg|_{V_2=0} \tag{11-44}$$

We observe that $1/A$ is an open-circuit voltage gain, that $-1/B$ is a short-circuit transfer admittance, that $1/C$ is an open-circuit transfer impedance, and that $-1/D$ is a short-circuit current gain. These are now all familiar functions, so that the determination of A, B, C, and D should present no problems.

The transmission parameters are useful in describing two-port networks which are connected in *cascade* (or in a chain arrangement). To justify this statement, consider the network of Fig. 11-9. The two cascaded networks are N_a and N_b. For these two networks, Eqs. (11-39) and (11-40) are, in matrix form,

$$\begin{bmatrix} V_{1a} \\ I_{1a} \end{bmatrix} = \begin{bmatrix} A_a & B_a \\ C_a & D_a \end{bmatrix} \begin{bmatrix} V_{2a} \\ -I_{2a} \end{bmatrix} \tag{11-45}$$

and

$$\begin{bmatrix} V_{1b} \\ I_{1b} \end{bmatrix} = \begin{bmatrix} A_b & B_b \\ C_b & D_b \end{bmatrix} \begin{bmatrix} V_{2b} \\ -I_{2b} \end{bmatrix} \tag{11-46}$$

For the composite network, N (indicated by the dashed line in Fig. 11-9), we have

$$\begin{bmatrix} V_1 \\ I_1 \end{bmatrix} = \begin{bmatrix} A & B \\ C & D \end{bmatrix} \begin{bmatrix} V_2 \\ -I_2 \end{bmatrix} \tag{11-47}$$

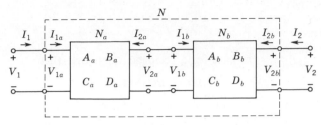

Fig. 11-9. Two networks, N_{11} and N_n connected in cascade to form the network N. Each network is described by the chain parameters, A, B, C, D.

Our objective is to determine A, B, C, D in this equation in terms of the transmission parameters of the previous two equations. This is accomplished by observing that $V_1 = V_{1a}$, $I_1 = I_{1a}$, $V_{2a} = V_{1b}$, $I_{2a} = -I_{1b}$, $I_{2b} = I_2$, and $V_{2b} = V_2$. If we use these conditions to eliminate voltages and currents with a subscript a or b, then we are able to express Eqs. (11-45) and (11-46) in the form of Eq. (11-47). This accomplished, we find that

$$\begin{bmatrix} A & B \\ C & D \end{bmatrix} = \begin{bmatrix} A_a & B_a \\ C_a & D_a \end{bmatrix}\begin{bmatrix} A_b & B_b \\ C_b & D_b \end{bmatrix} \tag{11-48}$$

which is the desired result. This result may be generalized for any number of two-ports connected in cascade, and we see that the overall transmission parameter matrix for cascaded two-port networks is simply the matrix product of the transmission matrices for each of the two-port networks in cascade.

In Eqs. (11-39) and (11-40), we expressed V_1 and I_1 in terms of V_2 and I_2. If we express V_2 and I_2 in terms of V_1 and I_1, then the equations are written

$$V_2 = A'V_1 - B'I_1 \tag{11-49}$$

and

$$I_2 = C'V_1 - D'I_1 \tag{11-50}$$

and the inverse transmission parameters are A', B', C', and D'. These equations apply for transmission in the opposite direction to that in Eqs. (11-39) and (11-40). The $A'B'C'D'$ parameters have properties which are similar to those outlined for the $ABCD$ parameters.

In studying the y and z parameters, we concluded that for reciprocal networks $y_{12} = y_{21}$ and $z_{12} = z_{21}$. In our discussion of the transmission parameters, we have made use of four parameters. Evidently, one of the four is superfluous and may be eliminated. Just as we have shown that the z functions could be expressed in terms of the y functions, in Eqs. (11-14) and (11-15), for example, so we may show that $A = z_{11}/z_{21}$, $B = \Delta_z/z_{21}$, $C = 1/z_{21}$, and $D = z_{22}/z_{21}$.

Observe that

$$AD - BC = \frac{z_{12}}{z_{21}} \tag{11-51}$$

If $z_{12} = z_{21}$, as is the case for a reciprocal network, then

$$AD - BC = 1 \tag{11-52}$$

so that if three of the transmission parameters are known, the fourth is determined. We may similarly show that for the inverse transmission parameters, reciprocity implies that

$$A'D' - B'C' = 1 \tag{11-53}$$

11-5. THE HYBRID PARAMETERS

The hybrid parameters that we study next find wide usage in electronic circuits, especially in constructing models for transistors. The properties of these parameters and their interpretation in terms of the two-port variables stem from the defining equations:

$$V_1 = h_{11}I_1 + h_{12}V_2 \tag{11-54}$$

and

$$I_2 = h_{21}I_1 + h_{22}V_2 \tag{11-55}$$

The h parameters are defined in terms of two of the variables by letting $I_1 = 0$ or $V_2 = 0$. There results:

$$h_{11} = \left.\frac{V_1}{I_1}\right|_{V_2=0} \tag{11-56}$$

$$h_{21} = \left.\frac{I_2}{I_1}\right|_{V_2=0} \tag{11-57}$$

$$h_{12} = \left.\frac{V_1}{V_2}\right|_{I_1=0} \tag{11-58}$$

and

$$h_{22} = \left.\frac{I_2}{V_2}\right|_{I_1=0} \tag{11-59}$$

From these equations, we see that h_{11} is the short-circuit input impedance, h_{21} is the short-circuit current gain, h_{12} is the open-circuit reverse voltage gain, and h_{22} is the open-circuit output admittance. These parameters are dimensionally mixed and for this reason are called "hybrid" parameters.

The usefulness of these parameters in representing transistors comes from the ease with which measurements for determining h_{11} and h_{21} are made under short-circuit conditions at port 2. It is relatively more difficult to make measurements with port 2 open-circuited.

The inverse hybrid parameters, or g parameters, are those in the equations

$$I_1 = g_{11}V_1 + g_{12}I_2 \qquad (11\text{-}60)$$

and

$$V_2 = g_{21}V_1 + g_{22}I_2 \qquad (11\text{-}61)$$

From these, we see that

$$g_{11} = \left.\frac{I_1}{V_1}\right|_{I_2=0} \qquad (11\text{-}62)$$

$$g_{21} = \left.\frac{V_2}{V_1}\right|_{I_2=0} \qquad (11\text{-}63)$$

$$g_{12} = \left.\frac{I_1}{I_2}\right|_{V_1=0} \qquad (11\text{-}64)$$

and

$$g_{22} = \left.\frac{V_2}{I_2}\right|_{V_1=0} \qquad (11\text{-}65)$$

Here g_{11} is the open-circuit input admittance, g_{21} is the open-circuit voltage ratio, g_{12} the short-circuit current ratio, and g_{22} is the short-circuit input impedance at port 2. Networks characterized by the h and g parameters which are equivalent to the general two-port network are shown in Fig. 11-10.

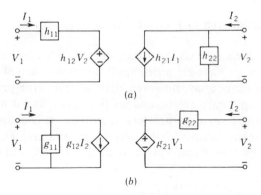

Fig. 11-10. Networks which are equivalent to the general two-port network in terms of (a) the h parameters, and (b) the g parameters.

EXAMPLE 4

A model for a transistor in the common-emitter connection is shown in Fig. 11-11. For this network, we see that

$$V_1 = (r_b + r_e)I_1 + \mu_{bc}V_2 \qquad (11\text{-}66)$$

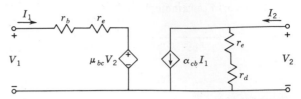

Fig. 11-11. Model of the common-emitter connected transistor used in Example 4.

and

$$I_2 = \alpha_{cb}I_1 + \frac{V_2}{r_e + r_d} \tag{11-67}$$

Using Eqs. (11-56) through (11-59), the hybrid h parameters are seen to have the following values:

$$h_{11} = r_b + r_e \tag{11-68}$$

$$h_{12} = \mu_{bc} \tag{11-69}$$

$$h_{21} = \alpha_{cb} \tag{11-70}$$

and

$$h_{22} = \frac{1}{r_e + r_d} \tag{11-71}$$

The simplicity of these results suggests the utility of the hybrid parameters in representing transistor networks.

11-6. RELATIONSHIPS BETWEEN PARAMETER SETS

In introducing the six sets of parameters in the previous sections of this chapter, we have suggested applications for each of the parameter sets. We cannot say, however, that all transistor problems are solved by using h parameters, and we frequently find it necessary to convert from one set of parameters to another. It is a simple matter to find the relationships of the sets of parameters. For example, comparing Eqs. (11-14) and (11-15) with Eqs. (11-18) and (11-19), we see that

$$\begin{bmatrix} z_{11} & z_{12} \\ z_{21} & z_{22} \end{bmatrix} = \frac{1}{\Delta_y} \begin{bmatrix} y_{22} & -y_{12} \\ -y_{21} & y_{11} \end{bmatrix} \tag{11-72}$$

All similar relationships between sets of parameters are summarized in Table 11-2. In this table, the matrices appearing in each of the rows are equivalent. Note that the equivalences involve a factor $\Delta_x = x_{11}x_{22} - x_{12}x_{21}$ where x is either z, y, T, T', h, or g.

The conditions under which a two-port network is reciprocal are given in Table 11-3 for the six sets of parameters. Also tabulated are the conditions under which a passive reciprocal two-port network is *symmetrical* in the sense that the ports may be exchanged without affecting the port voltages and currents.

Table 11-2. CONVERSION CHART

(Matrices in the same row in the table are equivalent)

$$\Delta_x = x_{11}x_{22} - x_{12}x_{21}$$

	[z]		[y]		[T]		[T']		[h]		[g]	
[z]	z_{11}	z_{12}	$\dfrac{y_{22}}{\Delta_y}$	$-\dfrac{y_{12}}{\Delta_y}$	$\dfrac{A}{C}$	$\dfrac{\Delta_T}{C}$	$\dfrac{D'}{C'}$	$\dfrac{1}{C'}$	$\dfrac{\Delta_h}{h_{22}}$	$\dfrac{h_{12}}{h_{22}}$	$\dfrac{1}{g_{11}}$	$-\dfrac{g_{12}}{g_{11}}$
	z_{21}	z_{22}	$-\dfrac{y_{21}}{\Delta_y}$	$\dfrac{y_{11}}{\Delta_y}$	$\dfrac{1}{C}$	$\dfrac{D}{C}$	$\dfrac{\Delta_{T'}}{C'}$	$\dfrac{A'}{C'}$	$-\dfrac{h_{21}}{h_{22}}$	$\dfrac{1}{h_{22}}$	$\dfrac{g_{21}}{g_{11}}$	$\dfrac{\Delta_g}{g_{11}}$
[y]	$\dfrac{z_{22}}{\Delta_z}$	$-\dfrac{z_{12}}{\Delta_z}$	y_{11}	y_{12}	$\dfrac{D}{B}$	$-\dfrac{\Delta_T}{B}$	$\dfrac{A'}{D'}$	$-\dfrac{1}{B'}$	$\dfrac{1}{h_{11}}$	$-\dfrac{h_{12}}{h_{11}}$	$\dfrac{\Delta_g}{g_{22}}$	$\dfrac{g_{12}}{g_{22}}$
	$-\dfrac{z_{21}}{\Delta_z}$	$\dfrac{z_{11}}{\Delta_z}$	y_{21}	y_{22}	$-\dfrac{1}{B}$	$\dfrac{A}{B}$	$-\dfrac{\Delta_{T'}}{B'}$	$\dfrac{D'}{B'}$	$\dfrac{h_{21}}{h_{11}}$	$\dfrac{\Delta_h}{h_{11}}$	$-\dfrac{g_{21}}{g_{22}}$	$\dfrac{1}{g_{22}}$
[T]	$\dfrac{z_{11}}{z_{21}}$	$\dfrac{\Delta_z}{z_{21}}$	$-\dfrac{y_{22}}{y_{21}}$	$-\dfrac{1}{y_{21}}$	A	B	$\dfrac{D'}{\Delta_{T'}}$	$\dfrac{B'}{\Delta_{T'}}$	$-\dfrac{\Delta_h}{h_{21}}$	$-\dfrac{h_{11}}{h_{21}}$	$\dfrac{1}{g_{21}}$	$\dfrac{g_{22}}{g_{21}}$
	$\dfrac{1}{z_{21}}$	$\dfrac{z_{22}}{z_{21}}$	$-\dfrac{\Delta_y}{y_{21}}$	$\dfrac{y_{11}}{y_{21}}$	C	D	$\dfrac{C'}{\Delta_{T'}}$	$\dfrac{A'}{\Delta_{T'}}$	$-\dfrac{h_{22}}{h_{21}}$	$-\dfrac{1}{h_{21}}$	$\dfrac{g_{11}}{g_{21}}$	$\dfrac{\Delta_g}{g_{21}}$
[T']	$\dfrac{z_{22}}{z_{12}}$	$\dfrac{\Delta_z}{z_{12}}$	$-\dfrac{y_{11}}{y_{12}}$	$-\dfrac{1}{y_{12}}$	$\dfrac{D}{\Delta_T}$	$\dfrac{B}{\Delta_T}$	A'	B'	$\dfrac{1}{h_{12}}$	$\dfrac{h_{11}}{h_{12}}$	$-\dfrac{\Delta_g}{g_{12}}$	$-\dfrac{g_{22}}{g_{12}}$
	$\dfrac{1}{z_{12}}$	$\dfrac{z_{11}}{z_{12}}$	$-\dfrac{\Delta_y}{y_{12}}$	$-\dfrac{y_{22}}{y_{12}}$	$\dfrac{C}{\Delta_T}$	$\dfrac{A}{\Delta_T}$	C'	D'	$\dfrac{h_{22}}{h_{12}}$	$\dfrac{\Delta_h}{h_{12}}$	$-\dfrac{g_{11}}{g_{12}}$	$-\dfrac{1}{g_{12}}$
[h]	$\dfrac{\Delta_z}{z_{22}}$	$\dfrac{z_{12}}{z_{22}}$	$\dfrac{1}{y_{11}}$	$-\dfrac{y_{12}}{y_{11}}$	$\dfrac{B}{D}$	$\dfrac{\Delta_T}{D}$	$\dfrac{B'}{A'}$	$\dfrac{1}{A'}$	h_{11}	h_{12}	$\dfrac{g_{22}}{\Delta_g}$	$-\dfrac{g_{12}}{\Delta_g}$
	$-\dfrac{z_{21}}{z_{22}}$	$\dfrac{1}{z_{22}}$	$\dfrac{y_{21}}{y_{11}}$	$\dfrac{\Delta_y}{y_{11}}$	$-\dfrac{1}{D}$	$\dfrac{C}{D}$	$-\dfrac{\Delta_{T'}}{A'}$	$\dfrac{C'}{A'}$	h_{21}	h_{22}	$-\dfrac{g_{21}}{\Delta_g}$	$\dfrac{g_{11}}{\Delta_g}$
[g]	$\dfrac{1}{z_{11}}$	$\dfrac{z_{12}}{z_{11}}$	$\dfrac{\Delta_y}{y_{22}}$	$\dfrac{y_{12}}{y_{22}}$	$\dfrac{C}{A}$	$-\dfrac{\Delta_T}{A}$	$\dfrac{C'}{D'}$	$-\dfrac{1}{D'}$	$\dfrac{h_{22}}{\Delta_h}$	$-\dfrac{h_{12}}{\Delta_h}$	g_{11}	g_{12}
	$\dfrac{z_{21}}{z_{11}}$	$\dfrac{\Delta_z}{z_{11}}$	$-\dfrac{y_{21}}{y_{22}}$	$\dfrac{1}{y_{22}}$	$\dfrac{1}{A}$	$\dfrac{B}{A}$	$\dfrac{\Delta_{T'}}{D'}$	$\dfrac{B'}{D'}$	$-\dfrac{h_{21}}{\Delta_h}$	$\dfrac{h_{11}}{\Delta_h}$	g_{21}	g_{22}

Table 11-3. SOME PARAMETER SIMPLIFICATIONS FOR PASSIVE, RECIPROCAL NETWORKS

Parameter	Condition for Passive Networks	Condition for Electrical Symmetry
z	$z_{12} = z_{21}$	$z_{11} = z_{22}$
y	$y_{12} = y_{21}$	$y_{11} = y_{22}$
$ABCD$	$AD - BC = 1$	$A = D$
$A'B'C'D'$	$A'D' - B'C' = 1$	$A' = D'$
h	$h_{12} = -h_{21}$	$\Delta_h = 1$
g	$g_{12} = -g_{21}$	$\Delta_g = 1$

337

11-7. PARALLEL CONNECTION OF TWO-PORT NETWORKS

We have discussed the cascade or tandem connection of two-port networks in Section 11-4. Another useful connection, called a *parallel* connection, is shown in Fig. 11-12. The short-circuit admittance functions are useful in characterizing parallel two-port networks as we will now see. For network A in Fig. 11-12(a), we have

$$I_{1a} = y_{11a}V_{1a} + y_{12a}V_{2a} \tag{11-73a}$$

and

$$I_{2a} = y_{12a}V_{1a} + y_{22a}V_{2a} \tag{11-73b}$$

Similarly for network B

$$I_{1b} = y_{11b}V_{1b} + y_{12b}V_{2b} \tag{11-74a}$$

and

$$I_{2b} = y_{12b}V_{1b} + y_{22b}V_{2b} \tag{11-74b}$$

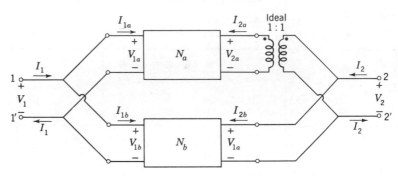

Fig. 11-12. The two networks N_a and N_b are shown in a parallel connection. The combined networks are described by Eqs. (11-77).

In proposing that the two networks be connected in parallel, we have assumed that making the connection will not alter the nature of the networks themselves. That this will not always be the case is illustrated by Fig. 11-13; in this case, the T-network shorts out the lowest resistor of the lattice network and so the network is altered. This difficulty is circumvented by the 1:1 turns-ratio ideal transformer which is shown in Fig. 11-12. An important special case arises when all of the networks to be connected in parallel have a common ground, as in Fig. 11-14. In this case, the ideal transformer is not required. Most of the applications of the ideas of this section will be applied to networks with a common ground, but, if this is not the case, then the ideal transformer is required or tests must be made to assure that the networks can be paralleled.[1]

[1]The necessary conditions are discussed in Norman Balabanian, Theodore A. Bickart, and S. Seshu, *Electrical Network Theory* (John Wiley & Sons, Inc., New York, 1969), pp. 174–76.

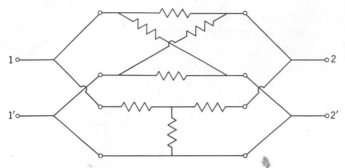

Fig. 11-13. Two resistive networks with a parallel connection, used to illustrate the necessity of the isolation provided by the 1:1 ideal transformer of Fig. 11-12.

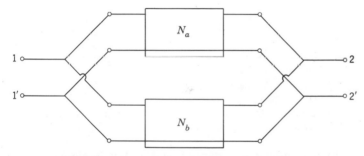

Fig. 11-14. The figure illustrates the manner in which the special case of a common ground in N_a and N_b is indicated.

Assuming that the parallel connection can be made, then the connection requires that

$$V_1 = V_{1a} = V_{1b} \quad \text{and} \quad V_2 = V_{2a} = V_{2b} \tag{11-75}$$

Further,

$$I_1 = I_{1a} + I_{1b} \quad \text{and} \quad I_2 = I_{2a} + I_{2b} \tag{11-76}$$

from Kirchhoff's current law. Combining these equations gives

$$I_1 = (y_{11a} + y_{11b})V_1 + (y_{12a} + y_{12b})V_2$$

and

$$I_2 = (y_{12a} + y_{12b})V_1 + (y_{22a} + y_{22b})V_2 \tag{11-77}$$

This result may be generalized for any number of networks connected in parallel; the individual short-circuit admittance functions are added to determine the overall short-circuit admittance function.

To illustrate the usefulness of this result, consider the network shown in Fig. 11-15(*a*). This is called a *twin-T* network; when properly designed, it is useful as a notch filter. It may be viewed as the parallel connection of two two-port networks as illustrated in (*b*) of the figure.

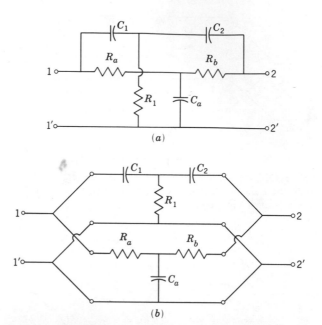

(a)

(b)

Fig. 11-15. The network of (a) is known as a *twin-T* network. It is topologically equivalent to the network shown in (b) which is the parallel connection of two *T*-networks.

Since the short-circuit admittance functions for the simple constituent networks are easily found, the overall short-circuit admittance functions may be routinely found by this artifice. To illustrate, call the top network A and the bottom network B. Then by inspection (with the output terminals shorted), we have

$$y_{11a} = \cfrac{1}{\cfrac{1}{C_1 s} + \cfrac{1}{\cfrac{1}{R_1} + C_2 s}} \tag{11-78}$$

and

$$y_{11b} = \cfrac{1}{R_a + \cfrac{1}{\cfrac{1}{R_b} + C_a s}} \tag{11-79}$$

If $R_1 = R_a = R_b = 1\ \Omega$ and $C_1 = C_2 = C_a = 1$ F, then

$$y_{11a} = \frac{s(s+1)}{2s+1} \quad \text{and} \quad y_{11b} = \frac{s+1}{s+2} \tag{11-80}$$

Then for the combined networks,

$$y_{11} = y_{11a} + y_{11b} = \frac{s^3 + 5s^2 + 5s + 1}{(2s+1)(s+2)} \tag{11-81}$$

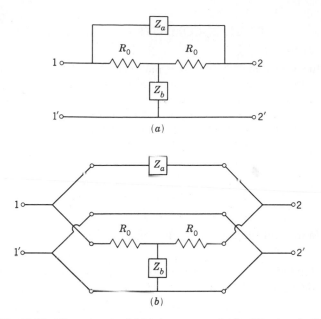

Fig. 11-16. The network of (*a*) is known as a *bridged-T* network. It is equivalent to the parallel connection of the two networks shown in (*b*) of the figure.

Similar computations may be made for the remaining network functions.

The network shown in Fig. 11-16(*a*) is known as a *bridged-T* network. It too may be recognized as two parallel two-port networks, as in (*b*) of the figure. These individual networks may be analyzed as was done for the twin-*T* network, and the short-circuit admittance functions added.

FURTHER READING

CHIRLIAN, PAUL M., *Basic Network Theory*, McGraw-Hill Book Company, New York, 1969. Chapter 8.

CLOSE, CHARLES M., *The Analysis of Linear Circuits*, Harcourt Brace Jovanovich, Inc., New York, 1966, Chapter 12.

DESOER, CHARLES A., AND ERNEST S. KUH, *Basic Circuit Theory*, McGraw-Hill Book Company, New York, 1969. Chapter 17.

HUELSMAN, LAWRENCE P., *Basic Circuit Theory with Digital Computations*, Prentice-Hall, Inc., Englewood Cliffs, N.J., 1972. See Chapter 9.

KARNI, SHLOMO, *Intermediate Network Analysis*, Allyn and Bacon, Inc., Boston, 1971. Chapter 4.

KIM, WAN H., AND HENRY E. MEADOWS, JR., *Modern Network Analysis*, John Wiley & Sons, Inc., New York, 1971. Chapter 5.

DIGITAL COMPUTER EXERCISES

In connection with the matrix multiplication of the *ABCD* parameter matrices for networks connected in cascade, see the exercises in references cited in Appendix E-3.1. The determination of the other parameters involves ordinary network analysis with the special condition that the one pair of network terminals be either open or shorted. These topics are considered in references cited in Appendix E-8.

PROBLEMS

In the problems to follow, all element values are in ohms, farads, or henrys.

11-1. Find the y and z parameters for the two simple networks shown in the figure if they exist.

11-2. For the two networks shown in the figure, find the z and y parameters if they exist.

11-3. Find the y and z parameters for the resistive network of the accompanying figure.

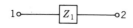

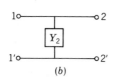

Fig. P11-1.

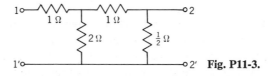

Fig. P11-3.

11-4. The network of the figure contains a current-controlled current source. For this network, find the y and z parameters.

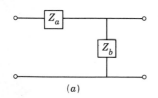

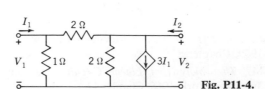

Fig. P11-4.

11-5. Find the y and z parameters for the resistive network containing a controlled source as shown in the accompanying figure.

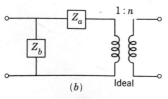

Fig. P11-2.

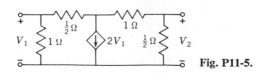

Fig. P11-5.

11-6. The accompanying figure shows a resistive network containing a single controlled source. For this network, find the y and z parameters.

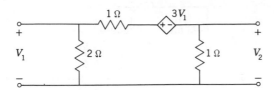

Fig. P11-6.

11-7. The network of the figure contains both a dependent current source and a dependent voltage source. For the element values given, determine the y and z parameters.

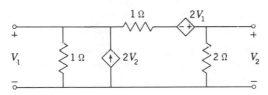

Fig. P11-7.

11-8. The accompanying network contains a voltage-controlled source and a current-controlled source. For the element values specified, determine the y and z parameters.

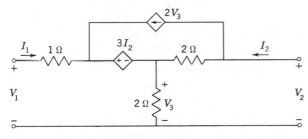

Fig. P11-8.

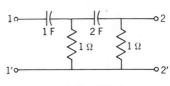

Fig. P11-9.

11-9. Find the y and z parameters for the RC ladder network of the figure.

11-10. The network of the figure is a bridged-T RC network. For the values given, find the y and z parameters.

11-11. Determine the $ABCD$ (transmission) parameters for the network of Prob. 11-10.

11-12. The accompanying figure shows a network with passive elements and two ideal transformers having $1:1$ turns-ratios. For the element values specified, determine the z parameters.

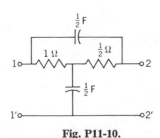

Fig. P11-10.

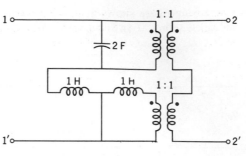

Fig. P11-12.

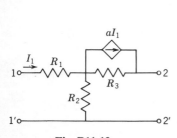

Fig. P11-13.

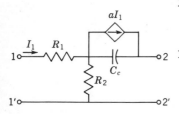

Fig. P11-14.

11-13. The network of the figure represents a certain transistor over a given range of frequencies. For this network, determine (a) the *h* parameters, and (b) the *g* parameters. Check your results using Table 11-2.

11-14. The network of the figure represents the transistor of Prob. 11-13 over a different range of frequencies. For this network, determine (a) the *h* parameters, and (b) the *g* parameters.

11-15. Show that the standard *T* section representation of a two-port network may be expressed in terms of the *h* parameters by the equations shown in the accompanying figure.

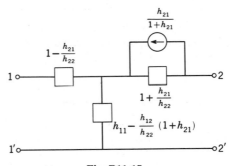

Fig. P11-15.

11-16. The network of the figure may be considered as a two-port network embedded in another resistive network. The resistive network is

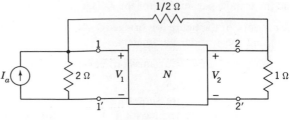

Fig. P11-16.

described by the following short-circuit admittances: $y_{11} = y_{22} = 2\,\mho$, $y_{21} = 2\,\mho$, and $y_{12} = 1\,\mho$. If I_a is a constant equal to 1 amp, find the voltages and the two ports of the network N, V_1 and V_2.

11-17. The network shown in the figure consists of a resistive T-and a resistive π-network connected in parallel. For the element values given, determine the y parameters.

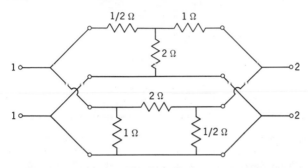

Fig. P11-17.

11-18. The resistive network shown in the figure is to be analyzed to determine the y parameters.

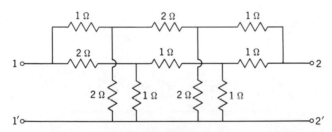

Fig. P11-18.

11-19. The accompanying figure shows two two-port networks connected in parallel. One two-port contains only a gyrator, and the other is a resistive network containing a single controlled source. For this network, determine the y parameters.

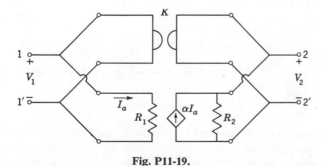

Fig. P11-19.

11-20. In the network of Fig. 11-16, let $Z_a = s/2$, $Z_b = 2/s$, and $R_0 = 1$. For these specific element values, determine the y parameters.

11-21. The network of the figure is of the type used for the so-called "notch filter." For the element values that are given, determine the y parameters.

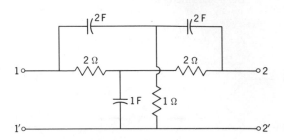

Fig. P11-21.

11-22. Let the element values for the network shown in Fig. 11-15 be as follows: $C_1 = C_2 = 1$ F, $R_1 = 1$ Ω, $R_a = R_b = 2$ Ω, $C_a = \frac{1}{2}$ F. Using these values, determine the y parameters.

11-23. The figure shows two networks as (a) and (b). It is asserted that one is the equivalent of the other. Is this assertion correct? Show reasoning. If it is, might one network have an advantage over the other as far as the calculation of network parameters is concerned?

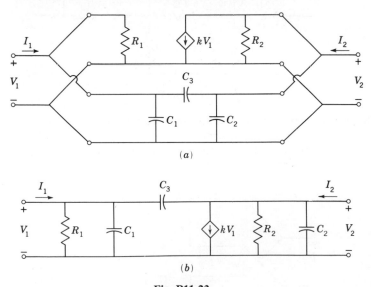

Fig. P11-23.

11-24. Two two-port networks are said to be equivalent if they have identical y or z parameters (or other of the characterizing parameters). In this problem, we wish to study the conditions under

which the π-network of (a) is equivalent to the T-network of (b).
Show that the two networks are equivalent if

$$Y_a = \frac{Z_2}{D}, \qquad Y_b = \frac{Z_3}{D}, \quad \text{and} \quad Y_c = \frac{Z_1}{D}$$

where

$$D = Z_1 Z_2 + Z_2 Z_3 + Z_3 Z_1$$

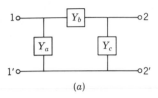
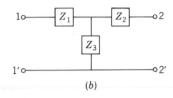

(a)   (b)

Fig. P11-24.

11-25. Derive equations similar to those given in Prob. 11-24 expressing
Z_1, Z_2, and Z_3 in terms of Y_a, Y_b, and Y_c. This result and that given
in Prob. 11-24 are used in obtaining a T-π *transformation*.

11-26. Apply the T-π transformation of Prob. 11-24 or 11-25 to the network
of the figure to obtain an equivalent (a) T-network, (b) π-network.

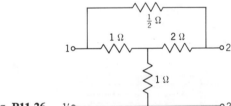

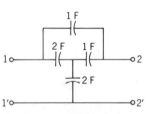

Fig. P11-26.

11-27. Apply the T-π transformation to obtain an equivalent (a) T-network
and (b) π-network for the capacitive network given in the figure.

Fig. P11-27.

11-28. Apply the T-π transformation as many times as is necessary to the
inductive ladder network shown in the accompanying figure in order
to determine the numerical values for the equivalent (a) T-network,
(b) π-network.

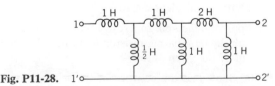

Fig. P11-28.

11-29. The network given in the figure is known as a lattice network; this
lattice is symmetrical in the sense that two arms of the lattice have
impedance Z_a and two have impedance Z_b. For this network, (a)
determine the z parameters, and (b) express Z_a and Z_b in terms of z
parameters.

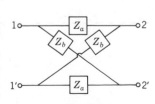

Fig. P11-29.

11-30. In this problem, we consider two-port networks having a symmetry property illustrated in (*a*) of the figure: If the network is divided at the dashed line, the two half networks have mirror symmetry with respect to the dashed line. The two half networks are connected by any number of wires as shown, and we will consider only the cases in which these wires do not cross. If a network meeting these specifications is *bisected* at the dashed line, then with the connecting wires open, the input impedance at either port is $Z_{1/2oc}$ as shown in (*b*). Similarly, with the connecting wires shorted, the impedance at either port is $Z_{1/2sc}$ as shown in (*c*). A theorem due to Bartlett states that these impedances are related to those given for the arms of the lattice in Prob. 11-29 by the equations

$$Z_a = Z_{1/2sc}, \qquad Z_b = Z_{1/2oc}$$

This is known as *Bartlett's bisection theorem*, and permits an equivalent lattice network to be found for any symmetrical network. Prove the theorem.

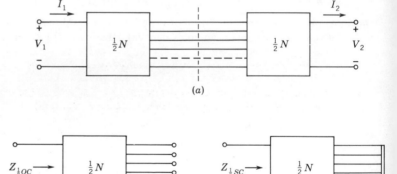

(*a*)

(*b*)   (*c*)

Fig. P11-30.

11-31. Apply the theorem of Prob. 11-30 to the network given in Prob. 10-2 with the terminating resistor at port 2 removed, and so obtain a lattice equivalent network.

11-32. Apply the theorem of Prob. 11-30 to the network of Prob. 10-31 with the terminating resistor R_0 removed to find the lattice equivalent of the given network.

11-33. (a) Show that the network of the accompanying figure satisfies the requirements described in Prob. 11-30. (b) Find the lattice equivalent of the network.

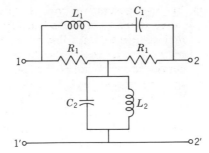

Fig. P11-33.

11-34. Find the lattice equivalent of the network of the accompanying figure making use of the results of Prob. 11-30.

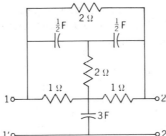

Fig. P11-34.

11-35. The network N in the accompanying figure may be described by the z parameters. Show that with port 2 open,

$$G_{12} = \frac{z_{21}}{z_{11}}$$

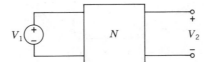

Fig. P11-35.

11-36. The network N in the figure is terminated at port 2 with a network having impedance $Z_L = 1/Y_L$. Show that

$$G_{12} = \frac{-y_{21}}{y_{22} + Y_L}$$

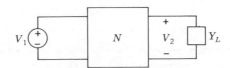

Fig. P11-36.

11-37. The network N of the figure is terminated at port 2 in impedance $Z_L = 1/Y_L$. Show that the transfer impedance for the combination is

$$Z_{12} = \frac{z_{21} Z_L}{z_{22} + Z_L}$$

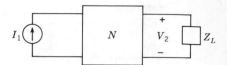

Fig. P11-37.

11-38. The figure shows two two-port networks connected in cascade. The two networks are distinguished by the subscripts a and b. Show that the combined network may be described by the equations

$$z_{12} = \frac{z_{12a} z_{12b}}{z_{11b} + z_{22a}}$$

and

$$y_{12} = \frac{-y_{12a} y_{12b}}{y_{11b} + y_{22a}}$$

for the transfer functions.

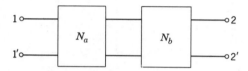

Fig. P11-38.

Sinusoidal Steady-State Analysis 12

12-1. THE SINUSOIDAL STEADY STATE

The sinusoid is a distinctive signal waveform. If a sinusoidal source is connected to a network of linear passive elements, then every voltage and current in that network will be sinusoidal *in the steady state*, differing from the source waveform only in amplitude and phase angle. This property follows from two observations:

(1) The sinusoid may be repeatedly differentiated or integrated and still be a sinusoid of the same frequency.
(2) The sum of a number of sinusoids of one frequency, but of arbitrary amplitude and phase, is a sinusoid of the same frequency. This summation property is important because analysis involves the application of Kirchhoff's voltage or current law.

Stated in other words, the class of signals described by the exponential function e^{st} has a steady-state value which is finite and nonzero only when $s = j\omega$ and $s = -j\omega$.

In addition to this mathematical distinction, the sinusoid is generated rather commonly in nature: a bottle bobbing in the water, a swinging pendulum, the shadow of a crank handle on a wheel—all these devices have motions which are sinusoidal in time. A sinusoidal voltage is generated by a conductor constrained to move in a circular path at right angles to a magnetic field; such generators produce the power used in our homes and in industry. Every electrical engineering laboratory is stocked with sinusoidal sources which may have frequency ranges from zero to gigahertz.

In this chapter, we develop the techniques that apply particularly to calculations for the sinusoidal steady state in terms of the general solutions we have previously obtained.[1] Consider first the equation

$$I(s) = Y(s)V(s) \qquad (12\text{-}1)$$

in which $I(s)$ is the response transform, $Y(s)$ is a network function, and $V(s)$ is an excitation transform. Let $V(s)$ be the transform of $v(t) = V_1 \sin \omega_1 t$ where V_1 and ω_1 are real positive constants so that

$$I(s) = \frac{\omega_1 V_1 Y(s)}{s^2 + \omega_1^2} \qquad (12\text{-}2)$$

Now $Y(s)$ is a quotient of polynomials, $Y = p/q$, where $q = (s - p_1)(s - p_2)\dots(s - p_n)$ where p_j are the simple poles of Y. The response transform may be expressed in partial fraction form as

$$I(s) = \frac{K_{-j\omega_1}}{s + j\omega_1} + \frac{K_{j\omega_1}}{s - j\omega_1} + \frac{K_1}{s + p_1} + \dots + \frac{K_n}{s + p_n} \qquad (12\text{-}3)$$

We may use the Heaviside method of Chapter 7 to evaluate the residues $K_{-j\omega_1}$ and $K_{j\omega_1}$; thus,

$$K_{-j\omega_1} = \frac{\omega_1 V_1 Y(-j\omega_1)}{-2j\omega_1} = \frac{V_1 |Y_1| e^{-j\phi}}{-2j} \qquad (12\text{-}4)$$

where we have defined $Y(-j\omega_1)$ to be given in polar form as $|Y_1| e^{-j\phi}$. Similarly, we see that

$$K_{j\omega_1} = \frac{\omega_1 V_1 Y(j\omega_1)}{2j\omega_1} = \frac{V_1 |Y_1| e^{j\phi}}{2j} \qquad (12\text{-}5)$$

with the identification in terms of ϕ made from the general property of complex variables, $[f(z^*)]^* = f(z)$. We now substitute Eqs. (12-4) and (12-5) into Eq. (12-3) and simplify the first two terms to the form

$$\frac{V_1 |Y_1|}{2j} \left(\frac{e^{j\phi}}{s - j\omega_1} - \frac{e^{-j\phi}}{s + j\omega_1} \right) \qquad (12\text{-}6)$$

having an inverse transform

$$V_1 |Y_1| \sin(\omega_1 t + \phi) \qquad (12\text{-}7)$$

Then the inverse transform of Eq. (12-3) becomes

$$i(t) = V_1 |Y_1| \sin(\omega_1 t + \phi) + K_1 e^{p_1 t} + \dots + K_n e^{p_n t} \qquad (12\text{-}8)$$

We recall from Chapter 10 that, for passive networks with nonzero loss, the real part of all p_j in this equation is negative, and so we see

[1] The elements of phasor algebra are reviewed in Appendix A.

that if we wait for a sufficiently long time after $t = 0$ only the first term in Eq. (12-8) will be important. This value we label as the steady-state response, as used in Chapter 6, so that

$$i_{ss} = V_1|Y_1|\sin(\omega_1 t + \phi) \tag{12-9}$$

Comparing this equation with that for the excitation, $v = V_1 \sin \omega_1 t$, we see that the excitation and response differ in magnitude by $|Y_1|$ and in phase by ϕ, these two quantities being the magnitude and phase of $Y(j\omega_1)$ from Eq. (12-5). Thus we see that the steady-state response for sinusoidal excitation is found by letting $s = j\omega$ in the network function relating excitation and response. This is an important result that is used in this chapter and the chapters that follow

12-2. THE SINUSOID AND $e^{\pm j\omega t}$

The sinusoidal waveform shown in Fig. 12-1 is described by the equation

$$v = V \sin(\omega t + \phi) \tag{12-10}$$

where V is the amplitude (or maximum value) of the sine wave, ω is the frequency in radians per sec, and ϕ is the phase angle of v with respect to the *reference*, $v' = V \sin \omega t$, shown as the dashed curve in Fig. 12-1. The *period* of v is the time interval between $\omega t = 0$ and $\omega t = 2\pi$ radians. Let the second value of time be $t = T$ so that the period is defined by $\omega T = 2\pi$ and

$$T = \frac{2\pi}{\omega} \text{ sec} \tag{12-11}$$

The quantity ϕ/ω shown in Fig. 12-1 represents the time displacement of v with respect to the reference v'. If ϕ is positive, v has zero value ϕ/ω sec earlier than v', and v is said to *lead* v'; conversely, if ϕ is negative, v passes through zero ϕ/ω sec later than v', and v *lags* v'.

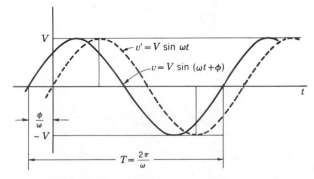

Fig. 12-1. A sine wave of amplitude V and phase angle ϕ. T is the period of the periodic function.

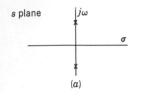

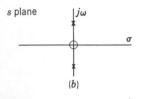

Fig. 12-2. Poles and zeros of the transform of (*a*) the sine function and (*b*) the cosine function.

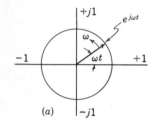

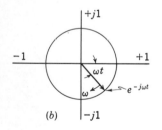

Fig. 12-3. Rotating phasors of unit amplitude (*a*) rotating in the positive direction, (*b*) in the negative direction. The complex plane shown in the figure is not to be confused with the *s* plane.

Using the relationship between the sine and cosine function, v of Eq. (12-10) may be written in the equivalent form:

$$v = V \cos\left[\omega t + \phi - (\pi/2)\right] = V \cos(\omega t + \beta) \qquad (12\text{-}12)$$

This equation may be written as the following sum of sine and cosine functions:

$$v = (V \cos \beta) \cos \omega t + (-V \sin \beta) \sin \omega t \qquad (12\text{-}13)$$

Thus the general form of the signal with which we shall be dealing in our study is the sum of the sine and cosine functions.

The transforms for the sine and cosine functions are

$$\mathcal{L}[\sin \omega t] = \frac{\omega}{s^2 + \omega^2}, \qquad \mathcal{L}[\cos \omega t] = \frac{s}{s^2 + \omega^2} \qquad (12\text{-}14)$$

indicating that sinusoidal signals have transforms with poles and zeros restricted to the imaginary axis as shown in Fig. 12-2. Poles at $\pm j\omega$ correspond to the time-domain factors $e^{j\omega t}$ and $e^{-j\omega t}$. This is also seen directly from the following equations:

$$\sin \omega t = \frac{e^{j\omega t} - e^{-j\omega t}}{2j} \qquad (12\text{-}15)$$

$$\cos \omega t = \frac{e^{j\omega t} + e^{-j\omega t}}{2} \qquad (12\text{-}16)$$

The term $e^{j\omega t}$ is commonly interpreted in terms of a unit rotating phasor rotating in the positive (or counterclockwise) direction; $e^{-j\omega t}$ likewise is interpreted as a unit rotating phasor rotating in the negative (clockwise) direction. The unit phasors are illustrated in Fig. 12-3. Now the sine, according to Eq. (12-15), is made up of the difference of two rotating unit phasors, rotating in opposite directions, divided by the factor $(2j)$. The construction of a sine wave in terms of these unit exponentials is illustrated in Fig. 12-4. The combination of the phasor $(e^{j\omega t}/2)$ and $(-e^{-j\omega t}/2)$ gives a phasor on the j axis. The factor $(1/j) = -j$ corresponds to a negative rotation of $90°$ $(-\pi/2$ radians). The sine is a real number; it has a value of zero when $\omega t = 0$, and a value of unity when $\omega t = \pi/2$. As ωt increases from 0 to 2π, the sine function is seen to have values between the limits of 1 and -1.

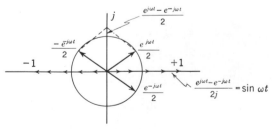

Fig. 12-4. The sine function described by two oppositely rotating phasors.

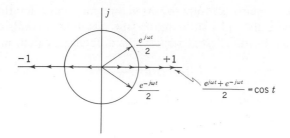

Fig. 12-5. The cosine function described by two oppositely rotating phasors.

The cosine may be similarly constructed in terms of exponential factors, as is illustrated in Fig. 12-5. The cosine is also a real number having a total range of values from $+1$ to -1. When $\omega t = 0$, the cosine has a value of unity; when $\omega t = \pi/2$, the cosine has zero value.

In the next two sections, we show how the exponential functions $e^{\pm j\omega t}$ may be exploited in computing the sinusoidal steady-state response.

12-3. SOLUTION USING $e^{\pm j\omega t}$

The determination of the steady-state component of the response for sinusoidal input was considered in Chapter 6 and the method of undetermined coefficients was introduced as a means of finding this part of the solution. The methods we study next have the same objective, but offer computational advantages. These methods were introduced into electrical engineering early in this century by Charles P. Steinmetz and others. The use of the exponential $e^{j\omega t}$ in determining the solution is especially popular in electromagnetic theory and in applications such as propagation and antennas where the excitation is sinusoidal and computations are made for steady-state conditions. We introduce this method by an example.

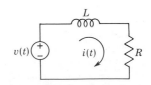

Fig. 12-6. *RL* series network for Example 1.

EXAMPLE 1

Consider the network of Fig. 12-6 in which the source is sinusoidal and the response, $i(t)$, is required for steady-state operation. For this network, $v = V \cos \omega t$, where V is a real positive constant, which may be written

$$V \cos \omega t = V\left(\frac{e^{j\omega t}}{2} + \frac{e^{-j\omega t}}{2}\right) \tag{12-17}$$

The generator $v(t) = V \cos \omega t$ is thus seen to be equivalent to *two* generators, one generating $(V/2)e^{j\omega t}$, the other generating $(V/2)e^{-j\omega t}$.

Using the principle of superposition, we may consider the driving forces separately and then combine the resulting currents to obtain the solution. For the first generator, the differential equation becomes

$$L\frac{di}{dt} + Ri = \frac{V}{2}e^{j\omega t} \qquad (12\text{-}18)$$

To solve this equation, we use the method of undetermined coefficients of Chapter 6. This method requires that the solution be of the form $i_{1ss} = Ae^{j\omega t}$, where A is the coefficient to be determined. Substituting i_{1ss} into Eq. (12-18), we obtain

$$j\omega LA + RA = \frac{V}{2} \qquad (12\text{-}19)$$

or

$$A = \frac{V/2}{R + j\omega L} \qquad (12\text{-}20)$$

Similarly, we may let $Be^{-j\omega t}$ be the steady-state solution of Eq. (12-18) with $e^{-j\omega t}$ replacing $e^{j\omega t}$ to give

$$B = \frac{V/2}{R - j\omega L} \qquad (12\text{-}21)$$

The solution for the steady state becomes

$$i_{ss} = \frac{V}{2}\left(\frac{e^{j\omega t}}{R + j\omega L} + \frac{e^{-j\omega t}}{R - j\omega L}\right) \qquad (12\text{-}22)$$

Algebraic reduction employing the identities of Eq. (12-15) and (12-16) gives

$$i_{ss} = \frac{V}{R^2 + \omega^2 L^2}(R\cos\omega t + \omega L \sin\omega t) \qquad (12\text{-}23)$$

This may be further reduced, using the method given in Eqs. (6-73) through (6-76) to obtain

$$i_{ss} = \frac{V}{\sqrt{R^2 + \omega^2 L^2}}\left[\cos\left(\omega t - \tan^{-1}\frac{\omega L}{R}\right)\right] \qquad (12\text{-}24)$$

which is the required solution.

In Example 1, we note that $B = A^*$ which suggests that all of the information required may be found with one exponential function, $e^{j\omega t}$. To show that this is indeed the case, we will employ the excitation

$$v = Ve^{j(\omega t + \phi)}$$
$$= V\cos(\omega t + \phi) + jV\sin(\omega t + \phi) \qquad (12\text{-}25)$$

This is simply an artifice employed to lead to useful results; there are no physical generators of complex numbers! Let the response due to the v of Eq. (12-25) be $i(t)$. Furthermore, let $i_c(t)$ be the response due to the excitation $V\cos(\omega t + \phi)$, and $i_s(t)$ be the response due to the excitation $V \sin(\omega t + \phi)$. We observe that $i(t)$ is related to $i_c(t)$ and $i_s(t)$ by

$$i(t) = i_c(t) + ji_s(t) \tag{12-26}$$

from Eq. (12-25) through superposition. Then

$$i_c = \text{Re}\,[i(t)] \tag{12-27}$$

and

$$i_s = \text{Im}\,[i(t)] \tag{12-28}$$

where "Re" denotes the *real part of* and "Im" the *imaginary part of* the function.

Now v of Eq. (12-25) may be written

$$v = Ve^{j\phi}e^{j\omega t} = \mathbf{V}e^{j\omega t} \tag{12-29}$$

where $\mathbf{V}$ is a complex number of magnitude V and phase ϕ. Note that due to the homogeneity aspect of the superposition property for linear networks, we may employ the excitation $e^{j\omega t}$ to find the solution and then multiply the result by $\mathbf{V}$. This discussion may be summarized in the form of two procedures as follows.

Procedure 1. The steady-state response to the excitation $V \cos(\omega t + \phi)$ may be found by determining the response to the excitation $e^{j\omega t}$, multiplying this response by $\mathbf{V} = Ve^{j\phi}$, and then determining the real part of this product.

Procedure 2. The steady-state response to the excitation $V \sin(\omega t + \phi)$ may be found by determining the response to the excitation $e^{j\omega t}$, multiplying this response by $\mathbf{V} = Ve^{j\phi}$, and then determining the imaginary part of this product.

EXAMPLE 2

Let us rework Example 1, using Procedure 1. For the excitation $e^{j\omega t}$, Eq. (12-18) becomes

$$L\frac{di}{dt} + Ri = e^{j\omega t} \tag{12-30}$$

for which the solution is

$$Ae^{j\omega t} = \frac{1}{R + j\omega L}e^{j\omega t} \tag{12-31}$$

For this problem, $\mathbf{V} = Ve^{j0}$, so that the steady-state response is

$$i_{ss} = \text{Re}\frac{Ve^{j\omega t}}{R + j\omega L} \tag{12-32}$$

Substituting $e^{j\omega t} = \cos \omega t + j \sin \omega t$ into this equation, we obtain, after algebraic reduction paralleling that for Example 1,

$$i_{ss} = \frac{V}{\sqrt{R^2 + \omega^2 L^2}}\left[\cos\left(\omega t - \tan^{-1}\frac{\omega L}{R}\right)\right] \qquad (12\text{-}33)$$

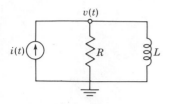

Fig. 12-7. *RL* parallel network for Example 3.

EXAMPLE 3

The network of Fig. 12-7 has a sinusoidal excitation $i_1 = I_1 \sin(\omega t + \phi)$ and we wish to determine the response node-to-datum voltage v in the steady state. The Kirchhoff current law gives

$$\frac{v}{R} + \frac{1}{L}\int_{-\infty}^{t} v \, dt = I_1 \sin(\omega t + \phi) \qquad (12\text{-}34)$$

We may convert this equation to a differential equation by differentiation, of course; instead, we consider the integral in the equation. Since the initial value found by integrating from $t = -\infty$ to $t = 0$ does not affect the steady-state solution, the lower limit does not enter into the evaluation, and we may write the equation in indefinite integral form. Thus the equation

$$\frac{v}{R} + \frac{1}{L}\int v \, dt = e^{j\omega t} \qquad (12\text{-}35)$$

has the solution

$$Ae^{j\omega t} = \frac{1}{(1/R) + (1/j\omega L)}e^{j\omega t} \qquad (12\text{-}36)$$

Since the excitation is expressed in terms of the sine, we use Procedure 2 which was previously outlined. Following this method, we have

$$v_{ss} = \text{Im}\left[I_1 e^{j\phi}\left(\frac{j\omega LR}{R + j\omega L}\right)e^{j\omega t}\right] \qquad (12\text{-}37)$$

Carrying out the algebraic steps yields the steady-state solution

$$v_{ss} = \frac{\omega LRI_1}{\sqrt{R^2 + \omega^2 L^2}}\sin\left(\omega t + \phi + \tan^{-1}\frac{R}{\omega L}\right) \qquad (12\text{-}38)$$

12-4. SOLUTION USING Re $e^{j\omega t}$ or Im $e^{j\omega t}$

A similar approach to that of the last section makes use of the identities

$$\cos \omega t = \text{Re } e^{j\omega t} \qquad (12\text{-}39)$$

and

$$\sin \omega t = \text{Im } e^{j\omega t} \qquad (12\text{-}40)$$

To illustrate the method, we will again consider the network of Example 1 shown in Fig. 12-6. We replace $V_1 \cos \omega t$ with Re $V_1 e^{j\omega t}$, giving the equation

$$L\frac{di}{dt} + Ri = \text{Re } V_1 e^{j\omega t} \tag{12-41}$$

Following the method of undetermined coefficients, the solution is assumed to be

$$i_{ss} = \text{Re } A e^{j\omega t} \tag{12-42}$$

Substituting into Eq. (12-41) gives

$$L\frac{d}{dt}[\text{Re } A e^{j\omega t}] + R[\text{Re } A e^{j\omega t}] = \text{Re } V_1 e^{j\omega t} \tag{12-43}$$

The algebraic operations with the complex quantities are next manipulated, using the following three rules: If $f - Ve^{j\omega t}$, we note that

$$\text{Re } (k_1 f_1 + k_2 f_2) = k_1 \text{ Re} f_1 + k_2 \text{ Re} f_2 \tag{12-44}$$

$$\text{Re}\frac{df}{dt} = \frac{d}{dt}\text{Re } f \tag{12-45}$$

and, finally, for problems other than this example, we will need to know that

$$\text{Re }\int f \, dt = \int \text{Re} f \, dt \tag{12-46}$$

Then Eq. (12-43) becomes

$$\text{Re}\left[A\left(L\frac{d}{dt} + R \right)e^{j\omega t} \right] = \text{Re } (V_1 e^{j\omega t}) \tag{12-47}$$

or

$$\text{Re } \{[A(j\omega L + R) - V_1]e^{j\omega t}\} = 0 \tag{12-48}$$

Since the equation must be satisfied for all t, it is necessary that

$$A(j\omega L + R) - V_1 = 0 \tag{12-49}$$

or

$$A = \frac{V_1}{j\omega L + R} \tag{12-50}$$

and A in Eq. (12-42) is determined. Then the solution is

$$i_{ss} = \text{Re}\left[\left(\frac{V_1}{j\omega L + R} \right)e^{j\omega t} \right] \tag{12-51}$$

This equation is identical with Eq. (12-32) of Example 2, and so the solution is that given by Eq. (12-33).

Observe that if the excitation is given in terms of the sine, then Im $e^{j\omega t}$ is used in Eq. (12-41) and the remaining steps parallel those given for the real part.

To generalize the method illustrated by the example, observe that

$$V_1\cos(\omega t + \phi) = \text{Re }[V_1 e^{j(\omega t + \phi)}] \tag{12-52}$$

$$= \text{Re }[\mathbf{V}e^{j\omega t}], \quad \text{if } \mathbf{V} = V_1 e^{j\phi} \tag{12-53}$$

The assumed solution is

$$i_{ss} = \text{Re}\,[\mathbf{I}e^{j\omega t}], \qquad \text{where } \mathbf{I} = I_1{}^{j\theta} \tag{12-54}$$

Note that the ratio

$$\frac{\mathbf{V}}{\mathbf{I}} = \frac{V_1}{I_1}e^{j(\phi-\theta)} = Z(j\omega) \tag{12-55}$$

Here $Z(j\omega)$ is the network function of our previous studies of the Laplace transformation with $s = j\omega$,

$$Z(j\omega) = \left.\frac{V(s)}{I(s)}\right|_{s=j\omega} \tag{12-56}$$

Equation (12-55) may also be written in the form

$$\mathbf{I}Z(j\omega) = \mathbf{V} \tag{12-57}$$

or

$$\mathbf{I} = \frac{\mathbf{V}}{Z(j\omega)} \tag{12-58}$$

Once $\mathbf{I}$ is determined, the solution is

$$i_{ss} = \text{Re}\,(\mathbf{I}e^{j\omega t}) \tag{12-59}$$

By methods previously discussed, this may be written

$$i_{ss} = \frac{V_1}{|Z|}\cos{(\omega t + \theta)} \tag{12-60}$$

where

$$\theta = \phi - \tan^{-1}\frac{\text{Im}\,Z(j\omega)}{\text{Re}\,Z(j\omega)} \tag{12-61}$$

EXAMPLE 4

A 1-Ω resistor is connected in series with a $\frac{1}{2}$-F capacitor and a voltage source $v = 10\cos{[2t + (\pi/4)]}$. We are required to find the steady-state current in the network by using Eqs. (12-60) and (12-61). From the given $v(t)$, we see that $V_1 = 10$ and $\phi = \pi/4$. Also, $Z(j2) = R + (1/j\omega C) = 1 - j1$. From this we see that the magnitude of $Z(j2)$ is $\sqrt{2}$, and the phase of $Z(j2)$ is $-\pi/4$. Substituting this information into Eq. (12-60), we obtain

$$i_{ss} = 7.07\cos\left(2t + \frac{\pi}{2}\right) = 7.07\sin{(2t + \pi)} \tag{12-62}$$

which is the required solution.

12-5. PHASORS AND PHASOR DIAGRAMS

In this section we generalize the results given by Eq. (12-57), employing the various methods used thus far in this chapter. Consider a network of L independent loops excited by sinusoidal sources, each operating at the same frequency. We are required to determine the steady-state response in this network. Suppose that L loop equations are written in transform form. In letting $s = j\omega$, where ω is the single frequency of excitation, the impedance functions become $Z_{ik}(j\omega)$, each loop voltage is represented by a phasor[2] $\mathbf{V}_i e^{j\omega t}$, and each current is also written in phasor form as $\mathbf{I}_k e^{j\omega t}$. Then for this network, we have

$$Z_{11}(j\omega)\,\mathbf{I}_1 + Z_{12}(j\omega)\,\mathbf{I}_2 + \ldots + Z_{1L}(j\omega)\,\mathbf{I}_L = \mathbf{V}_1$$

$$Z_{21}(j\omega)\,\mathbf{I}_1 + Z_{22}(j\omega)\,\mathbf{I}_2 + \ldots + Z_{2L}(j\omega)\,\mathbf{I}_L = \mathbf{V}_2 \qquad (12\text{-}63)$$

$$\cdot \quad \cdot \quad \cdot \quad \cdot \quad \cdot \quad \cdot \quad \cdot \quad \cdot \quad \cdot \quad \cdot \quad \cdot \quad \cdot$$

$$Z_{L1}(j\omega)\,\mathbf{I}_1 + Z_{L2}(j\omega)\,\mathbf{I}_2 + \ldots + Z_{LL}(j\omega)\,\mathbf{I}_L = \mathbf{V}_L$$

where the exponential $e^{j\omega t}$ has been cancelled in each of the L equations. When the required response phasor $\mathbf{I}_k$ is found through routine algebraic manipulations of complex numbers, it will be a phasor of the form

$$\mathbf{I}_k = I_k e^{j\theta_k} \qquad (12\text{-}64)$$

from which the response in the time domain may be written by using only the magnitude and phase information (in addition to the frequency, which is known),

$$i_k(t) = I_k \cos{(\omega t + \theta_k)} \qquad (12\text{-}65)$$

and the solution is complete.

This general method may be described by a flow chart similar to that of Fig. 7-1(*b*), which was used to introduce the Laplace transformation. Such a chart is shown in Fig. 12-8, which describes the method of solution that applies when all sources are sinusoidal, only the steady-state solution is desired, and all sources are operating at the same frequency. Note these steps:

(1) Express the network equations in transform form and let $s = j\omega$ in all network functions. This ω is the frequency of the sinusoidal sources.

(2) All initial-condition sources are set equal to zero, because initial conditions are not relevant to the determination of a steady-state solution.[3]

(3) All source voltages are described in terms of the cosine function

$$v_i(t) = V_i \cos{(\omega t + \phi_i)} \qquad (12\text{-}66)$$

from which the phasor is $\mathbf{V}_i = V_i e^{j\phi_i}$. Useful equations in changing sine functions to cosines are:

$$\sin{(\omega t + \alpha)} = \cos{\left(\omega t + \alpha - \frac{\pi}{2}\right)} \qquad (12\text{-}67)$$

---

[2]A phasor is also known as a *vector* or a *sinor*. See the discussion in Appendix A.

[3]We assume here that all poles of the network function are in the left half-plane.

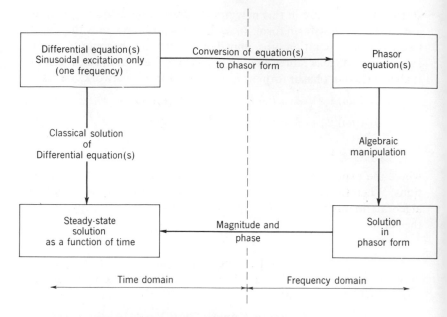

Fig. 12-8. Flow chart which describes the method of determining the steady-state response using phasors.

and

$$\cos(\omega t + \beta) = \sin\left(\omega t + \beta + \frac{\pi}{2}\right) \tag{12-68}$$

(4) The system of equations that results is of the form of Eqs. (12-63) which may be solved by usual methods (determinants, for example) to find I_k.

(5) The phasor I_k yields the required phase and magnitude information for the solution to be written in the time domain as[4]

$$i_k(t) = I_k \cos(\omega t + \theta_k) \tag{12-69}$$

and the solution is complete.

We now illustrate the application of these rules by a number of examples.

EXAMPLE 5

The network of Fig. 12-9 is described by the equation,

$$\mathbf{V_R} + \mathbf{V_L} + \mathbf{V_C} = Z(j\omega)\mathbf{I_1} = \mathbf{V_1} \tag{12-70}$$

where

$$Z(j\omega) = R + j\left(\omega L - \frac{1}{\omega C}\right) \tag{12-71}$$

---

[4]By changing steps 3 and 5, the rules may be written so that the solution is obtained in terms of the sine function.

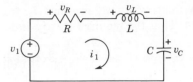

Fig. 12-9. Network of Example 5.

Since $v_1 = \cos \omega t$, we have $\mathbf{V}_1 = 1$, and

$$\mathbf{I}_1 = I_1 e^{j\theta_1} = \frac{\mathbf{V}_1}{\mathbf{Z}} = \frac{1}{R + j\left(\omega L - \dfrac{1}{\omega C}\right)} \qquad (12\text{-}72)$$

From this result, we see that

$$I_1 = \frac{1}{\sqrt{R^2 + \left(\omega L - \dfrac{1}{\omega C}\right)^2}} \qquad (12\text{-}73)$$

and

$$\theta_1 = \tan^{-1} \frac{\omega L - \dfrac{1}{\omega C}}{R} \qquad (12\text{-}74)$$

Finally, the time-domain expression is

$$i_1(t) = \mathbf{I}_1 \cos(\omega t + \theta_1) \qquad (12\text{-}75)$$

and Eqs. (12-73) to (12-75) describe the steady-state response.

EXAMPLE 6

For the network of Fig. 12-10, we are to find i_2 in the steady state if $v_1 = \cos 2t$. The impedance of each element and the voltage phasor are shown in Fig. 12-10(b). From the figure, we write the following loop equations:

$$\begin{aligned} j1\mathbf{I}_1 + j1\mathbf{I}_2 - 1 \\ j1\mathbf{I}_1 + (2 - j1)\mathbf{I}_2 = 0 \end{aligned} \qquad (12\text{-}76)$$

Eliminating $\mathbf{I}_1$, we find that

$$\mathbf{I}_2 = \frac{1}{-2 + j2} \qquad (12\text{-}77)$$

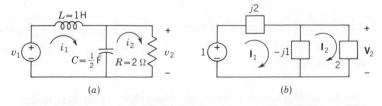

(a) (b)

Fig. 12-10. (a) The network to be analyzed in Example 6, and (b) an equivalent network of impedances and phasor functions.

From this result, we see that $I_2 = \sqrt{2}/4$ and that the phase is $-3\pi/4$. Thus the required steady-state current is

$$i_2 = \frac{\sqrt{2}}{2} \cos\left(2t - \frac{3\pi}{4}\right) \tag{12-78}$$

Phasors, being complex numbers, may be represented in the complex plane in the usual polar form as an arrow, with a length corresponding to the phasor magnitude, and an angle with respect to the positive real axis, which is the phase of the phasor. If the various phasor quantities pertaining to a given network are combined in such a way that one or both of the Kirchhoff laws is illustrated, the resultant figure is described as a *phasor diagram*. Such diagrams provide geometrical insight into the voltage and current relationships in a network, and they are especially helpful in visualizing such steady-state phenomena as resonance or in making magnitude plots, as we shall see.

In constructing a phasor diagram, we represent each sinusoidal voltage and current by a phasor of length equal to the maximum amplitude of the sinusoid,[5] and with an angular displacement from the positive real axis, which is the angle of the equivalent cosine function at $t = 0$. This use of the cosine is arbitrary, of course, and the student will have no difficulty formulating similar rules in terms of the sine function. Thus the first step is to express each voltage and current in terms of an equivalent cosine function. Equations useful in accomplishing this are Eqs. (12-67) and (12-68).

The next three examples illustrate the method for constructing phasor diagrams.

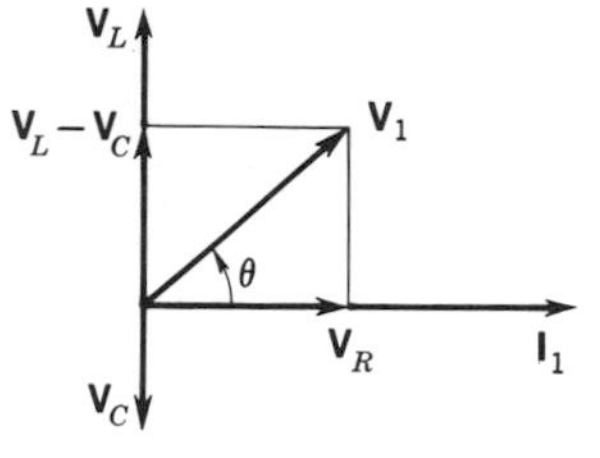

Fig. 12-11. Complete phasor diagram for the network of Fig. 12-9, as discussed in Example 7.

EXAMPLE 7

Consider once more the network shown in Fig. 12-9. Since each of the voltages is expressed in terms of one current, let us assume $\mathbf{I}_1$ as a *reference* in the sense that it is drawn at $\theta = 0$ in the complex plane. Now, since $\mathbf{V}_R = R\mathbf{I}_1$, $\mathbf{V}_R$ is in phase with $\mathbf{I}_1$ and is assigned an arbitrary length as in Fig. 12-11. Similarly, $\mathbf{V}_L = j\omega L\mathbf{I}_1$ and so is represented at $\theta = 90°$, while $\mathbf{V}_C = (1/j\omega C)\mathbf{I}_1$ and is at $\theta = -90°$. The phasor addition of the three voltages gives $\mathbf{V}_1$ which is displaced from $\mathbf{I}_1$ by θ as shown. We might have started our construction of the phasor diagram by using any other reference, but the completed diagram would have been the same as that shown in Fig. 12-11, except rotated such that the new reference was at $\theta = 0$.

[5]In Chapter 14, we show that it is sometimes useful to scale the magnitude by $1/\sqrt{2}$ to obtain *effective* or *rms* values.

EXAMPLE 8

For our second example, we will construct a phasor diagram for the network shown in Fig. 12-12(a); this network is the dual of that considered in Example 7. For this network, we select $\mathbf{V}$ as the reference (since other quantities are easily related to $\mathbf{V}$), and then determine each of the currents. Observe that $\mathbf{I}_C$ leads $\mathbf{V}$ by 90°, $\mathbf{I}_L$ lags $\mathbf{V}$ by 90°, and $\mathbf{I}_R$ is in phase with $\mathbf{V}$. Since specified values are not assigned, we represent each phasor by an arbitrary length, and so construct the phasor diagram shown in Fig. 12-12(b). The phasor addition of the three currents gives $\mathbf{I}$.

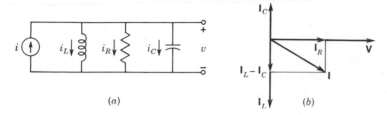

(a)   (b)

Fig. 12-12. Complete phasor diagram for the given network.

EXAMPLE 9

Consider the network of Fig. 12-10, used earlier for Example 6. For this network, we select $\mathbf{V}_2$ as the reference, which determines the position of $\mathbf{I}_R$ and $\mathbf{I}_C$ as shown in Fig. 12-13. The addition of these two currents is indicated as $\mathbf{I}_{RC}$ which must be the current through the inductor, $\mathbf{I}_1$. Now the inductor voltage leads the current by 90° (from the equation $\mathbf{V}_L = j\omega L\,\mathbf{I}_L$) and so has the position shown in the figure. Finally, the applied voltage $\mathbf{V}_1$ is the phasor addition of $\mathbf{V}_L$ and $\mathbf{V}_2$.

In constructing this phasor diagram, we have ignored the values specified for the network, and simply constructed the diagram with arbitrary lengths for each of the phasors. We can, of course, construct a phasor diagram to match specific element and source values by determining the value of the magnitude and phase of each phasor before beginning the diagram construction.

Fig. 12-13. Complete phasor diagram for the network of Fig. 12-10.

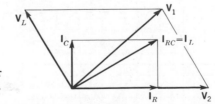

FURTHER READING

BALABANIAN, NORMAN, *Fundamentals of Circuit Theory*, Allyn and Bacon, Inc., Boston, 1961. Chapter 4.

CHIRLIAN, PAUL M., *Basic Network Theory*, McGraw-Hill Book Company, New York, 1969. Chapter 6.

CLOSE, CHARLES M., *The Analysis of Linear Circuits*, Harcourt Brace Jovanovich, Inc., New York, 1966. Chapter 5.

HUANG, THOMAS S., AND RONALD R. PARKER, *Network Theory: An Introductory Course*, Addison-Wesley Publishing Co., Inc., Reading, Mass., 1971. Chapter 10.

LEON, BENJAMIN J., AND PAUL A. WINTZ, *Basic Linear Networks for Electrical and Electronics Engineers*, Holt, Rinehart & Winston, Inc., New York, 1970. Chapter 4.

MANNING, LAURENCE A., *Electrical Circuits*, McGraw-Hill Book Company, New York, 1965. Chapter 6.

WING, OMAR, *Circuit Theory with Computer Methods*, Holt, Rinehart & Winston, Inc., New York, 1972. See Chapter 7.

DIGITAL COMPUTER EXERCISES

This chapter is devoted to a discussion of networks operating in the sinusoidal steady state. Analysis of large systems in this condition is straightforward but tedious if done with pencil and paper, and the computer can be used to advantage. See the references cited in Appendix E-8.3 for suggested exercises. In particular, see Chapters 9 and 10 of Huelsman, reference 7 in Appendix E-10, and Chapters 3 and 11 of Ley, reference 11 in Appendix E-10.

PROBLEMS

12-1. Let $v(t) = V_1 \cos \omega_1 t$ for Eq. (12-1) and carry out the derivation leading to a result similar to Eq. (12-9).

12-2. For the sinusoidal waveform of the figure, write an equation for $v(t)$ using numerical values for the magnitude, phase, and frequency.

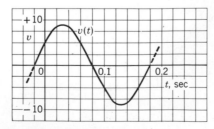

Fig. P12-2.

12-3. Starting with the rotating phasors, $e^{\pm j\omega t}$, show by a construction similar to that illustrated in Figs. 12-4 and 12-5 that

$$\sin^2 \omega t + \cos^2 \omega t = 1$$

12-4. Given the equation

$$\sin 377t + 3\sqrt{2} \sin \left(377t + \frac{\pi}{4}\right) = A \cos (377t + \theta)$$

determine A and θ.

12-5. Show that

$$\sum_{k=1}^{n} A_k \sin (\omega_1 t + \phi_k) = C \sin (\omega_1 t + \theta)$$

In other words, show that the sum of any number of sinusoids of arbitrary amplitude and phase angle but all of the same frequency is a sinusoid of the same frequency.

12-6. Using the equation of Prob. 12-5 with $n = 2$, determine C and θ in terms of A_1, A_2, ϕ_1, and ϕ_2.

12-7. Using the method of Section 12-3, solve the following differential equations for the steady-state solution (called the particular integral in Chapter 6):

(a) $\dfrac{di}{dt} + 2i = \sin 2t$

(b) $\dfrac{di}{dt} + i = \cos 3t$

(c) $\dfrac{di}{dt} + 3i = \cos (2t + 45°)$

(d) $\dfrac{d^2 i}{dt^2} + 2\dfrac{di}{dt} + i = 5 \sin (2t + 30°)$

(e) $\dfrac{d^2 i}{dt^2} + i = 2 \sin t$

12-8. Repeat Prob. 12-7 for the following differential equations, solving only for the steady-state solution:

(a) $\dfrac{d^2 i}{dt^2} + 2\dfrac{di}{dt} + 2i = 3 \cos (t + 30°)$

(b) $\dfrac{d^2 i}{dt^2} + 4i = 3 \cos (2t + 45°)$

(c) $\dfrac{di}{dt} + 2i = \sin 2t + \cos t$

12-9. The network of the figure has a sinusoidal voltage source and is operating in the steady state. Use the method of Section 12-3 to determine the steady-state current $i(t)$ if $v_1 = 2 \cos 2t$.

12-10. In the network of the figure, $i_1 = 3 \cos (t + 45°)$ and the network is operating in the steady state. Make use of the method of Section 12-3 to determine the node-to-datum voltage $v_1(t)$.

12-11. For the given network, find $v_a(t)$ in the steady state if $v_1 = 2 \sin 2t$. Make use of the method of Section 12-3.

12-12. In the resistive network shown in the figure, $v_1 = 2 \sin (2t + 45°)$ for all t. (a) Determine $i_a(t)$. (b) Determine $i_b(t)$.

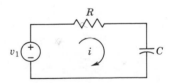

Fig. P12-9.

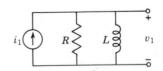

Fig. P12-10.

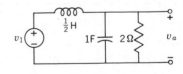

Fig. P12-11.

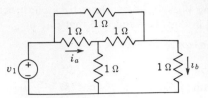

Fig. P12-12.

12-13. The network shown in the accompanying figure is operating in the steady state with sinusoidal voltage sources. If $v_1 = 2\cos 2t$ and $v_2 = 2\sin 2t$, determine the voltage $v_a(t)$.

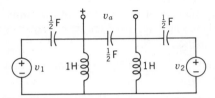

Fig. P12-13.

12-14. The inductively coupled network of the figure is operating in the sinusoidal steady state with $v_1(t) = 2\cos t$. If $L_1 = L_2 = 1$ H, $M = \frac{1}{4}$ H, and $C = 1$ F, determine the voltage $v_a(t)$.

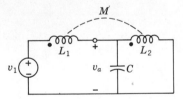

Fig. P12-14.

12-15. The network of the figure is operating in the sinusoidal steady state. In the network, it is determined that $v_a = 10\sin(1000t + 60°)$ and $v_b = 5\sin(1000t - 45°)$. The magnitude of the impedance of the capacitor is 10 Ω. Determine the impedance at the input terminals of the network N.

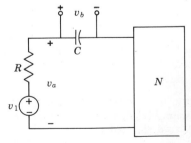

Fig. P12-15.

12-16. In the network shown, $v_1 = 10\sin 10^6 t$ and $i_1 = 10\cos 10^6 t$, and the network is operating in the steady state. For the element values given, determine the node-to-datum voltage $v_a(t)$.

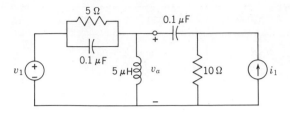

Fig. P12-16.

12-17. For the bridged-T network of the accompanying figure, $v_1 = 2 \cos t$ and the system is in the steady state. For this network, (a) determine $i_a(t)$, and (b) determine $i_b(t)$

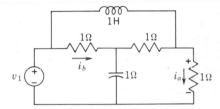

Fig. P12-17.

12-18. The network of the figure is operating in the steady state with $v_1 = 2 \sin 2t$ and $K_1 = -\frac{1}{3}$. Under these conditions, determine $i_2(t)$.

The following series of problems are intended to give practice in constructing phasor diagrams. The network shown in the figure for Prob. 12-19 is assumed to be operating in the sinusoidal steady state. In the element values given in the table, a double entry in column 1 implies a series connection, in column 2 a parallel connection. For each problem, (a) determine $v_1(t)$. (b) Draw a complete phasor diagram showing all voltages and all currents, as well as all relationships between the voltages and the currents.

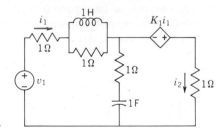

Fig. P12-18.

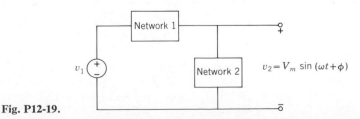

Fig. P12-19.

	Network 1	Network 2	V_m	ω	ϕ
12-19.	$R = 1$	$C = 2$	2	$\frac{1}{2}$	$-30°$
12-20.	$R = 2$	$C = 1$	10	2	$45°$
12-21.	$R = 20$	$C = \frac{1}{2}$	1	0.1	$0°$
12-22.	$R = 2$	$L = 2$	100	$\frac{1}{2}$	$30°$
12-23.	$L = \frac{1}{2}$	$R = 1$	10	$\frac{1}{2}$	$0°$
12-24.	$C = 2$	$R = 2$	3	1	$45°$
12-25.	$L = 3$	$C = 1$	10	$\frac{1}{2}$	$-45°$
12-26.	$C = 1$	$L = \frac{1}{2}$	1	2	$0°$
12-27.	$R = 1, C = 1$	$L = 2$	2	1	$30°$
12-28.	$R = 1, L = 2$	$C = \frac{1}{2}$	2	1	$45°$
12-29.	$L = 1, C = 2$	$R = 1$	10	$\frac{1}{2}$	$0°$
12-30.	$R = 1, C = 1$	$R = 1, L = 2$	10	1	$90°$
12-31.	$R = 3, L = 2$	$R = 1, C = \frac{1}{2}$	1	$\frac{1}{2}$	$0°$
12-32.	$L = 1, C = 2$	$R = 1, C = 1$	100	1	$-90°$
12-33.	$L = 1, C = 2$	$R = 1, L = 2$	1	1	$0°$

12-34. The network of the figure is operating in the sinusoidal steady state and it is known that $v_3 = 2 \sin 2t$. For the element values given, determine $\mathbf{V}_2/\mathbf{V}_1 = A e^{j\phi}$.

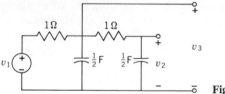

Fig. P12-34.

12-35. The network of the figure is adjusted so that $R_L = R_C = \sqrt{L/C}$. (a) Draw a complete phasor diagram showing all voltages and currents (and their relationships to each other) for the condition $|I_L| = |I_C|$. (b) Let the frequency for the condition of part (a) be ω_1. Draw a phasor diagram for a frequency $\omega_2 > \omega_1$. (c) Repeat part (b) for a frequency $\omega_3 < \omega_1$.

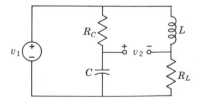

Fig. P12-35.

12-36. The network of the figure is adjusted so that $R_1 C_1 = R_2 C_2 = T$. Let the phase angle of v_2 with respect to v_1 be ϕ. (a) Show how ϕ varies with T. (b) For a fixed T, show how ϕ varies with ω. (c) For a fixed T, show how the maximum amplitude of v_2 is related to the maximum amplitude of v_1 as a function of ω.

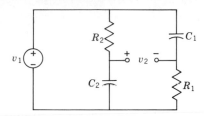

Fig. P12-36.

13 Frequency Response Plots

13-1. PARTS OF NETWORK FUNCTIONS

In this chapter, we restrict our study to network functions which describe networks in the sinusoidal steady state, with $s = j\omega$. Let $G(s)$ be any network function, either driving-point or transfer. Now $G(j\omega)$ is a complex function which may be written in rectangular coordinates as

$$G(j\omega) = R(\omega) + jX(\omega) \tag{13-1}$$

or in the polar form[1]

$$G(j\omega) = |G(j\omega)|e^{j\phi(\omega)} \tag{13-2}$$

Here $R(\omega) = \text{Re } G(j\omega)$ is the *real part* of the network function, $X(\omega) = \text{Im } G(j\omega)$ is the *imaginary part*, $|G(j\omega)|$ is the *magnitude*, and $\phi(\omega)$ is the *phase*. The parts of Eqs. (13-1) and (13-2) are related through the equations,[2]

$$\phi(\omega) = \tan^{-1}\frac{X(\omega)}{R(\omega)} \tag{13-3}$$

and

$$|G(j\omega)| = \sqrt{[R(\omega)]^2 + [X(\omega)]^2} \tag{13-4}$$

[1] A useful shorthand for this equation is $|G(j\omega)| \, \underline{/\phi(\omega)}$.

[2] There are, of course, other relationships of the four parts: Given any one of the parts, the other three may be determined analytically. See Chapter 8 of the author's *Introduction to Modern Network Synthesis* (John Wiley & Sons, Inc., New York, 1960).

372

These parts of network functions are important in network design for two good reasons: (1) As we will see later in Chapter 14, specifications from which networks are to be designed are usually given in terms of magnitude and phase—and less frequently in terms of real and imaginary parts. (2) Measurements of these parts of network functions are easily made by using standard instruments such as the cathode ray oscilloscope, a voltmeter-ammeter-wattmeter combination, the bridge, or the Q-meter. While our present concern is with analytical methods to expedite the determination of the parts of $G(j\omega)$, we must keep in mind that we may be required to check our results in the laboratory and to do this quickly and conveniently, using simple instruments.

The general network function of previous chapters becomes the following with $s = j\omega$:

$$G(j\omega) = \frac{a_0(j\omega)^n + a_1(j\omega)^{n-1} + \ldots + a_n}{b_0(j\omega)^m + b_1(j\omega)^{m-1} + \ldots + b_m} \quad (13\text{-}5)$$

or

$$G(j\omega) = \frac{a_0(j\omega - z_1)(j\omega - z_2)\ldots(j\omega - z_n)}{b_0(j\omega - p_1)(j\omega - p_2)\ldots(j\omega - p_m)} \quad (13\text{-}6)$$

For specified numerical values of the a and b coefficients or the poles and zeros, we wish to compute network functions for all values of ω. This will be found to be tedious in general. Fortunately, the amount of labor can be reduced in many cases by using methods which are discussed in this chapter.

13-2. MAGNITUDE AND PHASE PLOTS

There are a number of possibilities for coordinates for frequency response plots. The independent variable is always frequency ω—otherwise the plot would not be a *frequency* response—but the dependent variable may be any of the parts discussed in the last section; or it may be two of the associated parts, magnitude and phase in the polar coordinate interpretation, with frequency as the parameter. The last possibility is considered in the next section under the title of complex loci. The use of logarithmic coordinates in magnitude and phase plots is also treated separately in Section 13-5 as Bode plots. Here we consider a number of important features in the use of linear coordinates in plots of magnitude and phase with frequency.

It is usual to make use of the frequency range from 0 to ∞ rather than $-\infty$ to ∞. This is a natural choice since laboratory generators of sine waves produce only positive ω. In addition, the information for negative frequency is redundant because magnitude and real part are even functions, while phase and imaginary part are odd functions of

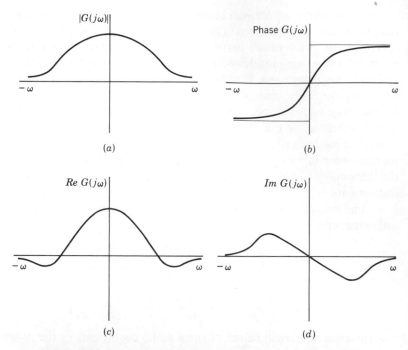

Fig. 13-1. Four parts of a network function for $s = j\omega$ (a) the magnitude, (b) the phase, (c) the real part, and (d) the imaginary part.

ω as illustrated in Fig. 13-1. We should note that for some purposes, the application of the Nyquist criterion[3] being one example, it is necessary to use the complete range of values of ω. Finally, we should note that, in comparison to plots of magnitude and phase, in engineering practice little use is made of the real or imaginary part frequency response, and for that reason such plots will not be considered here.

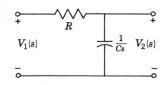

Fig. 13-2. Network for Example 1.

EXAMPLE 1

Figure 13-2 shows a two-port RC network for which the voltage-ratio transfer function is

$$G_{12}(s) = \frac{V_2(s)}{V_1(s)} = \frac{1}{RCs + 1} \tag{13-7}$$

For the sinusoidal steady state, G_{12} becomes

$$G_{12}(j\omega) = \frac{1}{j\omega RC + 1} = \frac{1/RC}{j\omega + (1/RC)} \tag{13-8}$$

[3]H. Nyquist, "Regeneration theory," *Bell System Tech. J.*, **11**, 126–147 (1932).

The asymptotic values of magnitude and phase can be determined by inspection of Eq. (13-8). For small ω, the magnitude is approximately 1 and the phase is $0°$. For large ω, the magnitude approaches 0 while the phase becomes nearly $-90°$. For one other frequency, $\omega = 1/RC$, magnitude and phase are especially easy to determine, the values being $1/\sqrt{2} \approx 0.707$ and $-45°$. If a number of other values of ω are selected, the magnitude and phase response curves shown in Fig. 13-3 may be drawn, where we have let $M = |G_{12}(j\omega)|$ for convenience.

To study the significance of the frequency $\omega = 1/RC$ in Example 1, let us digress by considering the network of Fig. 13-2 as a one-port network excited by a voltage source for which the current phasor is $\mathbf{I} = Y\mathbf{V}$ and the magnitude of the current is

$$|I(j\omega)| = |Y(j\omega)||V(j\omega)| \qquad (13-9)$$

We will show in Chapter 14 that the average power in the resistor of the RC combination is

$$P = \tfrac{1}{2}|I(j\omega)|^2 R \qquad (13-10)$$

Let ω_{hp} be the frequency at which $|I| = (1/\sqrt{2})|I|_{\max}$. This will, of course, be the frequency at which the admittance is $1/\sqrt{2}$ times its maximum value, since from Eq. (13-9) we see that $|I|$ varies directly with $|Y|$ when $|V|$ is constant. Let P_1 be the maximum power given by Eq. (13-10). At the frequency ω_{hp}, the average power in the resistor is

$$P_2 = \frac{1}{2}\left|\frac{I(j\omega)}{\sqrt{2}}\right|^2 R = \frac{1}{2}P_1 \qquad (13-11)$$

indicating that at this frequency the power in the resistor is *half* of its maximum value. The frequency ω_{hp} is called the *half-power frequency*. The name is applied to the frequency at which any magnitude of a network function, either driving-point or transfer, reduces to 0.707 of its maximum value.

Returning now to Eq. (13-7), we see that (1) the frequency $\omega = 1/RC$ is the half-power frequency of the frequency response plot, and (2) $1/RC$ is the reciprocal of the time constant T of the network (say, to a step input), being the negative of the pole location in the s plane, $\sigma = -1/RC$. Thus we have a simple connection between the two kinds of response for this network which is summarized by $T = 1/\omega_{hp}$. Now equations like Eq. (13-7) describe a number of two-element RC or RL networks. For such networks, Eq. (13-8) may be written in a standard form

$$G_{12}(j\omega) = \frac{1}{1 + j(\omega/\omega_{hp})} \qquad (13-12)$$

and for this a universal plot of magnitude and phase may be prepared. This will be done in Section 13-5 in logarithmic coordinates.

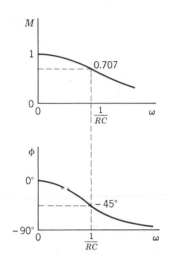

Fig. 13-3. The variation of magnitude and phase for the transfer function G_{12} of the network of Fig. 13-2.

13-3. COMPLEX LOCI

Given a network function $G(j\omega)$ having the general form of Eq. (13-5) or (13-6), we may determine the real and imaginary parts identified by Eq. (13-1), or the magnitude and phase of Eq. (13-2), as a function of ω. As ω is varied, a point representing $G(j\omega) = R(\omega) + jX(\omega)$ moves on the complex $G(j\omega)$ plane and is said to describe a *locus*. A complex locus is a plot of the rectangular coordinates of $G(j\omega)$, $R(\omega)$, and $X(\omega)$, or the polar coordinates $M(\omega)$ and $\phi(\omega)$—only one plot, but two interpretations—with ω as the parameter. Some of the features of such a locus are shown in Fig. 13-4. Each point on the locus may be thought of as a phasor extending from the origin to the point, of magnitude M and angle ϕ, and the locus is simply the path of the "tips" of the phasors.

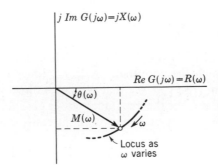

Fig. 13-4. The locus of the tip of the phasor of magnitude M and phase θ as ω varies.

Clearly, the complex loci we have just described contains all the information of the magnitude and phase plots of the last section, but no more. This being the case, we should point out that there are a number of reasons why we might prefer one over the other. The locus plot may be a more efficient or dramatic way of presenting information. In addition, use is made of the complex loci in applying the Nyquist criterion as we will see in Section 13-6. We now show the construction of complex loci by a number of examples.

EXAMPLE 2

As our first example, consider the network used in Example 1, that shown in Fig. 13-2. For $G_{12}(j\omega)$ of Eq. (13-8), the locus is that shown in Fig. 13-5 which starts at $1 + j0$ for $\omega = 0$ and traces a semicircular locus to the origin as ω approaches ∞. That the locus is a circle may be seen by rationalizing Eq. (13-8) to obtain:

$$\text{Re } G_{12} + j \text{ Im } G_{12} = \frac{1}{1 + R^2C^2\omega^2} - j\frac{RC\omega}{1 + R^2C^2\omega^2} \qquad (13\text{-}13)$$

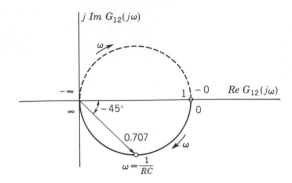

Fig. 13-5. Locus plot for the network of Fig. 13-2 shown solid for $\omega > 0$. The complete locus is a circle of radius $\frac{1}{2}$.

Eliminating the parameter ω, we obtain the following relation between Re G_{12} and Im G_{12}:

$$(\text{Re } G_{12} - \tfrac{1}{2})^2 + (\text{Im } G_{12})^2 = (\tfrac{1}{2})^2 \qquad (13\text{-}14)$$

which is the equation of a circle of radius $\frac{1}{2}$ and center at Re $G_{12} = \frac{1}{2}$, Im $G_{12} = 0$. The upper half of the circle in Fig. 13-5 is obtained for negative frequencies. Observe that the locus for negative frequencies is the *image* of that for positive frequencies, since the real part has even and the imaginary part odd symmetry. We will follow the practice of showing the locus for negative frequency as a dashed curve.

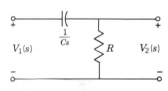

Fig. 13-6. *RC* network for Example 3.

EXAMPLE 3

If the resistor and capacitor of the *RC* network of Example 2 are interchanged, the resulting two-port network, shown in Fig. 13-6, has the transfer function

$$G_{12}(j\omega) = \frac{j\omega}{j\omega + (1/RC)} \qquad (13\text{-}15)$$

Following the same procedure as in Example 1, we determine the following points on the locus:

ω	$G_{12}(j\omega)$
0	0 at $+90°$
$1/RC$	0.707 at $+45°$
∞	1 at $0°$

The complete locus is shown in Fig. 13-7, together with the equivalent $M(\omega)$ and $\phi(\omega)$ plots.

Comparing the results in Examples 1 and 3, we see that the network of Fig. 13-2 has an output in which the phase *lags* that of

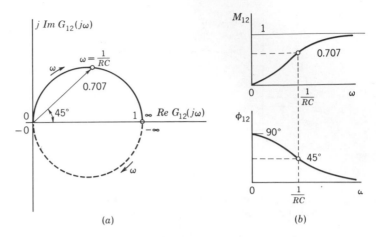

(a)   (b)

Fig. 13-7. Locus plot for the transfer function G_{12} of the network of Fig. 13-6, and the corresponding magnitude and phase plots.

the input for all ω, while in the network of Fig. 13-6, the phase of the output voltage *leads* the input for all ω. In some applications, these networks are called lag networks and lead networks, respectively. The magnitude variations of the two $G_{12}(j\omega)$ with ω are inverse; for a constant magnitude of input voltage, the output voltage of the network of Fig. 13-2 decreases with frequency, and for that of Fig. 13-6 increases with an increase in frequency.

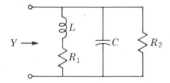

Fig. 13-8. Network analyzed in Example 4.

EXAMPLE 4

Our next example shows that, in some cases of network structures, subloci for parts of the network may be combined to determine a composite locus for the entire network. Consider the network shown in Fig. 13-8. This network may be regarded as the composite of three subnetworks: R_1L, C, and R_2. These subnetworks are connected in parallel so that the admittances may be added to find the driving-point admittance; thus

$$Y(j\omega) = Y_{RL}(j\omega) + Y_C(j\omega) + Y_R(j\omega) \qquad (13\text{-}16)$$

The loci for these three functions may be added point by point (a phasor addition) to give the composite locus for $Y(j\omega)$. For the first subnetwork,

$$Y_{RL}(j\omega) = \frac{1/L}{j\omega + (R_1/L)} \qquad (13\text{-}17)$$

meaning that the locus for Y_{RL} is that of Fig. 13-5, except for scale and half-power frequency, and is shown in (a) of Fig. 13-9. Now $Y_C = j\omega C$ so that the locus is confined to the imaginary axis while $Y_R = 1/R_2$, which does not vary with frequency. These loci are shown as (b) and

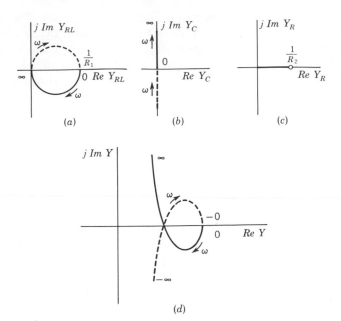

Fig. 13-9. Individual locus plots for (*a*) the R_1L branch, (*b*) the *C* branch, (*c*) the R_2 branch (the locus being a single point) and (*d*) the complete locus for *Y* found by adding the loci for the three parallel branches.

(*c*) of Fig. 13-9. The composite locus for the network of Fig. 13-8 is shown in Fig. 13-9(*d*) for one set of values of elements.

The method of this example applies when a driving-point immittance may be expressed as the sum of admittances or impedances. For more complicated networks, it may be necessary to calculate $G_{12}(j\omega)$ for a number of values of ω, using enough values of ω so that the locus plot may be made accurately.

13-4. PLOTS FROM *s*-PLANE PHASORS

Any network function $G(s)$ has the form of a quotient of poly-nomials, each of which may be expressed as the product of factors of the form

$$(s - s_r) \tag{13-18}$$

where s_r is either a pole or a zero. In the sinusoidal steady state, $s = j\omega$, and the typical factor becomes

$$(j\omega - s_r) \tag{13-19}$$

In the *s* plane, each of the terms $j\omega$ and s_r are phasors, and we are interested in the *difference* of these two phasors—which is also a

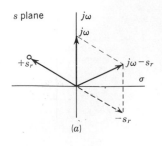

(a)

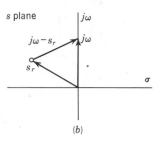

(b)

Fig. 13-10. (*a*) Construction of and (*b*) the direction of the phasor $(j\omega - s_r)$ which is from s_r to $j\omega$.

phasor. In general, s_r is complex while $j\omega$ is imaginary. The two phasors and their difference are shown in Fig. 13-10. In (*a*) of this figure, we see the component phasors and their sum; in (*b*) of Fig. 13-10 is shown a *string* phasor diagram which is better suited to our needs. The phasor difference $(j\omega - s_r)$ is seen to be a phasor directed *from s_r to $j\omega$*. Now as ω increases, the phasor $(j\omega - s_r)$ changes in length and angle. If a network function involves a number of such factors, some in the numerator and some in the denominator, the variation of the network function with ω may be determined by studying the variation of the individual factors, each changing in a pattern determined by the position of s_r with respect to the imaginary axis.

EXAMPLE 5

To illustrate the use of *s*-plane phasors in determining magnitude and phase variation with ω, consider the admittance of a series *RL* circuit which is

$$Y(s) = \frac{1}{L} \frac{1}{[s + (R/L)]} \tag{13-20}$$

This admittance has a pole at $s = -R/L$, and a zero at infinity. The construction of the phasor $[j\omega - (-R/L)]$ with respect to this pole location is shown in Fig. 13-11. Three steps in the change of the phasor position are shown in Fig. 13-12 as frequency increases from ω_1 to ω_2 to ω_3. Let

$$L\left(j\omega + \frac{R}{L}\right) = M(\omega)e^{j\phi(\omega)} \tag{13-21}$$

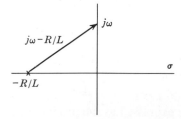

Fig. 13-11. Phasor representation of $(j\omega + R/L)$.

Then the admittance may be found from the equation

$$Y(j\omega) = \frac{1}{M(\omega)} e^{-j\phi(\omega)} \tag{13-22}$$

It can be seen that the magnitude changes from $1/R$ to 0 as ω changes from zero to infinity; similarly, the phase changes from $0°$ to $-90°$ as ω varies from zero to infinity. The complex locus representation of this variation is shown in Fig. 13-13.

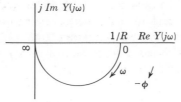

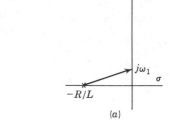

Fig. 13-13. The complex locus for $Y(j\omega)$ for values obtained by the technique illustrated in Fig. 13-12.

(a)

EXAMPLE 6

Consider next a series *RLC* circuit having the driving-point impedance

$$Z(s) = Ls + R + \frac{1}{Cs} \qquad (13\text{-}23)$$

and the admittance is the reciprocal of $Z(s)$; that is,

$$Y(s) = \frac{1}{L}\left[\frac{s}{s^2 + (Rs/L) + (1/LC)}\right] \qquad (13\text{-}24)$$

or

$$Y(s) = \frac{1}{L}\left[\frac{s}{(s - s_a)(s - s_a^*)}\right] \qquad (13\text{-}25)$$

where

$$s_a, s_a^* = -\frac{R}{2L} \pm j\sqrt{\frac{1}{LC} - \left(\frac{R}{2L}\right)^2} = -\zeta\omega_n \pm j\omega_n\sqrt{1 - \zeta^2} \qquad (13\text{-}26)$$

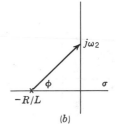

(b)

In this equation, we have followed the conventions of Chapter 6, where ω_n is the natural frequency of the network, and ζ is the damping ratio. In this example, we consider only the oscillatory case with $\zeta < 1$. For this case, the poles are located on a circle of radius ω_n, and a line from the pole to the origin makes an angle of $\theta = \cos^{-1}\zeta$ with the negative real axis. A single zero is located at the origin, as shown in Fig. 13-14.

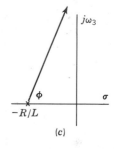

(c)

Fig. 13-12. Illustrating the change in the phasor length as ω increases where $\omega_1 < \omega_2 < \omega_3$. The variation of $Y(j\omega)$ is found from Eq. (13-22).

The frequency response is determined by letting ω assume several values, four of which are shown in Fig. 13-15. The magnitude and phase are then computed by expressing each factor in Eq. (13-25) in the form of Eq. (13-22), giving

$$Y(j\omega) = \frac{M_0}{M_a M_a^*} e^{j(\phi_0 - \phi_a - \phi_a^*)} \qquad (13\text{-}27)$$

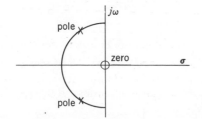

Fig. 13-14. The poles and zero of $Y(s)$ given by Eq. 13-24.

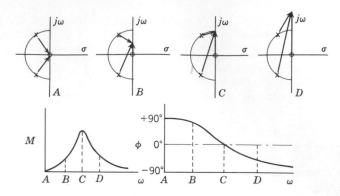

Fig. 13-15. Phasors which determine the frequency response shown for four different values of ω.

The maximum value of $|Y(j\omega)|$ evidently takes place near the frequency at which M_a has a minimum value. The frequency to cause M_a to have a minimum value is the frequency at the point of closest approach to one of the conjugate poles. In that frequency range, M_a changes rapidly, and at the same time M_a^* and M_0 are changing very slowly. The frequency corresponding to a maximum $|Y(j\omega)|$ is defined as the *frequency of resonance*. Since $I(j\omega)$ varies directly with $Y(j\omega)$, the frequency of resonance is also the frequency of maximum $I(j\omega)$.

The magnitude of $Y(j\omega)$ may be written in the form

$$|Y(j\omega)| = \frac{1}{\sqrt{R^2 + [\omega L - (1/\omega C)]^2}} \tag{13-28}$$

and from this equation it is seen that $|Y(j\omega)|$ has a maximum value of $(1/R)$ when

$$\omega L - \frac{1}{\omega C} = 0 \quad \text{or} \quad \omega = \frac{1}{\sqrt{LC}} = \omega_n \tag{13-29}$$

that is, resonance occurs at $\omega = \omega_n$ (and not at the point opposite the pole on the $j\omega$ axis). An enlarged view of the various phasors for the condition of resonance is shown in Fig. 13-16. The phase angles from the poles to $j\omega_n$ are marked ϕ_a and ϕ_a^*. The phasor from the zero to $j\omega_n$ is along the $j\omega$ axis and thus has a constant phase angle of $+90°$. The sum of ϕ_a and ϕ_a^* is equal to $90°$ because the triangle ABC is a right triangle (being inscribed in a semicircle). The total phase angle, which is

$$\phi_0 - \phi_a - \phi_a^* = +90° - (+90°) = 0° \tag{13-30}$$

thus has zero value. The phase angle of $Y(j\omega)$ is *zero degrees at resonance*. The magnitude of $Y(j\omega)$ at resonance is $1/R$, and $|I(j\omega)|$ at resonance is V/R.

The *circuit Q*, or simply Q, for an RLC series circuit is defined as

$$Q = \frac{\omega_n L}{R} = \frac{1}{2} \frac{\omega_n}{R/2L} \tag{13-31}$$

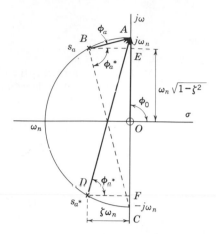

Fig. 13-16. Phasors drawn for the special condition of resonance.

Now the quantity $(R/2L)$ is the same as $\zeta\omega_n$ by Eq. (13-26). Thus the circuit Q is defined as

$$Q = \frac{1}{2}\frac{\omega_n}{\zeta\omega_n} = \frac{1}{2} \times \frac{\text{the length } OA}{\text{the length } EB} \qquad (13\text{-}32)$$

in terms of quantities shown in Fig. 13-16. The circuit Q can thus be taken directly from a scale plot of the poles and zeros of the immittance function for an RLC circuit. The circuit Q can alternatively be written in the form

$$Q = \frac{1}{2\zeta} \qquad (13\text{-}33)$$

$$= \frac{1}{2\cos\theta} \qquad (13\text{-}34)$$

where θ is the angle from the $-\sigma$ axis to the line OB (or OD) of Fig. 13-16. From these last three equations, two conclusions can be written:

(1) The closer poles s_a and s_a^* are to the $j\omega$ axis, the higher the Q. (This follows since Q varies inversely with the distance EB.)
(2) The value of Q varies inversely with damping ratio ζ. A high value of Q infers a low value of damping ratio. A series RLC circuit with low R thus has high Q.

Plots of the magnitude of $Y(j\omega)$ for various values of Q are shown in Fig. 13-17.

Another means of specifying the circuit Q is specification in terms of *half-power* frequencies. The current at resonance has the magnitude V/R. When the current has the magnitude

$$I = \frac{V}{\sqrt{2}\,R} \qquad (13\text{-}35)$$

the power will be *half* of that at resonance. At the half-power fre-

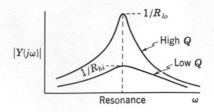

Fig. 13-17. The general shape of the response $|Y(j\omega)|$ for large and small values of Q.

quencies, the magnitude of the admittance $Y(j\omega)$ is $(1/\sqrt{2}\,R)$; this requires that

$$\sqrt{R^2 + [\omega L - (1/\omega C)]^2} = \sqrt{2}\,R \tag{13-36}$$

or

$$\left(\omega L - \frac{1}{\omega C}\right) = \pm R \tag{13-37}$$

This equation reduces to the form

$$\omega^2 \pm \frac{R}{L}\omega - \frac{1}{LC} = 0 \tag{13-38}$$

The values of ω that satisfy this equation are

$$\omega = \pm \frac{R}{2L} \pm \sqrt{(R/2L)^2 + (1/LC)} \tag{13-39}$$

or, in terms of damping ratio and undamped natural frequency,

$$\omega = \omega_n(\pm\zeta \pm \sqrt{\zeta^2 + 1}) \tag{13-40}$$

Let the two positive values of frequency given by Eq. (13-40) be ω_1 and ω_2.

$$\omega_1, \omega_2 = \omega_n(\sqrt{\zeta^2 + 1} \pm \zeta) \tag{13-41}$$

Two important results follow simply from Eq. (13-41). First, the product

$$\omega_1\omega_2 = \omega_n^2 \tag{13-42}$$

meaning that ω_n is the geometrical mean of the two frequencies ω_1 and ω_2. The differences of these two frequencies is defined as the *bandwidth B*, which is

$$B = \omega_2 - \omega_1 = 2\omega_n\zeta = \frac{\omega_n}{Q} \tag{13-43}$$

such that the bandwidth varies inversely with the circuit Q.

In many networks used as *selectors*, the damping ratio ζ is very small so that ζ^2 is negligible compared to 1. Under this assumption, the two half-power frequencies are

$$\omega_2 = \omega_n + \zeta\omega_n \tag{13-44}$$

and

$$\omega_1 = \omega_n - \zeta\omega_n \tag{13-45}$$

The quantity $(\zeta\omega_n)$ is the distance EB of Fig. 13-16, or the distance

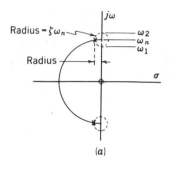

(a)

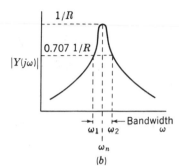

(b)

Fig. 13-18. The bandwidth is defined by the frequency interval, $\omega_2 - \omega_1$, and the bandwidth is defined by a circle of radius $\zeta\omega_n$ for the high Q case, as shown in the figure.

from the $j\omega$ axis to the pole s_a (or s_a^*). The location of the half-power frequencies is shown in Fig. 13-18: A circle of radius $\zeta\omega_n$ centered at $0 + j\omega_n$ crosses the imaginary axis at the half-power frequencies. As identified by Eq. (13-43), this range of frequencies is the bandwidth.

EXAMPLE 7

The final example of this section is chosen to show something new in complex locus plots. The network shown in Fig. 13-19(a) is known as a *symmetrical lattice*. Some of the properties of this particular kind of lattice were studied in Chapter 10. For our present discussion, we observe that through routine analysis the voltage-ratio transfer function may be determined to have the form

$$G_{12}(s) = \frac{V_2(s)}{V_1(s)} = \frac{s^2 - s + 1}{s^2 + s + 1} = \frac{(s - s_1)(s - s_1^*)}{(s - s_a)(s - s_a^*)} \qquad (13\text{-}46)$$

The two zeros have the values

$$s_1, s_1^* = +\frac{1}{2} \pm j\frac{\sqrt{3}}{2} \qquad (13\text{-}47)$$

and the poles are at

$$s_a, s_a^* = -\frac{1}{2} \pm j\frac{\sqrt{3}}{2} \qquad (13\text{-}48)$$

These poles and zeros are shown in Fig. 13-19(b); they are located on a

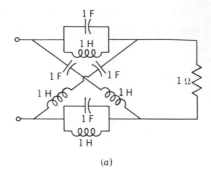

(a)

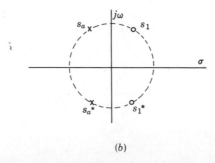

(b)

Fig. 13-19. (a) A symmetrical lattice, and (b) the poles and zeros of G_{12} for this network. These two poles and two zeros constitute a quad.

unit circle for this example and they are symmetrically located with respect to both the real and imaginary axes. Poles and zeros so located are said to form a *quad*. The left half-plane members of the quad are poles, and the right half-plane members are zeros.

As outlined in previous examples of this section, the frequency response may be found from phasors drawn from the poles and zeros to different points on the imaginary axis, as shown in Fig. 13-20. From this figure, we see that the magnitude of $(j\omega - s_1)$ is always equal to the magnitude of $(j\omega - s_a)$; likewise, the magnitude of $(j\omega - s_1^*)$ is always equal to that of $(j\omega - s_a^*)$. Thus we have discovered that

$$\left|\frac{V_2(j\omega)}{V_1(j\omega)}\right| = 1 \tag{13-49}$$

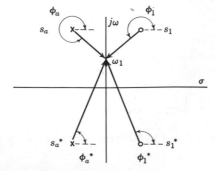

Fig. 13-20. Phasors used in determining the magnitude and phase response for G_{12} for the network of Fig. 13-19(a).

In other words, we have shown that the magnitude of the output voltage is *always* equal to the magnitude of the input voltage—*for all frequencies!* It is not surprising that such a network is known as an *all-pass* network.

In computing the phase, we determined positive contributions from the *zero phasors* and negative contributions from *pole phasors*:

$$\phi = \phi_1 + \phi_1^* - \phi_a - \phi_a^* \qquad (13\text{-}50)$$

where the angles are identified in Fig. 13-20. When the phasors move up the imaginary axis, we see that the phase of $G_{12}(j\omega)$ starts at $0°$ at zero frequency and becomes more negative as frequency increases while approaching $360°$. The phase shift characteristic for the transfer function is shown in Fig. 13-21, together with the complex locus plot.

In summary, this network has the property for sinusoidal inputs that everything which comes in goes out unaffected in magnitude but distorted in phase.[4] Such networks find applications in the compensation of telephone lines.

13-5. BODE DIAGRAMS

A logarithmic scale for the magnitude of network functions as well as the frequency variable was used extensively in the studies of H. W. Bode.[5] For that reason the logarithmic plots to be studied in this section are known as *Bode plots* or *Bode diagrams*. The natural logarithm of

$$G(j\omega) = M(\omega)e^{j\phi(\omega)} \qquad (13\text{-}51)$$

is the complex function

$$\ln G(j\omega) = \ln M(\omega) + j\phi(\omega) \qquad (13\text{-}52)$$

In this equation, $\ln M(\omega)$ is the *gain* or *logarithmic gain* in nepers, and ϕ is the *angle* function in radians. The usual unit for gain is the decibel (db); in this unit[6]

$$\text{Gain in decibels} = 20 \log M(\omega) \qquad (13\text{-}53)$$

Similarly, the usual unit for the angle ϕ is the degree. Conversion to these units may be accomplished by the equations:

$$\text{Number of decibels} = 8.68 \times \text{number of nepers} \qquad (13\text{-}54)$$

$$\text{Number of degrees} = 57.3 \times \text{number of radians} \qquad (13\text{-}55)$$

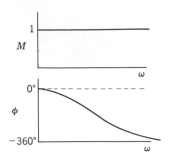

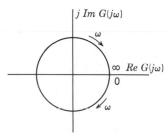

Fig. 13-21. The magnitude and phase response for the network of Fig. 13-19(*a*) showing "all-pass" characteristics.

[4]Some of my friends pass information along this same way!

[5]*Network Analysis and Feedback Amplifier Design* (D. Van Nostrand Co., Inc., Princeton, N.J., 1945), pp. 316 ff. or H. W. Bode, "Relations between attenuation and phase in feedback amplifier design," *Bell System Tech. J.*, 19, 421–454 (1940). Bode was a member of the technical staff at the Bell Telephone Laboratories and later joined the faculty at Harvard University.

[6]Throughout the book, *log* denotes the base 10 and *ln* the base $e = 2.718$.

At this point, we introduce a logarithmic frequency variable

$$u = \log \omega \quad \text{or} \quad \omega = 10^u \qquad (13\text{-}56)$$

In terms of this variable, two frequency intervals given by the equation

$$u_2 - u_1 = \log \omega_1 - \log \omega_2 = \log \frac{\omega_2}{\omega_1} \qquad (13\text{-}57)$$

find extensive use. These intervals are the *octave* for which $\omega_2 = 2\omega_1$, and the *decade* with $\omega_2 = 10\omega_1$. Slopes of straight lines in the coordinate system of Bode plots will be expressed in terms of these two intervals.

Let us next return to Eq. (13-6) for $G(j\omega)$, assuming for the time being that all poles and zeros are distinct (nonrepeated). If the first term in the numerator is modified by letting $z_1 = -1/T_1$, a real number, then we have

$$j\omega - z_1 = \frac{1}{T_1}(1 + j\omega T_1) \qquad (13\text{-}58)$$

The logarithm of this equation is

$$\ln (j\omega - z_1) = \ln \frac{1}{T_1} + \ln|1 + j\omega T_1| + j \tan^{-1} \omega T_1 \qquad (13\text{-}59)$$

The first two terms in the equation are expressed in decibels as

$$20 \log \frac{1}{T_1} + 20 \log |1 + j\omega T_1| \, \text{db} \qquad (13\text{-}60)$$

and the last term may be expressed in degrees as

$$\phi = 57.3 \tan^{-1} \omega T_1 \qquad (13\text{-}61)$$

All numerator factors will be of this form, with $z_i = -1/T_i$. For the denominator, we will let $p_1 = -1/T_a$, $p_2 = -1/T_b$, etc. Then the magnitude of $G(j\omega)$ in decibels is given by the summation

$$20 \log \frac{a_0 T_a T_b \dots T_m}{b_0 T_1 T_2 \dots T_n} + 20 \log |1 + j\omega T_1| + \dots + 20 \log |1 + j\omega T_n|$$
$$- 20 \log |1 + j\omega T_a| - \dots - 20 \log |1 + j\omega T_m| \, \text{db} \qquad (13\text{-}62)$$

and the angle in degrees is

$$57.3 \tan^{-1} \omega T_1 + 57.3 \tan^{-1} \omega T_2 + \dots + 57.3 \tan^{-1} \omega T_n$$
$$- 57.3 \tan^{-1} \omega T_a - \dots - 57.3 \tan^{-1} \omega T_m \qquad (13\text{-}63)$$

This result tells us that, aside from the first term in Eq. (13-62), which is a constant, all terms are of the same form, differing only in value of T_i—for both the magnitude and the phase. We may consider the second term in Eq. (13-62) and the first in Eq. (13-63) as typical and study these in detail. Once this is done, the total magnitude and phase functions will be found by a simple summation of such terms.

Magnitude of first-order factor. The expression for magnitude is

$$|G_1| = 20 \log|1 + j\omega T_1| = 20 \log (1 + \omega^2 T_1^2)^{1/2} \quad (13\text{-}64)$$

For small ω, that is, $\omega T_1 \ll 1$, G_1 approaches the straight-line asymptote

$$|G_1| \approx 20 \log 1 = 0 \text{ db} \quad (13\text{-}65)$$

which has zero magnitude and zero slope. For large ω, $\omega T_1 \gg 1$, $|G_1|$ approaches another straight-line asymptote

$$|G_1| \approx 20 \log \omega T_1 \text{ db} \quad (13\text{-}66)$$

The slope of the straight line in Bode plot coordinates is found by observing that the magnitude changes 6 db from $\omega T_1 = 1$ to $\omega T_1 = 2$ (or any other octave), giving a slope of 6 db/octave, or changes 20 db from $\omega T_1 = 1$ to $\omega T_1 = 10$ (or any other decade), corresponding to a slope of 20 db/decade.

The frequency $\omega = 1/T_1$ is called the *break point* or *break frequency*—and sometimes the *corner frequency*. This is a new name for a familiar quantity which has previously served as the reciprocal of the time constant, and also the half-power frequency. The value of G_1 at the break frequency is, from Eq. (13-64),

$$|G_1| = 20 \log 2 \text{ db} \approx 3 \text{ db} \quad (13\text{-}67)$$

Another convention will be introduced here. Plots of the straight line segments given by Eq. (13-65) and Eq. (13-66) constitute the *asymptotic curve*, which is distinguished from the true magnitude response given by Eq. (13-64). Values of both responses for a number of frequencies are given in Table 13-1. From these values, the plot shown in Fig. 13-22 may be made, and a technique for drawing the true magnitude response may be described.

From Table 13-1 and from Fig. 13-22, we see that a simple technique for drawing the true magnitude response is described in the following steps: (1) Knowing T, determine the break point. (2) From the break point, draw the two asymptotic lines, one of zero slope extending to lower frequencies, one of 6 db/octave slope extending to

Table 13-1.

ω	*True Magnitude*	*Asymptotic Magnitude*	*Difference*
$1/2T$	1 db	0 db	1 db
$1/1.31T$	2 db	0 db	2 db
$1/T$	3 db	0 db	3 db
$1.31/T$	4.3 db	2.3 db	2 db
$2/T$	7 db	6 db	1 db

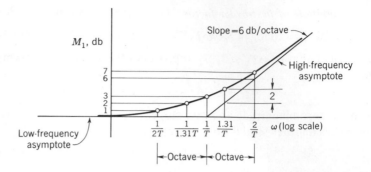

Fig. 13-22. The magnitude response given by Eq. 13-64 showing the actual curve and the low-frequency and high-frequency asymptotes. The five points on the actual curve are conveniently located in terms of the asymptotic straight lines.

higher frequencies. (3) At the break point, the true response is displaced 3 db from the intersection of asymptotes; an octave below and above the break point, the true curve is displaced 1 db from the asymptotic lines. If required, the 2-db difference points of Table 13-1 may be located. (4) If the factor comes from a pole rather than a zero in $G(j\omega)$, all values are negative, including the slope.

EXAMPLE 8

For this example, we will construct a Bode plot for the network function,

$$G(j\omega) = \frac{(1 + j\omega T_1)(1 + j\omega T_3)}{(1 + j\omega T_2)(1 + j\omega T_4)} \tag{13-68}$$

The break frequencies are first determined from $\omega_i = -1/T_i$, a transformation illustrated by an s-plane rotation in Fig. 13-23. Using Bode coordinates, the four asymptotic lines of ± 6 db/octave are drawn, as shown in Fig. 13-24(a). Finally, the responses are added as required by Eq. (13-62), giving the asymptotic response of Fig. 13-24(b). To this skeleton, the true response is determined by adding

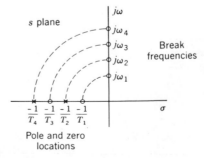

Fig. 13-23. Illustrating the simple relationship between pole and zero locations on the real axis and the break frequencies.

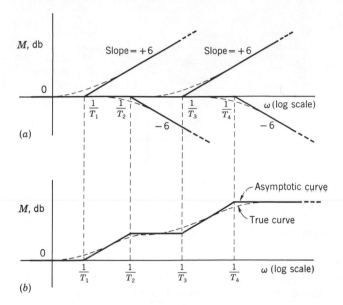

Fig. 13-24. (*a*) The individual responses for the four factors in Eq. (13-68) and (*b*) their addition to give the response for the magnitude of $G(j\omega)$.

the actual responses for each of the factors giving the result shown by the dashed line. Simple, isn't it?

Phase of first-order factor. From Eq. (13-63), the first term is

$$\phi_1 = 57.3 \tan^{-1} \omega T_1 \tag{13-69}$$

For this function, the values given in Table 13-2 are useful in making a plot such as the one of Fig. 13-25. One approximation to the phase curve that is sometimes used consists of a straight line of slope 45° per decade passing through 45° at $\omega = 1/T$ as shown by the dashed line in Fig. 13-25. The maximum difference between this approximation and

Table 13-2.

ω	$57.3 \tan^{-1} \omega T$
0	0°
$1/2T$	26.6°
$1/\sqrt{3}\,T$	30°
$1/T$	45°
$\sqrt{3}\,T$	60°
$2T$	63.4°
∞	90°

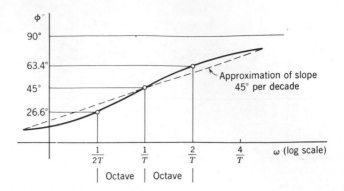

Fig. 13-25. The phase given by Eq. (13-69), showing three values which are useful in constructing the response. The dashed line represents a useful approximation.

the actual curve is less than 6° from $\omega = 0.1/T$ to $\omega = 10/T$. Values of ϕ_1 are multiplied by -1 if the factor arises from a pole in $G(s)$.

The factor s in $G(s)$—a pole or zero at the origin—has a magnitude response given by $|G_0| = 20 \log \omega$ db for all ω. The equation is multiplied by $+1$ for a zero, by -1 for a pole. The phase contribution from this factor is $\pm 90°$, using the same sign convention: $+1$ for a zero, -1 for a pole.

Second-order factors. If $G(s)$ has poles or zeros which are complex conjugates, then the factor $(1 + sT_i)$ is replaced by a second-order factor which may be written in the form

$$G_2 = 1 + 2\zeta sT + s^2T^2 \tag{13-70}$$

where ζ is the damping ratio of Chapter 6. From this, the magnitude and phase functions are

$$M = 10 \log [(1 - \omega^2T^2)^2 + (2\zeta\omega T)^2] \tag{13-71}$$

and

$$\phi = \tan^{-1} \frac{2\zeta\omega T}{1 - \omega^2T^2} \tag{13-72}$$

Plots of $M(\omega)$ and $\phi(\omega)$ from these equations with ζ as the parameter are shown in Fig. 13-26. From the curves and the equations, a number of useful observations may be made concerning the use of curves for second-order factors in constructing Bode diagrams:

(1) The low-frequency asymptote is the line 0 db, as was the case for the first-order factor.
(2) The high frequency asymptote is

$$|G_2| \approx 40 \log \omega T \, \text{db} \tag{13-73}$$

which is a straight line of slope 12 db/octave or 40 db/decade.

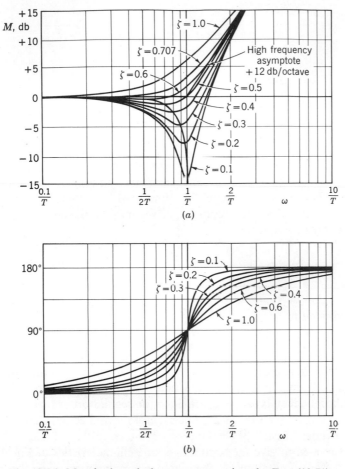

Fig. 13-26. Magnitude and phase responses given by Eqs. (13-71) and (13-72), for several values of ζ.

(3) The break frequency, the intersection of the low and high frequency asymptotes, is $\omega = 1/T$ for all ζ.

(4) At the break frequency, $G_2(j\omega)$ has the magnitude

$$M_2(1/T) = 20 \log 2\zeta \text{ db} \qquad (13\text{-}74)$$

A number of values of this equation which are useful in constructing Bode diagrams are given in Table 13-3.

(5) The phase varies from $0°$ at low frequencies to $180°$ at high frequencies, with the value $90°$ at the break frequency for all ζ. The phase characteristics change abruptly near the break frequency; the more abrupt, the smaller the value of ζ.

(6) All the equations given here are for zero factors. For pole factors, multiply all equations by -1.

Table 13-3.

ζ	$20 \log 2\zeta$ db	ζ	$20 \log 2\zeta$ db
0	$-\infty$	0.4	-2
0.05	-20	0.5	0
0.1	-14	0.6	$+1.5$
0.2	-8	0.707	$+3$
0.3	-4.5	1.0	$+6$

In constructing Bode plots, the various magnitude and phase curves may be plotted on a convenient size and type of semilog graph paper. Thin graph paper of the same kind may then be used as an overlay onto which the required curves may be traced at the appropriate break frequency.

EXAMPLE 9

Suppose that the transfer function for a two-port network is determined to be

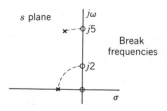

Fig. 13-27. Showing the relationship of pole locations to break frequencies for the $G(j\omega)$ of Example 9. The break frequency for the complex pole is found by drawing a circle through it, centered at the origin, to intersect the imaginary axis.

$$G(s) = \frac{s}{(1 + 0.5s)(1 + 0.12s + 0.04s^2)} \tag{13-75}$$

and it is required that Bode plots be made for the corresponding magnitude and phase. For the first-order factor, we note that the break frequency is $\omega_1 = 2$, while for the second-order factor, $\omega_2 = 5$ and $\zeta_2 = 0.3$. See Fig. 13-27. The asymptotic and actual curves for the magnitudes of each of the factors are shown in Fig. 13-28, and these are summed to give the total magnitude characteristics of Fig. 13-29. A similar summation of the phase characteristics gives the phase response also shown in Fig. 13-29 (dashed line).

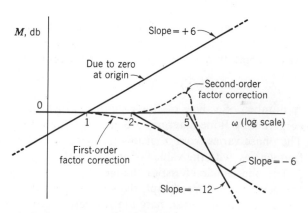

Fig. 13-28. Plots of the various factors in Eq. (13-75) used to determine the magnitude response.

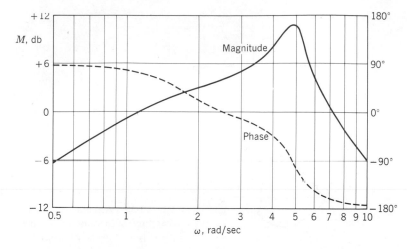

Fig. 13-29. The magnitude and phase response for the transfer function given by Eq. (13-75).

13-6. THE NYQUIST CRITERION

We next study a criterion having the same objective as the Routh-Hurwitz criterion of Chapter 10, namely the stability of the system under study. The Routh-Hurwitz criterion related directly to the roots of the characteristic equation of the system. The Nyquist criterion employs a different approach making use of sinusoidal steady-state concepts of this chapter and Chapter 12. It was originally formulated in 1932 by Harry Nyquist[7] of the Bell Telephone Laboratories. While we will present the Nyquist criterion as a method for the computational analysis of a system, as was the case for the Routh-Hurwitz criterion, it is important to observe that its usefulness in practice relates to the fact that it may be applied through routine sinusoidal measurements which may be made in the laboratory.[8]

The basic operation in applying the Nyquist criterion is a *mapping* from the s plane to the $F(s)$ plane. By the term mapping, we mean that a set of values of s (for example, s_1, s_2, and s_3) have, for a given $F(s)$, a corresponding set of values of $F(s)$, namely, $F(s_1)$, $F(s_2)$, and $F(s_3)$. These three—and an infinite number of other—points are shown in Fig. 13-30. Here an arbitrary contour in the s plane is "mapped" into a corresponding contour in the $F(s)$ plane. A specific example is shown in Fig. 13-31. The mapping is made for imaginary

[7]H. Nyquist, "Regeneration Theory," *Bell System Tech. J.*, **11**, 126–147 (1932).

[8]Kuh, Ernest S., and R. A. Rohrer, *Theory of Linear Active Networks* (Holden-Day, Inc., San Francisco, 1967), pp. 561–578. Spence, Robert, *Linear Active Networks* (John Wiley & Sons, Inc., New York, 1970), pp. 175–189.

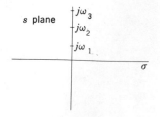

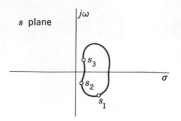

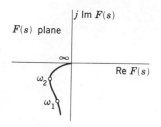

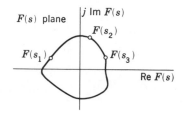

Fig. 13-30. An illustration of a mapping of a contour in the s plane to one in the $F(s)$ plane.

Fig.13-31. A mapping of points on the imaginary axis of the s plane to the $F(s)$ plane for $F(s) = 1/[s(sT + 1)]$.

values of s; that is, $s - j\omega$ for $\omega \geq 0$. The specific function is

$$F(s) = \frac{1}{s(sT + 1)} \qquad (13\text{-}76)$$

As another example of a mapping operation, not so difficult as the one given above, suppose that two functions are related by the equation

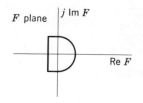

$$F(s) = P(s) + 1 \qquad (13\text{-}77)$$

that is, the function $P(s)$ plus a constant (unity) is equal to the function $F(s)$. A typical plot in the two planes is shown in Fig. 13-32. The transformation evidently moves the plot one unit to the left.

Next, suppose that $F(s)$ is factored to find its poles and zeros which are given in the equation

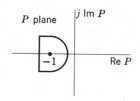

$$F(s) = K\frac{(s - s_1)(s - s_2)\dots(s - s_n)}{(s - s_a)(s - s_b)\dots(s - s_m)} \qquad (13\text{-}78)$$

where $s_1, s_2, \dots, s_n$ are the zeros and $s_a, s_b, \dots, s_m$ are the poles. These poles and zeros are displayed on a plot of the s plane shown in Fig. 13-33(a) (an arbitrary array for purposes of illustration). A single zero, s_1, is isolated in Fig. 13-33(b). This zero comes from the term $(s - s_1)$ in Eq. (13-78). At some particular frequency s_a, this term has a value $(s_a - s_1)$, which may be expressed in polar form as

Fig. 13-32. A mapping defined by the equation $F(s) = P(s) + 1$.

$$(s_a - s_1) = M_1 e^{j\phi_1} \qquad (13\text{-}79)$$

where M_1 is the magnitude and ϕ_1 is the phase angle of the phasor $(s_a - s_1)$. This magnitude and phase are shown on the s plane in Fig. 13-33(b). Any other term in Eq. (13-78) can be similarly expressed; for example,

$$(s_a - s_b) = M_b e^{j\phi_b} \qquad (13\text{-}80)$$

When all terms are so expressed, Eq. (13-78) takes the form which is a generalization of Eq. (13-27),

$$M_t e^{j\phi_t} = |F(s)| e^{j \text{ Ang } F(s)} = \frac{K M_1 M_2 M_3 M_4 \cdots}{M_a M_b M_c M_d \cdots} e^{j\phi_t} \quad (13\text{-}81)$$

where

$$\phi_t = \phi_1 + \phi_2 + \ldots - \phi_a - \phi_b - \ldots \quad (13\text{-}82)$$

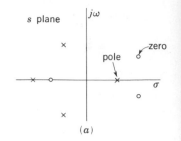

This last equation tells us that the total phase at some frequency s_a for the function $F(s)$ may be found by adding the phase of the "zero phasors" and subtracting the phase of the "pole phasors"; in other words,

$$\text{Ang } F(s) = \text{Ang}(s - s_1) + \text{Ang}(s - s_2)$$
$$+ \ldots - \text{Ang}(s - s_a) - \ldots \quad (13\text{-}83)$$

for any value of s.

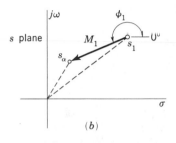

Figure 13-34 shows the s plane with two zeros, s_1 and s_2, and a mapping contour C (s_a is a point on the contour). Consider the effect of s_1 (ignoring s_2) as s_a moves along C in a *clockwise* direction. After one complete traversing of the closed contour C, the phase of the phasor term $(s - s_1)$ has increased by -2π radians. Next, consider the effect of ignoring s_1, as the same closed contour C is traversed in the same clockwise direction. There is *no* net gain in phase of the phasor term $(s - s_2)$. In summary, if the closed contour encircles a zero in traversing a closed path in the clockwise direction, the function changes in phase by -2π radians; if no zero is encircled, there is no change in phase. Exactly the same conclusion may be reached in the case of a pole except that the phase is changed by $+2\pi$ radians.

Fig. 13-33. (*a*) A configuration of poles and zeros in the s plane, and (*b*) a phasor drawn from a zero s_1 to the point s_α.

Suppose next that a contour is selected in the s plane of Fig. 13-33 such that p poles and Z zeros are encircled as the contour is traversed in a clockwise direction. The net change in the phase of the function $F(s)$ will be given by the equation

$$\Delta \text{ Ang } F(s) = 2\pi(P - Z) \text{ radians} \quad (13\text{-}84)$$

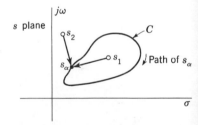

Return next to the mapping of the s plane into the $F(s)$ plane. Let us examine the behavior of the $F(s)$ plot in the complex plane as the closed contour in the s plane is traversed. An increase in the phase of $F(s)$ manifests itself in the $F(s)$ plane by an encirclement of the origin for every 2π-radian increase. Further, every zero encircled will cause one counterclockwise encirclement of the origin just as every pole will cause a clockwise encirclement. Should the contour not encircle any poles or zeros—or if it encircles equal numbers of poles and zeros—the contour in the $F(s)$ plane will not encircle the origin. In summary, if the closed contour C in the s plane encircles in a clockwise (or negative) direction P poles and Z zeros, the corresponding contour in the $F(s)$ plane encircles the origin $(P - Z)$ times in a counterclockwise (or positive) direction. Two examples are given in Fig. 13-35 to illustrate this conclusion.

Fig. 13-34. A contour in the s plane which encloses s_1 but not s_2.

The principles we have developed will next be applied to a

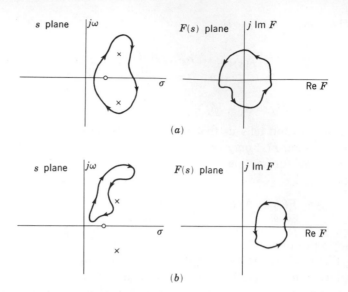

(a)

(b)

Fig. 13-35. Two mappings from the s plane to the $F(s)$ plane which illustrate the rule $\Delta\phi = 2\pi(P - Z)$.

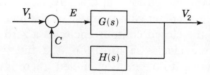

Fig. 13-36. A block diagram representation of a closed-loop system one part of which has the transfer function $G(s)$ and the other part $H(s)$.

closed-loop system represented by the block diagram of Fig. 13-36. The block-diagram algebra is defined such that the "input" to the block multiplied by the transfer function is equal to the "output" of the block. Then for the connections shown in the figure, we have the three equations

$$E(s) = V_1(s) - C(s) \tag{13-85}$$

$$C(s) = H(s)V_2(s) \tag{13-86}$$

and

$$V_2(s) = G(s)E(s) \tag{13-87}$$

From these three equations, we may eliminate $E(s)$ and $C(s)$ to determine $V_2(s)$ in terms of $V_1(s)$, arranged in the form of the overall transfer function

$$\frac{V_2(s)}{V_1(s)} = \frac{G(s)}{1 + G(s)H(s)} \tag{13-88}$$

The poles and zeros of the two functions $1 + GH$ and GH must be considered in the derivation of the Nyquist criterion. Let

$$1 + G(s)H(s) = \frac{P(s)}{Q(s)} = K\frac{(s - s_1)(s - s_2)\ldots(s - s_n)}{(s - s_a)(s - s_b)\ldots(s - s_m)} \tag{13-89}$$

$$G(s)H(s) = K'\frac{(s - s_\alpha)(s - s_\beta)\ldots(s - s_\omega)}{(s - s_a)(s - s_b)\ldots(s - s_m)} \tag{13-90}$$

398

Observe that the two functions have the *same poles*. In Eq. (13-89), the order of the polynominal $P(s)$ is n, and the order of $Q(s)$ is m. In deriving the Nyquist criterion, the orders are restricted to the case $n \leq m$, such that

$$\lim_{s \to \infty} G(s)H(s) = 0 \quad \text{or} \quad \text{a constant} \tag{13-91}$$

It is important to distinguish the various poles and zeros. They are tabulated as follows:

$s_1, s_2, \ldots, s_n$ are zeros of $[1 + G(s)H(s)]$.

$s_a, s_b, \ldots, s_m$ are poles of $[1 + G(s)H(s)]$.

$s_a, s_b, \ldots, s_m$ are also the poles of $G(s)H(s)$.

$s_a, s_\beta, \ldots, s_\omega$ are the zeros of $G(s)H(s)$.

The $s_1, s_2, \ldots, s_n$ roots are of vital concern to us because these are the zeros of the equation $1 + GH = 0$, which is the characteristic equation of the closed-loop system. These roots must *not have positive real parts* for the system they represent to be stable. Note that the *zeros* of $1 + GH$ are, by Eq. (16-88), the *poles* of V_2/V_1.

In studying stability, our specific interest is the zeros of the polynomial $1 + GH$ with positive real parts. This suggests that we choose a contour in the s plane to include the entire right half-plane as shown in Fig. 13-37. This contour will enclose all the zeros of interest. The contour is traced in the direction 1-2-3-4-1, starting at $s = -j\infty$, avoiding the origin ($s = 0$) for the time being, and continuing to $s = +j\infty$, thence on a circle of infinite radius to the point of beginning. This contour is traversed in a clockwise (or negative) direction. The contour in the s plane can be mapped in either the $1 + GH$ plane or the GH plane [the simple relationship between these mappings was considered in Eq. (13-77)]. If any poles or zeros of $1 + GH$ are encircled in the right half of the s plane, then (1) the locus in the $1 + GH$ plane will encircle the origin, or (2) the locus in the GH plane will encircle the point $-1 + j0$.

Let X = the number of zeros of $1 + GH$ with positive real parts, P = the number of poles of $1 + GH$ with positive real parts

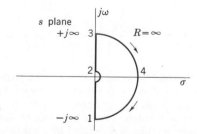

Fig. 13-37. A particular contour in the s plane which avoids the origin but encircles the entire right half of the s plane.

[also the number of poles of (GH) with positive real parts], $R =$ the net counterclockwise encirclements of the point $-1 + j0$ in the GH plane or the origin in the $1 + GH$ plane, then

$$R = P - X \tag{13-92}$$

Since X, the number of zeros of $1 + GH$ and the number of poles of V_2/V_1 with positive real parts, *must be equal to zero* for the system to be stable, the system with the characteristic equation $1 + GH = 0$ *is stable if and only if*

$$R = P \tag{13-93}$$

In most cases $P = 0$, and the criterion reduces to the requirement that $R = 0$ for stability.

To apply the Nyquist criterion, plot the $G(s)H(s)$ locus for the range of frequencies, $-\infty < \omega < \infty$. If R is the net counterclockwise encirclements[9] of the point $-1 + j0$ and P is the number of poles of $G(s)H(s)$ with positive real parts (and so in the right half-plane), the system is stable if and only if $R = P$.

We have thus far avoided any problem that might arise because of a pole of $G(s)H(s)$ at the origin or several poles at the origin. Actually, there is a practical matter involved in taking into account these poles at the origin deserving of special attention. To illustrate the problem, consider a transfer function

$$G(s)H(s) = \frac{K}{s(sT + 1)} \tag{13-94}$$

which is plotted in Fig. 13-38 for frequencies in the range $-\infty < \omega < +\infty$. The plot is complete except for one detail. The points $(+0)$ and (-0) should be joined together (as the same point). If the locus closes through the right half-plane, the system is stable, since $R = 0$; however, if the locus closes the other direction in the left half-plane, then $R = 1$ and the system is unstable. This is, as we see, a vital point.

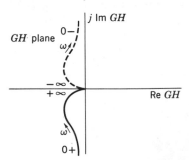

Fig. 13-38. A partial plot of $G(j\omega)H(j\omega)$ given by Eq. (13-94).

[9]To find the value of R, imagine a phasor with one end securely tied to the point $-1 + j0$ pointing away from this point. Let the end of this phasor trace the locus starting at $-\infty$ through -0 and $+0$ finally ending at $+\infty$. Count the net number of counterclockwise rotations of this phasor. This is the value of R. A clockwise rotation is designated by a negative number for R.

As s becomes small, only the pole at the origin has an effect on the transfer function $G(s)H(s)$. Thus for small s, the transfer function can be written

$$(Gs)H(s) = \frac{K}{s^n} \tag{13-95}$$

where n is the order (or multiplicity) of the poles at the origin. For the semicircular path shown in Fig. 13-39, the equation of the s-plane phasor locus is

$$(s - 0) = \delta c^{j\theta} \tag{13-96}$$

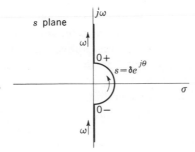

Fig. 13-39. An amplification of the region of the origin in the contour shown in Fig. 13-37 showing the contour which avoids the origin.

where δ is the radius of the semicircle and θ is the angle of the phasor $(s - 0)$ directed from the origin to a point on the circle. As $\delta \rightarrow 0$, the transfer function has the limiting value

$$\lim_{\delta \to 0} G(s)H(s) = \lim_{\delta \to 0} \frac{1}{\delta^n} e^{-jn\theta} = \infty \, e^{-jn\theta} \tag{13-97}$$

Hence as θ shown in Fig. 13-39 varies from $-\pi/2$ to $+\pi/2$, the phase of $G(s)H(s)$ ranges from $n\pi/2$ to $-n\pi/2$. In summary, the n poles at the origin in the transfer function $G(s)H(s)$ cause $n/2$ *clockwise* rotations at infinite radius of the phasor locus of $G(j\omega)H(j\omega)$.

Applying this rule to the example in Eq. (13-94), we see that $n = 1$ causes $\frac{1}{2}$ clockwise rotation of the phasor locus of $G(j\omega)H(j\omega)$ in going from $s = -0$ to $s = +0$. From this information, we are able to complete the plot of Fig. 13-38; the result is shown in Fig. 13-40.

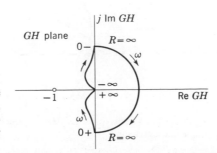

Fig. 13-40. The completion of the plot of $G(j\omega)H(j\omega)$ shown in Fig. 13-38 with attention to the closure corresponding to values near $s = 0$.

In making the Nyquist plot, only positive values of ω need be considered. Because the real part of $G(j\omega)H(j\omega)$ is even and the imaginary part odd, it follows that

$$\text{Im } G(-j\omega)H(-j\omega) = -\text{Im } G(+j\omega)H(+j\omega)$$
$$\text{Re } G(-j\omega)H(-j\omega) = +\text{Re } G(+j\omega)H(+j\omega)$$

(13-98)

The plot for negative values of ω can be made by reflecting the plot for positive frequency upon the real axis of the GH plane.

If the transfer function $G(s)H(s)$ has no poles in the right half-plane (and the Routh-Hurwitz criteria can be used to advantage in making this determination) such that $P = 0$ in Eq. (13-93), a rule of thumb may be used to advantage. Trace ("walk") from $\omega = 0$ to $\omega = +\infty$ on the Nyquist plot. If the point $-1 + j0$ is on the *right* at the point (ω) of nearest approach of $G(j\omega)H(j\omega)$ to $-1 + j0$, the system is *unstable*; if on the *left*, the system is *stable* ($P = 0$ only).

Suppose that a system is stable. We might next ask for some measure of relative stability. For the $P = 0$ case, an encirclement by the locus of $-1 + j0$ indicates an unstable system. If the locus goes through $-1 + j0$, then the system will exhibit sustained oscillations. If not encircled, then the system is stable. In the no-encirclement case, it seems possible that the further the locus plot misses $-1 + j0$, the more stable the system is. A measure of relative stability is given by two numbers that are frequently employed in the specifications of a system: the *gain margin* and *phase margin*.

In Fig. 13-41, we see that the locus crosses the 180° line a dis-

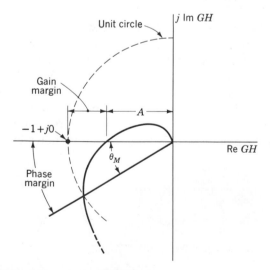

Fig. 13-41. A locus in the $G(s)H(s)$ plane used to define gain margin and phase margin.

tance A from the origin, where $A < 1$. We define the distance of the point A from -1 as the *gain margin*. It is the amount by which the gain would have to be increased to make the system unstable. This amount must be $1/A$ so that $1/A \times A = 1$. In logarithmic units, this is 20 log A db. If we draw a circle with the origin of the GH plane as the center and with a radius of 1, then the intersection of this circle with the locus determines a line drawn from the origin which defines the *phase margin* θ_M, as shown in the figure. A system with a gain margin of 2 and a phase margin of 30° is stable; another system with a gain margin of 4 and a phase margin of 45° is likely relatively more stable.

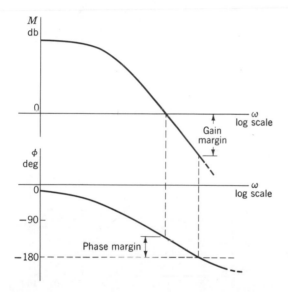

Fig. 13-42. The Bode plot corresponding to the locus shown in Fig. 13-41.

The Bode plot corresponding to Fig. 13-41 is shown as Fig. 13-42. Two important points on the Bode plots are (1) the point at which the locus crosses the 0 db line on the magnitude plot, and (2) the point at which the locus crosses the 180° line on the phase plot. The construction by which the gain margin and phase margin are determined on the Bode plot is shown by the dashed lines in Figs. 13-42 and 13-43.

Several examples will illustrate the application of the Nyquist criterion to the study of stability.

Figure 13-44 shows the plot of the transfer function $GH = K/(sT + 1)$ for all ω. The plot is that of a circle of diameter K. For all values of K, the locus does not encircle the point $-1 + j0$, and the

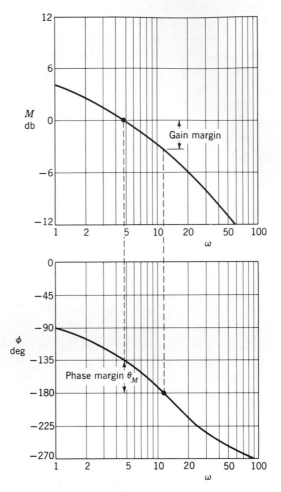

Fig. 13-43. A Bode plot used to illustrate the determination of gain margin and phase margin.

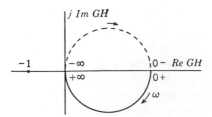

Fig. 13-44. The function $GH = K/(sT + 1)$ plotted for $s = j\omega$ for all values of ω.

system may be said to be *unconditionally stable*. There is nothing that can be done to make it unstable.

Figure 13-45 shows the locus of a transfer function which we will show is *conditionally stable*. For the particular plot that is shown

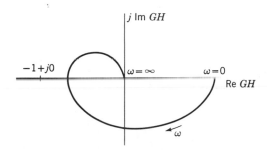

Fig. 13-45. The locus plot of a transfer function which represents a conditionally stable network (or system).

in the figure, there is no encirclement of $-1 + j0$ and so the system is stable. However, if K is increased, the locus will be magnified with respect to the coordinate system on which it is plotted. A larger value of K will cause the locus to pass through the point $-1 + j0$ and the system represented by the locus will be unstable in the sense that it has sustained oscillation. An even larger value of K will cause the point $-1 + j0$ to be completely encircled and the system will be unstable. Thus a conditionally stable system is one that is stable for some values of K and unstable for others.

The locus shown in Fig. 13-46 is similar to that of Fig. 13-45 but somewhat more complicated. The high-frequency asymptote is $180°$ rather than $270°$. As shown, the locus represents a stable system. However, with K increased (the locus magnified) the locus will encircle $-1 + j0$ and the system will be unstable. Observe that an even larger increase in K will result in no encirclements and therefore a stable system. The system represented by this plot is conditionally stable, with two ranges of gain causing stability and one range causing instability.

The locus plots of Fig. 13-47 represent a transfer function containing a factor like $(s - a)$ for a positive a, a pole or a zero in the right half of the s plane. We have come to associate a right half-plane pole with instability, but we shall next show that this is not necessarily the case with a closed-loop system. For the locus of Fig. 13-47(a), we observe that $R = 1$ and we are given that $P = 1$. Using the equation $R = P - X$, we see that $X = 0$ and the system is stable. A similar system locus is represented in Fig. 13-47(b). In this case, however, the

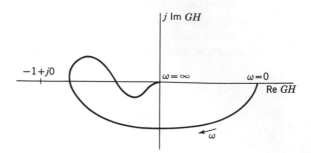

Fig. 13-46. The locus plot of a transfer function which represents a conditionally stable network (or system) which is more complex than that shown in Fig. 13-45.

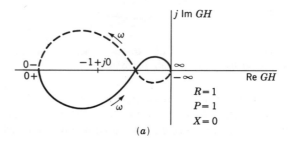

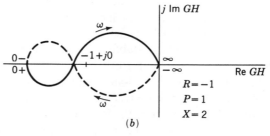

Fig. 13-47. Two locus plots for transfer functions containing a factor $(s - a)$ for a positive in $G(s)H(s)$.

point $-1 + j0$ is "inside" the loop closer to the origin. From the plot, we find that $R = -1$, and since $P = 1$, it follows that $X = 2$ and the system is unstable with two poles of the closed-loop transfer function in the right half-plane.

The locus plotted in Fig. 13-48 is different from those previously given as examples in that the low-frequency asymptote is $-90°$ rather than $0°$. This is caused by a pole at the origin in the transfer function GH; for example,

$$G(s)H(s) = \frac{K}{s(s + a)(s + b)} \qquad (13\text{-}99)$$

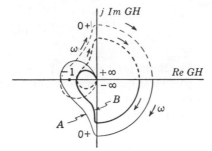

Fig. 13-48. Locus plot for $G(s)H(s)$ of the type given by Eq. (13-99).

Such transfer functions arise frequently in control systems due to the presence of an "integrator element" in the loop, but are less frequently seen in networks. The plot shows two loci. For a value of K that results in locus B, $R = 0$, and since $P = 0$, the system is stable. However, if K is increased to give the locus marked A, then $P = 0$ and $R = -2$, giving $X = 2$ and implying an unstable system. Thus this system is a *conditionally stable* system.

FURTHER READING

CHIRLIAN, PAUL M., *Basic Network Theory*, McGraw-Hill Book Company, New York, 1969. Chapter 6.

GRAY, PAUL E., AND CAMPBELL L. SEARLE, *Electronic Principles—Physics, Models and Circuits*, John Wiley & Sons, Inc., New York, 1969. Chapter 20.

KUH, ERNEST S., AND R. A. ROHRER, *Theory of Linear Active Networks*, Holden-Day, Inc., San Francisco, 1967. See the discussion of the Nyquist criterion in Section 4 of Chapter 11.

PERKINS, WILLIAM R., AND JOSÉ B. CRUZ, JR., *Engineering of Dynamic Systems*, John Wiley & Sons, Inc., New York, 1969. Chapter 8.

SCOTT, RONALD E., *Linear Circuits*, Addison-Wesley Publishing Co., Reading, Mass., 1960. Chapter 18.

THORNTON, RICHARD D., C.L. SEARLE, D.O. PEDERSON, R.B. ADLER, AND E.J. ANGELO, JR., *Multistage Transistor Circuits*, John Wiley & Sons, Inc., New York, 1965. Chapter 4.

DIGITAL COMPUTER EXERCISES

The subject of frequency response plots is a natural one for the use of the digital computer to avoid otherwise tedious computation. These topics are covered by references cited in Appendix E-9. Especially recommended are Chapters 9 and 10 of Huelsman, reference 7 of Appendix E-10, and Chapters 3 and 11 of Ley, reference 11 of Appendix E-10.

PROBLEMS

13-1. Sketch the (a) magnitude, (b) phase, (c) real part, and (d) imaginary part variation of the following network functions with ω for both $\omega > 0$ and $\omega < 0$:

(a) $1 + j2\omega$

(b) $\dfrac{1}{1 - j2\omega}$

(c) $\dfrac{(1 - 2\omega^2) + j\omega}{1 + j2\omega}$

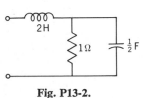

Fig. P13-2.

13-2. Consider the *RLC* one-port network shown in the figure. For this network, determine the driving-point functions $Z(j\omega)$ and $Y(j\omega)$. For each of these functions, plot the magnitude, phase, real part, and imaginary part as a function of frequency for $\omega > 0$ and $\omega < 0$.

13-3. For the two-port network of the figure, determine the voltage-ratio transfer function, $G_{12}(j\omega) = V_2(j\omega)/V_1(j\omega)$. Plot the variation of this function with ω for the two methods employed in Fig. 13-7.

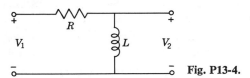

Fig. P13-3.

13-4. The two-port network of the figure shows an *RL* network. Repeat the plots specified in Prob. 13-3.

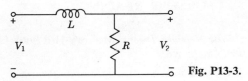

Fig. P13-4.

13-5. Repeat Prob. 13-3 for the *RC* two-port network shown in the accompanying figure.

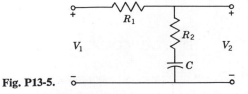

Fig. P13-5.

13-6. Show that the locus plot of Eq. (13-15) shown in Fig. 13-7 is a semicircle centered at $G_{12}(j\omega) = 0.5 + j0$ for the frequency range $0 < \omega < \infty$.

13-7. Consider the locus plot required in Prob. 13-5. Show that this locus is a circle for the frequency range, $-\infty < \omega < \infty$. Determine the center of the circle and its radius.

13-8. Consider the *RLC* series circuit shown in the figure. (a) Suppose that this network is connected to a sinusoidal voltage source. Plot the variation of the current magnitude and phase with frequency. (b) Suppose that the same network is connected to a current source of a sinusoidal waveform. Plot the variation of the voltage across the three elements using the same coordinates as in part (a). Element values are in ohms, farads, and henrys.

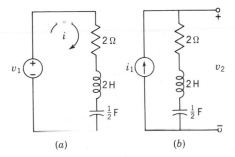

Fig. P13-8.

13-9. The figure shows a network which functions as a low-pass filter. For this network, determine the transfer function V_2/I_1 and plot the magnitude and phase as a function of frequency for this ratio.

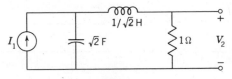

Fig. P13-9.

13-10. The network shown in the accompanying figure serves a similar function to that considered in Prob. 13-9, namely, it is a low-pass filter. For this network, determine the transfer function V_2/I_1 and plot the magnitude and phase as a function of frequency.

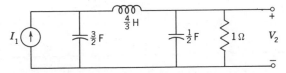

Fig. P13-10.

13-11. A network is analyzed and it is found that the transfer function is

$$\frac{V_2}{V_1} = \frac{1}{s^3 + 2s^2 + 2s + 1}$$

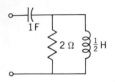

Fig. P13-12.

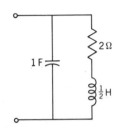

Fig. P13-13.

For this function, plot the magnitude and phase as a function of frequency for the range $0 < \omega < 4$.

13-12. For the RLC network shown in the figure, plot (a) the locus of the impedance function, and (b) the locus of the admittance function.

13-13. Plot (a) the admittance locus, and (b) the impedance locus for the RLC network shown in the figure.

13-14. The four-element network shown in the figure is to be analyzed to determine (a) the locus of the impedance of the network, and (b) the locus of the admittance function for the network.

13-15. For the network of the figure, plot (a) the locus of the impedance function, and (b) the locus of the admittance function.

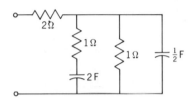

Fig. P13-15.

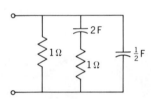

Fig. P13-14.

13-16. The RL network shown in (a) of the figure has element values such that the phase of the voltage measured with respect to the current is

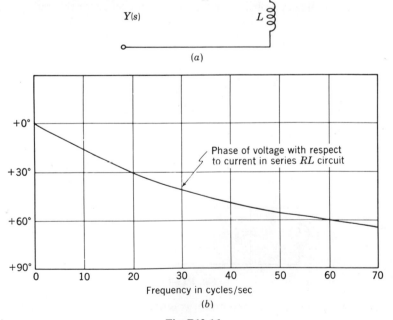

Fig. P13-16.

that shown in (*b*) of the figure. From this information, determine the pole and zero locations for $Y(s)$.

13-17. The figure shows the variation of the magnitude of the current with ω for an RLC series network with an applied sinusoidal voltage of constant magnitude. From the figure, determine the locations of the poles and zeros of the admittance of the network.

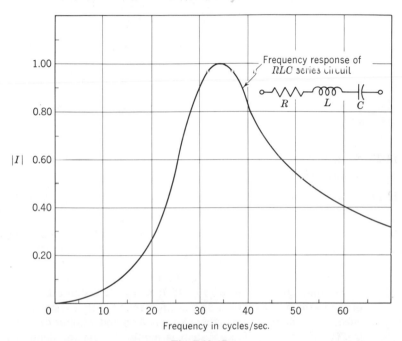

Frequency response of RLC series circuit

Frequency in cycles/sec.

Fig. P13-17.

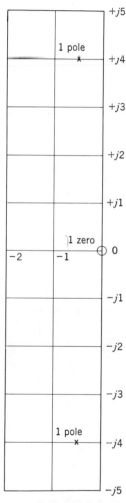

Fig. P13-18.

13-18. The pole-zero configuration shown in the figure represents the admittance function for the series RLC circuit. From the pole-zero configuration, determine: (a) the undamped natural frequency ω_n, (b) the damping ratio ζ, (c) the circuit Q, (d) the bandwidth (to the half-power points), (e) the actual frequency of oscillation of the transient response, (f) the damping factor of the transient response, (g) the frequency of resonance, (h) the parameter values (in terms of L if the values cannot otherwise be uniquely determined). (i) Sketch the magnitude of the admittance $|Y(j\omega)|$ as a function of frequency. (j) If the frequency scale is magnified by a factor of 1000, how do the values of the parameters, R, L, and C change?

13-19. The figure shows two configurations of poles and zeros for a certain transfer function. Use a graphical procedure to determine the variation of the magnitude of the network function for the two configurations. Superimpose the two plots on the same system of coordinates.

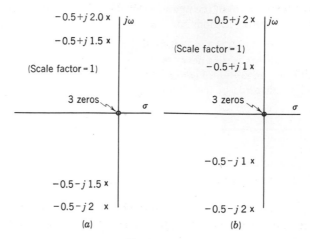

Fig. P13-19.

13-20. Show that the bandwidth B varies inversely with the circuit Q for a *series RLC* circuit.

13-21. Show that for an *RLC* series network the product of $|Y|_{max}$ and the bandwidth B equals $1/L$, where L is the inductance.

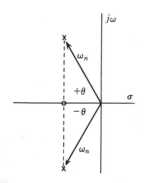

Fig. P13-22.

13-22. The two poles and zero shown in the s plane of the accompanying sketch are for the transfer function of a two-terminal-pair network, $G(s) = V_2(s)/V_1(s)$. The zero is on the real axis at a position to correspond with the same real part of the poles. The poles have positions corresponding to $\zeta = 0.707(\theta = 45°)$; ω_n is the distance from the origin to the pole as shown. In this problem, we will investigate the effect of the finite zero by computations with and without the zero. (a) The bandwidth of the system is modified from the definition given in the chapter as the range of frequencies from $\omega = 0$ to the half-power point. Compute the bandwidth of the system with the pole-zero configuration shown above; compute the bandwidth with the zero removed. In which case is the bandwidth greater and by what factor? A graphical construction is suggested.

13-23. The Q of a series *RLC* network at resonance is 10. The maximum amplitude of the current at resonance is 1 amp when the maximum amplitude of the applied voltage is 10 V. If $L = 0.1$ H, find the value of C in microfarads.

13-24. A coil under test may be represented by the model of L in series with R. The coil is connected in series with a calibrated capacitor. A sine wave generator of 10 V maximum amplitude and frequency $\omega = 1000$ radians/sec is connected to the coil. The capacitor is varied and it is found that the current is a maximum when $C = 100$ μF. Also, when $C = 12.5$ μF, the current is 0.707 of the maximum value. Find the Q of the coil at $\omega = 1000$ radians/sec.

13-25. The network of the figure is found to have the driving-point impedance

$$Z(s) = \frac{10^6(s+1)}{(s+1+j100)(s+1-j100)}$$

From this information, determine the values of R_1, R_2, L, and C.

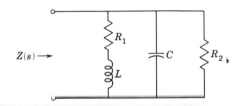

Fig. P13 26.

13-26. For the following network function, plot the straight-line asymptotic
magnitude response and the phase response. Use 4- or 5-cycle semilog
paper.

$$G(s) = \frac{100}{s(1+0.01s)(1+0.001s)}$$

13-27. Given the network function,

$$G(s) = \frac{(1+0.1s)(1+0.01s)}{(1+s)(1+0.001s)}$$

Plot the straight-line asymptotic magnitude response and the phase
response. Use 4- or 5-cycle semilog paper.

13-28. Plot the straight-line asymptotic magnitude response and phase angle
for the network function

$$G(s) = 100\frac{s^2}{(1+0.17s)(1+0.53s)}$$

Use 3- or 4-cycle semilog paper.

13-29. (a) Plot the straight-line asymptotic magnitude response, and (b)
determine the actual (or true) response for the network function

$$G(s) = 1000\frac{(1+0.25s)(1+0.1s)}{(1+s)(1+0.025s)}$$

(c) On the same coordinate system, plot the phase response. Use 4-
or 5-cycle semilog paper for the plotting.

13-30. Repeat Prob. 13-29 for the following network functions:

(a) $G(s) = 50\dfrac{(1+0.025s)}{s(1+0.05s)}$

(b) $G(s) = \dfrac{1000s}{(1+0.01s)(1+0.0025s)}$

(c) $G(s) = 180\dfrac{s(1+0.01s)}{(1+0.05s)(1+0.001s)}$

13-31. (a) Plot the straight-line asymptotic magnitude response, and
(b) determine the actual (or true) response curve for the network

function

$$G(s) = 120 \frac{(1 + 0.2s)}{s(s^2 + 2s + 10)}$$

(c) On the same coordinate system, plot the phase response. Use 3- or 4-cycle semilog paper.

13-32. Repeat Prob. 13-31 for the following network functions:

(a) $G(s) = 1000 \dfrac{s}{(1 + 0.001s)(1 + 4 \times 10^{-5}s + 10^{-8}s^2)}$

(b) $G(s) = \dfrac{100s}{(1 + s + 0.5s^2)(1 + 0.4s + 0.2s^2)}$

13-33. We are required to construct a network function $G(s)$ satisfying the following specifications: The asymptotic curve should have a low-frequency response of 0 db/octave slope, and the high-frequency response has a slope of -24 db/octave. The break frequency between these two slopes is at $\omega = 1$ radian/sec. At no frequency should the difference between the asymptotic and the true response exceed ± 1 db.

13-34. The figure shows two straight-line segments having slopes of $\pm n6$ db/octave. The low-and high-frequency asymptotes extend indefinitely, and the network function the response represents has first-order factors only. Find $G(s)$ and evaluate the constant multiplier of the function.

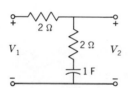

Fig. P13-36.

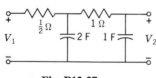

Fig. P13-37.

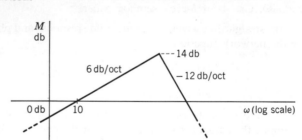

Fig. P13-34.

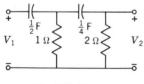

Fig. P13-38.

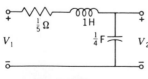

Fig. P13-39.

13-35. Repeat Prob. 13-34 if the response is changed only by the high-frequency asymptote having a slope of -18 db/octave.

13-36. For the two-port network shown in the figure, determine V_2/V_1 and plot the magnitude response (Bode plot) showing both asymptotic and true curves.

13-37. Prepare a Bode plot for the network function V_2/V_1 for the network shown in the accompanying figure.

13-38. Prepare a Bode plot for the voltage-ratio transfer function $G_{12} = V_2/V_1$ for the two-port network shown in the figure.

13-39. The figure shows an *RLC* network. For this two-port network, plot the transfer function $G_{12} = V_2/V_1$ showing both the asymptotic and true curves.

13-40. Consider the following transfer functions:

(a) $G(s)H(s) = K\dfrac{s-1}{s+1}$

(b) $G(s)H(s) = K\dfrac{s+1}{s-1}$

(c) $G(s)H(s) = \dfrac{K}{s(1 + 0.05s)}$

For each of these functions: (a) plot $G(j\omega)H(j\omega)$ in the complex GH plane from $\omega = 0$ to $\omega = \infty$ with $K = 1$. (b) Determine the range of values of K that will result in a stable system by means of the Nyquist criterion.

13-41. For the locus plot shown in Fig. 13-45, sketch the corresponding Bode plots for the magnitude and phase, making some assumption as to the frequency scale. Estimate the gain and phase margins and indicate these on the Bode plots.

13-42. Repeat Prob. 13-41 for the locus plot shown in Fig. 13-48.

13-43. Starting with the locus plot shown in the figure for Prob. 13-45, sketch the corresponding magnitude and phase plots using Bode coordinates. Make an assumption about the frequency scale along the locus. Indicate on the figure the gain and phase margins.

13-44. The Nyquist plot of the figure is made for a system for which $P = 0$. Analyze the system by applying the Nyquist criterion, indicating whether the system is stable, conditionally stable, or unstable.

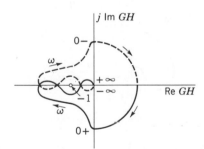

Fig. P13-44.

13-45. The locus plot is made for a system for which $P = 0$. It is given that $A = -0.75$, $B = -1.3$, and $C = -2$. Assuming that the plot is

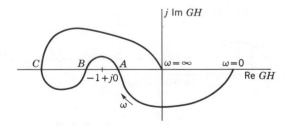

Fig. P13-45.

made for a gain K, what is the range of values of gain for which the system will be (a) stable, and (b) unstable.

13-46. Repeat Prob. 13-45 if $P = 1$.

13-47. The figure shows a locus plot made for a system for which $P = 0$. Is the system stable? Determine your answer to this question by applying the Nyquist criterion. Repeat if $P = 1$, $P = 2$.

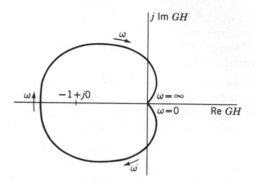

Fig. P13-47.

13-48. The locus plot shown in the figure is made for a system with $P = 2$, two poles with positive real parts. Apply the Nyquist criterion to this system to determine the stability of the system.

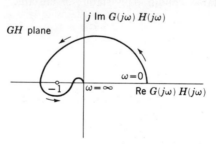

Fig. P13-48.

13-49. The locus plot of $G(j\omega)H(j\omega)$ shown in the figure is made for a system with $P = 0$. For this system, apply the Nyquist criterion to study the stability of the system.

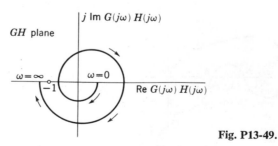

Fig. P13-49.

13-50. The accompanying figure shows a plot of the locus of $G(j\omega)H(j\omega)$ from $\omega = 0$ to $\omega = \infty$. From this plot determine everything you can about $G(s)H(s)$ as a quotient of polynomials in s.

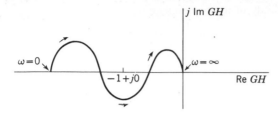

Fig. P13-50.

13-51. The figure shows the feedback system for which the Nyquist criterion has been developed. For this problem, let $H = 1$, and

$$G(s) = \frac{K}{(s - a)(s + 2)(s + 3)}$$

Make use of the Nyquist criterion to study this system for stability for the case $a = 1$.

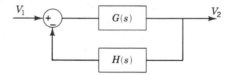

Fig. P13-51.

13-52. Repeat Prob. 13-51 if $a = 2$.

13-53. Repeat Prob. 13-51 if $a = 4$.

13-54. A system is described by the transfer functions which relate to the system of Fig. P13-51.

$$G(s) = \frac{10^5}{(s + 2)(s + 10)(s + 20)}$$

and $H = 1$. Make use of the Nyquist criterion to determine if this system is stable.

13-55. Repeat Prob. 13-54 for the given $G(s)$, but for a new feedback transfer function

$$H(s) = \frac{s + 20}{20}$$

This causes cancellation in the product $H(s)G(s)$ and is a form of *compensation* of a system to improve stability. Comment on the effectiveness of this compensation function.

13-56. The figure shows a model of a feedback amplifier. For this system, identify $G(s)$ and $H(s)$ as in Fig. P13-51 and express each as a quotient of polynomials in s. Is this system capable of oscillation? Make use of the Nyquist criterion in answering this question and in a general study of the system stability.

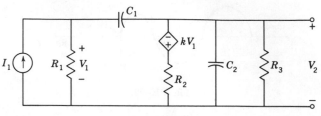

Fig. P13-56.

Input Power, Power Transfer, and Insertion Loss

14

In Chapter 1, the circuit parameters were introduced in terms of the basic concepts of energy and charge. The important role of energy and its time derivative, power, in network analysis and design is extended in this chapter, and applied to one- and two-port networks operating in the sinusoidal steady state. Such studies are important because energy costs money, and so the efficient transfer of energy from the source to the load (or sink) is vital to the engineer. In addition, we know that one limitation in the design of equipment—ranging in size from a huge motor in a rolling mill to a tiny silicon diode or transistor —is the amount of heat that can be dissipated without mechanical damage.

The order in which topics are considered in this chapter may be described by using the models of Fig. 14-1. We first consider power relationships for a general one-port network shown in (a), which may

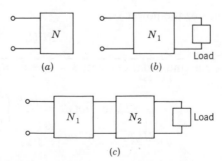

Fig. 14-1. General network configurations to be considered in this chapter: (a) a one-port network, (b) a two-port network terminated in a load, and (c) the network of (b) with another network N_2 inserted in cascade.

419

contain any number or connection of elements. In (*b*) of this figure, part of the network is identified as the *load* and the remaining network becomes a two-port; our problem here is to study power transfer from the source to the load. Finally, we extend our study of power transfer by studying the effect of inserting a new network in cascade with the two-port, as shown in (*c*) of Fig. 14-1, and determining the power loss (or gain) at the load due to this insertion.

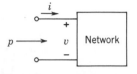

Fig. 14-2. One-port network with reference directions for *v* and *i* to define positive *p*.

14-1. ENERGY AND POWER

Consider the one-port network of Fig. 14-2. Energy and power relationships for this one-port, which we will write next, apply for elements which are linear or nonlinear, active or passive. The energy absorbed by the network from time t_1 to t_2 is

$$w = \int_{t_1}^{t_2} v(t)i(t)\,dt \quad \text{J (joules)} \tag{14-1}$$

The rate at which energy is being absorbed is the power given by

$$p = \frac{dw}{dt} = v(t)i(t) \quad \text{W (watts)} \tag{14-2}$$

as first derived in Eqs. (1-5) and (1-6). The convention for the reference direction for the flow of energy is shown in Fig. 14-2. For the voltage and current references shown, a positive p indicates a flow of energy into the network, negative p out of the network. The direction of flow may change with time, of course, and will depend only on the sign of p. If either the voltage or current reference is reversed, so is the reference for the flow of energy.

Next, let us specialize the general results of Eqs. (14-1) and (14-2) for the case of a single linear element in the network of Fig. 14-2, and also for the case in which voltage and current are sinusoidal and the network is in the steady state.

For the resistor, $v = Ri$, and we have for the energy absorbed

$$w_R = \int_{t_1}^{t_2} Ri^2(t)\,dt = \int_{t_1}^{t_2} \frac{v^2(t)}{R}\,dt \tag{14-3}$$

and for the power

$$p_R = Ri^2(t) = \frac{v^2(t)}{R} \tag{14-4}$$

For a sinusoidal current, $i = I_m \sin \omega t$, and with $t_1 = 0$, Eq. (14-3) gives

$$w_R = \int_0^{t_2} RI_m^2 \sin^2 \omega t\,dt = \frac{RI_m^2}{2}\int_0^{t_2}(1 - \cos 2\omega t)\,dt$$
$$= \frac{RI_m^2}{2}\left(t - \frac{\sin 2\omega t}{2\omega}\right) \quad \text{J} \tag{14-5}$$

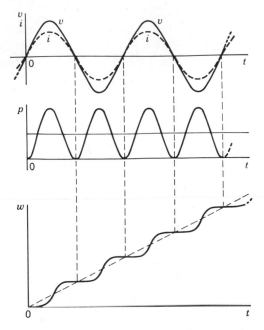

Fig. 14-3. The variation of $v, i, p,$ and w for the resistor with sinusoidal excitation.

for any time, $t = t_2$. Similarly, from Eq. (14-4), we have

$$p_R = RI_m^2 \sin^2 \omega t = \frac{RI_m^2}{2}(1 - \cos 2\omega t) \quad \text{W} \tag{14-6}$$

The variation with time of the four quantities, $v_R, i_R, w_R,$ and p_R, is shown in Fig. 14-3. Observe that p_R varies at twice the frequency of v_R and i_R, that both p_R and w_R are always positive, and that w_R increases to a very large value with increasing time.

The energy which enters the inductor for storage is found by substituting $v = L\, di/dt$ into Eq. (14-1), giving

$$w_L = \int_{t_1}^{t_2} L\frac{di}{dt} i\, dt = \int_{i_1}^{i_2} Li\, di \tag{14-7}$$
$$= \tfrac{1}{2}L(i_2^2 - i_1^2)\ \text{J}$$

where i_1 and i_2 are the currents at t_1 and t_2. If we let $i_1 = 0$ at $t_1 = 0$, and if $i_2 = I_m \sin \omega t$, then

$$w_L = \tfrac{1}{2}LI_m^2 \sin^2 \omega t = \tfrac{1}{4}LI_m^2(1 - \cos 2\omega t) \tag{14-8}$$

and the power is

$$p_L = (LI_m\omega \cos \omega t)(I_m \sin \omega t)$$
$$= LI_m^2\omega \sin \omega t \cos \omega t = \tfrac{1}{2}LI_m^2\omega \sin 2\omega t \quad \text{W} \tag{14-9}$$

Again, the four quantities—$v_L, i_L, w_L,$ and p_L—for the case of the

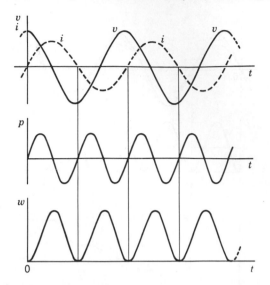

Fig. 14-4. The variation of $v, i, p,$ and w for the inductor with sinusoidal excitation.

inductor are shown in Fig. 14-4. Again, w_L and p_L are varying at twice the frequency of v_L and i_L, but this time the maximum value of w_L remains the same for each cycle.

Finally, the energy which enters the capacitor for storage is found by substituting $i = C\, dv/dt$ into Eq. (14-1) to obtain

$$w_C = \int_{t_1}^{t_2} C\frac{dv}{dt}v\, dt = \int_{v_1}^{v_2} Cv\, dv$$
$$= \tfrac{1}{2}C(v_2^2 - v_1^2) \quad \text{J} \tag{14-10}$$

where v_1 and v_2 are the voltages at t_1 and t_2. Again, we let $v_1 = 0$ at $t_1 = 0$, and $v_2 = V_m \sin \omega t$, so that

$$w_C = \tfrac{1}{2}CV_m^2 \sin^2 \omega t = \tfrac{1}{4}CV_m^2(1 - \cos 2\omega t) \tag{14-11}$$

and the power is

$$p_C = (V_m \sin \omega t)(CV_m \omega \cos \omega t)$$
$$= CV_m^2 \omega \sin \omega t \cos \omega t = \tfrac{1}{2}CV_m^2 \omega \sin 2\omega t \tag{14-12}$$

Time variations for $v_C, i_C, p_C,$ and w_C for the case of the capacitor are shown in Fig. 14-5. The relationships shown in this figure are similar to those found for the inductor.[1]

Figure 14-6 reminds us that Eqs. (14-1), (14-2), (14-3), (14-4), (14-7), and (14-10) are general and apply to the current and voltage at the driving terminals of any one-port network of Fig. 14-2. Most of

---

[1]In finding p_L, we let $i_2 = I_m \sin \omega t$, while in finding p_C, $v_2 = V_m \sin \omega t$. If we had let $i_2 = I_m \sin \omega t$ for both cases, then p_C in Eq. (14-12) would be multiplied by -1 and the $-$ sign in Eq. (14-11) would be $+$.

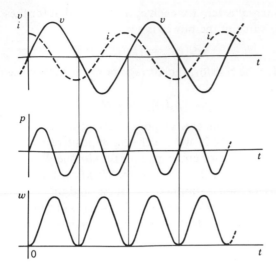

Fig. 14-5. The variation of $v, i, p,$ and w for the capacitor with sinusoidal excitation.

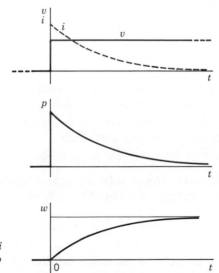

Fig. 14-6. Nonsinusoidal v and i and the corresponding plots of p and w.

the relationships that follow in this chapter are specialized to the sinusoidal steady state. Unless these *special* conditions are satisfied, the general relationships must be used!

Let us next compare the results obtained for the three elements under the restriction of the sinusoidal steady state. Note first that the energy and power variations for the inductor and capacitor have the same form. The average value of $p(t)$ over n periods of duration T seconds is the integral of $p(t)$ from $t = 0$ to $t = nT$ divided by nT.

Since the integral is zero for each cycle, we see that for these elements, the average value of the power is zero:

$$[p_L(t)]_{av} = [p_C(t)]_{av} = 0 \qquad (14\text{-}13)$$

However, for the resistor, the average power is half of the peak value and

$$[p_R(t)]_{av} = \tfrac{1}{2}RI_m^2 \qquad (14\text{-}14)$$

Energy is the integral of power. Using the convention for sign illustrated by Fig. 14-2, we see that, for positive p, energy is being supplied to the inductor and capacitor for storage. When p is negative, energy is being returned to the source. We know that no more energy can be returned than is supplied, of course, and the fact that w returns to zero value for every cycle of the voltage and current implies that the energy supplied is returned entirely every cycle for the inductor and capacitor. Thus,

$$\Delta w_L \text{ per cycle} = \Delta w_C \text{ per cycle} = 0 \qquad (14\text{-}15)$$

Now the resistor absorbs energy which is dissipated as heat (or converted to energy in some other form). If T is the period of v and i, then

$$\Delta w_R = R \int_0^T i^2(t)\, dt = \tfrac{1}{2}RI_m^2 T \qquad (14\text{-}16)$$

is the energy supplied each cycle.

The one-port network of Fig. 14-2 may contain any number of elements of each of the three kinds. Since energy is a scalar quantity and energy is conserved, the energy supplied to the one-port, w_t, must equal that stored plus that dissipated in the network. If there are n elements in the one-port, then

$$w_t = w_1 + w_2 + \ldots + w_n \qquad (14\text{-}17)$$

where any term w_j is the energy stored or dissipated for that element. The total power is found by differentiating Eq. (14-17), giving[2]

$$p_t = p_1 + p_2 + \ldots + p_n \qquad (14\text{-}18)$$

14-2. EFFECTIVE OR ROOT-MEAN-SQUARE VALUES

The effective value I_{eff} of a periodic current $i(t)$ is defined as the constant value of current which will produce the same power in a resistor as is produced on the average by the periodic current. The power in a resistor due to a constant current I is

$$p = I^2 R \qquad (14\text{-}19)$$

---

[2]This equation may be written compactly as $\sum_{k=1}^{n} v_k i_k$ or in the matrix form $v^T i$, where v and i are column matrices and T indicates transpose.

In the sinusoidal steady state, the average power in the resistor is given by Eq. (14-14) and is

$$P_{av} = [p_R(t)]_{av} = \tfrac{1}{2}RI_m^2 \tag{14-20}$$

Equating p and p_{av}, $I = I_{eff}$ is

$$I_{eff} = \frac{I_m}{\sqrt{2}} \tag{14-21}$$

Since $p = V^2/R$ and $V_m = RI_m$ for the resistor, we may also write

$$V_{eff} = \frac{V_m}{\sqrt{2}} \tag{14-22}$$

For a nonsinusoidal but periodic current, $i(t)$ of period T, the average power is

$$P_{av} = \left[\frac{1}{T} \int_{t_0}^{t_0+T} Ri^2(t)\, dt \right] \tag{14-23}$$

Again, equating P_{av} to p of Eq. (14-19), we may solve for $I = I_{eff}$, which is

$$I_{eff} = \left[\frac{1}{T} \int_{t_0}^{t_0+T} i^2(t)\, dt \right]^{1/2} \tag{14-24}$$

This equation may be regarded as the defining equation for the effective value of any periodic function. Thus the effective value of the voltage of period T is

$$V_{eff} = \left[\frac{1}{T} \int_{t_0}^{t_0+T} v^2(t)\, dt \right]^{1/2} \tag{14-25}$$

The operations in Eqs. (14-24) and (14-25) involve extracting the square *root* of the *mean* (average) of the *squared* function $i^2(t)$, and a descriptive name for the resulting value is *root-mean-square*, abbreviated rms. Thus we have equivalent notation which may be illustrated for the sine wave, using Eqs. (14-21) and (14-22):

$$I_{rms} = I_{eff} = \frac{I_m}{\sqrt{2}} \approx 0.707 I_m \tag{14-26}$$

and

$$V_{rms} = V_{eff} = \frac{V_m}{\sqrt{2}} \approx 0.707 V_m \tag{14-27}$$

One reason for the importance of the effective or rms value of a periodic function is that many voltmeters and ammeters are constructed to read these values. When the voltage at a wall plug is described as 110 V, it is implied that this is the rms value; the maximum value of the voltage is $\sqrt{2}$ greater or 156 V. A convention that facilitates the writing of an equation in terms of the rms value is illustrated by

$$v(t) = 110\sqrt{2} \sin(\omega t + \phi) \tag{14-28}$$

for the wall-plug voltage.

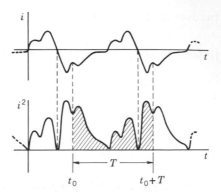

Fig. 14-7. Waveforms pertinent to the determination of the rms or effective value of a nonsinusoidal but periodic function.

The rms value of a nonsinusoidal but periodic $i(t)$ is determined with Eq. (14-24) by the operations described by Fig. 14-7. The figure shows $i(t)$ and the corresponding $i^2(t)$. The shaded area of the figure is the integral of i^2 over one period. If the number corresponding to this area is divided by the base T, then extracting the square root of this number yields the rms value for $i(t)$.

EXAMPLE 1

The waveform shown in Fig. 14-8 is that produced by a half-wave rectifier. It is not necessary to use Eq. (14-24) to determine the rms value for this current, because we see directly that the area of $i^2(t)$ over one period is half of that of the regular sine wave. Then, using Eq. (14-26), we have

$$I_{\text{rms}} = \frac{1}{\sqrt{2}}(I_{\text{rms}} \text{ of sine}) = \frac{I_m}{2} \qquad (14\text{-}29)$$

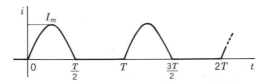

Fig. 14-8. The half-wave rectified sine wave for which the rms value is computed in Example 1.

EXAMPLE 2

The waveform of Fig. 14-9 may be described over the period from 0 to 2 by

$$i(t) = \begin{cases} t, & 0 < t < 1 \\ 1, & 1 < t < 2 \end{cases} \qquad (14\text{-}30)$$

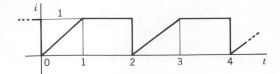

Fig. 14-9. Waveform considered in Example 2.

Substituting numerical values into Eq. (14-24) gives

$$I_{rms} = \left[\frac{1}{2} \left(\int_0^1 t^2 \, dt + \int_1^2 dt \right) \right]^{1/2} = \left[\frac{1}{2} \frac{1}{3} + 1 \right]^{1/2} \tag{14-31}$$
$$= 0.816 \text{ amp}$$

14-3. AVERAGE POWER AND COMPLEX POWER

We have shown that the average power for the resistor is

$$P_{av} = \frac{1}{2} I_m^2 R = \frac{1}{2} V_m^2 \frac{1}{R} \qquad \text{W} \tag{14-32}$$

This may now be expressed in terms of rms voltage and current as

$$P_{av} = I_{rms}^2 R = \frac{V_{rms}^2}{R} \qquad \text{W} \tag{14-33}$$

In this section, we determine a number of equivalent formulas for P_{av} and also generalize P_{av} into a complex power.

The one-port network of Fig. 14-2 may be characterized by a driving-point impedance, assuming it contains no independent sources. This impedance is

$$Z(j\omega) = R + jX = |Z| e^{j\theta_z} \tag{14-34}$$

Here we see that

$$R = \text{Re } Z(j\omega) = |Z| \cos \theta_z \tag{14-35}$$

so that Eq. (14-33) becomes

$$P_{av} = I_{rms}^2 |Z| \cos \theta_z = V_{rms} I_{rms} \cos \theta_z \tag{14-36}$$

which is an especially convenient expression for p_{av}. In this equation, $\cos \theta_z = $ pf is defined as the *power factor* with the convention that it be said to be *leading* if current leads voltage, *lagging* if current lags voltage. Thus a power factor of 0.8 lagging implies that the current lags the voltage by $\cos^{-1} 0.8 = 36.8°$.

We will next determine Eq. (14-36) from the voltage and current phasors, V and I, which are assumed *scaled to rms values* by division by $\sqrt{2}$. Let these phasors be

$$\mathbf{I} = I e^{j\alpha} \quad \text{and} \quad \mathbf{V} = V e^{j\beta} \tag{14-37}$$

so that

$$Z = \frac{\mathbf{V}}{\mathbf{I}} = \frac{V}{I} e^{j(\beta - \alpha)} \tag{14-38}$$

Comparing the phases of Eqs. (14-34) and (14-38), we make the identification

$$\beta - \alpha = \theta_Z \tag{14-39}$$

which is the angle of Eq. (14-36). Now, the angle of the product of $\mathbf{V}$ and $\mathbf{I}$ is $\beta + \alpha$. Since we require $\beta - \alpha$, the appropriate product is that of $\mathbf{V}$ and the conjugate of $\mathbf{I}$. Thus we see that Eq. (14-36) may be written

$$P_{av} = \text{Re } \mathbf{V} \mathbf{I}^* \quad \text{W} \tag{14-40}$$

Experience has shown that in making power calculations it is convenient to define an imaginary part of $\mathbf{V} \mathbf{I}^*$ which we identify as Q; thus

$$Q = \text{Im } \mathbf{V} \mathbf{I}^* \tag{14-41}$$

and the phasor sum is

$$\mathbf{S} = P_{av} + jQ = \mathbf{V} \mathbf{I}^* \tag{14-42}$$

where $\mathbf{S}$ is known as *complex* (or *phasor*) *power*. Since $Q = \text{Im } \mathbf{S} = \text{Im } \mathbf{V} \mathbf{I}^*$, it may be written in a form similar to Eq. (14-36), which is

$$Q = V_{rms} I_{rms} \sin \theta_Z \text{ vars} \tag{14-43}$$

This quantity is known as *reactive power*. The unit for reactive power is the *volt-ampere reactive or var*. A sign is assigned to Q to distinguish reactive power for the inductor and the capacitor: That for the inductor is positive and for the capacitor is negative. The unit for the magnitude of complex power of Eq. (14-42), which is

$$|\mathbf{S}| = \sqrt{P_{av}^2 + Q^2} = V_{rms} I_{rms} \text{ va} \tag{14-44}$$

is the *volt-ampere* or simply *va*. It is also known as apparent power, being the product of the reading of the voltmeter and the reading of the ammeter.

A number of other expressions are useful in solving power problems. From $\mathbf{I} = Y \mathbf{V}$,

$$\mathbf{S} = \mathbf{V} \mathbf{I}^* = \mathbf{V} \mathbf{V}^* Y^* = |\mathbf{V}|^2 \, Y^* \tag{14-45}$$

If we let $Y = G + jB$ so that $Y^* = G - jB$, we see that

$$P_{av} + jQ = |\mathbf{V}|^2 \, (G - jB) \tag{14-46}$$

Equating real and imaginary parts, we have

$$P_{av} = V_{rms}^2 G \quad \text{W} \tag{14-47}$$

and

$$Q = -V_{rms}^2 B \text{ vars} \tag{14-48}$$

Similarly,

$$\mathbf{S} = \mathbf{V} \mathbf{I}^* = \mathbf{I} \mathbf{I}^* Z = |\mathbf{I}|^2 Z \tag{14-49}$$

Since $Z = R + jX$, we obtain

$$P_{av} = I_{rms}^2 R \quad \text{W} \tag{14-50}$$

and

$$Q = I_{rms}^2 X \text{ vars} \tag{14-51}$$

It is understood, of course, that all voltages and currents are rms values, but the subscript rms is included in these equations for emphasis.

Another set of useful relationships follow for reasons similar to those given in arriving at Eq. (14-18) in Section 14-1. The average power is a scalar quantity which is always positive. The total average power supplied to the one-port network from the source is the sum of P_{av} for each element of the network. Thus

$$\text{Total } P_{av} = P_{av_1} + P_{av_2} + \ldots + P_{av_n} \tag{14-52}$$

if there are n elements in the network. Reactive power may similarly be summed, provided account is taken of the sign of each value. Then,

$$\text{Total } Q = Q_1 + Q_2 + \ldots + Q_n \tag{14-53}$$

The addition of the last two equations gives the general result

$$\text{Total } \mathbf{S} = \mathbf{S}_1 + \mathbf{S}_2 + \ldots + \mathbf{S}_n \tag{14-54}$$

EXAMPLE 3

The network shown in Fig. 14-10 is driven by a current source

$$i_1 = 5\sqrt{2} \sin 2t \tag{14-55}$$

and is in the steady state. Let us first determine the driving-point impedance of the one-port network, as shown in (b) of this figure. The result is $Z(j2) = 3 + (j/3)$. Then,

$$P_{av} = I_{rms}^2 \operatorname{Re} Z = 25 \times 3 = 75 \text{ W} \tag{14-56}$$

$$Q = I_{rms}^2 \operatorname{Im} Z = 25 \times \tfrac{1}{3} = 25/3 \text{ vars} \tag{14-57}$$

and

$$\mathbf{S} = 75 + j\frac{25}{3}, \qquad |\mathbf{S}| = 75.4 \text{ va} \tag{14-58}$$

so that

$$V_{rms} = \frac{75.4}{5} = 15.1 \text{ V} \tag{14-59}$$

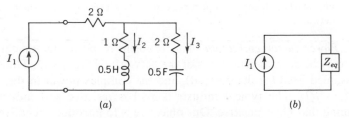

(a)   (b)

Fig. 14-10. Network for Example 3.

The power factor is

$$\text{pf} = \cos \theta_z = \cos \tan^{-1} \tfrac{1}{9} = 0.994 \text{ leading} \qquad (14\text{-}60)$$

Note that the analysis may be completed in terms of the summation of Eq. (14-54). The three branch currents in the network may be found by routine analysis. They are

$$I_1 = 5, \quad I_2 = \frac{10\sqrt{5}}{6}, \quad \text{and} \quad I_3 = \frac{10\sqrt{2}}{6} \text{ amp} \qquad (14\text{-}61)$$

The total average power is found by summing the power in the three resistors, giving

$$\text{Total } P_{\text{av}} = 50 + \frac{500}{36} + \frac{400}{36} = 75 \text{ W} \qquad (14\text{-}62)$$

Similarly, the reactive power sums as

$$\text{Total } Q = \frac{500}{36} - \frac{200}{36} = \frac{25}{3} \text{ vars} \qquad (14\text{-}63)$$

in agreement with Eqs. (14-56) and (14-57).

14-4. PROBLEMS IN OPTIMIZING POWER TRANSFER

In this section, we consider several problems encountered in optimizing the transfer of average power from the source to the load. We will assume that some portion of the network is identified as *the load* consisting of one or more elements, and that this load is connected to the source through a two-port network as shown in Fig. 14-11(*a*). In many cases, we will find it advantageous to consider the Thévenin equivalent of the network and source, as studied in Chapter 9 and shown in (*b*) of the figure.

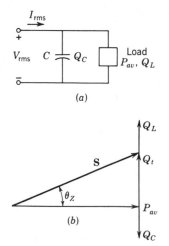

Fig. 14-12. (*a*) Network and (*b*) phasor diagram pertaining to power factor correction.

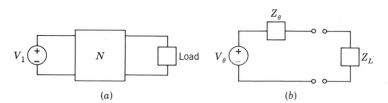

Fig. 14-11. A loaded two-port network and its Thévenin equivalent network.

Power factor correction. The problem of power factor correction may be described in terms of the network of Fig. 14-12. The load is assumed fixed (or determined), and the complex power in the load is $P_{\text{av}} + jQ_L$. The typical industrial load is resistive and inductive, meaning that Q_L is positive. Our objective is to introduce negative Q_C into the network to cancel part or all of Q_L. This is accomplished by a capacitor connected in parallel so that the terminal voltage at the load remains unchanged. Since

$$|\mathbf{S}| = \sqrt{P_{av}^2 + Q_t^2} = V_{rms}I_{rms} \tag{14-64}$$

where $Q_t = Q_L - Q_C$, we see that reducing Q_t will reduce the product $V_{rms}I_{rms}$ so that for a fixed terminal voltage the current to the combined load will be reduced. Viewed in another way, the power company charges for both P_{av} and Q_t, either directly or in terms of a penalty on rate for a large Q_t. It may be economically advantageous to reduce Q_t.

The power factor is a measure of the relative size of Q_t and P_{av}, since

$$\theta_Z = \tan^{-1}\frac{Q_t}{P_{av}} \tag{14-65}$$

The power factor is the cosine of this angle or

$$\text{pf} = \cos\theta_Z = \frac{P_{av}}{V_{rms}I_{rms}} = \frac{P_{av}}{\sqrt{P_{av}^2 + Q_t^2}} \tag{14-66}$$

Thus a power factor near unit value implies $|Q_t| \ll P_{av}$, whereas a power factor near zero implies $P_{av} \ll |Q_t|$. Using Eq. (14-66), we may determine the power factor before correction, and also determine the value of Q_t needed to produce the desired power factor. The problem is completed by finding $Q_C = Q_L - Q_t$, and since

$$Q_C = -V_{rms}^2 B_C = -V_{rms}^2 \omega C \tag{14-67}$$

we determine the value of the capacitor from

$$C = \frac{Q_C}{-V_{rms}^2 \omega} \tag{14-68}$$

A phasor diagram showing the components of complex power for the network of Fig. 14-12(*a*) is shown in (*b*) of the figure.

EXAMPLE 4

For a given load operating at a given voltage, $P_{av} = 500$ W and $Q = 500$ vars. The power factor is cos $\tan^{-1} 1 = 0.707$ lagging (from the positive sign for Q). We wish to correct the power factor to 0.9 lagging by the connection of a capacitor as in Fig. 14-12. From Eq. (14-66), Q_t must be 244 vars, meaning that $Q_C = 244 - 500 = -256$ vars must be supplied by the capacitor. The capacitance required may be determined from Eq. (14-68) when V_{rms} and ω are specified.

Impedance matching for maximum power transfer. We first redraw the network of Fig. 14-11(*b*) in the form shown in Fig. 14-13, using notation that will be followed for the remainder of the chapter.

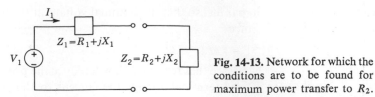

Fig. 14-13. Network for which the conditions are to be found for maximum power transfer to R_2.

Here V_1 and Z_1 are either the generator voltage and generator internal impedance, or the Thévenin equivalent of a more complicated two-port network (which may contain internal sources). We assume that the load impedance $Z_2 = R_2 + jX_2$ is of a nature such that R_2 and X_2 can be individually varied. We wish to determine the value for R_2 and X_2 that will result in the maximum transfer of power to the load. The phasor current is

$$\mathbf{I}_1 = \frac{\mathbf{V}_1}{Z_1 + Z_2} \tag{14-69}$$

having a magnitude squared which, when multiplied by R_2, gives the power in the load

$$P_2 = I_1^2 R_2 = \frac{V_1^2 R_2}{(R_1 + R_2)^2 + (X_1 + X_2)^2} \tag{14-70}$$

We first consider R_2 to be a constant, and determine the value of X_2 that will maximize P_2, the average power in R_2. Differentiating Eq. (14-70) gives

$$\frac{dP_2}{dX_2} = V_1^2 R_2 \left\{ \frac{-2(X_1 + X_2)}{[(R_1 + R_2)^2 + (X_1 + X_2)^2]^2} \right\} \tag{14-71}$$

For this derivative to be zero and so P_2 a maximum, requires that

$$X_2 = -X_1 \tag{14-72}$$

Substituting this value into Eq. (14-70) gives

$$P_2 = \frac{V_1^2 R_2}{(R_1 + R_2)^2} \tag{14-73}$$

Using this equation, we may verify routinely that $dP_2/dR_2 = 0$ requires that

$$R_2 = R_1, \qquad R_1 \text{ and } R_2 \text{ positive} \tag{14-74}$$

Thus maximum power transfer is accomplished when the reactive components of Z_1 and Z_2 cancel and when the real components are equal. Equations (14-72) and (14-74) may be combined into a single requirement, which is

$$Z_2 = Z_1^* \tag{14-75}$$

With impedances so adjusted, a *conjugate match* of impedances is said to exist.

Transformer adjustment for maximum power transfer. Consider the network of Fig. 14-14 which is similar to that of Fig. 14-13 with a

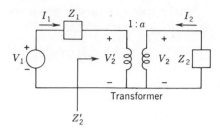

Fig. 14-14. In this network, the transformer may be adjusted to maximize the power to $R_2 = \operatorname{Re} Z_2$.

transformer inserted between the load and the two-port represented by the source V_1 and impedance Z_1. We will consider the transformer to be *ideal* in the sense that magnetizing and leakage inductance and internal losses are neglected. The voltages and currents of the two windings are related in such a way that $V_1 I_1 = -V_2 I_2$. Our problem is to maximize the power to the load, if only the transformation ratio of the transformer is adjustable.

Let a be the ratio of secondary to primary voltage for the transformer, meaning that $V_2 = aV'_2$. The driving-point impedance of the transformer terminated in Z_2 is

$$Z'_2 = \frac{1}{a^2} Z_2 \tag{14-76}$$

Let us write Z'_2 in the form

$$Z'_2 = |Z'_2| e^{j\theta} = |Z'_2| \cos\theta + |Z'_2| \sin\theta \tag{14-77}$$

Then the average power in R'_2 may be written in a form similar to Eq. (14-70):

$$P_2 = I_1^2 |Z'_2| \cos\theta = \frac{V_1^2 |Z'_2| \cos\theta}{(R_1 + |Z'_2| \cos\theta)^2 + (X_1 + |Z'_2| \sin\theta)^2} \tag{14-78}$$

Now only the magnitude $|Z'_2|$ is adjustable by changing a. If we form $dP_2/d|Z'_2|$ and set this to zero, we obtain the condition necessary for maximum P_2. This is, after some algebraic simplification,

$$|Z'_2| = \sqrt{R_1^2 + X_1^2} = |Z_1| \tag{14-79}$$

Then from Eq. (14-76), we have

$$\frac{1}{a^2}|Z_2| = |Z_1| \tag{14-80}$$

meaning that the choice of the transformation ratio

$$a = \sqrt{\left|\frac{Z_2}{Z_1}\right|} \tag{14-81}$$

causes the average power in the load to be maximized. This adjustment, however, gives a value of P_2 which is less than that obtained for a conjugate match for nonreal Z_1 and Z_2.

Maximizing efficiency of power transfer. We define the per cent efficiency of power transfer as

$$\eta_t = \frac{P_2}{P_s}\, 100\% = \frac{P_2}{P_1 + P_2}\, 100\% \qquad (14\text{-}82)$$

where P_1 is the average power of R_1, P_2 is the average power of R_2 (the load), and P_s is the power supplied by the source. Clearly η_t is a maximum when P_1 is as small as possible.

Recall that with a conjugate match, $R_1 = R_2$ so that under this condition $\eta_t = 50\%$, meaning that P_s is shared equally by the two resistors. Such an efficiency of transmission may be permitted in communication systems, but it would not be tolerated in a power system. In power systems involving large blocks of power, R_2 is large compared to R_1, meaning that $P_1 \ll P_2$ such that the efficiency is high.

14-5. INSERTION LOSS

To begin our study of the important subject of insertion loss, consider the network shown in Fig. 14-14 in which a transformer has been "inserted" between the load and the source. We will consider only the resistive case by letting $Z_1 = R_1$ and $Z_2 = R_2$. From Eq. (14-81), we may determine the transformation ratio of the transformer. Thus, when $a^2 = R_2/R_1$, the resistors appear to the source V_s to be equal and we have maximized the average power in R_2. This maximum value for power in R_2 is denoted by $P_{2\text{max}}$ and may be regarded as the *power available* to R_2 in the sense that P_2 may have this value under the best condition, but it may not be larger. The value of available power is

$$P_{2\text{max}} = \left(\frac{V_s}{2}\right)^2 \frac{1}{R_1} = \frac{V_s^2}{4R_1} \qquad (14\text{-}83)$$

Without the transformer of Fig. 14-14, we have the network shown in Fig. 14-15(*a*). For this network,

$$V_{20} = \frac{R_2}{R_1 + R_2} V_s \qquad (14\text{-}84)$$

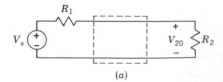

(*a*)

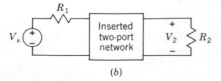

(*b*)

Fig. 14-15. The resistor network shown in (*a*) is modified by the insertion of a two-port network (which is usually lossless or *LC*).

and the power to the load R_2 is

$$P_{20} = \frac{V_{20}^2}{R_2} = \frac{V_s^2 R_2}{(R_1 + R_2)^2} \tag{14-85}$$

Combining Eqs. (14-83) and (14-85) gives the result

$$P_{20} = \frac{4R_1 R_2}{(R_1 + R_2)^2} P_{2\max} \tag{14-86}$$

Note that if $R_1 = R_2$, then $P_{20} = P_{2\max}$ in agreement with the results of Section 14-4.

Finally, we consider the network of Fig. 14-15(b) with a two-port network N inserted between R_1 and R_2. If the voltage at the load is V_2 (distinguished from V_{20}), then

$$P_2 = \frac{V_2^2}{R_2} \tag{14-87}$$

The insertion of the two-port network N causes the average power in R_2 to be reduced, meaning that there is a loss due to this insertion. Note that we do not ignore the possibility of a gain in the power to N, but gain will be regarded as negative loss! As a conceptual aid, we will regard the insertion of N as in Fig. 14-15(b) to cause power that might have reached R_2 to be "reflected." By definition, this reflected average power is

$$P_R = P_{2\max} - P_2 \tag{14-88}$$

The ratio of P_R to $P_{2\max}$ is appropriately called the *reflection coefficient squared*; thus

$$\frac{P_R}{P_{2\max}} = \frac{P_{2\max} - P_2}{P_{2\max}} = 1 - \frac{P_2}{P_{2\max}} = |\rho|^2 \tag{14-89}$$

where $|\rho|^2$ is always positive and

$$\rho(j\omega)\rho(-j\omega) = |\rho(j\omega)|^2 \tag{14-90}$$

and $\rho(j\omega)$ is the reflection coefficient. Substituting Eqs. (14-87) and (14-83) into Eq. (14-89), we have

$$|\rho|^2 = 1 - \frac{4R_1}{R_2}\left(\frac{V_2}{V_s}\right)^2 \tag{14-91}$$

where V_2 and V_s are rms magnitudes of $\mathbf{V}_2$ and $\mathbf{V}_s$.

Similarly, the ratio of P_2 to $P_{2\max}$ is known as the *transmission coefficient* squared (what else?) and is

$$|t(j\omega)|^2 = \frac{P_2}{P_{2\max}} = \frac{4R_1}{R_2}\left(\frac{V_2}{V_s}\right)^2 \tag{14-92}$$

Combining Eqs. (14-91) and (14-92) gives

$$|\rho(j\omega)|^2 + |t(j\omega)|^2 = 1 \tag{14-93}$$

In other words we return to our defining statement:

$$\frac{\text{Power reflected} + \text{power transmitted}}{\text{Power available}} = 1 \tag{14-94}$$

The effect of inserting the two-port network N in Fig. 14-15(*b*) is characterized by the ratio of P_{20} to P_2, the ratio of power *before* to *after* insertion. We define[3]

$$\frac{P_{20}}{P_2} = e^{2\alpha} \tag{14-95}$$

to be the *insertion power ratio* and α to be the *insertion loss* in nepers. In terms of the network variables,

$$\frac{P_{20}}{P_2} = \left(\frac{R_2}{R_1 + R_2}\right)^2 \left[\frac{V_s(j\omega)}{V_2(j\omega)}\right]^2 \tag{14-96}$$

or

$$\frac{P_{20}}{P_2} = \frac{4R_1R_2}{(R_1 + R_2)^2} \frac{1}{|t(j\omega)|^2} \tag{14-97}$$

The usual practice is to express insertion loss in decibels. Then,

$$\alpha = 10 \log \frac{P_{20}}{P_2} \text{ db} \tag{14-98}$$

or, from Eq. (14-96),

$$\alpha = 20 \log \frac{R_2}{R_1 + R_2}\left(\frac{V_s}{V_2}\right) \text{ db} \tag{14-99}$$

This result is important in the design of filters or equalizers for transmission systems such as telephone lines.

EXAMPLE 5

Consider the network of Fig. 14-16(*a*) for which we wish to determine the insertion loss. By routine analysis, we find that

$$\frac{V_2}{V_s} = \frac{\frac{1}{2}}{s^3 + 2s^2 + 2s + 1} \tag{14-100}$$

Substituting this result into Eq. (14-96), the insertion power ratio is

$$\frac{P_{20}}{P_2} = (1 - 2\omega^2)^2 + (2\omega - \omega^3)^2 = 1 + \omega^6 \tag{14-101}$$

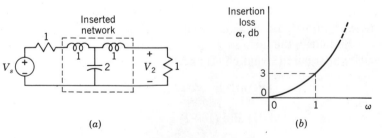

(a)   (b)

Fig. 14-16. (*a*) The network considered in Example 5 and (*b*) the insertion loss computed in Eq. (14-102).

---

[3]In network theory, we adopt a pessimistic point of view and regard positive α as loss, negative α as gain. In amplifier theory, the opposite convention is used.

and so the insertion loss is determined from Eq. (14-98) to be

$$\alpha = 10 \log (1 + \omega^6) \text{ db} \qquad (14\text{-}102)$$

which is shown plotted as a function of ω in Fig. 14-16(b).

14-6. TELLEGEN'S THEOREM

We next return to a consideration of instantaneous power studied in Section 14-1 and complex power as considered in Section 14-3 to introduce Tellegen's theorem.[4] This theorem appeared in 1952, which is surprisingly late considering its fundamental importance. As we shall see, Tellegen's theorem has a surprising number of applications in the study of electric circuits.

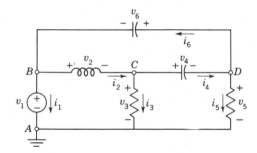

Fig. 14-17. Network used in Example 6 to illustrate Tellegen's theorem.

Consider the network shown in Fig. 14-17. In this network, arbitrary reference directions have been selected for all of the branch currents, and the corresponding branch voltage is indicated, with the positive reference direction at the tail of the current arrow. For this network, we will *select* a set of branch voltages, the only requirement being that the voltages must satisfy Kirchhoff's voltage law. We will then *select* a set of currents for the branches, without any consideration of the previous choice of branch voltages, but with the requirement that the Kirchhoff current law be satisfied at each node. We will then demonstrate that these arbitrarily chosen voltages and currents satisfy the equation

$$\sum_{k=1}^{b} v_k i_k = 0 \qquad (14\text{-}103)$$

if Kirchhoff's voltage and current laws have been satisfied.

[4]This result was announced by B. D. H. Tellegen in "A General Network Theorem with Applications," *Philips Research Reports*, 7, 259–269 (1952). Tellegen (1900–   ) is with Philips Research Laboratories, Eindhoven, The Netherlands, and has retired as Professor at the Technological University of Delft.

EXAMPLE 6

In the network of Fig. 14-17, let us choose $v_1 = 4$ and $v_2 = 2$. Applying Kirchhoff's voltage law around the loop $ABCA$, we see that $v_3 = 2$ is required. Similarly, around loop $ACDA$, we choose $v_4 = 3$, and then we are required to let $v_5 = -1$. Around the loop $BCDB$, the values we have selected for v_2 and v_4 require that $v_6 = -5$. We next apply Kirchhoff's current law successively to nodes B, C, and D. At node B, we let $i_1 = 2$, $i_2 = 2$, and then it is required that $i_6 = 4$. At node C, we select $i_3 = 4$ and then it is required that $i_4 = -2$. At node D, i_4 and i_6 have been selected, so we must let $i_5 = -6$. Carrying out the operations of Eq. (14-103), we have

$$(4)(2) + (2)(2) + (2)(4) + (3)(-2)$$
$$+ (-1)(-6) + (-5)(4) = 0 \qquad (14\text{-}104)$$

verifying Eq. (14-103) for this example. This information is summarized in Table 14-1, along with another set of currents, marked i'_k, which satisfy Kirchhoff's current law for the same network. We observe from the table that Eq. (14-103) is satisfied for the summation of $v_k i'_k$.

To demonstrate that this summation works in general, consider the network of Fig. 14-17 redrawn as Fig. 14-18. We will consider this

Table 14-1.

				Element		
Item	1	2	3	4	5	6
v_k	4	2	2	3	-1	-5
i_k	2	2	4	-2	-6	4
$v_k i_k$	8	4	8	-6	6	-20
i'_k	-1	3	2	1	-1	2
$v_k i'_k$	-4	6	4	3	1	-10

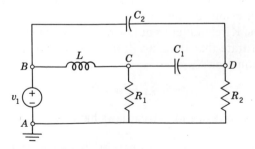

Fig. 14-18. Network used to show the method of general proof of Tellegen's theorem.

specific network and then note that the things we have done for this network will lead to the same result in general. In this study, we will make use of node-to-datum voltages and currents with a double subscript to indicate direction; for example, for the inductor between nodes B and C, $v_2 i_2 = (v_B - v_C)i_{BC}$. Summing a similar product for each of the six elements, we have

$$\sum_{k=1}^{6} v_k i_k = v_B i_{AB} + (v_B - v_C)i_{BC} + v_C i_{CA} + (v_C - v_D)i_{CD}$$
$$+ v_D i_{DA} + (v_D - v_B)i_{DB} \qquad (14\text{-}105)$$

This equation is next rearranged by factoring with respect to the node-to datum voltages. Thus

$$\sum_{k=1}^{6} v_k i_k = v_B(i_{AB} + i_{BC} - i_{DB}) + v_C(-i_{BC} + i_{CA} + i_{CD})$$
$$+ v_D(-i_{CD} + i_{DA} + i_{DB}) \qquad (14\text{-}106)$$
$$= v_B(\text{KCL at node } B) + v_C(\text{KCL at node } C)$$
$$+ v_D(\text{KCL at node } D) = 0 \qquad (14\text{-}107)$$

where we have used KCL to indicate Kirchhoff's current law. Each product vanishes because each Kirchhoff current law summation equals zero. While this has been a specific example, the procedure used is identical to that used for a general proof. This is a procedure which is used to establish Tellegen's theorem.

Suppose that we are given a network like that of Fig. 14-18 made up of both active and passive elements. Using the two Kirchhoff laws, we may *analyze* the network and find all voltages and currents for the branches. These voltages and currents differ from those previously used in that they are the actual voltages and currents resulting from a given set of excitations. For this situation, the product terms $v_k i_k$ are recognized as the *instantaneous power* for the kth branch of the network. Equation (14-103) tells us that the summation of instantaneous powers for the b branches in the network must be equal to zero.

As discussed in Chapter 1, instantaneous power is equal to the instantaneous rate at which energy is being supplied or removed, $p(t) = dw(t)/dt$. Equation (14-103) tells us that this summation must be zero, that energy must be supplied at a rate which is just equal to the rate at which energy is dissipated in resistors and stored in inductors and capacitors. If the network is divided into the two parts illustrated in Fig. 14-19, one part with all of the energy sources and the other part with all of the passive elements, then we can say that the power delivered by the independent sources of the network must equal the sum of the power absorbed (dissipated or stored) in all of the other branches of the network. All of this is implied by the equation

$$\sum_{k=1}^{b} v_k i_k = 0 \qquad (14\text{-}108)$$

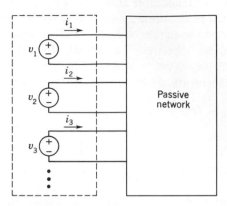

Fig. 14-19. The general network is divided into an active and a passive part for a discussion of the conservation of instantaneous power and complex power.

We turn next to complex power, which was discussed in Section 14-3. There it was shown that if we consider linear networks operating in the sinusoidal steady state, then complex power is given by Eq. (14-42), which is

$$\mathbf{S} = \mathbf{V}\,\mathbf{I}^* = P_{av} + jQ \tag{14-109}$$

In analyzing the network in the sinusoidal steady state, each time-varying voltage v_k is replaced by a phasor voltage $\mathbf{V}_k$; similarly, time-varying currents i_k are replaced by phasors $\mathbf{I}_k$. If Kirchhoff's voltage law applies to the set of voltages v_k, then it applies to the derived phasor voltage set $\mathbf{V}_k$; similarly, if Kirchhoff's current law applies to the set of currents i_k, it applies to the set of phasor currents $\mathbf{I}_k$ as well as to the conjugates of these currents $\mathbf{I}_k^*$. Then Eq. (14-108) may be written

$$\sum_{k=1}^{b} \mathbf{V}_k \mathbf{I}_k^* = \sum_{k=1}^{b} \mathbf{S}_k = 0 \tag{14-110}$$

where all phasor voltages and currents are rms values. Applying this concept to the network divided into two parts of Fig. 14-19, then Eq. (14-110) implies that the sum of the complex power of the sinusoidal sources must equal that delivered to the passive elements of the network. This may be interpreted as the *conservation of complex power*.

The discussion concerning Tellegen's theorem thus far has been for linear networks. We should observe that since the theorem depends only on the two Kirchhoff laws it is applicable to a very general class of lumped networks composed of elements that are linear or nonlinear, passive or active, time-invariant or time-varying. This generality is one of the reasons that Tellegen's theorem is a powerful tool. Some of the variations on the theorem are as follows:

(1) Given two networks, N_1 and N_2, having the same graph with the same reference directions assigned to the branches in the two networks, but with different element values and kinds. Let v_{1k} and i_{1k} be the voltages and currents in N_1, and v_{2k} and i_{2k} similarly be the voltages and currents in N_2, where all voltages and currents satisfy the appropriate Kirchhoff law. Then by Tellegen's theorem

$$\sum_{k=1}^{b} v_{k1}i_{k2} = 0 \quad \text{and} \quad \sum_{k=1}^{b} v_{k2}i_{k1} = 0 \qquad (14\text{-}111)$$

(2) We see from Eq. (14-111) that voltage and the current in the product which is summed for all elements can be very different, the only requirement being that the two Kirchhoff laws be satisfied. For example, if t_1 and t_2 are two different times of observation, it still follows that

$$\sum_{k=1}^{b} v_k(t_1)i_k(t_2) = 0 \qquad (14\text{-}112)$$

We begin to see how remarkable this theorem of Tellegen really is!

FURTHER READING

CHIRLIAN, PAUL M., *Basic Network Theory*, McGraw-Hill Book Company, New York, 1969. Chapter 6.

CRUZ, JOSÉ B., AND M. E. VAN VALKENBURG, *Signals in Linear Circuits*, Houghton-Mifflin Co., Boston, 1974. Chapter 6.

DIRECTOR, S. W., *Circuit Theory—The Computational Approach*, John Wiley & Sons, Inc., New York, 1974, Chapters 5 and 6.

DESOER, CHARLES A., AND ERNEST S. KUH, *Basic Circuit Theory*, McGraw-Hill Book Company, New York, 1969. See the discussion of Tellegen's theorem.

FRIEDLAND, B., OMAR WING, AND R. B. ASH, *Principles of Linear Networks*, McGraw-Hill Book Company, New York, 1961. Chapter 7.

HAYT, WILLIAM H., JR., AND JACK E. KEMMERLY, *Engineering Circuit Analysis*, McGraw-Hill Book Company, New York, 1971. Chapter 11.

MANNING, LAURENCE A., *Electrical Circuits*, McGraw-Hill Book Company, New York, 1965. Chapter 8.

WING, OMAR, *Circuit Theory with Computer Methods*, Holt, Rinehart & Winston, Inc., New York, 1972. See Chapter 7.

DIGITAL COMPUTER EXERCISES

The power analysis of a network requires the determination of all voltages and currents in the network. This is accomplished by ordinary network analysis, as treated in Appendices E-8.3 and E-8.4. Such analysis also provides a basis for network studies making use of Tellegen's theorem, as outlined in Director, reference 5 of Appendix E-10.

PROBLEMS

14-1. The figure shows the waveform of a voltage v_1 which is applied to the terminals of a one-port network, with reference directions as shown in Fig. 14-2. In this problem, we will consider various possibilities for the current to the one-port network. (a) For $i_1(t)$ shown

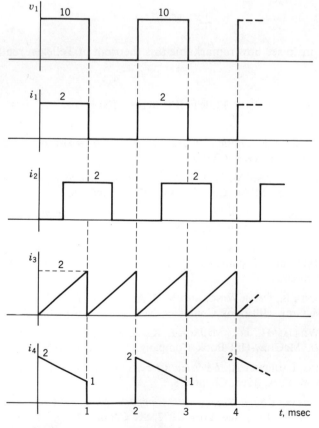

Fig. P14-1.

in the figure, plot $p(t)$, $w(t)$, and determine the energy absorbed by the network per cycle. (b) Repeat part (a) for i_2. (c) Repeat part (a) for i_3. (d) Repeat part (a) for i_4.

14-2. The waveform of the figure represents current with $K = 10$ amp, in a resistor $R = 10\,\Omega$. (a) For this waveform, sketch $p_R(t)$ from $t = 0$ to $t = 3T$ when $a = T/2$. (b) Sketch the energy $w_R(t)$ over the same interval specified in (a). (c) Compute the energy absorbed per cycle as a function of a.

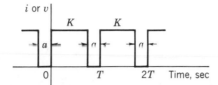

Figs. P14-2 to 5.

14-3. The waveform of the figure represents current with $K = 10$ amp in a capacitor $C = 2\,\mu F$. Let $a = T/2$, and sketch $p_C(t)$ from $t = 0$ to $t = 3T$. (b) Under the same conditions as (a), sketch $w_C(t)$. Assume that $v_C(0) = 0$.

14-4. The waveform of the figure represents voltage with $K = 100$ V of an inductor $L = 2$ mH. Let $a = T/3$ and sketch $p_L(t)$ from $t = 0$ to $t = 3T$ if $i_L(0) = 0$. (b) Under the same conditions as (a), sketch $w_L(t)$.

14-5. Determine the effective value of the current waveform of the figure as a function of a and K. Check your result for $a = 0$ and $a = T$.

14-6. The waveform of the figure is known as a sawtooth. Determine the effective value of $v(t)$.

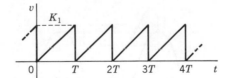

Fig. P14-6.

14-7. The waveform of the figure consists of a train of isosceles triangles. For this waveform, determine the effective value of $v(t)$.

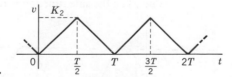

Fig. P14-7.

14-8. The waveform of the figure is similar to that of Prob. 14-7 except that the duration of the triangular waveform is less than the period. Determine the effective value of this $v(t)$ for $k_1 < \frac{1}{2}$. Compare your result to that of Prob. 14-7 for $k_1 = \frac{1}{2}$.

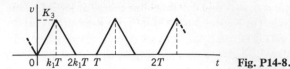

Fig. P14-8.

14-9. The waveform for this problem is similar to that of Prob. 14-7 except that the triangles are no longer isosceles unless $k_2 = \frac{1}{2}$. Determine the effective value of this $v(t)$ for $0 < k_2 < 1$.

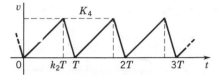

Fig. P14-9.

14-10. The waveform of the figure is identified as a *full-wave rectified* waveform which is described by the equation $v(t) = V_m |\cos \omega t|$. For this $v(t)$, determine the effective value.

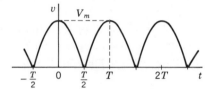

Fig. P14-10.

14-11. The waveform of the figure is known as a *half-wave rectified* waveform being a cosine function when the function is positive and zero when the function is negative. For this waveform, determine the effective value.

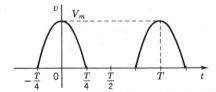

Fig. P14-11.

14-12. The waveform of the figure is derived from a sine function and has zero value when the sine function is negative and also from $t = 0$ to

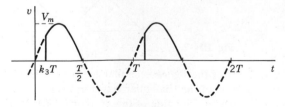

Fig. P14-12.

$t = k_3 T$, $k_3 < \frac{1}{2}$, and for the corresponding interval of each period. Find the effective value of this $v(t)$ as a function of k_3.

14-13. The waveform of the figure is known as a *fractional sine wave*. It is derived by adding a negative constant to the sine wave and then defining it to be nonzero only when the resulting function has a positive value. For this waveform, determine the effective value.

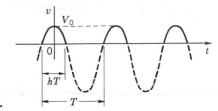

Fig. P14-13.

14-14. The figure shows a train of cosine-squared pulses which are derived by squaring the half-wave rectified waveform of Prob. 14-11. Determine the effective value of this $v(t)$.

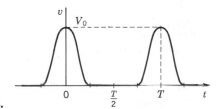

Fig. P14-14.

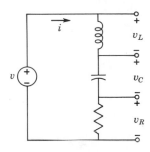

Fig. P14-15.

14-15. The series-connected RLC network shown in the figure is operating in the sinusoidal steady state. If the effective value of the source voltage is 1 V, the effective value of the current is 1 amp, $i(t)$ lags $v(t)$ by $45°$, $L = 1$ H, and $\omega = 2$ radians/sec, determine the effective value of the voltage across each of the three elements.

14-16. A voltmeter is used to make measurements of the effective value of the various voltages in the RLC network of the accompanying figure, and the following values are recorded: (a) The voltage from a to c is 20 V. (2) The voltage from b to d is 9 V. (3) The voltage from a to d is 15 V. Find all of the possible voltmeter readings for measurements made for each of the three elements.

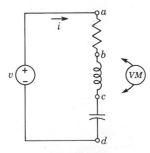

Fig. P14-16.

14-17. A sinusoidal source of voltage with an effective value of 5 V is connected to a series RLC network. When $C = \frac{1}{5}$ F, the effective value of the current is 1 amp, and the average power is 3 W. With the same voltage source, operating at the same voltage and frequency, connected to the network but with the capacitor charged such that $C = \frac{1}{45}$ F, the effective value of the current and the average power are the same. Find the value of L in henrys.

14-18. Consider the one-port network shown in Fig. 14-2. For this network, suppose that $v = 150 \cos \omega t$ V and $i = 5 \cos(\omega t + 60°)$ amp. Find: (a) the instantaneous power, $p(t)$, (b) P_{av} in watts, (c) Q in vars, and (d) S in va.

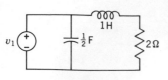

Fig. P14-20.

14-19. Repeat Prob. 14-18 if $v = 100 \sin (\omega t - 30°)$ V and $i = 10 \cos (\omega t + 45°)$ amp.

14-20. The network of the figure is operated in the sinusoidal steady state with the element values given and $v_1 = 100 \cos 2t$. Determine (a) the complex power generated by the source, (b) the effective current in each of the passive elements, and (c) the complex power for each of the passive elements in the network.

14-21. In the network of the figure, $i_1 = 10^{-3} \sin 3000t$. For the element values given, determine the quantities specified in Prob. 14-20.

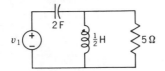

Fig. P14-22.

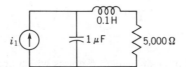

Fig. P14-21.

14-22. In the network shown in the figure, $v_1 = 10 \cos 2t$ V. For the element values specified, determine the quantities required in Prob. 14-20.

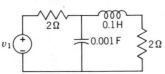

Fig. P14-23.

14-23. In the network of the figure, $v_1 = 440\sqrt{2} \cos 377t$ and the passive element values are as specified on the figure. For this network, determine the quantities required in Prob. 14-20.

14-24. In the network of the figure, $v_1 = (\sqrt{2}/20) \cos 100t$ and it is given that $k = 30$ and $R_L = 1000$ Ω. Compute the average power in the load resistor R_L.

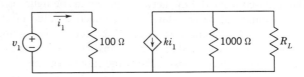

Fig. P14-24.

14-25. The current source connected to the network shown in the figure is described by the equation,

$$i_1 = 5\sqrt{2} \sin 1000t$$

Determine the effective current in each element and also the complex power for each element in the network. Find the total complex power for the network.

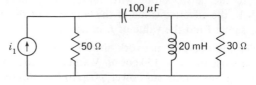

Fig. P14-25.

14-26. The average power to the plant shown in the accompanying figure is 250 kW and the power factor is 0.707 lagging. The generator is sinusoidal in waveform and the effective value is 2300 V. Determine the value of the capacitance C such that: (a) the power factor is 0.866 lagging, (b) the power factor is 1.0, and (c) the power factor is 0.866 leading. Let $\omega = 377$ radians/sec.

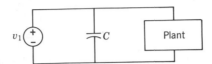

Fig. P14-26.

14-27. The plant of the figure for Prob. 14-24 has a rating of 100 kva with a power factor of 0.8 lagging. The voltage of the system has an effective value of 2000 V. Determine the value of capacitance C necessary to correct the power factor of the system to: (a) 0.9 lagging, (b) 1.0, (c) 0.9 leading. Let $\omega = 377$ radians/sec.

14-28. A series-connected RL network is connected to a sinusoidal voltage source and the system is operating in the steady state. If $R = 5\,\Omega$, and $\omega L = 10\,\Omega$, what must be the capacitance of a capacitor connected in parallel with the RL combination to produce a unity power factor for the RLC network if $\omega = 377$ radians/sec?

14-29. For the given network, determine the value of R_L that will cause the power in R_L to have a maximum value. What will be the value of power under this condition?

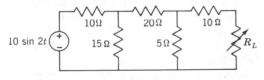

Fig. P14-29.

14-30. For the network of the figure, determine the impedance Z_x such that maximum power is transferred from the source to the load of impedance Z_x.

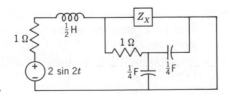

Fig. P14-30.

14-31. Consider the network for Prob. 9-31. Let the 1-Ω resistor in series with the controlled voltage source be considered as the load. What must be the new value for this load for maximum power transfer with $K_1 = -3$ and $\omega = 1$ radian/sec?

14-32. For the network of the figure, it is given that $v_1 = 2\sqrt{2}\,\sin 2t$. For the element values given, find the value of C that will cause maximum power in the 1-Ω load.

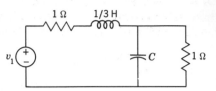

Fig. P14-32.

14-33. For the given network, let $C_1 = \frac{1}{10}$ F. What value of C_2 will result in the maximum power delivery to the 1-Ω resistor load? Let $v_1 = \cos 2t$.

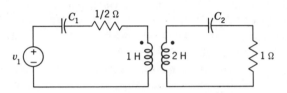

Fig. P14-33.

14-34. Consider the network given for Prob. 14-33. In this case, let $C_2 = \frac{1}{10}$ F and C_1 be variable. What value of C_1 will result in maximum power delivery to the load which is the 1-Ω resistor? Again, let $v_1 = \cos 2t$.

14-35. Consider a system operating in the sinusoidal steady state consisting of a current source connected to two subnetworks in parallel, N_a and N_b. Let these networks be characterized by their admittance functions, which are $Y_a = G_a + jB_a = |Y_a|\,e^{j\phi_a}$ and $Y_b = G_b + jB_a = |Y_b|\,e^{j\phi_b}$. The problem to be considered is the maximizing of the power to the subnetwork N_b under a number of different constraints. Consider the following cases:
(a) G_b and B_b may be varied.
(b) The magnitude of Y_b may be varied.
(c) G_a and B_a may be varied.
(d) The magnitude of Y_a may be varied.

14-36. Consider Example 5, but with a different form for the transfer function V_2/V_s than that used in Eq. (14-100). For the two denominator polynomials (a) $s^2 + 2s + 1$ and (b) $s^3 + 3s^2 + 3s + 1$, determine an expression for the insertion loss similar to that given as Eq. (14-102).

14-37. Consider the LC two-port networks shown in the accompanying figure. Each network is to be inserted between the two 1-Ω resistors as in Fig. 14-16(a). For each of the three networks, (a) determine an

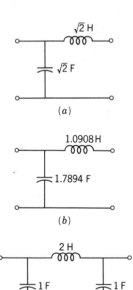

Fig. P14-37.

expression for the insertion loss, and (b) plot α in db as a function of ω.

14-38. The *LC* two-port networks shown in the figure are to be inserted between two 1-Ω resistors as shown in Fig. 14-16(*a*). For each of the four networks, (a) determine an expression for the insertion loss α, and (b) plot α in db as a function of ω as was done in Fig. 14-16(*b*).

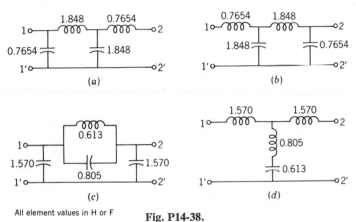

All element values in H or F **Fig. P14-38.**

14-39. The connection shown in (*a*) of the figure represents a *Y connection* of voltage sources. These voltages are said to be a three-phase

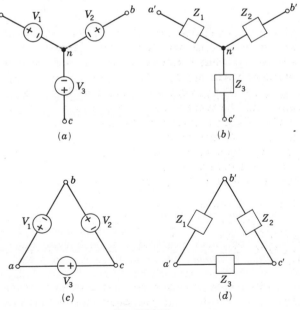

Fig. P14-39.

sequence if $V_1 = Ve^{j0}$, $V_2 = Ve^{-j2\pi/3}$, and $V_3 = Ve^{-j4\pi/3}$. In (b) of the figure is shown the corresponding *Y-connected* load. The load is said to be balanced if $Z_1 = Z_2 = Z_3 = Z$. For this problem, connect a to a', b to b', c to c', and n to n' and assume that the current reference direction is from the unprimed to the primed quantity.

(a) For a balanced load, find the current in each line, showing both the magnitude and phase with respect to V_1.

(b) Show that the line current from n to n', the so-called current in the neutral, is equal to zero.

(c) Show that the magnitude of the line-to-line voltages (excluding the neutral) is $\sqrt{3}\ V$ and find the phase of each with respect to V_1.

(d) If V_L and I_L are line quantities and θ is the power factor of the phase loads, show that the total power to the three-phase load is

$$P_T = \sqrt{3}\ V_L I_L \cos\theta$$

14-40. In (c) and (d) of the figure for Prob. 14-39 is shown the *delta connection* of voltages and loads. Assume that the voltages and impedances are defined as in Prob. 14-39. As in Prob. 14-39, connect unprimed to primed terminals, and repeat parts (a), (c), and (d).

14-41. For this problem, we make use of the three-phase voltages shown in (a) of Fig. P14-39(a) connected to the delta-connected load of (d) of the same figure. Connect a to a', b to b' and c to c' and let the reference direction for the currents be from the unprimed to the primed terminals. For the voltages and impedances as defined in Prob. 14-39, repeat parts (a), (c), and (d).

14-42. Consider the system described in Prob. 14-39. Let $V = 120$ V and $Z = 5 + j5\,\Omega$. Find the power in each load and the total power into the three-phase load.

14-43. Consider the system described in Prob. 14-41 with the addition that there is resistance in the wires connecting the unprimed and primed quantities, $R_W = 0.5\,\Omega$. If $V = 120$ V, $Z_1 = 4 + j2$, $Z_2 = 4 - j2$, and $Z_3 = 4\,\Omega$, find the total power delivered by the three-phase voltage sources.

14-44. Repeat Prob. 14-43 for the values given, but with the delta-delta connection described in Prob. 14-40.

14-45. The figure illustrates the manner in which a wattmeter is connected to a passive network to measure the average power to the network. Power is sometimes measured in a three-phase system using the *two-wattmeter method*. Let two wattmeters be connected into a three-phase system such that each has its current coils in a line and the voltage coils connected from line to line. Show that the method is general and that the total power is the sum of the two wattmeter readings.

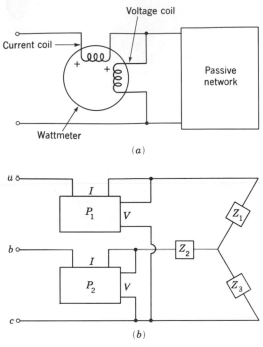

Fig. P14-45.

14-46. Apply the two-wattmeter method of Prob. 14-45 to the three-phase system described in Prob. 14-43 and determine the reading of each meter.

14-47. Apply the two-wattmeter method of Prob. 14-45 to the three-phase system described in Prob. 14-44 and determine the reading of each of the two meters.

15 Fourier Series and Signal Spectra

15-1. FOURIER SERIES

In Chapter 8, we studied time-domain responses in networks subjected to periodic inputs such as that shown in Fig. 15-1. There we found that the response could be determined as the inverse Laplace transform of the product of the transform of the input signal and the appropriate network function. In this chapter, we consider only the steady-state response to a periodic input which ideally started at the beginning of time and is destined to last forever. In addition, we are interested in the input signal and the network response in terms of their *frequency content*. The notion of frequency content of periodic signal waveforms is especially useful in engineering problems, as we will see, and is the basis for much of the specialized language or jargon in which electrical engineers communicate with each other.

The French mathematician, J. B. J. Fourier (1758–1830), while studying problems in the flow of heat (electrical applications were scarce in 1822), showed that arbitrary periodic functions could be represented by an infinite series of sinusoids of harmonically related frequencies. A number of words we have used in this statement require elaboration. A signal $f(t)$ is said to be periodic of period T if $f(t) = f(t + T)$ for all t. Several signal waveforms that fulfill this requirement are shown in Fig. 15-2. In (a) we have a train of pulses, in (b) a train of half sine waves (said to be half-wave rectified), and in (c) a continuous (but otherwise unrecognizable) periodic signal. Of special

452

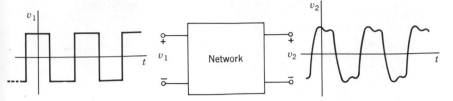

Fig. 15-1. Illustrating the type of problem considered in this chapter: a square wave input to a two-port network causes a periodic response, $v_2(t)$.

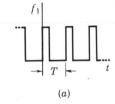

(a)

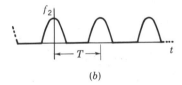

(b)

Fig. 15-2. Three periodic wave forms, each with the period T identified.

(c)

interest to us are the sinusoids

$$f_1(t) = \cos \frac{2n\pi}{T} t = \cos n\omega_0 t \qquad (15\text{-}1)$$

and

$$f_2(t) = \sin \frac{2n\pi}{T} t = \sin n\omega_0 t \qquad (15\text{-}2)$$

where n is any integer (or zero). Each frequency of the sinusoids, $n\omega_0 = 2n\pi/T$, is said to be the *nth harmonic* of the *fundamental frequency*, ω_0. Thus a periodic wave will be described in terms of its fundamental frequency, its second harmonic, third harmonic, etc., and each of these frequencies is simply related to the period T.

If $f(t)$ is periodic and satisfies the *Dirichlet conditions* to be discussed in Section 15-4, then the *Fourier series* is

$$f(t) = a_0 + a_1 \cos \omega_0 t + a_2 \cos 2\omega_0 t + \ldots + a_n \cos n\omega_0 t + \ldots$$
$$+ b_1 \sin \omega_0 t + \ldots + b_n \sin n\omega_0 t + \ldots \qquad (15\text{-}3)$$

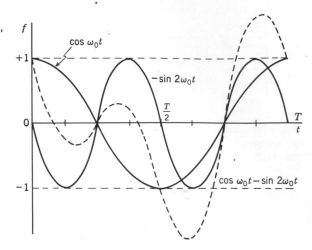

Fig. 15-3. An example of the combination of two sinusoids of harmonically related frequencies to give a nonsinusoidal periodic function shown by the dashed line.

The summation of two such terms to give a periodic function is shown in Fig. 15-3 for $a_1 = 1$, $b_2 = -1$, and all other coefficients equal to zero. *Fourier analysis* consists of two operations: (1) The determination of the values of the coefficients $a_0, a_1, \ldots, b_1, b_2, \ldots$, and (2) a decision as to the number of terms to include in a *truncated* series such that the partial sum will represent the function within allowable error. If convergence of the series is rapid, only a few terms will suffice.

The series of Eq. (15-3) may be written in a number of apparently different although equivalent forms, one of which is obtained by recognizing that for all n,

$$a_n \cos n\omega_0 t + b_n \sin n\omega_0 t = c_n \cos (n\omega_0 t + \theta_n) \qquad (15\text{-}4)$$

where

$$c_n = \sqrt{a_n^2 + b_n^2} \quad \text{and} \quad \theta_n = -\tan^{-1}\frac{b_n}{a_n} \qquad (15\text{-}5)$$

a result first used in Chapter 6, Eqs. (6-127) and (6-128).

Combining pairs of terms in Eq. (15-3) gives the equivalent form for the Fourier series

$$f(t) = c_0 + c_1 \cos (\omega_0 t + \theta_1) + \ldots$$
$$+ c_n \cos (n\omega_0 t + \theta_n) + \ldots \qquad (15\text{-}6)$$

with $c_0 = a_0$ and all other c_n and θ_n defined by Eqs. (15-5). The coefficient c_n is the *amplitude* and θ_n the *phase* of the nth harmonic.

Observe that if we know that a Fourier series is to be constructed in the form of Eq. (15-6), then the set of numbers c_n and θ_n contains all the needed information. Plots by which this information may be displayed are shown in Fig. 15-4. The plot of c_n as a function of n or

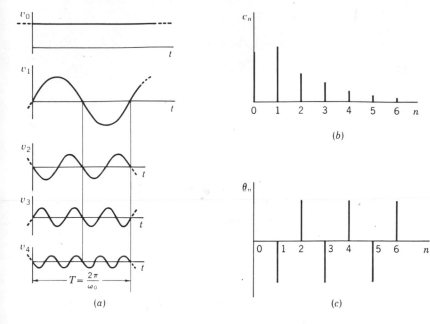

Fig. 15-4. (a) The terms in the Fourier series are shown as a function of time. The wave forms of (a) may be described by the amplitude spectrum of (b) together with the phase spectrum of (a).

$n\omega_0$ (the two being related by simple linear scaling) is known as the amplitude spectrum; the plot of θ_n as a function of n or $n\omega_0$ is the phase spectrum. Later in the chapter, we refine this definition to distinguish between the line (or discrete) spectra we have just discussed and continuous spectra defined for all frequencies. This chapter emphasizes the study of spectra as a means for signal representation and the solution of circuit problems.

15-2. EVALUATION OF FOURIER COEFFICIENTS[1]

The evaluation of the a and b coefficients in Eq. (15-3) is accomplished by using simple integral equations which may be derived from using the *orthogonality property* of the set of functions involved, namely, cos $n\omega_0 t$ and sin $m\omega_0 t$ with integer values for n and m. These functions are *orthogonal* over the interval from t_0 to $t_0 + T$ for any t_0. We will often use the value $t_0 = 0$ or $t_0 = -T/2$, but with the understanding that any period may be used, we will replace t_0 to $t_0 + T$ with 0 to T in the equations to follow.

[1]For a summary of numerical methods for evaluating Fourier coefficients see *Reference Data for Radio Engineers*, 5th ed. (Howard W. Sams & Co., Inc. Indianapolis, Ind., 1970).

First, observe that

$$\int_0^T \sin m\omega_0 t \, dt = 0, \qquad \text{all } m \tag{15-7}$$

and

$$\int_0^T \cos n\omega_0 t \, dt = 0, \qquad \text{all } n \neq 0 \tag{15-8}$$

since the average value of a sinusoid over m or n complete cycles in the period T is zero. The following three cross-product terms are also zero for the stated relationships of m and n:

$$\int_0^T \sin m\omega_0 t \cos n\omega_0 t \, dt = 0, \qquad \text{all } m, n \tag{15-9}$$

$$\int_0^T \sin m\omega_0 t \sin n\omega_0 t \, dt = 0, \qquad m \neq n \tag{15-10}$$

and

$$\int_0^T \cos m\omega_0 t \cos n\omega_0 t \, dt = 0, \qquad m \neq n \tag{15-11}$$

Nonzero values for the integrals result when m and n are equal; thus

$$\int_0^T \sin^2 m\omega_0 t \, dt = \frac{T}{2}, \qquad \text{all } m \tag{15-12}$$

and

$$\int_0^T \cos^2 n\omega_0 t \, dt = \frac{T}{2}, \qquad \text{all } n \tag{15-13}$$

Our evaluation procedure follows a pattern described by the three steps: (1) Multiply Eq. (15-3) on both sides by a suitable factor, (2) integrate the resulting expression term by term over the interval of time from 0 to T, and (3) make use of Eqs. (15-7) through (15-13) to evaluate the integrals, most of which will be zero. Applying this procedure to a_0, the multiplying factor is 1 and the integral equation is

$$\int_0^T f(t) \, dt = a_0 \int_0^T dt + \int_0^T f_1(t) \, dt \tag{15-14}$$

where $f_1(t)$ is written compactly as the summation

$$f_1(t) = \sum_{n=1}^{\infty} (a_n \cos n\omega_0 t + b_n \sin n\omega_0 t) \tag{15-15}$$

This particular division is made because the first term on the right-hand side of Eq. (15-14) has the value $a_0 T$ while every term in the infinite summation of $f_1(t)$, when integrated from 0 to T, has zero value by Eqs. (15-7) and (15-8). Equating the left-hand side of Eq. (15-14) to $a_0 T$, we have

$$a_0 = \frac{1}{T} \int_0^T f(t) \, dt \tag{15-16}$$

indicating that a_0 is simply the average value of $f(t)$ over a period, sometimes also known as the d-c value of the signal.

The multiplying factor for evaluating a_n is cos $n\omega_0 t$, and the product of this factor and Eq. (15-3) integrated from 0 to T is

$$\int_0^T f(t) \cos n\omega_0 t \, dt = \int_0^T a_0 \cos n\omega_0 t \, dt + \int_0^T f_1(t) \cos n\omega_0 t \, dt$$

$$(15\text{-}17)$$

In this case, all the terms on the right-hand side of this equation are zero, except the one of the form of Eq. (15-13), which has the value $T/2$. Thus we have for the coefficient[2] a_n,

$$a_n = \frac{2}{T} \int_0^T f(t) \cos n\omega_0 t \, dt \qquad (15\text{-}18)$$

Following the pattern we have established, Eq. (15-3) is multiplied by sin $n\omega_0 t$ and integrated to give

$$\int_0^T f(t) \sin n\omega_0 t \, dt = \int_0^T a_0 \sin n\omega_0 t \, dt + \int_0^T f_1(t) \sin n\omega_0 t \, dt$$

$$(15\text{-}19)$$

and the only nonzero integral is the one from the second term in the right-hand side of this equation of the form of Eq. (15-12), which has the value $T/2$. Thus the value of b_n is given by the equation

$$b_n = \frac{2}{T} \int_0^T f(t) \sin n\omega_0 t \, dt \qquad (15\text{-}20)$$

and all Fourier coefficients have been evaluated. The use of the three equations, Eqs. (15-16), (15-18), and (15-20), are illustrated by the following examples.

EXAMPLE 1

Figure 15-5 shows a square-wave voltage signal which we wish to represent by a Fourier series. This waveform is written:

$$v(t) = \begin{cases} V, & 0 < t < T/4 \\ -V, & T/4 < t < 3T/4 \\ V, & 3T/4 < t < T \end{cases} \qquad (15\text{-}21)$$

By inspection of the square wave, we see that the average value over one period is zero so that $a_0 = 0$ is obtained without using Eq. (15-16).

---

[2]Some authors write the first term in Eq. (15-3) as $a_0/2$ so that Eq. (15-18) reduces to Eq. (15-16) for $n = 0$. With this convention, a_0 is twice the average value of $f(t)$ over one period.

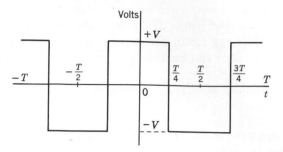

Fig. 15-5. A square or rectangular wave of amplitude $2V$ and period $T = 2\pi/\omega_0$.

The value of a_1 may be obtained with Eq. (15-18) with $n = 1$; thus

$$a_1 = \frac{2}{T}\left(V\int_0^{T/4} \cos\omega_0 t\, dt - V\int_{T/4}^{3T/4} \cos\omega_0 t\, dt + V\int_{3T/4}^{T} \cos\omega_0 t\, dt\right)$$

(15-22)

$$= \frac{2V}{\omega_0 T}\left[\sin\frac{\omega_0 T}{4} - \left(\sin\frac{3\omega_0 T}{4} - \sin\frac{\omega_0 T}{4}\right) + \left(\sin\omega_0 T - \sin\frac{3\omega_0 T}{4}\right)\right]$$

(15-23)

Since $\omega_0 T = 2\pi$, we obtain for a_1

$$a_1 = \frac{V}{\pi}(1 + 2 + 1) = \frac{4V}{\pi}$$

(15-24)

Applying the same procedure for all n, we find that

$$a_n = \begin{cases} \dfrac{+4V}{n\pi}, & n = 1, 5, 9, \ldots \\[2ex] \dfrac{-4V}{n\pi}, & n = 3, 7, 11, \ldots \\[2ex] 0, & n = \text{even integers} \end{cases}$$

(15-25)

$$b_n = \quad 0, \qquad \text{all } n$$

Thus the Fourier series is

$$v(t) = \frac{4V}{\pi}\left(\cos\omega_0 t - \tfrac{1}{3}\cos 3\omega_0 t + \tfrac{1}{5}\cos 5\omega_0 t - \tfrac{1}{7}\cos 7\omega_0 t + \ldots\right)$$

(15-26)

This sum of harmonically related voltages is equal to the square wave, as illustrated by Fig. 15-6, and for the multi-generator network of (b) the total response may be determined by superposition. The amplitude and phase spectra for the square wave are shown in Fig. 15-7.

From Eqs. (15-25) and (15-26), we observe that most of the Fourier coefficients are zero. We next show that this is a consequence of certain symmetries in the signal, and that this information can be determined directly from the waveform, thus avoiding the necessity of evaluating integrals which must have zero value.

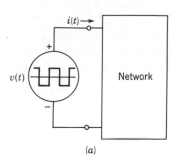

(a)

Fig. 15-6. By means of the Fourier series, a periodic function as voltage source may be replaced by n sinusoidal voltage sources in series, each having an amplitude given by the Fourier coefficients and each at harmonically related frequencies, $n\omega_0$ where n is an integer.

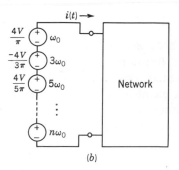

(b)

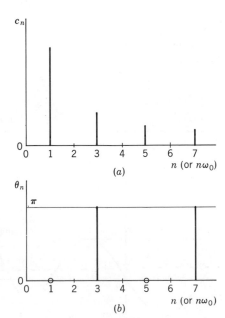

(a)

Fig. 15-7. The magnitude and phase spectra for the waveform of Fig. 15-5. The plots are shown as a function of n, but may be frequency scaled and plotted as a function of $n\omega_0$ as indicated on the figure.

(b)

15-3. WAVEFORM SYMMETRIES AS RELATED TO FOURIER COEFFICIENTS

In this section, we study a number of interesting properties of periodic signals, and also derive rules which will simplify the evaluation of Fourier coefficients. A function $f(t)$ satisfying the condition

$$f(t) = f(-t) \qquad (15\text{-}27)$$

is said to be an *even function*. Even functions of special interest to us are the functions cos $n\omega_0 t$ and the constant a_0, although there are many others such as $|\sin t|$, t^n for n even, and the square wave of Fig. 15-8(a). Similarly, if $f(t)$ satisfies the condition

$$f(t) = -f(-t) \qquad (15\text{-}28)$$

it is an *odd function*. Our interest is in the odd functions sin $n\omega_0 t$; other odd functions are t^n for n odd, $\theta = \tan^{-1} \omega$ in Chapter 13, and the triangular wave of Fig. 15-8(b).

The elementary functions used as examples in the discussion so far are either even or odd. If a number of even functions are summed, the result is even, and the sum of odd functions is likewise odd:

$$\text{Sum of even functions} = \text{even function} \qquad (15\text{-}29)$$

$$\text{Sum of odd functions} = \text{odd function} \qquad (15\text{-}30)$$

However, if one odd function is added to an even summation, the result is neither even nor odd, but may be said to have an even and an odd part:

$$f(t) = f_e(t) + f_o(t) \qquad (15\text{-}31)$$

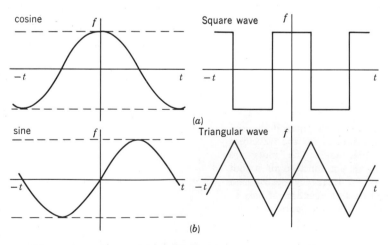

Fig. 15-8. Example of (a) even functions and (b) odd functions.

where

$$f_e = \text{Ev } f(t) \qquad (15\text{-}32)$$

and

$$f_o = \text{Od } f(t) \qquad (15\text{-}33)$$

Now

$$f(-t) = f_e(-t) + f_o(-t) \qquad (15\text{-}34)$$

and from the results of Eqs. (15-27) and (15-28),

$$f(-t) = f_e(t) - f_o(t) \qquad (15\text{-}35)$$

Adding and subtracting Eqs. (15-31) and (15-35), we obtain

$$f_e(t) = \tfrac{1}{2}[f(t) + f(-t)] \qquad (15\text{-}36)$$

and

$$f_o(t) = \tfrac{1}{2}[f(t) - f(-t)] \qquad (15\text{-}37)$$

We next apply these results to the equations for the Fourier coefficients. Since we are interested in comparing $f(t)$ and $f(-t)$, we choose our interval for integration from $-T/2$ to $T/2$ (corresponding to $t_0 = -T/2$ in Section 15-2). Then Eq. (15-18) may be written

$$a_n = \frac{2}{T}\left[\int_{-T/2}^{T/2} f_e \cos n\omega_0 t\, dt + \int_{-T/2}^{T/2} f_o \cos n\omega_0 t\, dt\right] \qquad (15\text{-}38)$$

and from Eq. (15-20),

$$b_n = \frac{2}{T}\left[\int_{-T/2}^{T/2} f_e \sin n\omega_0 t\, dt + \int_{-T/2}^{T/2} f_o \sin n\omega_0 t\, dt\right] \qquad (15\text{-}39)$$

Since function products are involved in these integrals, we recognize that

$$\text{Odd function} \times \text{odd function} = \text{even function} \qquad (15\text{-}40)$$

$$\text{Even function} \times \text{even function} = \text{even function} \qquad (15\text{-}41)$$

$$\text{Even function} \times \text{odd function} = \text{odd function} \qquad (15\text{-}42)$$

and that for any even function f_e

$$\int_{-t_0}^{t_0} f_e(t)\, dt = 2\int_0^{t_0} f_e(t)\, dt \qquad (15\text{-}43)$$

while for any odd function f_o

$$\int_{-t_0}^{t_0} f_o(t)\, dt = 0 \qquad (15\text{-}44)$$

for any t_0, although we will use $t_0 = T/2$. In terms of these results, we next consider a number of important symmetries that may exist in periodic functions.

1. $f(t)$ is an even function. From Eq. (15-29) we know that $f_o = 0$, implying that $b_n = 0$. This may also be seen from Eq. (15-39) with $f_o = 0$ since $f_e \sin n\omega_0 t$ is odd for all n and the integral is zero by

Eq. (15-44). Applying the same conditions to Eq. (15-38) for a_n, we obtain from Eq. (15-43):

$$a_n = \frac{4}{T} \int_0^{T/2} f(t) \cos n\omega_0 t \, dt \qquad (15\text{-}45)$$

and

$$a_0 = \frac{2}{T} \int_0^{T/2} f(t) \, dt \qquad (15\text{-}46)$$

Thus the Fourier series expansion of an even periodic function contains only cosine terms and a constant.

2. $f(t)$ is an odd function. By similar analysis, Eq. (15-30) tells us that $f_e = 0$ so that $a_n = 0$ and $a_0 = 0$. This is also seen from Eq. (15-38) where $f_e = 0$ and $f_o \cos n\omega_0 t$ is odd for all n, so that by Eq. (15-44) $a_n = 0$. Applying these conditions to Eq. (15-39) for b_n, we obtain by using Eq. (15-43):

$$b_n = \frac{4}{T} \int_0^{T/2} f(t) \sin n\omega_0 t \, dt \qquad (15\text{-}47)$$

In summary, *the Fourier series expansion of an odd periodic function contains only sine terms.*

3. Half-wave symmetry. This symmetry is described by the condition

$$f(t) = -f\left(t \pm \frac{T}{2}\right) \qquad (15\text{-}48)$$

and is illustrated in Fig. 15-9, showing that the waveform with t increasing from $-T/2$ to 0 is the negative of the waveform with t

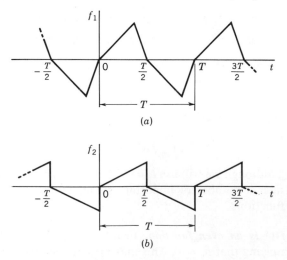

Fig. 15-9. Illustrations of half-wave symmetry.

increasing from 0 to $T/2$. Clearly this waveform is neither even nor odd, so it must be both. To obtain expressions for the Fourier coefficients, consider the equations:

$$a_n = \frac{2}{T} \int_{-T/2}^{T/2} f(t) \cos n\omega_0 t\, dt \qquad (15\text{-}49)$$

and

$$b_n = \frac{2}{T} \int_{-T/2}^{T/2} f(t) \sin n\omega_0 t\, dt \qquad (15\text{-}50)$$

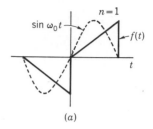

(a)

Now the four integrands in these equations are neither even nor odd for all n. We next outline a method by which we may show that the integrands are always odd for n even, and always even for n odd.

Figure 15-10 shows a waveform $f(t)$ with half-wave symmetry and also $\sin n\omega_0 t$ for $n = 1$ in (a) and for $n = 2$ in (c); the two waveforms are to be multiplied together and then integrated as specified by Eq. (15-50). As far as the value of the integral is concerned, the functions shown from $-T/2$ to 0 may be turned end for end without changing the integral. The waveforms are so modified in (b) and (d). For that of (b), we have an odd times an odd function giving an even function, so that by Eq. (15-43) the value of b_1 is twice Eq. (15-50) with the limits changed from 0 to $T/2$. Now, for the even value $n = 2$, (d) shows that we have an even times an odd function giving an odd function so that by Eq. (15-44) the value is zero. The same method may be applied to Eqs. (15-49) and (15-50) for all n, giving the result

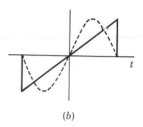

(b)

$$a_0 = 0, \qquad a_n = b_n = 0, \qquad n \text{ even} \qquad (15\text{-}51)$$

and

$$a_n = \frac{4}{T} \int_0^{T/2} f(t) \cos n\omega_0 t\, dt \qquad n \text{ odd} \qquad (15\text{-}52)$$

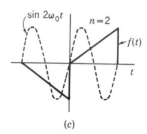

(c)

$$b_n = \frac{4}{T} \int_0^{T/2} f(t) \sin n\omega_0 t\, dt, \qquad n \text{ odd} \qquad (15\text{-}53)$$

In summary, *the Fourier series expansion of a periodic function with half-wave symmetry contains only odd harmonics.* The student will have no difficulty distinguishing between odd functions and odd harmonics. For example, $b_4 \sin 4\omega_0 t$ is an even harmonic but an odd function. What about $a_4 \cos 4\omega_0 t$?

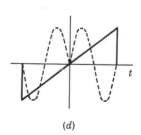

(d)

Fig. 15-10. Waveforms employed to illustrate the conclusion that periodic functions with half-wave symmetry contain only odd harmonics.

EXAMPLE 2

A triangular or sawtooth waveform is shown in Fig. 15-11. This is an odd function, so the symmetry conditions require that $a_n = 0$ for all n, including $n = 0$. To find the b_n coefficients, we substitute

$$v(t) = \begin{cases} \dfrac{4V}{T} t, & 0 < t < \dfrac{T}{4} \\[2ex] -\dfrac{4V}{T} t + 2V, & \dfrac{T}{4} < t < \dfrac{3T}{2} \end{cases} \qquad (15\text{-}54)$$

Table 15-1. SUMMARY OF SYMMETRY CONDITIONS FOR PERIODIC WAVEFORMS

Name of Symmetry	Condition	Illustration	Property	a_0	$a_n (n \neq 0)$	b_n
Even	$f(t) = f(-t)$		Cosine terms only	*	$\dfrac{4}{T} \displaystyle\int_0^{T/2} f(t) \cos n\omega_0 t \, dt$	0
Odd	$f(t) = -f(-t)$		Sine terms only	0	0	$\dfrac{4}{T} \displaystyle\int_0^{T/2} f(t) \sin n\omega_0 t \, dt$
Half-wave	$f(t) = -f\left(t \pm \dfrac{T}{2}\right)$		Odd n only	0	$\dfrac{4}{T} \displaystyle\int_0^{T/2} f(t) \cos n\omega_0 t \, dt$	$\dfrac{4}{T} \displaystyle\int_0^{T/2} f(t) \sin n\omega_0 t \, dt$

* $a_0 = \dfrac{2}{T} \displaystyle\int_0^{T/2} f(t) \, dt$, $\omega_0 = \dfrac{2\pi}{T}$. T is the minimum period.

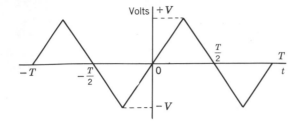

Fig. 15-11. Triangular waveform analyzed in Example 2.

into Eq. (15-47) from which we find that

$$v(t) = \frac{8V}{\pi^2}\left(\sin \omega_0 t - \frac{1}{3^2}\sin 3\omega_0 t + \frac{1}{5^2}\sin 5\omega_0 t - \ldots\right) \qquad (15\text{-}55)$$

Observe that the amplitude of the coefficients decreases more rapidly with n for this waveform than for that in Fig. 15-5 described by Eq. (15-26).

EXAMPLE 3

The signal waveform shown in Fig. 15-12(a) has none of the symmetries we have discussed in this section, and so must have both even and odd parts as in Eq. (15-31). Using Eqs. (15-36) and (15-37), the parts of the signal are determined as shown in (b) and (c) of this figure. These are the waveforms of Examples 1 and 2, so that the Fourier series is simply the sum of Eqs. (15-26) and (15-55). This illustrates a most useful technique in signal analysis.

Returning to the triangular waveform of Fig. 15-11, observe that the choice of the reference time $t = 0$ determines whether $v(t)$ is an even or an odd function. As shown, it is odd; if shifted to the right by $t = T/4$ units, it becomes even. Now the choice of a reference time seems somewhat arbitrary and it is surprising, at first, at least, to find that in one case we get a cosine series and in the other a sine series! What happens when we shift the reference time $t = 0$, and what guidelines may we use in selecting this time?

First, the shifting of a waveform upward or downward with respect to the $f = 0$ or horizontal axis is accomplished by changing a_0. If $f_2 = K_0 + f_1$, then in the Fourier series for f_1, K_0 combines only with a_0. Hence, the addition of K_0 to a signal f_1 shifts the resulting waveform f_2 upward K_0 units for positive values, downward K_0 units for negative values. Conversely, *the selection of the $f = 0$ or horizontal axis affects only a_0.*

A shift in the $t = 0$ or reference time axis or a time displacement of the waveform has a different effect. For a typical term in Eq. (15-6),

$$f_n(t) = c_n \cos(n\omega_0 t + \theta_n) \qquad (15\text{-}56)$$

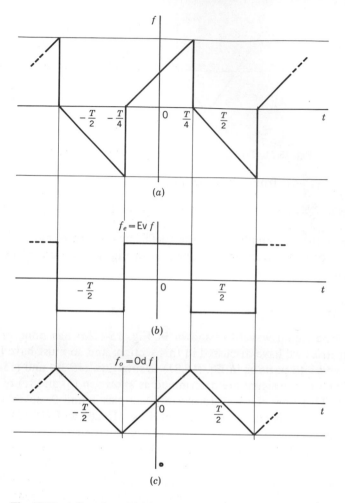

Fig. 15-12. A function which is neither even or odd may be resolved into an even part, f_e, and an odd part f_0 such that $f = f_e + f_0$.

a shift in the time axis so that $t' = 0$ when $t = \tau$ causes t in Eq. (15-56) to be replaced by $t' + \tau$, so that the shifted term is

$$f_n(t') = c_n \cos [n\omega_0(t' + \tau) + \theta_n]$$
$$= c_n \cos (n\omega_0 t' + \phi_n) \qquad (15\text{-}57)$$

where

$$\phi_n = \theta_n + n\omega_0\tau \qquad (15\text{-}58)$$

Observe that c_n, one term of the magnitude spectrum, is the same for $f_n(t')$ as for $f_n(t)$ but that the phase spectrum term is changed from θ_n to ϕ_n. We now see that for the triangular wave which started this discussion a sine series or cosine series might represent this function, because the sine and cosine functions are related by a 90° phase shift.

In analyzing a waveform for which a reference time must be selected, we can select the reference time in such a way that the determination of the Fourier coefficients is made easy, preferably by making the function either even or odd. However, *the selection of the t = 0 or reference time does not affect the magnitude spectrum, but does determine the phase spectrum.*

15-4. CONVERGENCE IN TRUNCATED SERIES

The conditions under which it is possible to write the Fourier series for a periodic function as in Eq. (15-3) are known as the *Dirichlet conditions*, after the mathematician Dirichlet who first found them. They require that in each period the function: (1) have a finite number of discontinuities, (2) possess a finite number of maxima and minima, and (3) be absolutely convergent,

$$\int_0^T |f(t)|\, dt < \infty \qquad (15\text{-}59)$$

For our discussion, we assume a strong form of the conditions and exclude impulse functions and derivatives of impulse functions.[3] Also excluded are such functions as $\sin(1/t)$ or $t^2 \sin(1/t)$, which are excluded by requirements (1) and (2).

We will first study the consequences of *truncating* the infinite Fourier series by which we mean that we will study finite series in which all terms after the nth are dropped. Let this partial sum be $s_n(t)$. The *truncation error* is the difference between $f(t)$ and $s_n(t)$:

$$\epsilon_n = f(t) - s_n(t) \qquad (15\text{-}60)$$

A useful measure or figure of merit which is used in advanced studies is the *mean-square-error* which is

$$E_n = \frac{1}{T} \int_0^T [\epsilon_n(t)]^2\, dt \qquad (15\text{-}61)$$

The so-called *least squares* property of Fourier series states that the truncated Fourier series s_n minimizes the value of E_n found by Eq. (15-61) in the sense that no smaller E_n may be found for another series with the same number of terms.[4] To illustrate, let us use the Fourier series for a square wave given by Eq. (15-26) with the simplification that $V = \pi/4$. For this series, the first three of the truncated

[3]For an example of the kinds of operations possible when impulses and doublets are permitted, see F. F. Kuo, *Network Analysis and Synthesis*, 2nd ed. (John Wiley & Sons, Inc., New York, 1966), pp. 58–63.

[4]For a detailed discussion, see E. A. Guillemin, *Mathematics of Circuit Analysis* (John Wiley & Sons, Inc., New York, 1949), pp. 482–496.

series are

$$s_1 = \cos \omega_0 t \qquad (15\text{-}62)$$

$$s_2 = \cos \omega_0 t - \tfrac{1}{3} \cos 3\omega_0 t \qquad (15\text{-}63)$$

and

$$s_3 = \cos \omega_0 t - \tfrac{1}{3} \cos 3\omega_0 t + \tfrac{1}{5} \cos 5\omega_0 t \qquad (15\text{-}64)$$

If we compare the third of these with another series

$$p_3 = d_1 \cos \omega_0 t + d_3 \cos 3\omega_0 t + d_5 \cos 5\omega_0 t \qquad (15\text{-}65)$$

then we know that there are no values of d_1, d_3, d_5 which minimize the error, in the least-squares sense, better than the values used in Eq. (15-64). The same statement applies to p_3 with $d_5 = 0$ compared to s_2, and to p_3 with $d_3 = d_5 = 0$ compared to s_1.

From this discussion, we know that the error will be minimized, but we have no general method for insuring that it will be less than some prescribed value other than by trial and error, using more terms until specifications are met.

The Dirichlet conditions permit a finite number of discontinuities in each period of $f(t)$ of the form of the jumps shown in Fig. 15-13. This is important to us because we often use the Fourier series representations of such functions, the square wave being one example. What value will the Fourier series $f(t)$ or the truncated $s_n(t)$ assume at the discontinuity? Assume that the function $f(t)$ is discontinuous at t_1 with different limits to the right and to the left of t_1. Let these values be distinguished as $f(t_1-)$ and $f(t+)$. The value at $t = t_1$ will be

$$f(t_1) = \frac{f(t_1+) + f(t_1-)}{2} \qquad (15\text{-}66)$$

such that

$$f(t_1) - f(t_1-) = f(t_1+) - f(t_1) \qquad (15\text{-}67)$$

meaning that the series assumes a value midway between the two values of the function. These values are shown as dots in Fig. 15-13.

Of course, it is too much to expect that s_n, being only a partial sum of the Fourier series, should pass through the three points $f(t_1-)$, $f(t_1)$, and $f(t_1+)$; this would require infinite slope. We should expect that there will be considerable error near the discontinuity, which

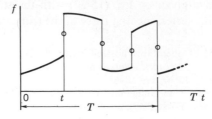

Fig. 15-13. Discontinuities in $f(t)$ with the circles indicating the value given by the truncated series at the discontinuity.

indeed turns out to be the case. The effect is known as the *Gibbs phenomena*,[5] after the first man to investigate it, Sir Willard Gibbs.

The characteristic overshoot and damped oscillatory decay of the Gibbs phenomena is shown in Fig. 15-14. The amount of overshoot is 9 per cent of the total jump as $n \rightarrow \infty$. The nature of the oscillations does change with n, the radian frequency increasing at the same time that the time interval for decay decreases.

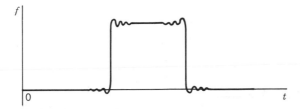

Fig. 15-14. The overshoot and oscillatory decay which characterizes the Gibbs phenomena.

In comparing the Fourier series for the square wave and the triangular wave in Example 3, we observed that the series for the triangular wave diminished more rapidly with increasing n than for the square wave. This is one example of the more general case which we discuss next.

Consider the waveforms shown in Fig. 15-15. These waveforms are members of one family which we studied in Chapter 8 in that f_b is the derivative of f_a, f_c of f_b, and f_d of f_c. We wish to study the relationship of the frequency content of the waveforms in (a), (b), and (c), excluding only (d) because it contains impulse functions. Intuitively, we expect that more terms in a truncated Fourier series will be necessary to represent b than a, and c than b, because of the decreasing "smoothness" of the waveforms f_a, f_b, and f_c. And, of course, this expectation matches our findings in the comparison of Example 3.

The law covering the manner in which the Fourier coefficients diminish with increasing n is conveniently expressed in terms of the number of times a function must be differentiated to produce a jump discontinuity.[6] Let that derivative be the kth. Then the following inequalities hold for the Fourier coefficients:

$$|a_n| \leq \frac{M}{n^{k+1}}, \qquad |b_n| \leq \frac{M}{n^{k+1}} \qquad (15\text{-}68)$$

where M is a constant which depends on $f(t)$ but not on n. Clearly, the equality sign in these equations represents the upper bound of

[5]E. A. Guillemin, *op. cit.*, pp. 485–496.

[6]I. S. Sokolnikoff and R. M. Redheffer, *Mathematics of Physics and Modern Engineering* (McGraw-Hill, Book Co., New York, 1958), p. 211.

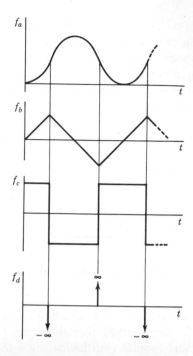

Fig. 15-15. Four waveforms which are related through differentiation in going from top to bottom or through integration in going from bottom to top.

coefficient decrease with n. The application of Eqs. (15-68) to the waveforms of Fig. 15-15 (and other waveforms of these types) is summarized in Table 15-2. We see that the smallest rate of decrease of the coefficients is $1/n$, which is a consequence of our having excluded impulse functions from consideration.

The rates of decrease of the Fourier coefficients with n can also be expressed in terms of the familiar 6 db/octave slopes of Chapter 13 by expressing $|a_n|$ or $|b_n|$ in db and plotting n (or $n\omega_0$) on a logarithmic

Table 15-2

Condition			Law* describing upper bound of decrease of $|a_n|$ and $|b_n|$ with n
Jump in:	*Impulse in:*	*Example*	
$f(t)$	$f'(t)$	Square wave	$\dfrac{1}{n}$
$f'(t)$	$f''(t)$	Triangular wave	$\dfrac{1}{n^2}$
$f''(t)$	$f'''(t)$	Parabolic waveform	$\dfrac{1}{n^3}$
f^{k-1}	$f^k(t)$	—	$\dfrac{1}{n^k}$

*This is also the ratio $|a_n/a_1|$ and $|b_n/b_1|$ if all quantities are nonzero.

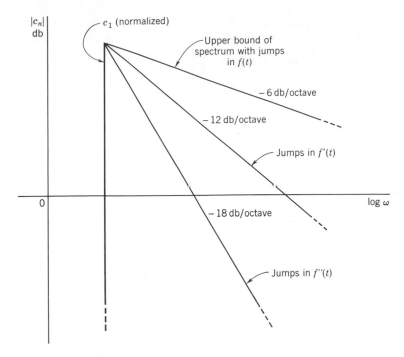

Fig. 15-16. The upper bound for the magnitude of the lines in the spectrum in db, plotted as a function of log ω. The figure illustrates the conclusion that this upper bound decreases at the rate of $6(n + 1)$ db/octave, where n is the number of times the function must be differentiated to produce a jump discontinuity.

scale. Figure 15-16 shows a plot of $20 \log |a_n|$ db with $\log \omega$. The upper bounds for the slopes for the first three cases of Table 15-2 are seen to be -6, -12, and -18 db/octave.

In terms of the notions introduced in this discussion, we may anticipate the magnitude spectrum for a given time-domain waveform, and, in reverse, we may anticipate the time-domain waveform from the magnitude spectrum. Such a facility is a real asset to any engineer! If a waveform contains jumps, we know that its magnitude spectrum will contain lines which diminish only as $1/n$. A waveform in which the slope changes abruptly at some point during the period, but without jumps, will require at most a $1/n^2$ law. A waveform with intricate parabolic wiggles will need at most a $1/n^3$ law. As the wiggles become smaller and smaller, we finally arrive at the smoothest of all waveforms, the sinusoid itself.[7] Examples to illustrate these remarks are shown in Fig. 15-17.

[7]Mason and Zimmermann describe waveforms and their spectra in terms of measures they call *content, variation, and wiggliness.* See S. J. Mason and H. J. Zimmermann, *Electronic Circuits, Signals, and Systems* (John Wiley & Sons, Inc., New York, 1960), pp. 235–242.

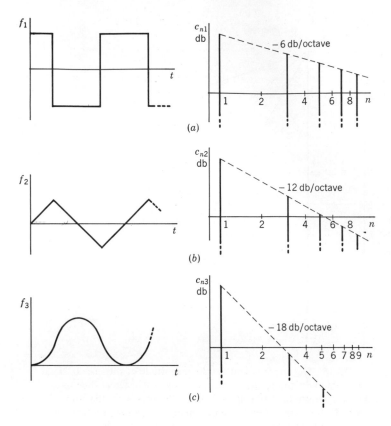

Fig. 15-17. Three examples which illustrate the general conclusion shown in Fig. 15-16. Each part of the figure shows a waveform and its magnitude spectrum with a normalized magnitude for $n = 1$.

15-5. EXPONENTIAL FORM OF THE FOURIER SERIES

The Fourier series can be expressed in an equivalent form in terms of the exponentials, $e^{\pm jn\omega_0 t}$. Suppose that the terms in the series are grouped together in the form

$$f(t) = a_0 + \sum_{n=1}^{\infty} (a_n \cos n\omega_0 t + b_n \sin n\omega_0 t) \qquad (15\text{-}69)$$

As in Chapter 6, the cosine and sine may be expressed in terms of the exponentials as

$$\cos n\omega_0 t = \tfrac{1}{2}(e^{jn\omega_0 t} + e^{-jn\omega_0 t}) \qquad (15\text{-}70)$$

and

$$\sin n\omega_0 t = \tfrac{1}{2j}(e^{jn\omega_0 t} - e^{-jn\omega_0 t}) \qquad (15\text{-}71)$$

Substituting these equations into Eq. (15-69), there results

$$f(t) = a_0 + \sum_{n=1}^{\infty} \left(a_n \frac{e^{jn\omega_0 t} + e^{-jn\omega_0 t}}{2} + b_n \frac{e^{jn\omega_0 t} - e^{-jn\omega_0 t}}{2j} \right) \qquad (15\text{-}72)$$

In order to simplify this equation, like exponential terms are grouped. Noting that $1/j = -j$, our equation becomes

$$f(t) = a_0 + \sum_{n=1}^{\infty} \left[\left(\frac{a_n - jb_n}{2} \right) e^{jn\omega_0 t} + \left(\frac{a_n + jb_n}{2} \right) e^{-jn\omega_0 t} \right] \qquad (15\text{-}73)$$

To simplify this expression, we next introduce a new coefficient to replace the a and b coefficients. By definition,

$$\tilde{c}_n = \frac{a_n - ib_n}{2}, \quad \tilde{c}_{-n} = \frac{a_n + jb_n}{2}, \quad \text{and} \quad \tilde{c}_0 = a_0 \qquad (15\text{-}74)$$

The new form for Eq. (15-73) is

$$f(t) = \tilde{c}_0 + \sum_{n=1}^{\infty} \left(\tilde{c}_n e^{jn\omega_0 t} + \tilde{c}_{-n} e^{-jn\omega_0 t} \right) \qquad (15\text{-}75)$$

We are now in a position to understand better all the maneuvering we have just been through. Letting n range through values from 1 to ∞ in this equation is equivalent to letting n range from $-\infty$ to $+\infty$ (including zero) in a compact equation,

$$f(t) = \sum_{n=-\infty}^{\infty} \tilde{c}_n e^{jn\omega_0 t} \qquad (15\text{-}76)$$

Here we have the exponential form of the Fourier series. The coefficients $\tilde{c}_n$ can easily be evaluated in terms of a_n and b_n, which we already know. Then

$$\tilde{c}_n = \frac{1}{T} \int_0^T f(t) \cos n\omega_0 t \, dt - \frac{j}{T} \int_0^T f(t) \sin n\omega_0 t \, dt$$

$$= \frac{1}{T} \int_0^T f(t)(\cos n\omega_0 t - j \sin n\omega_0 t) \, dt \qquad (15\text{-}77)$$

$$= \frac{1}{T} \int_0^T f(t) e^{-jn\omega_0 t} \, dt$$

This equation for $\tilde{c}_n$ holds whether n is positive as we have assumed, negative, or zero, as can be shown by exactly the same procedure. The tilde is used to distinguish between the real c_n of Section 15-1 and the complex $\tilde{c}_n$ of this discussion. If we let

$$\tilde{c}_n = |\tilde{c}_n| e^{j\phi_n}, \quad \tilde{c}_{-n} = \tilde{c}_n^* = |\tilde{c}_n| e^{-j\phi_n} \qquad (15\text{-}78)$$

then

$$|\tilde{c}_n| = \tfrac{1}{2}\sqrt{a_n^2 + b_n^2} = \tfrac{1}{2} c_n \qquad (15\text{-}79)$$

and

$$\phi_n = \tan^{-1} - \frac{b_n}{a_n} \qquad (15\text{-}80)$$

for all n except $n = 0$ when $\tilde{c}_0 = a_0$ which is real and is the average or d-c value of $f(t)$. This difference between $\tilde{c}_n$ and c_n makes a difference in the spectra from the ordinary and the exponential form of the Fourier series, but the difference is simply a scale factor of $\frac{1}{2}$ for all lines except the one for $n = 0$. We distinguish between the two cases by the tilde and also by plotting both positive and negative n for spectra obtained from the exponential form of the series.

EXAMPLE 4

The sweep voltage waveform shown in Fig. 15-18 is represented over one cycle by the equation of a straight line, $v = (V/T)t$ which may be substituted into Eq. (15-77) to obtain the Fourier $\tilde{c}_n$ coefficients:

$$\tilde{c}_n = \frac{1}{T} \int_0^T \frac{V}{T} t e^{-jn\omega_0 t} dt \tag{15-81}$$

$$\tilde{c}_n = \begin{cases} \dfrac{jV}{2n\pi}, & n \neq 0 \\ \dfrac{V}{2}, & n = 0 \end{cases} \tag{15-82}$$

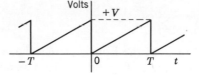

Fig. 15-18. Sweep voltage waveform for Example 4.

This result, which shows that $\tilde{c}_n$ diminishes with $1/n$, might have been anticipated from the jump discontinuity in the waveform. Substituting these values for $\tilde{c}_n$ into Eq. (15-76), we obtain the following form for the exponential Fourier series:

$$v(t) = \ldots - \frac{jV}{6\pi} e^{-j3\omega_0 t} - \frac{jV}{4\pi} e^{-j2\omega_0 t} - \frac{jV}{2\pi} e^{-j\omega_0 t} + \frac{V}{2}$$

$$+ \frac{jV}{2\pi} e^{j\omega_0 t} + \frac{jV}{4\pi} e^{j2\omega_0 t} + \ldots \tag{15-83}$$

If we wish to reduce this result to the alternate form of Fourier series, the a and b coefficients may be found from the equations which follow from the definitions of Eq. (15-74).

$$a_n = \tilde{c}_n + \tilde{c}_{-n}, \qquad b_n = j(\tilde{c}_n - \tilde{c}_{-n}), \qquad a_0 = \tilde{c}_0 \tag{15-84}$$

From these equations, $a_n = 0$, $a_0 = V/2$, and $b_n = -V/n\pi$, and the Fourier series becomes

$$v(t) = V\left[\frac{1}{2} - \frac{1}{\pi}\left(\sin \omega t + \frac{1}{2} \sin 2\omega t + \frac{1}{3} \sin 3\omega t + \ldots \right) \right] \tag{15-85}$$

The line spectra derived from Eq. (15-83) is shown in Fig. 15-19.

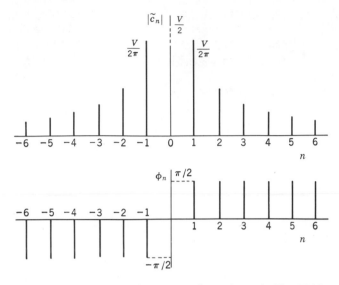

Fig. 15-19. Line spectra for the waveform shown in Fig. 15-18.

We may interpret each term in Eq. (15-83), or the more general Eq. (15-76), as a rotating phasor, except for $\tilde{c}_0$ which is stationary. Terms of the form $e^{jn\omega_0 t}$ rotate in the counterclockwise direction, $e^{-jn\omega_0 t}$ in the clockwise direction; $\tilde{c}_n$ and $\tilde{c}_{-n}$ determine the position of the phasor at $t = 0$; see Fig. 15-20. This infinite summation of spinning phasors, rotating only at speeds which are multiples of the fundamental frequency ω_0, sum to give $f(t)$. If the series is truncated, then the phasors sum to s_n which approximates f. To illustrate, consider the series for Example 4 in Eq. (15-83). For this series, $\phi_n = -\phi_{-n}$, meaning that the imaginary components of like frequency terms will cancel, but the real parts will add. This being the case, we can split the series in two, including the constant term $V/2 - (V/4) + (V/4)$, and sum only for positive n. This accomplished, the real part of the sum of the phasors will be half of the amplitude of the function. Figure 15-21 shows such a plot for Example 4, simplified by letting $V = 2\pi$. We

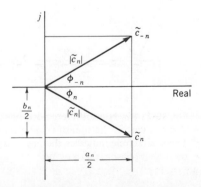

Fig. 15-20. Two rotating phasors having positions $\tilde{c}_{-n}$ and $\tilde{c}_n$ at $t = 0$.

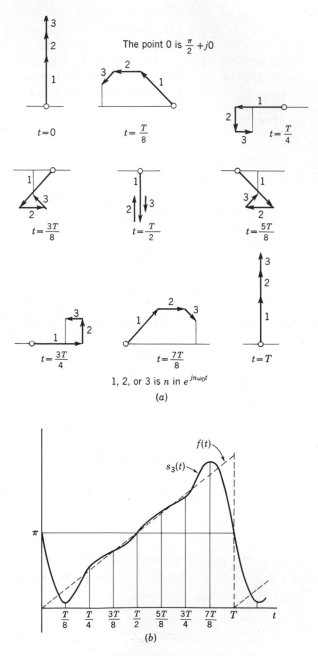

The point 0 is $\frac{\pi}{2} + j0$

$t = 0$ $t = \frac{T}{8}$ $t = \frac{T}{4}$

$t = \frac{3T}{8}$ $t = \frac{T}{2}$ $t = \frac{5T}{8}$

$t = \frac{3T}{4}$ $t = \frac{7T}{8}$ $t = T$

1, 2, or 3 is n in $e^{jn\omega_0 t}$

(a)

(b)

Fig. 15-21. (*a*) Nine different positions of the three rotating phasors. The real part of the sum of these phasors is the amplitude of $s_3(t)$. The small circle indicates the position $(\pi/2) + j0$, and the number with each phasor is n in $e^{jn\omega_0 t}$.

truncate the series at $n = 3$; the value of $\bar{c}_0$ is $\pi/2$, the remaining magnitudes are 1, $\frac{1}{2}$, and $\frac{1}{3}$, and the phasors are all 90°. Nine different positions of the rotating phasors are shown in Fig. 15-21(a) on page 476, and the values so determined from the real part summation are shown as $s_3(t)$ along with $f(t)$ in (b) of this figure. By adding more and more of the spinning phasors, of smaller length but increased speed of rotation, the phasor sum will spiral about in such a way that the real part increases linearly with time!

15-6. STEADY-STATE RESPONSE TO PERIODIC SIGNALS

We are now prepared to discuss the problem given at the beginning of this chapter. In terms of Fig. 15-1, what will be the steady-state response for a periodic input which may be expressed in the form of a Fourier series? We again recall that this topic was treated in Chapter 8 and that we are here considering an alternative method of solution. We will consider two approaches to this subject.

1. Direct phasor approach. This approach is based on the method of Chapter 12, using superposition. We know that the phasor representing the response is equal to the product of the network function and the phasor representing the excitation. Consider the driving-point case with

$$\mathbf{I} = Y\mathbf{V} \tag{15-86}$$

If $\mathbf{V}$ can be resolved into a number of other phasors such that

$$\mathbf{V} = \mathbf{V}_1 + \mathbf{V}_2 + \ldots + \mathbf{V}_n \tag{15-87}$$

then

$$\mathbf{I} = Y\mathbf{V}_1 + Y\mathbf{V}_2 + \ldots + Y\mathbf{V}_n \tag{15-88}$$

One term of this sum may be written as

$$\mathbf{I}_k = Y(jk\omega_0)\,\mathbf{V}_k \tag{15-89}$$

where Y is found at the frequency of the phasor $\mathbf{V}_k$ which is $k\omega_0$. We then transform from the frequency domain to the time domain to give the i_k corresponding to $\mathbf{I}_k$,

$$i_k(t) = |\mathbf{I}_k|\cos\left(k\omega_0 t + \theta_k\right) \tag{15-90}$$

and the response will be the summation of terms like Eq. (15-90) for all k. This is, of course, the *steady-state* response only. In this method, we are in effect replacing one excitation by an infinite number of sources, as shown in Fig. 15-6, and then applying the principle of superposition to obtain the steady-state response.

EXAMPLE 5

Consider the problem of determining the steady-state response when a square-wave source excites a series RL circuit. The Fourier series for the square-wave is given by Eq. (15-26); for simplicity, let $\omega_0 = 1$ radian/sec and $V = \pi/4$ so that the series is

$$v(t) = \cos t - \tfrac{1}{3}\cos 3t + \tfrac{1}{5}\cos 5t + \dots \qquad (15\text{-}91)$$

Using the method of Section 12-5, the phasors corresponding to the first three terms are

$$\mathbf{V}_1 = 1e^{j0°}, \qquad \mathbf{V}_3 = \tfrac{1}{3}e^{-j180°}, \qquad \mathbf{V}_5 = \tfrac{1}{5}e^{j0°} \qquad (15\text{-}92)$$

If we let $R = 1\ \Omega$ and $L = 1$ H, then for this problem

$$Y(jn) = \frac{1}{1+jn} \qquad (15\text{-}93)$$

and from this equation, we determine the values

$$Y(j1) = \frac{1}{\sqrt{2}}e^{-j45°}$$

$$Y(j3) = \frac{1}{\sqrt{10}}e^{-j71.6°} \qquad (15\text{-}94)$$

and

$$Y(j5) = \frac{1}{\sqrt{26}}e^{-j78.8°}$$

From Eq. (15-89), the current phasors are

$$\mathbf{I}_1 = 0.707\, e^{-j45°}$$
$$\mathbf{I}_3 = 0.105\, e^{-251.6°} \qquad (15\text{-}95)$$

and

$$\mathbf{I}_5 = 0.039\, e^{-j78.8°}$$

Then the total current has the following first three terms:

$$i(t) = 0.707 \cos (t - 45°) + 0.105 \cos (3t - 251.6°)$$
$$+ 0.039 \cos (5t - 78.8°) + \dots \qquad (15\text{-}96)$$

As the number of terms in this series becomes very large, the response $i(t)$ becomes that shown in Fig. 15-22.

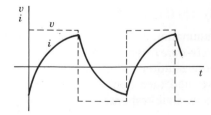

Fig. 15-22. The excitation v and the response i computed in Example 5.

2. Spectrum product approach. The second approach makes use of the amplitude and phase spectra of the excitation, together with that for the network function. By analogy to Eq. (15-86), we see that

$$\begin{pmatrix} \text{amplitude spectrum} \\ \text{of response} \end{pmatrix} = \begin{pmatrix} \text{magnitude of} \\ \text{network function} \end{pmatrix}$$
$$\times \begin{pmatrix} \text{amplitude spectrum} \\ \text{of excitation} \end{pmatrix} \quad (15\text{-}97)$$

and also that

$$\begin{pmatrix} \text{phase spectrum} \\ \text{of response} \end{pmatrix} = \begin{pmatrix} \text{phase of net-} \\ \text{work function} \end{pmatrix} + \begin{pmatrix} \text{phase spectrum} \\ \text{of excitation} \end{pmatrix} \quad (15\text{-}98)$$

In some cases, the response spectrum so obtained is sufficient for a solution to a problem, especially if the estimating methods of the last section are used. If the time response in the form of Eq. (15-96) is required, then the phasor methods of Chapter 12 may be used. If we let $\bar{c}_n = \mathbf{V}_n$ in Eq. (15-76), the excitation may be written

$$v(t) = \sum_{n=-\infty}^{\infty} \mathbf{V}_n e^{jn\omega_0 t} \quad (15\text{-}99)$$

Multiplying each $\mathbf{V}_n$ by the appropriate admittance, as in Eq. (15-89), we obtain for the steady-state response

$$i(t) = \sum_{n=-\infty}^{\infty} Y(jn\omega_0)\mathbf{V}_n e^{jn\omega_0 t} \quad (15\text{-}100)$$

This result may be simplified to the form of Eq. (15-96) by combining pairs of exponentials with the same frequency $n\omega_0$. Equivalent forms of Eq. (15-100) may be written in terms of the Re or Im parts with n extending from 0 to ∞, using the methods of Section 12-4.

EXAMPLE 6

Consider the square wave of Fig. 15-5 with the origin shifted to the left by $T/4$ to be at the jump discontinuity. Let $\omega_0 = 1$ and $V = \pi/8$ so that the Fourier series in the exponential form of Eq. (15-99) is given by

$$|\mathbf{V}_n| = \frac{1}{4|n|}, \qquad n = \pm 1, \pm 3, \pm 5, \ldots$$

$$\phi_n = -90°, \qquad n = 1, 3, 5, \ldots \quad (15\text{-}101)$$

and

$$\phi_n = +90°, \qquad n = -1, -3, -5, \ldots$$

The line spectra for the amplitude and phase are shown in Fig. 15-23(*a*). In (*b*) of this figure are shown the magnitude and phase of $Y(j\omega)$. Applying Eqs. (15-97) and (15-98), we obtain the amplitude and phase spectra of the response which is shown in Fig. 15-23(*c*). The values of Y in Eq. (15-94) may be used together with the values of V_n

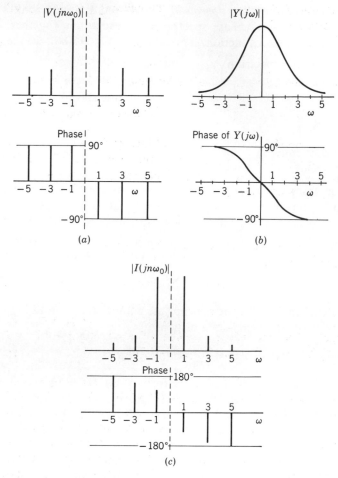

Fig. 15-23. (*a*) The line spectra for the voltage, (*b*) the continuous spectra of the admittance function, and (*c*) the line spectra of the current found by multiplying the two magnitude spectra and adding the two phase spectra.

in Eq. (15-101) to determine $i(t)$ in the form of Eq. (15-100). The result may be reduced algebraically to the form of Eq. (15-96).

15-7. THE POWER SPECTRUM OF PERIODIC SIGNALS

To determine the effective or rms value of a nonsinusoidal but periodic function, we begin with the Fourier series representation of the function, written in the exponential form of Eq. (15-76):

$$f(t) = \sum_{n=-\infty}^{\infty} \tilde{c}_n e^{jn\omega_0 t} = \tilde{c}_0 + \sum_{n=1}^{\infty} 2|\tilde{c}_n| \cos (n\omega_0 t + \phi_n) \quad (15\text{-}102)$$

Substituting $f(t)$ in this form into the defining equation for effective value, given as Eq. (14-24),

$$F_{eff}^2 = \frac{1}{T} \int_0^T f^2 \, dt \qquad (15\text{-}103)$$

where we have selected the reference time $t_0 = 0$, we have

$$F_{eff}^2 = \frac{1}{T} \int_0^T [\bar{c}_0 + \sum_{n=1}^\infty 2|\bar{c}_n| \cos(n\omega_0 t + \phi_n)]^2 \, dt \qquad (15\text{-}104)$$

The evaluation of this infinite sum of integrals is not as formidable as it first appears. Due to the orthogonality properties of such integrals discussed earlier in this chapter, in particular using Eqs. (15-11) and (15-13), the final result is

$$F_{eff}^2 = \bar{c}_0^2 + \sum_{n=1}^\infty 2|\bar{c}_n|^2 \qquad (15\text{-}105)$$

The last term in this equation can be written in two other forms by using the relationship of Eqs. (15-78),

$$|\bar{c}_n|^2 = \bar{c}_n \bar{c}_n^* = \bar{c}_n \bar{c}_{-n} \qquad (15\text{-}106)$$

Now, in Eq. (15-102), $2|\bar{c}_n|$ is the maximum value of the cosine function. If we now change notation to let $f(t)$ be a current (or a voltage), then the effective value of each component is $I_{eff\,n} = I_{max\,n}/\sqrt{2} = \sqrt{2}\,|\bar{c}_n|$ and Eq. (15-105) becomes

$$I_{eff}^2 = I_0^2 + I_{eff\,1}^2 + I_{eff\,2}^2 + \cdots \qquad (15\text{-}107)$$

where a typical term $I_{eff\,k}^2$ is the square of the effective value of the kth harmonic component. In other words, *the effective (or rms) value of a periodic function is the square root of the sum of the squares of the effective values of the harmonic components of the function and I_0^2.*

We next turn to a determination of the power in a network when the excitation is periodic but nonsinusoidal. Using the notation of Eqs. (14-37) and (14-39) for voltage and current, the Fourier series for the voltage is

$$v(t) = V_0 + V_1 \cos(\omega_0 t + \beta_1) + V_2 \cos(2\omega_0 t + \beta_2) + \cdots \qquad (15\text{-}108)$$

and the Fourier series for the current is

$$i(t) = I_0 + I_1 \cos(\omega_0 t + \alpha_1) + I_2 \cos(2\omega_0 t + \alpha_2) + \cdots \qquad (15\text{-}109)$$

where V_0, V_1, I_0, I_1, etc., are maximum values of the harmonic components of the voltage and current. The instantaneous power is $p(t) = v(t)i(t)$. However, our interest is in *average power*, which is given by Eq. (14-23) and, again setting $t_0 = 0$:

$$P_{av} = \frac{1}{T} \int_0^T v(t)i(t) \, dt \qquad (15\text{-}110)$$

Substituting Eqs. (15-108) and (15-109) into this equation to determine P_{av}, we are again able to make use of the orthogonality relationships of Section 15-2 to obtain great simplification. It turns out that P_{av} may be written as a summation, a typical term of which is

$$P_{av\,k} = \frac{V_k I_k}{2} \cos \theta_k = V_{eff\,k} I_{eff\,k} \cos \theta_k \qquad (15\text{-}111)$$

where, as in Eq. (14-39), $\theta_k = \beta_k - \alpha_k$. The summation of terms of the form of Eq. (15-111) may be written

$$P_{av} = P_{av\,1} + P_{av\,2} + P_{av\,3} + \dots \qquad (15\text{-}112)$$

where the typical term $P_{av\,k}$ is the average power in the kth harmonic component. *Thus the total power for periodic but nonsinusoidal voltage and current is the sum of the average power for the harmonic components.* There are no contributions to the average power from the current at one frequency and the voltage at another. This is, of course, the basis for constructing a power spectrum much the same as we have already constructed the spectrum for signals.

As was the case for signal spectra, there are two conventions for plotting power spectra. The first is shown in Fig. 15-24(a). It is constructed from Eq. (15-112) using a line of length $P_{av\,k}$ at the discrete frequency $k\omega_0$. Of course, all $P_{av\,k}$ are positive so that there is only a magnitude spectrum. Double-sided spectra relate to the single-sided spectra by assigning half of $P_{av\,k}$ to the positive frequency $k\omega_0$ and half to the negative frequency $-k\omega_0$, as shown in Fig. 15-24(b). Such spectra are commonly used in related advanced studies such as the power density spectra of random signals.

Plots of the power spectrum for a given load are useful in visualizing the distribution of average power as a function of frequency. A method by which such spectra might be measured is illustrated in Fig. 15-25. In (a) of this figure is shown an ideal filter terminated

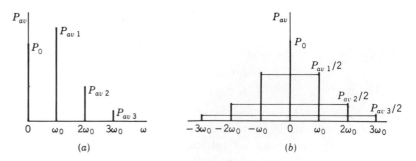

Fig. 15-24. Examples of the two conventions, used in plotting the power spectrum: (a) single-sided, and (b) double-sided spectra.

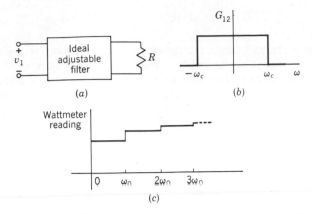

(a) (b)

(c)

Fig. 15-25. (a) A method by which the power spectrum might be measured employing an ideal adjustable filter having the magnitude characteristics shown in (b) with ω_c adjustable. (c) The power in R as ω_c increases.

in a resistive load R and driven by a voltage source. The ideal filter has the property that frequency components smaller than ω_c are transmitted, but components larger than ω_c are rejected, as shown by the magnitude characteristic in (b) of the figure. If the cut-off frequency of the filter is varied from a small value to a large value, the power in the load, as measured with a wattmeter, will vary as shown in Fig. 15-25(c) with step increases at the harmonically related frequencies.

FURTHER READING

CHIRLIAN, PAUL M., *Basic Network Theory*, McGraw-Hill Book Company, New York, 1969. Chapter 6.

CLOSE, CHARLES M., *The Analysis of Linear Circuits*, Harcourt Brace Jovanovich, New York, 1966. Chapter 9.

CRUZ, JOSE B., JR., AND M. E. VAN VALKENBURG, *Signals in Linear Circuits*, Houghton Mifflin Company, Boston, Mass., 1974. Chapter 17.

GABEL, ROBERT A., AND RICHARD A. ROBERTS, *Signals and Linear Systems*, John Wiley & Sons, Inc., New York, 1973. See Chapter 5.

HAYT, WILLIAM H., JR., AND JACK E. KEMMERLY, *Engineering Circuit Analysis*, 2nd ed., McGraw-Hill Book Company, New York, 1971. See Chapter 17.

MANNING, LAURENCE A., *Electrical Circuits*, McGraw-Hill Book Company, New York, 1965. Chapter 20.

DIGITAL COMPUTER EXERCISES

Fourier analysis may be carried out quickly and accurately with computer programs that are generally available. The references which treat computer methods of Fourier analysis are cited in Appendix E-2.3. Specific references that will prove to be useful are Huelsman, reference 7 of Appendix E-10, beginning on page 25, Chapter 5 of Ley, reference 11 in Appendix E-10, and Chapter 3 of Seely, Tarnoff, and Holstein, reference 14 of Appendix E-10.

PROBLEMS

15-1. You are given the following Fourier series:

$$v = \left[\frac{1}{4} - \frac{1}{\pi}\left(\sin \omega_0 t + \frac{1}{2}\sin 2\omega_0 t + \frac{1}{3}\sin 3\omega_0 t + \dots\right)\right]$$

In other problems in this chapter, you will be asked to determine such a series, but here you are given this result. Draw the amplitude and phase spectrum for this $v(t)$ patterned after the plot shown in Fig. 15-4(b) and (c).

15-2. Repeat Prob. 15-1 for the Fourier series

$$v(t) = \frac{1}{\pi}\left(1 + \frac{\pi}{2}\cos \omega_0 t + \frac{2}{3}\cos 2\omega_0 t - \frac{2}{15}\cos 4\omega_0 t + \dots\right)$$

15-3. Repeat Prob. 15-1 for the Fourier series given in the text as Eq. (15-26). Observe that you are not asked to derive this equation, but merely to use it in plotting the magnitude and phase spectra.

15-4. Repeat Prob. 15-1 for the Fourier series given in Eq. (15-55).

15-5. Repeat the amplitude and phase plots as in Prob. 15-1 for the following Fourier series:

$$v(t) = \frac{4V}{\pi}\left(\sin 2\pi t + \frac{1}{3}\sin 6\pi t + \frac{1}{5}\sin 10\pi t + \dots\right)$$

15-6. Repeat Prob. 15-1 for the series

$$v(t) = \frac{V}{2} - \frac{V}{\pi}\left(\sin 2\pi t + \frac{1}{2}\sin 4\pi t + \frac{1}{3}\sin 6\pi t + \dots\right)$$

15-7. Repeat Prob. 15-1 for the Fourier series
$$v(t) = V[1 + 2(\cos 2\pi t + \cos 4\pi t + \cos 6\pi t + \dots)]$$

15-8. Determine the Fourier coefficients for the pulse train shown in the accompanying figure for $a = T/2$ and $a = T/4$.

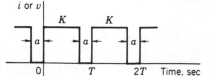

Fig. P15-8.

15-9. The waveform of the figure is known as a sawtooth. For this waveform, determine the Fourier coefficients, and plot the amplitude and phase spectra.

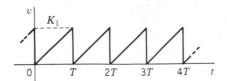

Fig. P15-9.

15-10. The waveform shown in the figure consists of a train of isosceles triangles. For this waveform, determine the Fourier coefficients and plot the corresponding amplitude and phase spectra.

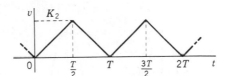

Fig. P15-10.

15-11. The waveform shown in the figure is trapezoidal in form and of period $T = 4a$. For this periodic function, determine the Fourier coefficients and plot the corresponding amplitude and phase spectra.

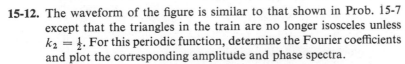

Fig. P15-11.

15-12. The waveform of the figure is similar to that shown in Prob. 15-7 except that the triangles in the train are no longer isosceles unless $k_2 = \frac{1}{2}$. For this periodic function, determine the Fourier coefficients and plot the corresponding amplitude and phase spectra.

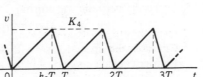

Fig. P15-12.

15-13. The figure shows a full-wave rectifier waveform which is the magnitude of the cosine function. For this waveform, determine the Fourier coefficients and plot the corresponding amplitude and phase spectra.

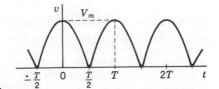

Fig. P15-13.

15-14. The waveform of this problem is similar to that of Prob. 15-13, being a plot of a cosine function when the cosine is positive and zero when the cosine is negative. For this periodic waveform, determine the Fourier coefficients and plot the corresponding amplitude and phase spectra.

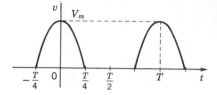

Fig. P15-14.

15-15. The figure shows a train of cosine-squared pulses derived from the waveform of Prob. 15-14 by squaring that function. For this periodic waveform, determine the Fourier coefficients and plot the corresponding amplitude and phase spectra.

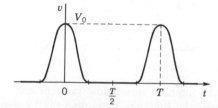

Fig. P15-15.

15-16. Given the periodic waveform

$$v(t) = t^2, \qquad 0 < t < T$$

For the special case $T = 1$, determine the Fourier coefficients and plot the amplitude and phase spectra.

15-17. Given the periodic waveform

$$v(t) = e^{kt}, \qquad 0 < t < T$$

Consider the special case with $k = 1$ and $T = 1$. Determine the Fourier coefficients and plot the amplitude and phase spectra.

15-18. Repeat Prob. 15-17 with $k = -1$ and $T = 1$.

15-19. Given the periodic waveform

$$v(t) = \cosh 2\pi t, \qquad 0 < t < 1$$

determine the Fourier series and plot the amplitude and phase spectra.

15-20. The input to a harmonic generator is a periodic waveform

$$v(t) = Ce^{-\alpha t}, \qquad 0 < t < 1, \quad T = 1$$

with $\alpha > 0$. The output is to be derived through a filter which will pass only one frequency. If α can be adjusted, what value of α will give the maximum output through the filter for the second harmonic? the third harmonic? the fifth harmonic?

15-21. For the following signal waveforms, make use of Eq. (15-36) and Eq. (15-37) to determine the even and odd parts of the given functions:

(a) $v(t) = 1/(1 + t)$

(b) $v(t) = t \sin t$

(c) $v(t) = b_0 + b_1 t + b_2 t^2$

(d) $v(t) = e^{j\omega t}$

(e) $v(t) = te^{-t}$

(f) $v(t) = 1/(1 + t^2)$

15-22. The accompanying figure shows waveforms which are periodic but are shown for only one period. For each waveform, make use of Eqs. (15-36) and (15-37) to determine the even and odd parts of the given function, and plot these two parts as in Fig. 15-12.

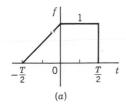

(a)

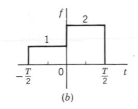

(b)

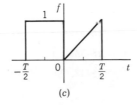

(c)

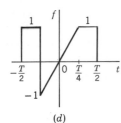

(d)

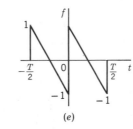

(e)

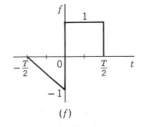

(f)

Fig. P15-22.

15-23. The figure shows a function $f(t)$ from $t = 0$ to $t = T/2$ where T is the period of the waveform. You are required to construct four $f(t)$ in the time interval between $t = T/2$ and $t = T$ such that the period function meets the following specifications:

(a) $a_0 = 0$

(b) $b_n = 0$ for all n

(c) $a_n = 0$ for all n

(d) a_n and b_n are present for odd n only.

For the following problems, determine the coefficients for the exponential form of the Fourier series, and plot the magnitude and phase spectra for the waveforms.

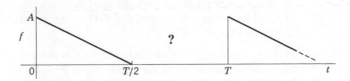

Fig. P15-23.

Problem	Use waveform of Prob.
15-24.	Example 1, Fig. 15-5
15-25.	Example 2, Fig. 15-11
15-26.	15-8
15-27.	15-9
15-28.	15-10
15-29.	15-11
15-30.	15-12
15-31.	15-13
15-32.	15-14
15-33.	15-15

15-34. The input to a device is a sine wave

$$v_{in}(t) = V_m \sin \omega t$$

as shown in (a) of the figure. There is an output of the device only when v_{in} exceeds a constant value, V_g, and then the output is the same as the input, as shown in (b) of the figure. The two functions, v_{in} and v_{out}, are periodic, of course. For v_{out}, determine the Fourier series in exponential form and plot the magnitude and phase spectra.

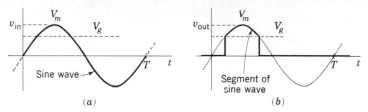

Fig. P15-34.

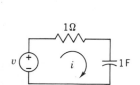

Fig. P15-35.

15-35. For the network shown in the figure, $v(t)$ is a periodic waveform for which $\omega_0 = 1$ radian/sec. For the following waveforms, determine the magnitude and phase spectrum of the periodic $i(t)$: (a) the square wave of Fig. 15-5, (b) the triangular wave of Fig. 5-11, and (c) the sweep voltage waveform of Fig. 15-18.

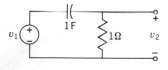

Fig. P15-36.

15-36. For the network of the figure, $v_1(t)$ is the input voltage to the two-port network which is periodic with $\omega_0 = 1$ radian/sec. For the following three waveforms for $v_1(t)$, determine the magnitude and the phase spectrum of the periodic $v_2(t)$: (a) the square wave of Fig. 15-5, (b) the triangular wave of Fig. 5-11, and (c) the triangular wave of Fig. 15-18.

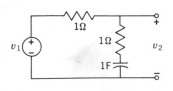

Fig. P15-37.

15-37. In the two-port network of the figure, $v_1(t)$ is a periodic voltage and the system is operating in the steady state. For the following waveforms for $v_1(t)$ with $\omega_0 = 1$ radian/sec, determine the magnitude and phase spectrum of the periodic $v_2(t)$: (a) the square wave of Fig.

15-5, (b) the triangular wave of Fig. 15-11, and (c) the triangular wave of Fig. 15-18.

15-38. For the network of the figure, $v_1(t)$ is the input voltage and $v_a(t)$ and $v_b(t)$ are output voltages. We are considering the case of $v_1(t)$ as a periodic function, and wish to determine the magnitude and phase spectra for the periodic outputs. Do this for the input

$$v_1(t) = t^2, \qquad 0 < t < 1, \quad T = 1$$

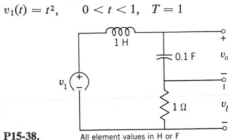

Fig. P15-38.   All element values in H or F

15-39. Repeat Prob. 15-38 if

$$v_1(t) = e^{-t}, \qquad 0 < t < 1, \quad T = 1$$

15-40. Repeat Prob. 15-38 if

$$v_1(t) = \sum_{-\infty}^{\infty} \delta(t - n)$$

which is an impulse train of infinite extent.

15-41. For the one-port network of Fig. 14-2, it is given that

$$i = 10 \cos t + 5 \cos(2t - 45°)$$
$$v = 2 \cos(t + 45°) + \cos(2t + 45°) + \cos(3t - 60°)$$

(a) What is the average power to the network? (b) Plot the power spectrum.

15-42. In the network of the figure, it is given that

$$v_1 = 120 \sin \omega_0 t + 50 \cos 3\omega_0 t$$
$$v_2 = 80 \sin(\omega_0 t + 30°) + 10 \sin 3\omega_0 t$$
$$i_1 = 15 \sin(\omega_0 t - 30°)$$
$$i_2 = 0.5 \sin(3\omega_0 t - 45°)$$

(a) Determine the equation for $v_3(t)$. (b) Determine the equation for $i_3(t)$. (c) Compute the average power to the passive network. (d) Plot the power spectrum for this system.

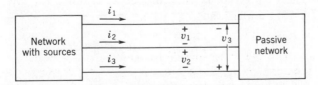

Fig. P15-42.

16 Fourier Integral and Continuous Spectra

In this chapter, we extend the Fourier methods of Chapter 15 to include the case of the signal which occurs once in some finite time interval and is never repeated, and so introduce the *Fourier integral*. This is accomplished by letting the period T become infinite in the Fourier series. The line spectrum then becomes a *continuous spectrum*.

16-1. SPECTRUM ENVELOPE FOR A RECURRING PULSE

Figure 16-1 shows the waveform of a periodic pulse of magnitude V_0 and duration a. The coordinates are selected so that

$$f(t) = \begin{cases} V_0, & -a/2 < t < a/2 \\ 0, & -T/2 < t < -a/2,\ a/2 < t < T/2 \end{cases} \quad (16\text{-}1)$$

In radar applications, it is useful to define the ratio a/T as the *duty factor*. We are interested in the spectrum of $f(t)$ as the duty factor becomes smaller and smaller, approaching 0 in the limit.

The Fourier coefficients may be determined for this $f(t)$ from Eq. (15-77) for all n; thus,

$$\bar{c}_n = \frac{1}{T} \int_{-a/2}^{a/2} V_0 e^{-jn\omega_0 t}\, dt \quad (16\text{-}2)$$

$$\bar{c}_n = \frac{V_0}{n\pi} \left(\frac{e^{jn\omega_0 a/2} - e^{-jn\omega_0 a/2}}{2j} \right) = V_0 \frac{\omega_0 a}{2\pi} \left(\frac{\sin(n\omega_0 a/2)}{n\omega_0 a/2} \right) \quad (16\text{-}3)$$

490

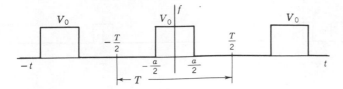

Fig. 16-1. A periodic pulse of amplitude V_0, duration a, and period T.

Now, since $T = 2\pi/\omega_0$, the equation finally may be written

$$\tilde{c}_n = V_0 \frac{a}{T} \frac{\sin(n\omega_0 a/2)}{n\omega_0 a/2} \tag{16-4}$$

If we let $x = n\omega_0 a/2$, then Eq. (16-4) is in the form $(\sin x)/x$ which is familiar in the literature of mathematics, and is shown plotted in Fig. 16-2. Note from the plot that the function has the value 1 when $x = 0$, corresponding to the case $n = 0$ in Eq. (16-4). Note also that $\tilde{c}_n$ has both positive and negative values, corresponding to a phase ϕ_n of either $0°$ or $180°$.

Now Eq. (16-4) has values only for the discrete frequencies $n\omega_0$; these are the values of frequency at which there will be lines in the spectrum for $\tilde{c}_n$. The *envelope* of $\tilde{c}_n$ is a *continuous* function found by replacing $n\omega_0$ with ω, and so is

$$\text{Envelope of } \tilde{c}_n = V_0 \frac{a}{T} \frac{\sin(\omega a/2)}{(\omega a/2)} \tag{16-5}$$

This envelope plays an important role in our study of this chapter.

We next examine $\tilde{c}_n$ as the ratio a/T changes. We do this first for a duty factor of $\frac{1}{2}$ and then for $\frac{1}{5}$, finally generalizing for $1/N$. For $a/T = \frac{1}{2}$, the envelope for $\tilde{c}_n$ is

$$\text{Envelope of } \tilde{c}_n = \frac{V_0}{2} \frac{\sin(\omega a/2)}{\omega a/2} \tag{16-6}$$

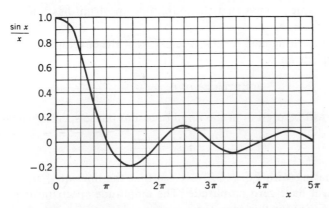

Fig. 16-2. The $(\sin x)/x$ of Eq. 16-5 with $x = \omega a/2$.

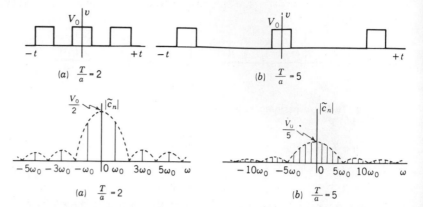

(a) $\frac{T}{a} = 2$ (b) $\frac{T}{a} = 5$

(a) $\frac{T}{a} = 2$ (b) $\frac{T}{a} = 5$

Fig. 16-3. Two periodic pulses are shown for (a) $T/a = 2$, and (b) $T/a = 5$. Below the waveforms are shown the corresponding line spectra for the amplitude.

which is shown as the dashed curve in Fig. 16-3(a). The envelope has zero values when

$$\frac{\omega a}{2} = \pm\pi, \pm2\pi, \pm3\pi, \ldots \tag{16-7}$$

or at the frequencies

$$\omega = \frac{\pm2\pi}{a}, \frac{\pm4\pi}{a}, \frac{\pm6\pi}{a}, \ldots \tag{16-8}$$

Now the fundamental frequency in $f(t)$ is $\omega_0 = 2\pi/T$, so that for $T = 2a$, $\tilde{c}_n$ has values at

$$n\omega_0 = \frac{n\pi}{a} \tag{16-9}$$

Comparing Eqs. (16-8) and (16-9), we see that the even-ordered harmonics have zero value, except for $\tilde{c}_0$. This is illustrated in Fig. 16-3(a) which shows that the amplitude spectrum has lines at $\pm n\omega_0$ only for n odd.

Repeating next for the case $a/T = \frac{1}{5}$, the envelope for the amplitude spectrum is

$$\text{Envelope of } \tilde{c}_n = \frac{V_0}{5}\frac{\sin(\omega a/2)}{\omega a/2} \tag{16-10}$$

Lines will be present in the amplitude and phase spectra when

$$n\omega_0 = \frac{2n\pi}{5a} \tag{16-11}$$

Comparing with Eq. (16-8), we see that the lines at $\pm5\omega_0$, $\pm10\omega_0$, etc., will have zero amplitude. The amplitude spectrum is shown in Fig. 16-3(b) for this case.

In the general case, the spacing between lines will be, for a duty factor of $a/T = 1/N$,

$$\Delta\omega = (n+1)\omega_0 - n\omega_0 = \omega_0 \qquad (16\text{-}12)$$

and the amplitude of the lines will be

$$|\tilde{c}_n| = \frac{V_0}{N}\left|\frac{\sin(n\omega_0 a/2)}{n\omega_0 a/2}\right| \qquad (16\text{-}13)$$

where

$$\omega_0 = \frac{2\pi}{Na} \qquad (16\text{-}14)$$

As N becomes large, say, 1000 or 10,000, the number of lines in a given frequency interval will become large, the separation of the lines will become very small, but the amplitude of each line, given by Eq. (16-13), will also become very small. More frequency components will be required to make up the shorter pulse, but the amplitude of the required frequency components will be smaller. We are now ready to consider the situation in the limit $N \longrightarrow \infty$.

16-2. THE FOURIER INTEGRAL AND TRANSFORM

In terms of the pulse which was the example of the last section, the limit $N \longrightarrow \infty$ is the same as the limit for the period $T \longrightarrow \infty$, since $a/T = 1/N$. Thus, in Fig. 16-1, the pulse width a remains fixed as $T \longrightarrow \infty$, and this may be interpreted as the case of a single pulse or a nonrecurring pulse.

We begin this study by considering the product of T and $\tilde{c}_n$ of Eq. (15-77):

$$\tilde{c}_n T = \int_{-T/2}^{T/2} f(t)e^{-jn\omega_0}\,dt \qquad (16\text{-}15)$$

Since $\tilde{c}_n$ is known to be complex, let the product be given the notation

$$F(jn\omega_0) = \tilde{c}_n T = 2\pi\frac{\tilde{c}_n}{\Delta\omega} \qquad (16\text{-}16)$$

which is obtained using Eq. (16-12). Then

$$\tilde{c}_n = \frac{F(jn\omega_0)\,\Delta\omega}{2\pi} \qquad (16\text{-}17)$$

Returning once more to the example of the last section, we noted that as the amplitude of the lines in the amplitude spectrum became small, the spacing between the lines also became small. In fact, if we divide Eq. (16-5) for $\tilde{c}_n$ by Eq. (16-12) for $\Delta\omega$, we obtain

$$\text{Envelope of } \frac{\tilde{c}_n}{\Delta\omega} = a\frac{V_0}{2\pi}\frac{\sin(\omega a/2)}{\omega a/2} \qquad (16\text{-}18)$$

which is independent of N and so of T, and consequently will not be changed as $T \to \infty$. This property, which we have illustrated by means of the pulse, holds for all $f(t)$ which are nonrecurring. Since from Eq. (16-17), $\bar{c}_n/\Delta\omega = F(j\omega n_0)/2\pi$, the conclusions reached for Eq. (16-18) hold for $F(j\omega n_0)$. Then substituting $\bar{c}_n$ from Eq. (16-17) into $f(t)$ given by Eq. (15-76), we obtain

$$f(t) = \lim_{T \to \infty} \sum_{n=-\infty}^{\infty} \frac{F(j\omega n_0)}{2\pi} e^{-jn\omega_0 t} \Delta\omega \qquad (16\text{-}19)$$

Now, as $T \to \infty$, we know from integral calculus that $\Delta\omega \to d\omega$. The product $n\omega_0 = n\Delta\omega \to \omega$, which is the frequency variable. Then if we interpret the limit of the sum in Eq. (16-19) as the integral, we have for $f(t)$ of Eq. (16-19)

$$f(t) = \frac{1}{2\pi} \int_{-\infty}^{\infty} F(j\omega)e^{j\omega t}\, d\omega \qquad (16\text{-}20)$$

$F(j\omega)$ is found directly from Eq. (16-15) by making the same substitutions that were made in Eq. (16-19), giving

$$F(j\omega) = \int_{-\infty}^{\infty} f(t)e^{-j\omega t}\, dt \qquad (16\text{-}21)$$

subject to the sufficient but not necessary condition

$$\int_{-\infty}^{\infty} |f(t)|\, dt < \infty \qquad (16\text{-}22)$$

which will be discussed in Section 16-4. Equations (16-20) and (16-21) constitute the *Fourier transform pair* in the same sense that $F(s)$ and $f(t)$ were Laplace transform pairs in Chapter 7. Thus,

$$F(j\omega) = \mathfrak{F}f(t) \qquad (16\text{-}23)$$

and

$$f(t) = \mathfrak{F}^{-1}F(j\omega) \qquad (16\text{-}24)$$

In terms of Eq. (16-21), $F(j\omega)$ is the *Fourier integral* of $f(t)$, and from Eq. (16-23) it is also the *Fourier transform* of $f(t)$. In other words, the Fourier integral is also called the Fourier transform.

Another terminology for $F(j\omega)$ is derived by a comparison of Eq. (16-21) and $\bar{c}_n$ of Eq. (15-77). Suppose that $f(t)$ is a pulse of arbitrary shape and duration $a < T$. Then the limits of integration for the two equations are identical and

$$F(j\omega) = \int_{-a/2}^{a/2} f(t)e^{-j\omega t}\, dt \qquad (16\text{-}25)$$

and

$$\bar{c}_n = \int_{-a/2}^{a/2} f(t)e^{-jn\omega_0 t}\, dt \qquad (16\text{-}26)$$

Comparing these two equations, we see that the only difference in $F(j\omega)$ and $\bar{c}_n(jn\omega_0)$ is that one is a continuous function and the other is not, being defined only for discrete values of frequency. Thus it is reasonable that in

$$F(j\omega) = |F(j\omega)| e^{j\phi(\omega)} \tag{16-27}$$

$|F(j\omega)|$ is known as the *continuous amplitude spectrum* and $\phi(\omega)$ is the *continuous phase spectrum* for a nonrecurring $f(t)$. The two kinds of spectra we have studied are thus distinguished by the names *line* or *discrete* and *continuous*. Concerning continuous spectra, we may make the following summarizing statements:

(1) The shape of the continuous amplitude and phase spectra for a nonrecurring $f(t)$ is identical with the envelopes of the amplitude and phase line spectra for the same pulse recurring.

(2) All frequencies are present in the continuous amplitude spectrum in the sense that $F(j\omega)$ is defined for all $\omega(-\infty < \omega < \infty)$; the amplitude of any frequency component is vanishingly small, being $F(j\omega) \, d\omega/2\pi$ by Eq. (16-17).

In terms of the synthesis of a pulse by the addition of frequency components, the continuous spectrum requires *all* frequency components. In terms of analysis, we know that a single pulse will contain all frequency components. This is confirmed by the common experience that a pulse of lightning will result in a burst of static on all receivers from low-frequency radio to television. In fact, the recurring pulse from a dirty sparkplug in an automobile does a fair job of transmitting all frequency components!

Examples of continuous spectra are given in the next two sections.

16-3. APPLICATION IN NETWORK ANALYSIS

The Fourier transform of a pulse waveform has served to introduce the notion of the new transform. Based on Eqs. (16-20) and (16-21), additional transform pairs may be computed for useful waveforms in a manner similar to that used in Chapter 7 for the Laplace transform. As the next example, consider the exponential waveform described by the equation

$$i_1(t) = I_0 e^{-kt}, \qquad t \geq 0 \tag{16-28}$$

which has zero value for negative t and satisfies Eq. (16-22) providing $k > 0$. The waveform is shown in Fig. 16-4. From Eq. (16-21), the corresponding Fourier transform is

$$I_1(j\omega) = I_0 \int_{-\infty}^{\infty} e^{-kt} u(t) e^{-j\omega t} \, dt \tag{16-29}$$

$$= \frac{I_0 e^{-(k+j\omega)t}}{-(k+j\omega)} \Bigg|_0^{\infty} = \frac{I_0}{j\omega + k} \tag{16-30}$$

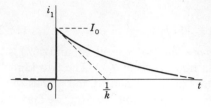

Fig. 16-4. Exponential waveform described by Eq. (16-28).

Hence

$$I_1(j\omega) = \mathfrak{F}[i_1(t)] = \frac{I_0}{j\omega + k} \qquad (16\text{-}31)$$

and the magnitude and phase are

$$|I_1(j\omega)| = \frac{I_0}{\sqrt{\omega^2 + k^2}} \qquad (16\text{-}32)$$

and

$$\theta(j\omega) = -\tan^{-1}\left(\frac{\omega}{k}\right) \qquad (16\text{-}33)$$

These two equations are plotted in Fig. 16-5; they are the continuous spectra for the exponential function described by Eq. (16-28). Because of the uniqueness property of the Fourier transform, the inverse transform is

$$\mathfrak{F}^{-1}\frac{I_0}{j\omega + k} = I_0 e^{-kt} u(t) \qquad (16\text{-}34)$$

One of the classical uses of the Fourier transforms is similar to that we have already studied in connection with the Laplace transform, the solution of a system described by a differential equation. To illustrate this use, we first determine the Fourier transform of the first derivative of a time function. From Eq. (16-20), we see that

$$\frac{df(t)}{dt} = \frac{1}{2\pi}\int_{-\infty}^{\infty} j\omega F(j\omega)e^{j\omega t}\, d\omega \qquad (16\text{-}35)$$

This result follows because the only time function within the integral is $e^{j\omega t}$, and we have made use of the usual procedure of interchanging the order of differentiation and integration. Comparing this equation with Eq. (16-20), we observe that $j\omega F(j\omega)$ has replaced $F(j\omega)$, and so

$$\mathfrak{F}\left[\frac{df(t)}{dt}\right] = j\omega F(j\omega) \qquad (16\text{-}36)$$

To continue, let us apply this result to the network shown in Fig. 16-6(a). Let the current source be described by the equation

$$i_1(t) = 2e^{-t}u(t) \qquad (16\text{-}37)$$

For the element values given in Fig. 16-6, the network is described by

$$\frac{1}{2}j\omega V_2(j\omega) + V_2(j\omega) = I_1(j\omega) = \frac{2}{j\omega + 1} \qquad (16\text{-}38)$$

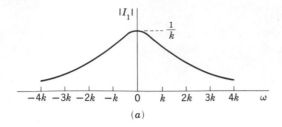

(a)

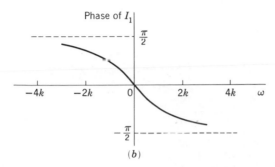

(b)

Fig. 16-5. Magnitude and phase spectra for the exponential wave-form of Eq. (16-28).

By algebraic manipulation

$$V_2(j\omega) = \frac{4}{(j\omega + 1)(j\omega + 2)} \tag{16-39}$$

Following the pattern we employed in Chapter 7, we expand this function by partial fractions to give

$$V_2(j\omega) = \frac{4}{j\omega + 1} - \frac{4}{j\omega + 2} \tag{16-40}$$

We are now in a position to make use of the inverse Fourier transform, and the necessary transform pair is given in Eq. (16-34). Thus

$$v_2(t) = 4e^{-t} - 4e^{-2t}, \qquad t \geq 0 \tag{16-41}$$

and zero value for $t < 0$. This result is shown in Fig. 16-6(b).

To illustrate a second approach made possible through the Fourier transform, we rearrange Eq. (16-38) in the form

$$Y(j\omega)V_2(j\omega) = I_1(j\omega) \tag{16-42}$$

or since $Z(j\omega) = 1/Y(j\omega)$,

$$V_2(j\omega) = Z(j\omega)I_1(j\omega) \tag{16-43}$$

where

$$Z(j\omega) = \frac{2}{j\omega + 2} \tag{16-44}$$

and

$$I_1(j\omega) = \frac{2}{j\omega + 1} \tag{16-45}$$

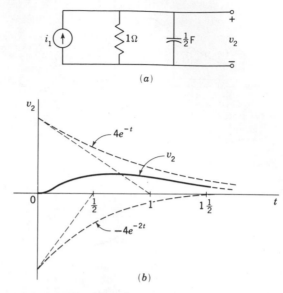

Fig. 16-6. (a) An RC network in which $i_1(t)$ is described by Eq. (16-37) and $v_2(t)$ is to be found. (b) Plots of the two parts of Eq. (16-41) and the response $v_2(t)$.

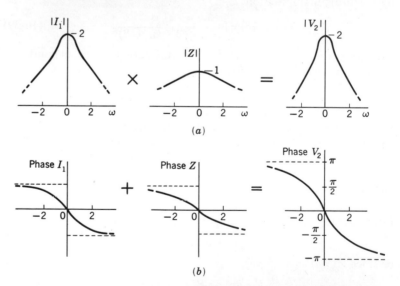

Fig. 16-7. (a) The magnitude and (b) the phase spectrum of $V_2(j\omega)$ as determined by Eqs. (16-46) and (16-47).

From Eq. (16-43), we see that the response transform, $V_2(j\omega)$, may be considered as the product of two other transforms, $Z(j\omega)$ and $I_1(j\omega)$. Now since each of these has a relatively simple form as shown in Fig. 16-5, the two spectral representatives may be multiplied together to determine the total response. As illustrated in Fig. 16-7, the magnitude and phase of the response $V_2(\omega)$ may be found by multiplying magnitudes and adding phases; thus, combining the factors in Eq. (16-43) in the usual manner for complex numbers,

$$|V_2(j\omega)| = |Z(j\omega)||I_1(j\omega)| \qquad (16\text{-}46)$$

and

$$\text{Phase of } V_2(j\omega) = \text{phase of } Z(j\omega) + \text{phase of } I_1(j\omega) \qquad (16\text{-}47)$$

The result is the same as that computed from Eq. (16-39), which is

$$|V_2(j\omega)| = \frac{4}{[(2-\omega^2)^2 + (3\omega)^2]^{1/2}} \qquad (16\text{-}48)$$

and

$$\text{Phase of } V_2(j\omega) = -\tan^{-1}\frac{3\omega}{2-\omega^2} \qquad (16\text{-}49)$$

However, the visualization of the response spectrum as being the result of the multiplication of a system-characterizing spectrum and the excitation spectrum has conceptual advantages in engineering applications.

16-4. SOME USEFUL FOURIER TRANSFORMS

We have mentioned earlier that a sufficient but not necessary condition for a Fourier transform to exist is that it satisfy the Dirichlet conditions stated in Section 15-4, the most important of which is the requirement that

$$\int_{-\infty}^{\infty} |f(t)|\, dt < \infty \qquad (16\text{-}50)$$

This requirement is not satisfied by many of the most familar waveforms used in earlier chapters, such as the step function, the ramp function, and periodic functions like the sine wave. We shall show in this section that if the impulse function is allowed to exist in the Fourier transform, then this limitation, Eq. (16-50), may be removed.

One of the simplest Fourier transforms is that for the unit impulse function. From the defining equation,

$$\mathcal{F}[\delta(t)] = F(j\omega) = \int_{-\infty}^{\infty} \delta(t)e^{-j\omega t}\, dt \qquad (16\text{-}51)$$

The evaluation of this integral is easily found in terms of the scanning function studied in Chapter 8 as Eq. (8-115), which is

$$\int_{-\infty}^{\infty} f(t)\delta(t - t_0)\, dt = f(t_0) \qquad (16\text{-}52)$$

Hence, the integral of Eq. (16-51) has the value

$$\mathcal{F}[\delta(t)] = 1 \quad \text{and} \quad \mathcal{F}^{-1}[1] = \delta(t) \qquad (16\text{-}53)$$

the second result following because of the uniqueness property of

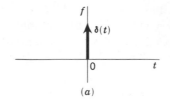

Fig. 16-8. An impulse response $\delta(t)$ shown in (a) has a constant magnitude spectrum as shown in (b).

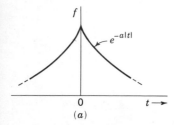

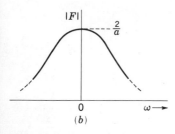

Fig. 16-9. The impulse function of (a), described by Eq. (16-54), has the magnitude spectrum shown in (b) and given by Eq. (16-55).

Fourier transforms. Thus the Fourier transform of an impulse is a simple constant, as shown in Fig. 16-8.

As the next example, consider an exponential function defined for both positive and negative t as shown in Fig. 16-9(a),

$$f(t) = e^{-a|t|}, \qquad \text{all } t \tag{16-54}$$

The corresponding Fourier transform is found from the defining equation and is

$$F(j\omega) = \int_{-\infty}^{0} e^{at} e^{-j\omega t}\, dt + \int_{0}^{\infty} e^{-at} e^{-j\omega t}\, dt$$

$$= \frac{1}{a - j\omega} + \frac{1}{a + j\omega} = \frac{2a}{\omega^2 + a^2} \tag{16-55}$$

This is a real function, with zero phase contribution for all ω, and is shown as a continuous spectrum in Fig. 16-9(b).

Let us make use of the last example to compute the Fourier transform for a more difficult time function: $f(t) = 1$. We accomplish this by starting from $f(t)$ given in Eq. (16-54) and take the limiting case as a approaches 0. In the form of an equation,

$$\mathcal{F}[1] = \lim_{a \to 0} \int_{-\infty}^{\infty} e^{-a|t|} e^{-j\omega t}\, dt \tag{16-56}$$

$$= \lim_{a \to 0} \frac{2a}{\omega^2 + a^2} \tag{16-57}$$

Now this limit has the value 0 for all a except $\omega = 0$. For $\omega = 0$, we may use l'Hospital's rule and differentiate both numerator and denominator before letting a approach zero. Thus

$$\lim_{a \to 0} \frac{2}{2a} = \infty \tag{16-58}$$

indicating an impulse function at $\omega = 0$. The magnitude of the impulse function is found by integrating $F(j\omega)$ for all ω to give the area contained by the function; thus

$$\int_{-\infty}^{\infty} \frac{2a}{\omega^2 + a^2} \, d\omega = 2\pi \tag{16-59}$$

Then we have determined that the Fourier transform of the constant one is an impulse of strength 2π, or

$$\mathcal{F}[1] = 2\pi\delta(\omega) \tag{16-60}$$

as illustrated by Fig. 16-10.

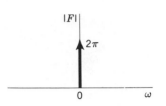

Fig. 16-10. The magnitude spectrum of the constant 1 is an impulse at $\omega = 0$, given by Eq. (16-60).

This result may be generalized to find the Fourier transform of $e^{j\omega_0 t}$, where ω_0 is a constant. Since we have just shown that

$$\int_{-\infty}^{\infty} 1 e^{-j\omega t} \, dt = 2\pi\delta(\omega) \tag{16-61}$$

then it follows that

$$\mathcal{F}[e^{j\omega_0 t}] = \int_{-\infty}^{\infty} 1 e^{j\omega_0 t - j\omega t} \, dt = 2\pi\delta(\omega - \omega_0) \tag{16-62}$$

Since the cosine and sine functions may be expressed in terms of the exponential factors $e^{\pm j\omega_0 t}$ as

$$\cos \omega_0 t = \frac{e^{j\omega_0 t} + e^{-j\omega_0 t}}{2} \quad \text{and} \quad \sin \omega_0 t = \frac{e^{j\omega_0 t} - e^{-j\omega_0 t}}{2j} \tag{16-63}$$

we may use Eq. (16-62) to determine the Fourier transforms of the cosine and sine functions.

$$\cos \omega_0 t = \pi[\delta(\omega - \omega_0) + \delta(\omega + \omega_0)] \tag{16-64}$$

$$\sin \omega_0 t = -j\pi[\delta(\omega - \omega_0) - \delta(\omega + \omega_0)] \tag{16-65}$$

Thus both functions are expressed in terms of two impulse functions as shown for the magnitude of $F(j\omega)$ in Fig. 16-11. The phase angles associated with the two functions do differ, of course, as shown in Eqs. (16-64) and (16-65).

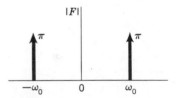

Fig. 16-11. The magnitude spectrum for the sine and cosine functions of time.

A function that finds frequent use in describing signal waveforms is the signum function, abbreviated sgn (t), which is shown in Fig. 16-12 and is defined as

$$\begin{aligned} \text{sgn } (t) &= +1, & t > 0 \\ &= 0, & t = 0 \\ &= -1, & t < 0 \end{aligned} \tag{16-66}$$

The direct evaluation of the Fourier transform for this function

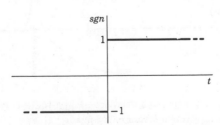

Fig. 16-12. The signum function sgn (t), defined by Eq. (16-66).

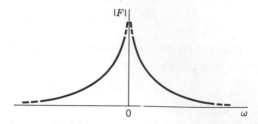

Fig. 16-13. The magnitude spectrum of the signum function.

presents difficulties because the integral of Eq. (16-50) diverges. As before, this function is expressed as a limiting case of other functions and the Fourier transform computed in this manner. As before, we make use of Eq. (16-54) and let a approach zero. Then

$$
\begin{aligned}
\mathfrak{F}[\text{sgn}(t)] &= \lim_{a \to 0} \int_{-\infty}^{\infty} e^{-a|t|} \, \text{sgn}\,(t) e^{-j\omega t} \, dt \\
&= \lim_{a \to 0} \left[\int_{-\infty}^{0} - e^{(a-j\omega)t} \, dt + \int_{0}^{\infty} e^{-(a+j\omega)t} \, dt \right] \quad (16\text{-}67) \\
&= \lim_{a \to 0} \left[\frac{-1}{a - j\omega} + \frac{1}{a + j\omega} \right] = \frac{2}{j\omega}
\end{aligned}
$$

The magnitude spectrum for the signum function is shown in Fig. 16-13. It is a plot of $|F(j\omega)| = 2/\omega$ and so has infinite value at $\omega = 0$. The associated phase function is a constant, $\phi(j\omega) = -\pi/2$ for $\omega > 0$ and $+\pi/2$ for $\omega < 0$.

The signum function makes it possible to evaluate the Fourier transform of the unit step function, another function that does not satisfy Eq. (16-50). We observe that

$$
u(t) = \tfrac{1}{2}[1 + \text{sgn}\,(t)] \quad (16\text{-}68)
$$

Since we know the Fourier transform of each of the two parts of this equation, we can write

$$
\mathfrak{F}[u(t)] = \pi\delta(\omega) + \frac{1}{j\omega} \quad (16\text{-}69)
$$

The magnitude of $F(j\omega)$ for the unit step is shown in Fig. 16-14; it is the combination of the rectangular hyperbola and the impulse

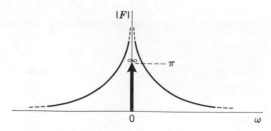

Fig. 16-14. The magnitude spectrum of a unit step function, $u(t)$, found from Eq. (16-69).

function of strength π to $\omega = 0$. This is indeed an unusual spectrum for such a simple time function, but the unusual nature is the price paid for finding the transforms of functions that do not satisfy the Dirichlet conditions.

Before other examples, let us consider a generalized periodic function for which we wish to compute the Fourier transform. Any periodic function will fail the condition of Eq. (16-50), of course, and so we anticipate that the result will contain impulse functions. We begin with the expression for the Fourier series for $f(t)$ expressed by Eq. (15-76) which is repeated here.

$$f(t) = \sum_{n=-\infty}^{\infty} \tilde{c}_n e^{jn\omega_0 t} \qquad (16\text{-}70)$$

The Fourier transform of this $f(t)$ is

$$\mathfrak{F}[f(t)] = F(j\omega) = \mathfrak{F}\left[\sum_{n=-\infty}^{\infty} \tilde{c}_n e^{jn\omega_0 t} \right]$$

$$= \sum_{n=-\infty}^{\infty} \tilde{c}_n \mathfrak{F}[e^{jn\omega_0 t}] \qquad (16\text{-}71)$$

Now from Eq. (16-62) we know the Fourier transform of the exponential function of this last equation, and so we may write for any periodic function

$$F(j\omega) = 2\pi \sum_{n=-\infty}^{\infty} \tilde{c}_n \delta(\omega - n\omega_0) \qquad (16\text{-}72)$$

This is an important result which is frequently used in applications of Fourier transform theory.

We will next apply Eq. (16-72) to a specific problem. In Chapter 8, we studied the impulse train which was periodic but started at $t = 0$. For this example, let the impulse train extend from $-\infty$ to ∞ and be of period T with unit strength for each impulse, as shown in Fig. 16-15. Then Eq. (8-72) becomes

$$\text{Ш}_T(t) = \sum_{n=-\infty}^{\infty} \delta(t - nT) \qquad (16\text{-}73)$$

The coefficient $\tilde{c}_n$ may be found by considering the periodic function over one period; it is

$$\tilde{c}_n = \frac{1}{T} \int_{-T/2}^{T/2} \delta(t) e^{jn\omega_0 t} \, dt = \frac{1}{T} \qquad (16\text{-}74)$$

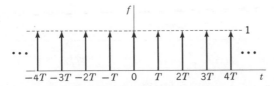

Fig. 16-15. An impulse train defined for all t, given by Eq. (16-73).

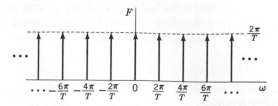

Fig. 16-16. The spectrum of an impulse train in the time domain is the impulse train in the frequency domain shown in the figure, given by Eq. (16-75).

because the integral has a value of 1, as we found in connection with the derivation of Eq. (16-51). Substituting this result into Eq. (16-72), we see that

$$\mathfrak{F}\text{III}_T(t) = \frac{2\pi}{T} \sum_{n=-\infty}^{\infty} \delta(\omega - n\omega_0) = \frac{2\pi \text{III}_{\omega_0}}{T}(\omega) \qquad (16\text{-}75)$$

where $\omega_0 = 2\pi/T$. This impulse train is shown in Fig. 16-16. This illustrates an interesting relationship between the time and frequency domains; in this case, an impulse train in the time domain has as its Fourier transform an impulse train in the frequency domain.

16-5. THE RELATIONSHIP OF FOURIER AND LAPLACE TRANSFORMS

That there is a relationship between the Fourier and Laplace transforms must have become abundantly clear by now. In Table 16-1, we have summarized the defining equations for the two transform pairs—from Eqs. (7-1) and (7-10) and Eqs. (16-20) and (16-21). The significant differences are few:

(1) The $j\omega$ in the Fourier transform has the same position as s in the Laplace transform.
(2) The limits of integration in the direct transform are different, one being one-sided and the other two-sided.
(3) The contours of integration in the inverse transform equations are different, one being along the imaginary axis and the other displaced by σ_1.

Table 16-1. COMPARISON OF LAPLACE AND FOURIER TRANSFORM PAIRS

	Laplace	*Fourier*
Direct transform	$F(s) = \int_0^{\infty} f(t)e^{-st}\,dt$	$F(j\omega) = \int_{-\infty}^{\infty} f(t)e^{-j\omega t}\,dt$
Inverse transform	$f(t) = \frac{1}{2\pi j}\int_{\sigma_1-j\infty}^{\sigma_1+j\infty} F(s)e^{st}\,ds$	$f(t) = \frac{1}{2\pi}\int_{-\infty}^{\infty} F(j\omega)e^{j\omega t}\,d\omega$

(4) The Laplace transform of $f(t)$ is identical with the Fourier transform of $f(t)$ multiplied by the convergence factor $e^{-\sigma t}$ if $f(t) = 0$ for $t < 0$; that is,

$$\mathcal{L}f(t) = \mathcal{F}[f(t)e^{-\sigma t}] \tag{16-76}$$

One advantage of this unification of the two subjects is that the tabulation of properties for the Laplace transform, such as that of

Table 16-2. SOME PROPERTIES OF FOURIER TRANSFORMS

Name:	*If $\mathcal{F}f(t) = F(j\omega)$, then·*
Definition	$F(j\omega) = \int_{-\infty}^{\infty} f(t)e^{-j\omega t}\,dt$
	$f(t) = \dfrac{1}{2\pi}\int_{-\infty}^{\infty} F(j\omega)e^{j\omega t}\,d\omega$
Superposition	$\mathcal{F}[af_1(t) + bf_2(t)] = aF_1(j\omega) + bF_2(j\omega)$
Simplification if:	
$\quad$(a) $f(t)$ is even	$F(j\omega) = 2\int_0^{\infty} f(t)\cos\omega t\,dt$
$\quad$(b) $f(t)$ is odd	$F(j\omega) = 2j\int_0^{\infty} f(t)\sin\omega t\,dt$
Negative t	$\mathcal{F}f(-t) = F^*(j\omega)$
Scaling:	
$\quad$(a) time	$\mathcal{F}f(at) = \dfrac{1}{|a|}F\left(\dfrac{j\omega}{a}\right)$
$\quad$(b) magnitude	$\mathcal{F}af(t) = aF(j\omega)$
Differentiation	$\mathcal{F}\left[\dfrac{d^n}{dt^n}f(t)\right] = (j\omega)^n F(j\omega)$
Integration	$\mathcal{F}\left[\int_{-\infty}^{t} f(x)\,dx\right] = \dfrac{1}{j\omega}F(j\omega) + \pi F(0)\delta(\omega)$
Time Shifting	$\mathcal{F}f(t - a) = F(j\omega)e^{-j\omega a}$
Modulation	$\mathcal{F}f(t)e^{j\omega_0 t} = F[j(\omega - \omega_0)]$
	$\mathcal{F}f(t)\cos\omega_0 t = \tfrac{1}{2}\{F[j(\omega - \omega_0)] + F[j(\omega + \omega_0)]\}$
	$\mathcal{F}f(t)\sin\omega_0 t = \tfrac{1}{2}j\{F[j(\omega - \omega_0)] - F[j(\omega + \omega_0)]\}$
Time Convolution	$\mathcal{F}^{-1}[F_1(j\omega)F_2(j\omega)] = \int_{-\infty}^{\infty} f_1(\tau)f_2(t - \tau)\,d\tau$
Frequency Convolution	$\mathcal{F}[f_1(t)f_2(t)] = \dfrac{1}{2\pi}\int_{-\infty}^{\infty} F_1(j\lambda)F_2[j(\omega - \lambda)]\,d\lambda$

Table 8-1, apply to Fourier transforms with $s = j\omega$. A separate summary is given in Table 16-2.

16-6. BANDWIDTH AND PULSE DURATION

A transmission system is said to be *ideal* or *distortionless* if the output waveform is the same as the input waveform, except that the magnitude is scaled by a constant F_o and the waveform is delayed a sec.

$$v_2(t) = F_o v_1(t - a) \tag{16-77}$$

If $F_o > 1$, the system is said to have gain; if $F_o < 1$, the system has loss. From Eq. (8-11), we have

$$V_2(s) = F_o V_1(s) e^{-sa} \tag{16-78}$$

so that in the sinusoidal steady state

$$\frac{\mathbf{V_2}}{\mathbf{V_1}} = G_{12}(j\omega) = F_o e^{-j\omega a} \tag{16-79}$$

Thus the ideal system has a $|G_{12}| = F_o$, which is a constant, and a phase which decreases linearly with ω. These ideal characteristics are shown in Fig. 16-17.

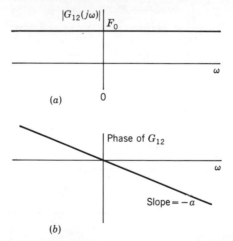

Fig. 16-17. The (*a*) magnitude and (*b*) phase characteristics for an ideal transmission system.

Now physical systems such as two-port networks or telephone lines are not ideal in the sense that they may be described by Eq. (16-79). Instead, the magnitude of $G_{12}(j\omega)$ decreases with ω, and the phase is not a linear function of ω. Actual magnitude and phase variations are more nearly like those shown in Fig. 16-18. For such systems, another ideal characteristic may be described. It is the *band-limited characteristic* or *ideal low-pass filter* having the characteristic shown in Fig. 16-19. The ideal low-pass filter does not exist in

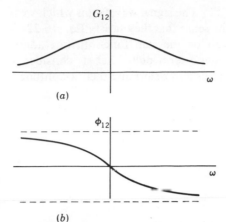

Fig. 16-18. The (a) magnitude and (b) phase characteristics for a typical network designed to approximate the characteristics of Fig. 16-17.

nature any more than that shown in Fig. 16-17, but it can be approximated by telephone transmission systems which consist of, say, 400 amplifiers in cascade. Pulses that pass through band-limited systems are distorted, as illustrated by Fig. 16-20. The signal distortion depends on both the bandwidth of the transmission system (or filter, if you prefer) and the pulse width of the signal, as we shall see.

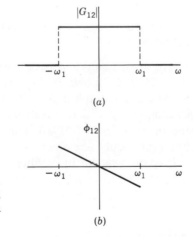

Fig. 16-19. The (a) magnitude and (b) phase characteristics for a band-limited system.

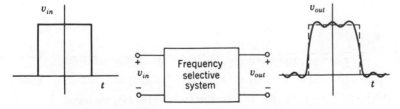

Fig. 16-20. Illustrating the distortion that results when a pulse is the input voltage for a frequency selective system and the output voltage is recorded.

The signal waveforms which we next consider are *time* limited in the sense that they satisfy Eq. (16-22). We thus exclude the ramp function or the step function, but admit waveforms described by the generic term pulse. Let us consider a number of different kinds of pulses and determine their amplitude spectra.

EXAMPLE 1

Let $f(t)$ be a pulse of magnitude V_0 and width a as shown in Fig. 16-21(a). From our study of Section 16-1 which involved this waveform, we may use Eqs. (16-5) and (16-16) with $n\omega_0 \longrightarrow \omega$ to write

$$F(j\omega) = V_0 a \frac{\sin(\omega a/2)}{\omega a/2} \qquad (16\text{-}80)$$

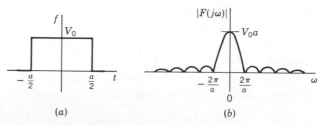

$$(a) \qquad\qquad (b)$$

Fig. 16-21. The (a) signal waveform and (b) continuous amplitude spectrum for Example 1.

The first zero of this equation occurs when $\omega a/2 = \pi$ or at the frequency $\omega = 2\pi/a$, as shown in Fig. 16-21(b). The most important part of the continuous amplitude spectrum, in the sense that it includes frequency components with relatively large values, is that part between $\omega = 0$ and the first zero of $F(j\omega)$. This is a qualitative statement, of course, but it turns out to be one that is useful in engineering practice. Then we may say that an ideal filter with cut-off frequency at ω_1, defined by Fig. 16-19(a), will pass most of the energy of the pulse if $\omega_1 = 2\pi/a$. Then we observe that

$$\omega_1 a = 2\pi \qquad (16\text{-}81)$$

or that the product of the bandwidth and the pulse width is a constant.

EXAMPLE 2

For our second example, consider an exponential pulse

$$f(t) = V_0 e^{-t/a}, \qquad t \geq 0 \qquad (16\text{-}82)$$

which has zero value for $t < 0$. From Eq. (16-21), we have

$$F(j\omega) = V_0 \int_0^\infty e^{-t/a} e^{-j\omega t}\, dt = \frac{V_0 a}{1 + j\omega a} \qquad (16\text{-}83)$$

In this case, neither $f(t)$ nor $F(j\omega)$ vanish, and so some new measure of "width" is required. The time constant a is a measure of the width of $f(t)$; a line tangent to $f(t)$ at $t = 0$ intersects the line $f = 0$ at $t = a$. A measure of the width of $F(j\omega)$ is provided from our studies of Chapter 13. The "half power" frequency, or the frequency at which $|F(j\omega)| = 0.707F(0)$, is found by equating real and imaginary parts of the denominator of Eq. (16-83), or $\omega = 1/a$. Equating this frequency to the cut-off frequency of the ideal filter, we have

$$\omega_1 a = 1 \tag{16-84}$$

which is of the same form as Eq. (16-81). See Fig. 16-22.

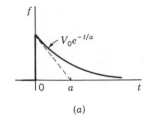

(a)

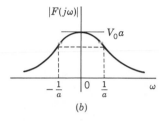

(b)

Fig. 16-22. The (a) signal waveform and (b) continuous amplitude spectrum for Example 2.

EXAMPLE 3

The triangular pulse shown in Fig. 16-23(a) has a spectrum found from

$$F(j\omega) = V_0 \int_{-a/2}^{a/2} \left(1 - \frac{2}{a}|t|\right) e^{-j\omega t}\, dt \tag{16-85}$$

Evaluating this integral gives

$$F(j\omega) = \frac{V_0 a}{2} \frac{\sin^2(\omega a/4)}{(\omega a/4)^2} \tag{16-86}$$

which is shown in Fig. 16-23(b). The first zero of $F(j\omega)$ occurs when $\omega = 4\pi/a$. Equating this to ω_1 for the low-pass filter, we have

$$\omega_1 a = 4\pi \tag{16-87}$$

which again is similar to Eq. (16-81).

From these three examples, as well as from others that will be found in the homework problems at the end of the chapter, we see that there is a basis for the conceptually useful statement[1]

$$\text{(Bandwidth required)} \times \text{(pulse duration)} = \text{constant} \tag{16-88}$$

In other words, *the bandwidth required for transmission is inversely proportional to the duration of the pulse being transmitted.*

It is interesting to test Eq. (16-88) with the impulse because it has zero width, and so represents a limiting case. From Eq. (16-53),

$$F(j\omega) = \int_{-\infty}^{\infty} \delta(t) e^{-j\omega t}\, dt = 1 \tag{16-89}$$

which is shown in Fig. 16-24. Thus $F(j\omega)$ is a constant for the impulse function, indicating that it contains all frequencies with equal ampli-

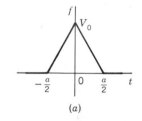

(a)

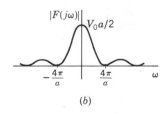

(b)

Fig. 16-23. The (a) signal waveform and (b) continuous amplitude spectrum for Example 3.

---

[1]For an advanced discussion of this subject in terms of rms duration, see Athanasios Papoulis, *The Fourier Integral and Its Applications* (McGraw-Hill, Inc., New York, 1962), pp. 62–64.

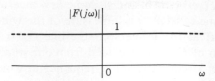

Fig. 16-24. The continuous amplitude spectrum for a unit impulse.

tudes and has infinite bandwidth. Equation (16-88) is satisfied in the sense that we are told that infinite bandwidth is required to transmit a pulse of zero width!

16-7. BANDWIDTH AND RISE TIME

In this section, we are interested in the network response when a step input is applied to a band-limited network (or ideal filter). Rather than study this problem directly, we employ the technique of Section 8-2 and first study the impulse response of the network from which the step response may be found by integration. In terms of this approach, we may formulate our problem as follows: When $\delta(t)$ is applied to an ideal transmission system having the characteristics of Fig. 16-17 the response is $\delta(t - a)$, where a is the time delay of the system. Now, when $\delta(t)$ is applied to the band-limited system, the response will not be an impulse (without infinite bandwidth). What is the impulse response, $v_2(t)$?

Fig. 16-25. A two-port network N assumed to have the response characteristics of Fig. 16-19.

In terms of Fig. 16-25 with the network N having the frequency characteristics of Fig. 16-19 so that ω_1 is the bandwidth (for positive ω), we have

$$V_2(j\omega) = G_{12}(j\omega)V_1(j\omega) \qquad (16\text{-}90)$$

and if $v_1(t) = \delta(t)$, then $V_1(j\omega) = 1$ as found in Eq. 16-89. Then the response is given in terms of the inverse Fourier transform equation

$$v_2(t) = \frac{1}{2\pi} \int_{-\infty}^{\infty} G_{12}(j\omega)e^{j\omega t}\, d\omega \qquad (16\text{-}91)$$

Here $G_{12}(j\omega)$ is described by Fig. 16-19 and is

$$G_{12}(j\omega) = \begin{cases} F_o e^{-j\omega a}, & -\omega_1 < \omega < \omega_1 \\ 0, & |\omega| > \omega_1 \end{cases} \qquad (16\text{-}92)$$

Since the magnitude of $G_{12}(j\omega)$ is zero for $|\omega| > \omega_1$, we replace the infinite limits of Eq. (16-91) with $\pm\omega_1$, and we have

$$v_2(t) = \frac{F_o}{2\pi} \int_{-\omega_1}^{\omega_1} e^{j\omega(t-a)}\, d\omega \qquad (16\text{-}93)$$

If we let $t - a = x$, then we obtain the result

$$v_2(t) = \frac{F_o \omega_1}{\pi} \frac{\sin \omega_1 x}{\omega_1 x} \qquad (16\text{-}94)$$

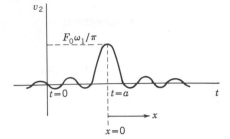

Fig. 16-26. The impulse response of the network of Fig. 16-25 having frequency characteristics of Fig. 16-19.

which has its maximum value when $x = 0$ or $t = a$. This value is

$$v_2(a) = \frac{F_o \omega_1}{\pi} \tag{16-95}$$

and the impulse response is shown in Fig. 16-26. We need not feel concern because there is a response even before the application of the impulse to the network. The low-pass amplitude characteristic we have assumed, that of Fig. 16-19, is not realizable by physical components, and so it is not surprising that we should obtain a result which predicts effect before cause. In real systems, *causality* insures that effect follow cause, and realizability may be determined from the elegant formulation of the Paley-Wiener criterion.[2]

To determine the step response, we integrate the impulse response of Eq. (16-94) to obtain

$$v_s(t) = \frac{F_o}{\pi} \int_{-\infty}^{t} \frac{\sin \omega_1 x}{\omega_1 x} d\omega_1 x = \frac{F_o}{\pi} \text{Si } \omega_1 x \Big|_{-\infty}^{t} \tag{16-96}$$

Here Si is the sine integral function[3] of the mathematical literature. One of the interesting properties of this function is that

$$\int_{-\infty}^{\infty} \frac{\sin x}{x} dx = 2 \int_{0}^{\infty} \frac{\sin x}{x} dx = \pi \tag{16-97}$$

which may be used to determine the final value of the step response:

$$v_s(\infty) = F_o \tag{16-98}$$

The relationship between $(\sin x)/x$ and Si x is shown in Fig. 16-27 for positive x. This is useful in determining the step function response from Eq. (16-96) which is shown in Fig. 16-28. Observe that the maximum slope on the plot of Si x and also of $v_s(t)$ occurs when $(\sin x)/x$ has its maximum value; the derivative of the Si x function is $(\sin x)/x$, of course.

The definition of *rise time* makes use of the maximum value of the slope of $v_s(t)$. This occurs when $t = a$, and hence is the numerical

[2] An excellent elementary treatment is given in F. F. Kuo, *Network Analysis and Synthesis*, 2nd ed., (John Wiley & Sons, Inc., New York, 1966), pp. 291–294.

[3] E. A. Guillemin, *The Mathematics of Circuit Analysis* (John Wiley & Sons, Inc., New York, 1949), pp. 491–496.

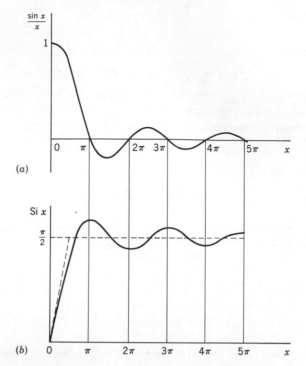

Fig. 16-27. (*a*) the (sin x)/x function in relationship to (*b*) the Si x or sine integral function.

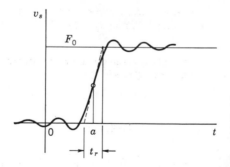

Fig. 16-28. The step function response corresponding to the impulse response of Fig. 16-24. In the figure, t_r is the rise time of Eq. (16-99).

value of the impulse response at $t = a$, $v_2(a)$. This has been determined and is given by Eq. (16-95). Now the straight line of slope $v_2(a)$ passing through $v_s(a)$ intersects the lines $v_s = 0$ and $v_s = F_o$ at the limits of the range of t which define the rise time. This simple geometrical construction which defines the rise time, t_r, is shown in Fig. 16-28. Using this definition and Eq. (16-95), the rise time is

$$t_r = \frac{F_o}{F_o \omega_1 / \pi} = \frac{\pi}{\omega_1} \qquad (16\text{-}99)$$

such that

$$\omega_1 t_r = \pi \qquad (16\text{-}100)$$

We can now make a statement similar to that of Eq. (16-88):

$$(\text{Bandwidth}) \times (\text{rise time}) = \text{constant} \qquad (16\text{-}101)$$

In other words, *the rise time of the response of a band-limited network to a step input is inversely proportional to the bandwidth.* We should note that there are other definitions of rise time, but for each, Eq. (16-101) applies.

FURTHER READING

CLOSE, CHARLES M., *The Analysis of Linear Circuits*, Harcourt Brace Jovanovich, Inc., New York, 1966. Chapter 9.

COOPER, GEORGE R., AND CLARE D. MCGILLEM, *Methods of Signal and System Analysis*, Holt, Rinehart & Winston, Inc. New York, 1967. Chapter 5.

GABEL, ROBERT A., AND RICHARD A. ROBERTS, *Signals and Linear Systems*, John Wiley & Sons, Inc., New York, 1973. See Chapter 5.

LATHI, B. P., *Signals, Systems, and Communication*, John Wiley & Sons, Inc., New York, 1965. Chapter 4.

MANNING, LAURENCE A., *Electrical Circuits*, McGraw-Hill Book Company, New York, 1965. Chapter 21.

RUSTON, HENRY, AND JOSEPH BORDOGNA, *Electric Networks: Functions, Filters, Analysis*, McGraw-Hill Book Company, New York, 1966. Chapter 8.

SCHWARZ, RALPH J., AND BERNARD FRIEDLAND, *Linear Systems*, McGraw-Hill Book Company, New York, 1965. Chapter 5.

DIGITAL COMPUTER EXERCISES

An appropriate exercise for the topics of this chapter will be the use of or the implementation of a computer program for the fast Fourier transform (FFT). References on this subject are given in Appendix E-5.2. Additional references in journals that are normally available to electrical engineers include the following:

GAE Subcommittee on Measurement Concepts, "What is the Fast Fourier Transform?" *Proceedings of the IEEE*, **55**, 1664–1674 (October, 1967).

G. Bergland, "A Guided Tour of the Fast Fourier Transform," *IEEE Spectrum*, **6**, 41–52 (July, 1969).

PROBLEMS

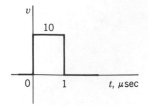

Fig. P16-1.

16-1. (a) The waveform shown in the figure is a rectangular pulse of 1 μsec duration and an amplitude of 10 V. Sketch the continuous amplitude spectrum for this waveform indicating the frequencies at which the envelope of the spectrum has zero value. (b) Repeat part (a) if the pulse duration is 0.01 μsec.

16-2. The waveform of the figure shows two pulses each of 1 μsec duration and spaced 1 μsec apart. Sketch the continuous amplitude spectrum for this waveform. Make use of the results of Prob. 16-1.

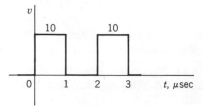

Fig. P16-2.

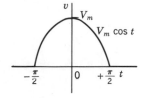

Fig. P16-3.

16-3. The figure shows a cosine pulse, $v = V_m \cos t$ which is zero for all time except $-\pi/2 \le t \le \pi/2$. (a) Determine the Fourier transform of this $v(t)$. (b) Sketch the continuous amplitude spectrum and the continuous phase spectrum for this $v(t)$.

16-4. The pulse shown in the figure is a cosine-squared function derived from the function of Prob. 16-3. Repeat Prob. 16-3 for this $v(t)$.

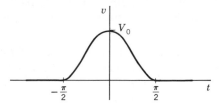

Fig. P16-4.

16-5. Repeat Prob. 16-3 for the pulse shown in the figure.

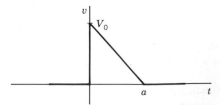

Fig. P16-5.

16-6. The figure shows a stepped pulse which is symmetrical about $t = 0$ and of duration T. For this waveform, determine the continuous amplitude spectrum and the continuous phase spectrum. Sketch both spectra.

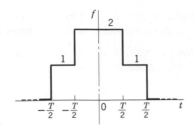

Fig. P16-6.

16-7. The waveform shown in the figure is made up of straight-line segments and has symmetry with respect to $t = 0$. Determine the continuous amplitude and phase spectra for this waveform, and sketch each spectrum.

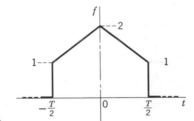

Fig. P16-7.

16-8. Determine the Fourier transform for the time functions (a) $f(t) = tu(t)$ and (b) $\cosh tu(t)$ if they exist.

16-9. Determine the Fourier transform for a signal waveform described by the equation

$$f(t) = [1 + m \cos \omega_1 t] \cos \omega_0 t, \qquad \text{for all } t$$

and sketch the amplitude and phase spectra.

16-10. A simple network consists of an inductor of $\frac{1}{2}$ H connected in series with a 1-Ω resistor and the combination driven by a voltage source, $v_1(t)$. The current in each of the elements is $i_2(t)$. If v_1 is the waveform shown in the figure for Prob. 16-6, determine the amplitude and phase spectrum for $i_2(t)$.

16-11. Repeat Prob. 16-10 for the waveform given in Prob. 16-7.

16-12. Repeat Prob. 16-10 for the waveform for $v_1(t)$ given in Prob. 16-3.

16-13. Verify the result shown for the Fourier transform for negative t given in Table 16-2.

16-14. Verify the result given for time scaling by the factor a in Table 16-2.

16-15. Verify the result given for the Fourier transform of the definite integral given in Table 16-2.

16-16. Verify the result given in Table 16-2 for the time shifting (advance or delay) of $f(t)$ to $f(t - a)$ with respect to the Fourier transform.

16-17. Consider the waveform given by the equation

$$v = V_0 e^{-\alpha t} \sin \omega_0 t, \qquad t \geq 0$$

and $v = 0$ for $t < 0$. Determine the continuous amplitude spectrum and the continuous phase spectrum for this damped oscillation.

16-18. Repeat Prob. 16-17 for the critically damped exponential pulse given by the equation

$$v(t) = V_0 \frac{t}{t_0} e^{-(t/t_0)}, \qquad t \geq 0$$

for which $v(t) = 0$ for $t < 0$.

16-19. Repeat Prob. 16-17 for a Gaussian pulse defined by the equation

$$v = V_0 e^{-(t/t_0)^2}, \qquad \text{for all } t$$

16-20. In the network of the figure, i_1 has the waveform shown in Prob. 16-6. For the element values given, determine the amplitude and phase spectra for the currents i_a and i_b.

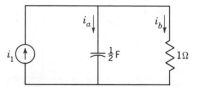

Fig. P16-20.

16-21. Repeat Prob. 16-20 if i_1 is the signal waveform described in Prob. 16-7.

16-22. In the network of the figure, v_1 has the waveform shown in Prob. 16-6. For the element values given, determine the magnitude and phase spectra for the voltages v_a and v_b.

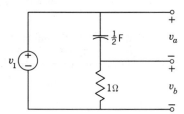

Fig. P16-22.

16-23. Repeat Prob. 16-21 if v_1 has the waveform which is shown in Prob. 16-7.

16-24. A system described by the voltage-ratio transfer function $G(s) = V_2(s)/V_1(s)$. If $G(s) = 3/(s + 2)$ and $v_1(t) = 5e^{-3t} u(t)$, make use of Fourier transforms in finding $V_2(j\omega)$ and sketch the magnitude and phase spectra.

16-25. Repeat Prob. 16-24 for the same $G(s)$ but for $v_1(t) = 3 \sin (2t + 45°)$.

16-26. Repeat Prob. 16-24 using the same $v_1(t)$ but the transfer function $G(s) = k_1 e^{-2s}$.

16-27. Repeat Prob. 16-24 using the $v_1(t)$ given in Prob. 16-25 with the transfer function given in Prob. 16-26 for $k_1 = 10$.

16-28. A system is described by the voltage-ratio transfer function $G(s) = V_2(s)/V_1(s)$. For this system, it is given that

$$G(s) = \frac{1}{1 + Ts}$$

If $v_1(t)$ is the waveform shown in Prob. 16-6, determine the amplitude and phase spectra for $v_2(t)$.

16-29. Repeat Prob. 16-28 if $v_1(t)$ is the waveform shown in Prob. 16-7.

16-30. The object of this problem is to show that the conclusion of Eq. (16-30), i.e., that the product of the bandwidth and the pulse duration is a constant, applies to a number of specific pulse waveforms. In each case, it is necessary to take some measure of the pulse duration, and also to select some measure of the bandwidth such as the frequency range to the first zero of the envelope of the continuous amplitude spectrum. For this problem, consider the cosine pulse given in Prob. 16-3. Show that the product of bandwidth and pulse duration is a constant and determine that constant.

16-31. Repeat Prob. 16-30 for the cosine-squared pulse described in Prob. 16-4.

16-32. Repeat Prob. 16-30 for the Gaussian pulse described in Prob. 16-19. In this case, the choice of time $t = t_0$ to measure pulse duration is suggested.

16-33. Repeat Prob. 16-30 for the triangular pulse of Prob. 16-5.

16-34. Repeat Prob. 16-30 for the exponential pulse described in Prob. 16-6.

16-35. The study of the relationship between bandwidth and rise time in Section 16-6 makes use of the idealized G_{12} characteristics of Fig. 16-19. In this problem, study this relationship for the transfer function

$$G_{12} = \frac{1}{1 + Ts}$$

for which the bandwidth is usually taken as being defined by the half-power frequency, $\omega = 1/T$. For this transfer function, show that the product of bandwidth and rise time is a constant and determine that constant.

16-36. This problem is similar to Prob. 16-35 except that here we consider the transfer function

$$G_{12} = \frac{\omega_0^2}{s^2 + 2\zeta\omega_0 s + \omega_0^2} \qquad \zeta \leq 1$$

Let the bandwidth be defined by the frequency $\omega = \omega_0$ and then normalize the frequency such that $\omega_0 = 1$. (a) Show that the product of rise time and bandwidth is a constant and determine the constant. (b) For $\zeta = \frac{1}{2}$, determine the overshoot that results when the input is a step function.

16-37. Consider the network of Prob. 15-36. For the following pulse waveforms for $v_1(t)$, determine the continuous amplitude spectrum and the continuous phase spectrum of $v_2(t)$. (a) Let $v_1(t)$ be described as shown in Fig. 16-22(a). (b) Let v_1 be given by the waveform shown in Fig. 16-23(a). (c) Let v_1 be given by the waveform shown in Prob. 16-3.

16-38. Repeat Prob. 16-37 for the three pulse waveforms indicated but using the network shown in Prob. 15-37.

Appendix A: Algebra of Complex Numbers (Phasors)

A-1. DEFINITIONS

The complex number $\mathbf{A} = a + jb$ is represented in the complex plane as shown in Fig. A-1(a). If we indicate the real part by Re and the imaginary part by Im, then

$$\text{Re } \mathbf{A} = a, \quad \text{Im } \mathbf{A} = b \qquad \text{(A-1)}$$

Observe that if $\mathbf{B} = c + jd$ and $\mathbf{B} = \mathbf{A}$, then $a = c$ and $b = d$. In polar or exponential form, the same complex number is written

$$\mathbf{A} = A e^{j\theta} \qquad \text{(A-2)}$$

and represented as in Fig. A-1(b). The magnitude and phase of A are written as

$$|\mathbf{A}| = A, \quad \arg \mathbf{A} = \theta \qquad \text{(A-3)}$$

Equations useful for converting from one form to the other are derived starting from Euler's identity,

$$e^{\pm j\theta} = \cos\theta \pm j\sin\theta \qquad \text{(A-4)}$$

Then Eq. (A-2) becomes

$$\mathbf{A} = A\cos\theta + jA\sin\theta \qquad \text{(A-5)}$$

Comparing this equation with Eq. (A-1), we have

$$a = A\cos\theta, \quad b = A\sin\theta \qquad \text{(A-6)}$$

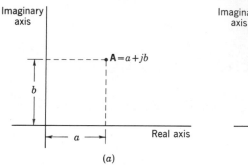 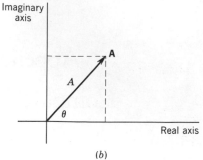

$$(a) \qquad\qquad\qquad (b)$$

Fig. A-1.

Squaring the two equations and adding, we find that

$$A^2 = a^2 + b^2 \tag{A-7}$$

Dividing b by a, we have

$$\theta = \tan^{-1}\left(\frac{b}{a}\right) \tag{A-8}$$

A-2. ADDITION AND SUBTRACTION

If $\mathbf{A} = a + jb$ and $\mathbf{B} = c + jd$, then

$$\mathbf{A} + \mathbf{B} = \mathbf{B} + \mathbf{A} = \mathbf{C} \tag{A-9}$$

and

$$\mathbf{C} = (a + c) + j(b + d) \tag{A-10}$$

In other words, the sum of two complex numbers is another complex number which is found by summing real parts and imaginary parts. The addition and subtraction operations are shown in the complex plane in Fig. A-2.

EXAMPLE 1

If $\mathbf{A} = 1 + j2$ and $\mathbf{B} = 1 - j1$, then

$$\mathbf{A} + \mathbf{B} = \mathbf{C} = 2 + j1 \tag{A-11}$$

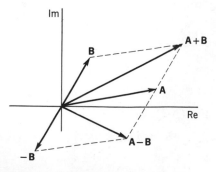

Fig. A-2.

and

$$A - B = D = 0 + j3 \qquad \text{(A-12)}$$

A-3. MULTIPLICATION

Using the complex numbers **A** and **B** of the last section, their product is

$$AB = BA = (a + jb)(c + jd) = (ac - bd) + j(ad + bc) \qquad \text{(A-13)}$$

showing that the product of two complex numbers is another complex number. In the polar or exponential form, we have

$$A = Ae^{j\theta_a}, \qquad B = Be^{j\theta_b} \qquad \text{(A-14)}$$

and

$$C = AB = ABe^{j(\theta_a + \theta_b)} \qquad \text{(A-15)}$$

so that

$$|C| = AB \quad \text{and} \quad \arg C = \theta_a + \theta_b \qquad \text{(A-16)}$$

The product of two complex numbers is shown in the complex plane in Fig. A-3. Observe that the multiplication of complex numbers is more easily accomplished in the polar or exponential form than in the complex form.

We are frequently required to multiply a complex number **A** by its conjugate **A***. By definition, if $A = a + jb$, then $A^* = a - jb$. Observe that

$$AA^* = (a + jb)(a - jb) = a^2 + b^2 \qquad \text{(A-17)}$$

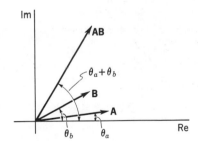

Fig. A-3.

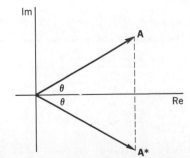

Fig. A-4.

showing that the product is a real number. In exponential form,

$$\mathbf{A} = Ae^{j\theta}, \quad \text{then} \quad A^* = Ae^{-j\theta} \tag{A-18}$$

as shown in Fig. A-4. Then

$$\mathbf{AA}^* = A^2 = a^2 + b^2 \tag{A-19}$$

in agreement with Eq. (A-7).

A-4. DIVISION

If $\mathbf{CB} = \mathbf{A}$, then for complex numbers

$$\mathbf{C} = \frac{\mathbf{A}}{\mathbf{B}} = \frac{\mathbf{A}}{\mathbf{B}} \frac{\mathbf{B}^*}{\mathbf{B}^*} = \frac{\mathbf{AB}^*}{|\mathbf{B}|^2} \tag{A-20}$$

in which multiplication by $\mathbf{B}^*/\mathbf{B}^*$ is known as *rationalization*. Then

$$\frac{\mathbf{A}}{\mathbf{B}} = \frac{(a + jb)(c - jd)}{c^2 + d^2} = \frac{ac + bd}{c^2 + d^2} + j\frac{bc - ad}{c^2 + d^2} \tag{A-21}$$

In the polar or exponential form

$$\frac{\mathbf{A}}{\mathbf{B}} = \frac{Ae^{j\theta_a}}{Be^{j\theta_b}} = \frac{A}{B} e^{j(\theta_a - \theta_b)} \tag{A-22}$$

so that

$$\left|\frac{\mathbf{A}}{\mathbf{B}}\right| = \frac{A}{B} \quad \text{and} \quad \arg \frac{\mathbf{A}}{\mathbf{B}} = \theta_a - \theta_b \tag{A-23}$$

These relationships are illustrated in the complex plane in Fig. A-5.

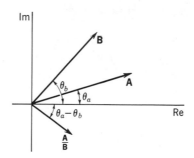

Fig. A-5.

EXAMPLE 2

Suppose that it is required to divide $1 + j\sqrt{3}$ by $j1$ and to express the result in the form $a + jb$. Then

$$\frac{1 + j\sqrt{3}}{j1} \frac{-j1}{-j1} = \sqrt{3} - j1 = a + jb \tag{A-24}$$

In exponential form

$$\frac{2e^{j60°}}{1e^{j90°}} = 2e^{-j30°} = 2\left(\frac{\sqrt{3}}{2} - j\frac{1}{2}\right) = \sqrt{3} - j1 \tag{A-25}$$

In obtaining this result, use was made of Eq. (A-5).

A-5. LOGARITHM OF A COMPLEX NUMBER

The logarithm of a complex number is found by expressing that number in exponential form and also observing that $\theta = \theta + k2\pi$ for any integer value of k with θ in radians. Then

$$\ln \mathbf{A} = \ln Ae^{j\theta} = \ln Ae^{j(\theta + k2\pi)} = \ln A + j(\theta + k2\pi) \qquad \text{(A-26)}$$

The value with $k = 0$ is known as the *principal value*.

EXAMPLE 3

We are required to find $\ln (3 + j4)$ in the form $a + jb$. Then

$$\ln (3 + j4) = \ln 5 + j(\tan^{-1}\tfrac{4}{3} + k2\pi), \qquad k = 0, 1, \ldots \qquad \text{(A-27)}$$

A-6. ROOTS AND POWERS OF COMPLEX NUMBERS

Roots and powers of complex numbers are found by making use of the law of exponents. Thus

$$\mathbf{A}^n = (Ae^{j\theta})^n = A^n e^{jn\theta} \qquad \text{(A-28)}$$

If we substitute $1/m$ for n and also add $k2\pi$ to θ, we have

$$A^{1/m} = +\sqrt[m]{A}\, e^{(j\theta + k2\pi)/m} \qquad \text{(A-29)}$$

where k is an integer, and $k = 0$ yields the *principal value*. This result is known as *De Moivre's theorem*.

EXAMPLE 4

It is required to determine the three roots of $+1 + j0$. Using Eq. (A-29), these are:

$$1e^{j0°}, \quad 1e^{j120°}, \quad \text{and} \quad 1e^{j240°} \qquad \text{(A-30)}$$

as shown in Fig. A-6.

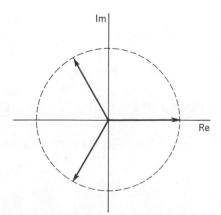

Fig. A-6.

PROBLEMS

A-1. Express the following complex numbers in exponential form:

$$2 - j2, \quad 3 - j4, \quad 1 + j\sqrt{3}, \quad -1 - j1$$

A-2. Express the following as complex numbers, $a + jb$:

$$2e^{j45°}, \quad 5e^{-j30°}, \quad 10e^{-j120°}, \quad 2e^{j105°}$$

A-3. Give the following complex numbers: $\mathbf{A} = 1 + j1$, $\mathbf{B} = 3 + j2$, $\mathbf{C} = -2 + j1$, $\mathbf{D} = -1 - j2$. Determine the following quantities showing each complex number and the parallelogram indicating addition or subtraction in the complex plane.

(a) $\mathbf{A} + \mathbf{B} + \mathbf{C}$ (c) $\mathbf{A} - \mathbf{B} + \mathbf{C} - \mathbf{D}$

(b) $2\mathbf{A} - \mathbf{B}$ (d) $2\mathbf{B} + \mathbf{C} - 3\mathbf{D}$

A-4. Using the complex numbers specified in Prob. A-3, determine the following quantities expressing each in the form of a complex number and also in exponential form. Show each quantity in the complex plane.

(a) $\mathbf{AB}$ (d) $\mathbf{AB^*}$

(b) $\mathbf{ABC}$ (e) $(\mathbf{AB})^*$

(c) $(\mathbf{A} + \mathbf{C})(\mathbf{B} + \mathbf{D})$ (f) $(\mathbf{A} - \mathbf{A^*})/\mathbf{BB^*}$

A-5. Using the complex numbers specified in Prob. A-3, determine the following quantities expressing each in the form of a complex number and also in exponential form. Show each quantity in the complex plane.

(a) $\mathbf{A/B}$ (d) $\mathbf{AB^*/CD^*}$

(b) $\mathbf{C/D}$ (e) $(\mathbf{A} + \mathbf{B})/(\mathbf{C} + \mathbf{D})$

(c) $\mathbf{A/D}$ (f) $\mathbf{A/C}(\mathbf{B} + \mathbf{D})$

A-6. If

$$\frac{1}{\cos\theta + j\sin\theta} = a + jb$$

determine a and b.

A-7. Determine the following quantities in the form of a complex number, $a + jb$.

(a) $\ln(j1)$ (c) $\ln(1 + j\omega)$

(b) $\ln(-1 - j1)$ (d) $\ln(1 - \omega^2 + j2\zeta\omega)$

A-8. Find all roots indicated in the following expressions and plot the roots in the complex plane.

(a) $(-1)^{1/4}$ (d) $(-j1)^{1/3}$

(b) $(1 + j0)^{1/4}$ (e) $(j32)^{1/5}$

(c) $(j1)^{1/3}$

A-9. Evaluate the following in either complex or exponential form:

$$(3 + j4)^{j6}, \quad (2 - j2)^{1+j1}$$

Appendix B: Matrix Algebra

B-1. DEFINITIONS

The rectangular array of numbers of functions

$$A = \begin{bmatrix} a_{11} & a_{12} & \cdots & a_{1n} \\ a_{21} & a_{22} & \cdots & a_{2n} \\ \cdot & \cdot & \cdot & \cdot \\ a_{m1} & a_{m2} & \cdots & a_{mn} \end{bmatrix} \tag{B-1}$$

is known as a *matrix* of order (m, n) or an $m \times n$ matrix. Double bars $\| \ \|$ and parentheses ( ) are also used in place of the square brackets. The numbers or functions a_{ij} are the *elements* of the matrix, with the first subscript indicating row position and the second subscript column position. We also use the notation $[a_{ij}]_{m \times n}$ to identify a matrix of order $m \times n$ whose elements are a_{ij}.

A number of matrices are given special names. A matrix of one column but any number of rows is known as a *column* matrix or a *vector*. A matrix of order (n, n) is a *square matrix of order n*. The main or principal diagonal of a square matrix consists of the elements a_{11}, $a_{22}, \ldots, a_{nn}$. A square matrix in which all elements except those on the principal diagonal are zero is known as a *diagonal matrix*. Further, if all elements of a diagonal matrix have the value 1, the matrix is known as a *unit or identity matrix, U*. If all elements of a matrix are zero, $a_{ij} = 0$, the matrix is known as a *zero* matrix 0. If $a_{ij} = a_{ji}$ in a matrix, it is known as a *symmetrical* matrix.

The three matrices

$$A = \begin{bmatrix} 1 & -1 & 2 \\ -1 & 0 & 3 \end{bmatrix}, \qquad B = \begin{bmatrix} 1 & 2 \\ -1 & 1 \\ 0 & 2 \end{bmatrix}, \qquad C = \begin{bmatrix} 1 \\ 2 \\ -1 \end{bmatrix} \qquad \text{(B-2)}$$

illustrate matrices of orders 2×3 for A, 3×2 for B and 3×1 for C. Matrix C is a column matrix or a vector. The matrix

$$U = \begin{bmatrix} 1 & 0 & 0 \\ 0 & 1 & 0 \\ 0 & 0 & 1 \end{bmatrix} \qquad \text{(B-3)}$$

is square, diagonal, unit, and also symmetrical.

In the next section, we shall be concerned with the *equality of matrices*. If two matrices are equal, then they have the same order and have equal corresponding elements.

B-2. ADDITION AND SUBTRACTION OF MATRICES

The *sum* of two matrices of the same order is found by adding the corresponding elements. Thus if the elements of A are a_{ij} and of B are b_{ij} and if $C = A + B$, then

$$c_{ij} = a_{ij} + b_{ij} \qquad \text{(B-4)}$$

Clearly, $A + B = B + A$ for matrices.

If a matrix A is multiplied by a constant α, then every element of A is multiplied by α:

$$\alpha A = [\alpha a_{ij}] \qquad \text{(B-5)}$$

In particular, if $\alpha = -1$, then

$$-A = [-a_{ij}] \qquad \text{(B-6)}$$

From this, we see that to subtract B from A, we multiply all elements of B by -1 and add. For example,

$$A - B = \begin{bmatrix} 1 & 2 & 1 \\ -1 & 0 & 1 \end{bmatrix} - \begin{bmatrix} 2 & 1 & 1 \\ 2 & 3 & -1 \end{bmatrix} = \begin{bmatrix} -1 & 1 & 0 \\ -3 & -3 & 2 \end{bmatrix} \qquad \text{(B-7)}$$

B-3. MULTIPLICATION OF MATRICES

The multiplication of matrices A and B is defined only if the number of columns of A is equal to the number of rows of B. Thus, if A is of order $m \times n$ and B is of order $n \times p$, then the product AB is a matrix C of order $m \times p$; thus

$$A_{m \times n} B_{n \times p} = C_{m \times p} \qquad \text{(B-8)}$$

The elements of C are found from the elements of A and B by multiplying the ith row of A and the jth column of B and summing these products to give c_{ij}. In equation form

$$c_{ij} = a_{i1}b_{1j} + a_{i2}b_{2j} + \ldots + a_{ip}b_{pj} = \sum_{k=1}^{p} a_{ik}b_{kj} \qquad \text{(B-9)}$$

Matrix multiplication is not commutative in general, i.e.,

$$AB \neq BA, \qquad \text{usually} \qquad \text{(B-10)}$$

even when BA is defined [meaning that the orders specified by Eq. (B-8) are satisfied].

EXAMPLE 1

The equations of Chapter 11,

$$z_{11}I_1 + z_{12}I_2 = V_1$$
$$z_{21}I_1 + z_{22}I_2 = V_2 \qquad \text{(B-11)}$$

are written in the following matrix form:

$$\begin{bmatrix} z_{11} & z_{12} \\ z_{21} & z_{22} \end{bmatrix} \begin{bmatrix} I_1 \\ I_2 \end{bmatrix} = \begin{bmatrix} V_1 \\ V_2 \end{bmatrix} \qquad \text{(B-12)}$$

as may be verified by applying Eq. (B-9).

EXAMPLE 2

Several matrix multiplications will illustrate the use of Eq. (B-9):

$$\begin{bmatrix} 1 & 1 \\ 1 & 1 \end{bmatrix} \begin{bmatrix} 1 & -2 \\ -1 & 2 \end{bmatrix} = \begin{bmatrix} 0 & 0 \\ 0 & 0 \end{bmatrix} \qquad \text{(B-13)}$$

$$\begin{bmatrix} 1 & 0 & 2 \\ 1 & -1 & 3 \end{bmatrix} \begin{bmatrix} 1 & 0 & 2 \\ -1 & 2 & -1 \\ 1 & 2 & 0 \end{bmatrix} = \begin{bmatrix} 3 & 4 & 2 \\ 5 & 4 & 3 \end{bmatrix} \qquad \text{(B-14)}$$

Finally, we note that

$$UA = AU = A \qquad \text{(B-15)}$$

if U and A are both square and of order n.

B-4. OTHER DEFINITIONS

A number of additional definitions will be used in discussing the solution of a matrix equation.

The *transpose* of a matrix A is A^T and is formed by interchanging the rows and columns of A. For example, if

$$A = \begin{bmatrix} 1 & 2 & -1 \\ 0 & -1 & 1 \end{bmatrix}, \quad \text{then} \quad A^T = \begin{bmatrix} 1 & 0 \\ 2 & -1 \\ -1 & 1 \end{bmatrix} \qquad \text{(B-16)}$$

The *determinant* of a square matrix has elements which are the elements of the matrix. In other words,

$$\det A = \det [a_{ij}] = |a_{ij}| \qquad \text{(B-17)}$$

For example,

$$\det \begin{bmatrix} 1 & -1 \\ 3 & 2 \end{bmatrix} = \begin{vmatrix} 1 & -1 \\ 3 & 2 \end{vmatrix} = 5 \qquad \text{(B-18)}$$

The cofactor A_{ij} of the element of a matrix a_{ij} is defined for a square matrix A. It has the value $(-1)^{i+j}$ times the determinant formed by deleting the ith row and the jth column in $\det A$.

The *adjoint matrix* of a square matrix A is formed by replacing each element a_{ij} by the cofactor A_{ij} and transposing. Thus

$$\text{adjoint of } A = [A_{ij}]^T \qquad \text{(B-19)}$$

Finally, we define the *inverse matrix* of A, A^{-1}, as the adjoint matrix divided by the determinant of A:

$$A^{-1} = \frac{\text{adjoint of } A}{\det A}, \qquad \det A \neq 0 \qquad \text{(B-20)}$$

EXAMPLE 3

Let it be required to determine A^{-1} given

$$A = \begin{bmatrix} 1 & 2 \\ -1 & 1 \end{bmatrix} \qquad \text{(B-21)}$$

Note that

$$\det A = 3, \qquad [A_{ij}] = \begin{bmatrix} 1 & 1 \\ -2 & 1 \end{bmatrix}, \qquad [A_{ij}]^T = \begin{bmatrix} 1 & -2 \\ 1 & 1 \end{bmatrix} \qquad \text{(B-22)}$$

Hence

$$A^{-1} = \frac{1}{3} \begin{bmatrix} 1 & -2 \\ 1 & 1 \end{bmatrix} = \begin{bmatrix} \frac{1}{3} & -\frac{2}{3} \\ \frac{1}{3} & \frac{1}{3} \end{bmatrix} \qquad \text{(B-23)}$$

B-5. MATRIX SOLUTION OF SIMULTANEOUS LINEAR EQUATIONS

In application to network analysis, we are concerned with equations formulated on the basis of one of the Kirchhoff laws. These equations have the form

$$ZI = V \qquad \text{(B-24)}$$

where Z is a square matrix of order n, I is a column matrix with elements $I_1, I_2, \ldots, I_n$, and V is a column matrix of elements V_1, $V_2, \ldots, V_n$. It was shown in Chapter 6 that I_j may be found using Cramer's rule as

$$I_j = \frac{1}{\det Z} \sum_{i=1}^{n} V_i Z_{ij} \qquad \text{(B-25)}$$

where Z_{ij} is the cofactor of z_{ij} in $\det Z$.

To obtain this same result using matrices, we multiply Eq. (B-24) by Z^{-1} and make use of the relationship $Z^{-1}Z = U$. Thus

$$Z^{-1}ZI = UI = I \tag{B-26}$$

and the solution of Eq. (B-24) is

$$I = Z^{-1}V \tag{B-27}$$

This may also be written in the form

$$I = \frac{\text{adjoint of } Z}{\det Z} V \tag{B-28}$$

which is seen to be in the same form as Eq. (B-25) but generalized for all I.

EXAMPLE 4

A network of resistors and batteries is described by the Kirchhoff voltage equations written in matrix form

$$\begin{bmatrix} 2 & -1 \\ -1 & 3 \end{bmatrix} \begin{bmatrix} I_1 \\ I_2 \end{bmatrix} = \begin{bmatrix} 2 \\ 1 \end{bmatrix} \tag{B-29}$$

As in Example 3, we find that

$$Z^{-1} = \frac{1}{5} \begin{bmatrix} 3 & 1 \\ 1 & 2 \end{bmatrix} \tag{B-30}$$

so that by Eq. (B-27),

$$\begin{bmatrix} I_1 \\ I_2 \end{bmatrix} = \frac{1}{5} \begin{bmatrix} 3 & 1 \\ 1 & 2 \end{bmatrix} \begin{bmatrix} 2 \\ 1 \end{bmatrix} = \begin{bmatrix} \frac{7}{5} \\ \frac{4}{5} \end{bmatrix} \tag{B-31}$$

In the analysis of networks, especially by the computer, other methods are employed rather than the evaluation of Z^{-1}. In Section 3-7, we studied the Gauss elimination method, and showed how to obtain the solution of a matrix equation by successive steps of triangularization and back substitution. The result is that the matrix equation

$$ZI = V \tag{B-32}$$

is reduced to the form

$$\begin{bmatrix} c_{11} & & & & \\ & c_{22} & & \mathbf{0} & \\ & & c_{33} & & \\ & & & \ddots & \\ & \mathbf{0} & & & c_{nn} \end{bmatrix} \begin{bmatrix} I_1 \\ I_2 \\ I_3 \\ \vdots \\ I_n \end{bmatrix} = \begin{bmatrix} k_1 V_1 \\ k_2 V_2 \\ k_3 V_3 \\ \vdots \\ k_n V_n \end{bmatrix} \tag{B-33}$$

where the square matrix has non-zero terms only on the principal diagonal. The solution is, directly,

$$I_1 = \frac{k_1}{c_{11}} V_1 \tag{B-34}$$

or, in general,

$$I_i = \frac{k_1}{c_{11}} V_i, \qquad i = 1, \ldots, n \tag{B-35}$$

A basic operation in the computer solution of matrix equations like Eq. (B-32) is *LU factorization* where L is a square matrix with non-zero elements on and below the principal diagonal, and U is a square matrix with non-zero elements on and above the principal diagonal, each matrix being in $n \times n$, and

$$LU = Z \tag{B-36}$$

Substituting this equation into Eq. (B-32) we have

$$LUI = V \tag{B-37}$$

If we identify $UI = y$, then we may solve

$$Ly = V \tag{B-38}$$

by forward Gauss elimination for y. Once y is known, we solve

$$UI = y \tag{B-39}$$

for I by the same straightforward Gauss elimination procedure. The advantage of this method is that if V changes in Eq. (B-32), then Z does not have to be refactored.

Finally, the computation of L and U is straightforward. If we let the elements of Z be z_{ij}, then

$$\begin{bmatrix} z_{11} & z_{12} & z_{13} \\ z_{21} & z_{22} & z_{23} \\ z_{31} & z_{32} & z_{33} \end{bmatrix} = \begin{bmatrix} l_{11} & 0 & 0 \\ l_{21} & l_{22} & 0 \\ l_{31} & l_{32} & l_{33} \end{bmatrix} \begin{bmatrix} 1 & u_{12} & u_{13} \\ 0 & 1 & u_{23} \\ 0 & 0 & 1 \end{bmatrix} \tag{B-40}$$

$$= \begin{bmatrix} l_{11} & l_{11}u_{12} & l_{11}u_{13} \\ l_{21} & l_{21}u_{12} + l_{22} & l_{21}u_{13} + l_{22}u_{23} \\ l_{31} & l_{31}u_{12} + l_{32} & l_{31}u_{13} + l_{32}u_{23} + l_{33} \end{bmatrix} \tag{B-41}$$

The solution of these equations for the l's and u's is routine.

For further reading on this subject, including a consideration of special techniques that may be employed when Z is *sparse*, see Donald A. Calahan, *Computer-Aided Network Design*, McGraw-Hill Book Co., New York, 1972, or George Forsythe and Cleve B. Moler, *Computer Solution of Linear Algebraic Systems*, Prentice-Hall, Inc., Englewood Cliffs, N.J., 1967.

PROBLEMS

B-1. Given the three matrices

$$A = \begin{bmatrix} 1 & -2 & 0 \\ 2 & 1 & 1 \end{bmatrix}, \qquad B = \begin{bmatrix} 2 & -1 & 0 \\ 1 & 1 & 1 \\ -1 & 0 & -1 \end{bmatrix}, \qquad C = \begin{bmatrix} 1 & 0 \\ 1 & 1 \\ 0 & -1 \end{bmatrix}$$

Determine (a) AB, (b) BC, and (c) ABC.

B-2. For B of Prob. B-1, find det B, B^T and B^{-1}.

B-3. For the matrix

$$D = \begin{bmatrix} 3 & -1 & -1 \\ -1 & 3 & -1 \\ -1 & -1 & 3 \end{bmatrix}$$

find det D, D^T, and D^{-1}.

B-4. For the matrix

$$E = \begin{bmatrix} 6 & -3 & -2 \\ -3 & 7 & -2 \\ -2 & -2 & 5 \end{bmatrix}$$

find det E, E^T, and E^{-1}.

B-5. Show that (a) $(A^T)^T = A$, and (b) $(AB)^T = B^T A^T$.

B-6. Using matrices, solve the following simultaneous algebraic equations and check your solutions.

(a) $\begin{bmatrix} 3 & 2 \\ 2 & 5 \end{bmatrix} \begin{bmatrix} I_1 \\ I_2 \end{bmatrix} = \begin{bmatrix} 2 \\ 1 \end{bmatrix}$

(b) $\begin{bmatrix} 3 & -2 \\ -2 & 3 \end{bmatrix} \begin{bmatrix} I_1 \\ I_2 \end{bmatrix} = \begin{bmatrix} 3 \\ -1 \end{bmatrix}$

(c) $\begin{bmatrix} 3 & -1 & -1 \\ -1 & 3 & -1 \\ -1 & -1 & 3 \end{bmatrix} \begin{bmatrix} I_1 \\ I_2 \\ I_3 \end{bmatrix} = \begin{bmatrix} 5 \\ 4 \\ -2 \end{bmatrix}$

B-7. By carrying out the operations shown in Eqs. (B-40) and (B-41), show that

$$\begin{bmatrix} 4 & -2 & -1 \\ -2 & \frac{17}{2} & -2 \\ -1 & -2 & 11 \end{bmatrix} = \begin{bmatrix} 0 & 0 & 0 \\ -2 & \frac{15}{2} & 0 \\ -1 & -\frac{5}{2} & \frac{119}{12} \end{bmatrix} \begin{bmatrix} 1 & -\frac{1}{2} & -\frac{1}{4} \\ 0 & 1 & -\frac{1}{3} \\ 0 & 0 & 1 \end{bmatrix}$$

by LU factorization.

Appendix C: Scaling

C-1. AN EXAMPLE OF SCALING

We will introduce the subject of scaling by considering the two networks given in (a) and (b) in Fig. C-1. For the parallel RLC network, the magnitude of the impedance is

$$|Z| = \frac{1}{[(1/R)^2 + (\omega C - 1/\omega L)^2]^{1/2}} \tag{C-1}$$

The variation of impedance magnitude with ω as determined by this equation is shown in (c) for the network of (a) and in (d) for the network of (b). We observe that these two plots differ in only two respects: The frequency scale of (d) is magnified by a factor of 10^3 and at corresponding frequencies (related by 10^3) the magnitude variation of $|Z_2|$ is 600 times that in (c) for $|Z_1|$. With these two factors taken into account, the plots have the same shape.

From the example, we can see that such scaling will be important in analysis. Networks having element values like those given in Fig. C-1(b) may be scaled to have values like that of (a) before analysis is undertaken. In the opposite direction, a network designed with element values such as those of (a) may be scaled to values such as those of (b) to meet any specifications as to range of frequencies and general level of impedance.

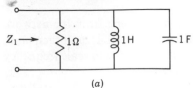
(a)

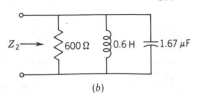
(b)

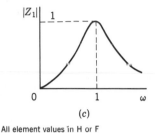
(c)

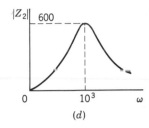
(d)

All element values in H or F

Fig. C-1.

C-2. FREQUENCY AND MAGNITUDE SCALING

Consider the problem of changing the scale of the frequency and at the same time changing the element values in a network such that the magnitude of the impedance remains constant. Let the frequency ω be scaled to the new value ω/ω_0, where ω_0 is a constant, in such a way that $Z(j\omega)$ will have the same magnitude as $Z(j\omega/\omega_0)$. From the equations

$$|Z_L| = \omega L = \frac{\omega}{\omega_0}\omega_0 L \qquad (C-2)$$

and

$$|Z_C| = \frac{1}{\omega C} = \frac{1}{(\omega/\omega_0)\omega_0 C} \qquad (C-3)$$

we see that if ω becomes ω/ω_0, then it is necessary that both L and C be multiplied by ω_0 in order that the impedance magnitude remain constant. Only L and C need be considered since $Z_R = R$ is frequency-invariant.

As a separate problem, consider the changes that must be made in element values in order that the magnitude of the impedance be changed by the same amount for every value of ω. From the impedance expressions, $Z_R = R$, $Z_L = j\omega L$, and $Z_C = 1/j\omega C$, we see that if Z is to be multiplied by a constant K, then it is necessary that R and L be multiplied by K and that C be divided by K.

By using both sets of rules just discussed, we may scale both magnitude and frequency. To illustrate the use of the rules, *up scaling* the magnitude and frequency as was done in going from the network of (a) to that of (b) in Fig. C-1 will be considered first, and then *down scaling* both magnitude and frequency. It is possible to scale frequency one direction and magnitude the other, of course.

Suppose that we wish to scale frequency so that ω_1 is scaled to ω_1/ω_0 for $\omega_0 > 1$, and at the same time the magnitude is to be scaled from a nominal value Z_0 (say the impedance at some frequency) to Z_0/K_1 for $K_1 > 1$. Then the element values for the *down-scaled network* may be computed from the equations

$$R_{ds} = \frac{R}{K_1} \tag{C-4}$$

$$L_{ds} = \frac{\omega_0}{K_1} L \tag{C-5}$$

and

$$C_{ds} = \omega_0 K_1 C \tag{C-6}$$

where R, L, and C are the element values of the network to be scaled.

Suppose that we now wish to scale frequency such that ω_2 is scaled to $\omega_0 \omega_2$, $\omega_0 > 1$, and at the same time the magnitude is to be increased from the nominal value (at some frequency) Z_0 to $K_2 Z_0$, $K_2 > 1$. Then the element values for the *up-scaled network* may be computed from the equations

$$R_{su} = K_2 R \tag{C-7}$$

$$L_{su} = \frac{K_2}{\omega_0} L \tag{C-8}$$

and

$$C_{su} = \frac{1}{\omega_0 K_2} C \tag{C-9}$$

to be scaled. Clearly one of the two sets of equations will suffice if K and ω_0 are allowed to have values less than 1.

The magnitude scaling procedure we have described applies to either driving-point or transfer impedances (Z or Z_{12}) and *inversely* to driving-point and transfer admittances (Y or Y_{12}). It is important to know that the magnitude scaling of impedance does *not* affect the magnitude of G_{12} or α_{12} or quantities derived from the voltage ratio or current ratio such as insertion loss.[1] This statement will be illustrated by an example.

[1]M. E. Van Valkenburg, *Introduction to Modern Network Synthesis* (John Wiley & Sons, Inc., New York, 1960), p. 53.

EXAMPLE 1

For our first example, we will derive the element values of the network of (b) in Fig. C-1 by up scaling the network of (a). Using Eqs. (C-7) to (C-9), with $\omega_0 = 10^3$ and $K_2 = 600$, we have

$$R_{su} = 600 \times 1 = 600 \; \Omega \qquad (C\text{-}10)$$

$$L_{su} = \left(\frac{600}{1000}\right)1 = 0.6 \; H \qquad (C\text{-}11)$$

and

$$C_{su} = \frac{1}{(600 \times 10^3)} = 1.67 \; \mu F \qquad (C\text{-}12)$$

EXAMPLE 2

The network and insertion loss characteristic shown in Fig. C-2 are taken from a handbook used in the design of filters.[2] A filter is to be designed in which the terminating resistors have the value of 600 Ω and the band of frequencies for which the insertion loss is 1 db or less is to extend from 0 to 12.0 kilocycles/sec. It is known that if this specification is satisfied then the 35 db minimum loss at higher frequencies is sufficient. This design may be completed using Eqs. (C-7)

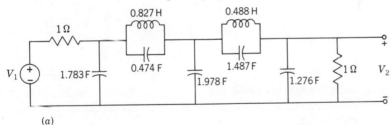

(a)

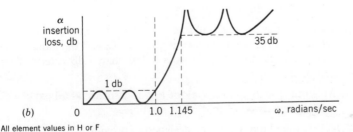

(b)

All element values in H or F

Fig. C-2.

[2]Deverl S. Humpherys, *The Analysis, Design and Synthesis of Electrical Filters* (Prentice-Hall, Inc., Englewood Cliffs, N.J., 1970).

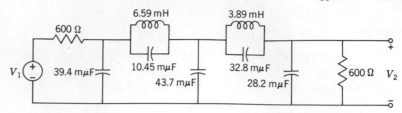

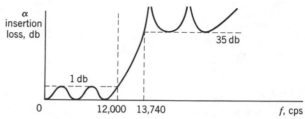

All element values in H or F

Fig. C-3.

to (C-9) with $K_2 = 600$ and $\omega_0 = 2\pi \times 12,000$. The scaled network and its insertion loss are shown in Fig. C-3. Observe that the magnitude of the insertion loss is not changed by the scaling operations.

PROBLEMS

C-1. Scale the network of (a) in Fig. C-1 such that the magnitude of Z_2 in (d) is 5000 Ω, at its maximum, and that this maximum occurs at 10^6 radians/sec.

C-2. Repeat Prob. C-1 such that the maximum magnitude of Z_2 is 0.001 Ω at $\omega = 0.1$ radian/sec.

C-3. Based on the network of Fig. C-2, design a network such that the terminating resistors have values of 5000 Ω and the band of frequencies for which the insertion loss is less than 1 db extends from 0 to 25,000 cps.

C-4. Magnitude scale the network of Fig. C-2 such that all passive element values are between 1 and 10, if possible.

C-5. It is required that a filter be designed to provide 35 db of insertion loss for all frequencies above 16.4 kilocycles/sec, and 1 db of insertion loss for most of the frequency range from 0 to 16.4 kilocycles/sec. The terminating resistors are to be of value 72 Ω. Design the network showing all element values.

Appendix D: Table of Inverse Laplace Transforms

The following table is given in a form devised by John A. Aseltine. In it, inverse Laplace transforms are given in terms of s-plane geometry for rational $F(s)$ with one, two, and three poles. In using the table, arrows are drawn from pole or zero to pole or zero as indicated. When the arrow so drawn is directed to the right, it is taken as positive, when directed to the left it is negative. When a line has arrows on both ends, it is always taken to be positive. [From John A. Aseltine, *Transform Method in Linear System Analysis*, copyright 1958 by McGraw-Hill, Inc. Used by permission.]

One Pole		
$F(s)$	s plane	$f(t)$
$\dfrac{1}{s+a}$		e^{-at}
Two Poles		
$\dfrac{1}{(s+a_1)(s+a_2)}$		$\dfrac{1}{A}(e^{-a_1 t} - e^{-a_2 t})$

Two Poles (*continued*)

$F(s)$	s plane	$f(t)$
$\dfrac{(s+b)}{(s+a_1)(s+a_2)}$		$\dfrac{1}{A}(B_1 e^{-a_1 t} - B_2 e^{-a_2 t})$
$\dfrac{1}{(s+a)^2}$		te^{-at}
$\dfrac{(s+b)}{(s+a)^2}$		$(Bt+1)e^{-at}$
$\dfrac{1}{(s+a)^2 + \beta^2}$		$\dfrac{1}{\beta} e^{-at} \sin \beta t$
$\dfrac{s+b}{(s+a)^2 + \beta^2}$		$\dfrac{B}{\beta} e^{-at} \sin(\beta t + \theta)$

Three Poles

$\dfrac{1}{(s+a_1)(s+a_2)(s+a_3)}$		$\dfrac{1}{A_{21}A_{31}} e^{-a_1 t} - \dfrac{1}{A_{21}A_{32}} e^{-a_2 t} + \dfrac{1}{A_{31}A_{32}} e^{-a_3 t}$
$\dfrac{s+b}{(s+a_1)(s+a_2)(s+a_3)}$		$\dfrac{B_1}{A_{21}A_{31}} e^{-a_1 t} - \dfrac{B_2}{A_{21}A_{32}} e^{-a_2 t} + \dfrac{B_3}{A_{31}A_{32}} e^{-a_3 t}$
$\dfrac{s^2}{(s+a_1)(s+a_2)(s+a_3)}$		$\dfrac{a_1^2}{A_{21}A_{31}} e^{-a_1 t} - \dfrac{a_2^2}{A_{21}A_{32}} e^{-a_2 t} + \dfrac{a_3^2}{A_{31}A_{32}} e^{-a_3 t}$

Three Poles (*Continued*)

$F(s)$	s plane	$f(t)$
$\dfrac{1}{(s+a_1)^2(s+a_2)}$		$\dfrac{1}{A^2}\left[e^{-a_2t}+(At-1)e^{-a_1t}\right]$
$\dfrac{s+b}{(s+a_1)^2(s+a_2)}$		$\dfrac{1}{A^2}\left[B_2e^{-a_2t}+(AB_1t-B_2)e^{-a_1t}\right]$
$\dfrac{s^2}{(s+a_1)^2(s+a_2)}$		$\dfrac{1}{A^2}\left[a_2^2e^{-a_2t}+(a_1^2At-a_1A-a_1a_2)e^{-a_1t}\right]$
$\dfrac{1}{(s+a_2)[(s+\alpha)^2+\beta^2]}$		$\dfrac{1}{A^2}e^{-a_2t}+\dfrac{1}{A\beta}e^{-\alpha t}\sin(\beta t-\theta)$
$\dfrac{s+b}{(s+a_1)[(s+\alpha)^2+\beta^2]}$		$\dfrac{B_1}{A^2}e^{-a_1t}+\dfrac{B}{\beta A}e^{-\alpha t}\sin(\beta t-\theta)$
$\dfrac{s^2}{(s+a_1)[(s+\alpha)^2+\beta^2]}$		$\dfrac{a_1^2}{A^2}e^{-a_1t}+\dfrac{B^2}{\beta A}e^{-\alpha t}\sin(\beta t-\theta_1-2\theta_2)$

Appendix E: Digital Computer Exercises (Elementary Numerical Methods)

This appendix is provided for the use of the references to descriptions of computer exercises suitable for collateral study with the subject matter of the various chapters. Specific reference to the appropriate exercises is made in a section found at the end of each chapter of the book.

Since the exercises suggested will be carried out at the computer center to which the reader has access, it is strongly recommended that contact be made with that center to determine a listing or catalog of subroutines that are available. Individual use of the computer will depend on the use of these subroutines, which may or may not be similar to those described in the references that are cited.

In the listings to follow, an abbreviation is used for the textbook references as given on pp. 542–543, and a page number indicating the location of the appropriate material.

E-1. NUMERICAL METHODS FOR FINDING ROOTS

E-1.1 Roots of Quadratic: MCC, 126; SEE, 59.

E-1.2 Newton-Raphson Method: GUP, 300; HU2, 867; KIN, 520; LEY, 139, 179; MCC, 44; SEE, 61; WNG, 216.

E-2. NUMERICAL INTEGRATION

E-2.1 Numerical Integration Methods: GUP, 19; HU1, 15; HU2, 173; KIN, 236; MCC, 142; LEY, 233; SEE, 73; WNG, 184.

540

E-2.2 Evaluation of an Indefinite Integral: GUP, 22; HU2, 185.

E-2.3 Fourier Analysis: GUP, 409; HU2, 794; MCC, 182; LEY, 252; SEE, 80.

E-2.4 Convolution and Superposition Integrals: GUP, 267; HU2, 285; KIN, 224.

E-3. MATRIX FORMULATION AND MATRIX OPERATIONS

E-3.1 Multiplication of Matrices: GUP, 114; MCC, 103; LEY, 458.

E-3.2 Matrix Formulation of Equations: HU2, 71; LEY, 473.

E-4. SOLVING SIMULTANEOUS LINEAR ALGEBRAIC EQUATIONS

E-4.1 Gauss Elimination Method: GUP, 132; HU2, 94; KIN, 272; LEY, 288; MCC, 154; STE, 282.

E-4.2 Sets of Simultaneous Equations: HU1, 100; HU2, 81; MCC, 116; SEE, 65.

E-4.3 Solution of State Equations: HU2, 81; KIN, Ch. 3; LEY, 768.

E-4.4 Simultaneous Equations Involving Impedance and Admittance: GUP, 425; HU2, 431, 444; WNG, 403, 419.

E-4.5 Implicit Integration: WNG, 467.

E-5. ALGEBRAIC OPERATIONS

E-5.1 Inverse Laplace Transforms (Determination of Residues) HU1, 118; HU2, 592; KIN, Ch. 9; LEY, 776; MCC, 196; STE, 120.

E-5.2 Fast Fourier Transform: KIN, 460; STE, 149.

E-6. SOLUTION OF ORDINARY DIFFERENTIAL EQUATIONS

E-6.1 Numerical Solution of First-Order Equations: GUP, 243; HU1, 63; HU2, 271; KIN, 115.

E-6.2 *RLC* Series Circuit Solution: MCC, 28, 53; SEE, 153.

E-6.3 Numerical Solution of Higher-Order Differential Equations (Runge-Kutta Method, etc.): GUP, 238; HU1, 79, 86; HU2, 364; KIN, 176; LEY, 749; SEE, 75.

E-7. SOLUTION OF NONLINEAR DIFFERENTIAL EQUATIONS

E-7.1 Solution of Nonlinear Differential Equations: HU1, 43, 70; WNG, 219.

E-7.2 Solution for Networks with Time-Varying Element: HU1, 61.

E-7.3 Solution for Distributed Networks: HU1, 167; WNG, 381.

E-8. NETWORK ANALYSIS

E-8.1 Solutions for Ladder Network: GUP, 141; LEY, 79; WNG, 132.

E-8.2 Solutions for Resistor Networks: HU1, 164; KIN, 269; STE, 286; WNG, 275, 286.

E-8.3 Solutions for Networks in the Sinusoidal Steady State: HU1, 131; HU2, 110; KIN, 351; SEE, 37; STE, 294.

E-8.4 General Network Analysis Programs: BOW, entire book; JEN, entire book.

E-9. COMPUTER DETERMINATION OF MAGNITUDE AND PHASE PLOTTING

E-9.1 Evaluation of Magnitude and Phase: GUP, 366, 371; HU1, 133; KIN, 404.

E-9.2 Graph of Solution: GUP, 195; HU1, 27; HU2, 863, 871; LEY, 108; MCC, 108.

E-9.3 Graph of Solution, Frequency Domain: HU2, 488; WNG, 369, 372, 511.

E-9.4 Graph of Solution, Time Domain: GUP, 195.

E-9.5 Special Plotting Techniques: (Bode Plots, Root-Locus Plots): HU1, 140, 149.

E-10. REFERENCES

1. BOW   BOWERS, JAMES C., AND STEPHEN R. SEDORE, *SCEPTRE: A Computer Program for Circuit and System Analysis*, Prentice-Hall, Inc., Englewood Cliffs, N.J., 1971, 455 pp.

2. CA1   CALAHAN, DONALD A., *Computer-Aided Network Design*, McGraw-Hill Book Company, New York, 1972, 350 pp.

3. CA2   CALAHAN, DONALD A., ALAN B. MACNEE, AND E. L. McMAHON, *Computer-Oriented Circuit Analysis*, Holt, Rinehart & Winston, Inc., New York, 1974. To be published.

4. DER   DERTOUZOS, MICHAEL L., MICHAEL ATHANS, RICHARD N. SPANN, AND SAMUEL J. MASON, *Systems, Networks and Computation: Basic Concepts*, McGraw-Hill Book Company, New York, 1972, 514 pp.

5. DIR   DIRECTOR, S. W., *Circuit Theory—The Computational Approach*, John Wiley & Sons, Inc., New York, 1974. To be published.

6. GUP   GUPTA, SOMESHWAR C., JON W. BAYLESS, AND BEHROUZ PEIKARI, *Circuit Analysis with Computer Applications to Problem Solving*, Intext Educational Publishers, Scranton, Pa., 1972, 546 pp.

7. HU1   HUELSMAN, LAWRENCE P., *Digital Computations in Basic Circuit Theory*, McGraw-Hill Book Company, New York, 1968, 203 pp.

8. HU2   HUELSMAN, LAWRENCE P., *Basic Circuit Theory with Digital Computations*, Prentice-Hall, Inc., Englewood Cliffs, N.J., 1972, 896 pp.

9. JEN   JENSEN, RANDALL W., AND MARK D. LIEBERMAN, *IBM Electronic Circuit Analysis Program: Techniques and Applications*, Prentice-Hall, Inc., Englewood Cliffs, N.J., 1968, 401 pp.

10. KIN   KINARIWALA, B.K., FRANKLIN F. KUO, AND NAI-KUAN TSAO, *Linear Circuits and Computation*, John Wiley & Sons, Inc., New York, 1973, 598 pp.

11. LEY   LEY, B. JAMES, *Computer Aided Analysis and Design for Electrical Engineers*, Holt, Rinehart & Winston, Inc., New York, 1970, 852 pp.

12. MCC   MCCRACKEN, DANIEL D., *FORTRAN With Engineering Applications*, John Wiley & Sons, Inc., New York, 1967, 237 pp.

13. RAM   RAMEY, ROBERT L., AND EDWARD J. WHITE, *Matrices and Computers in Electronic Circuit Analysis*, McGraw-Hill Book Co., New York, 1971, 390 pp.

14. SEE   SEELY, SAMUEL, NORMAN H. TARNOFF AND DAVID HOLSTEIN, *Digital Computers in Engineering*, Holt, Rinehart and Winston, Inc., New York, 1970, 329 pp.

15. STE   STEIGLITZ, KENNETH, *An Introduction to Discrete Systems*, John Wiley & Sons, Inc., New York, 1974, 318 pp.

16. WNG   WING, OMAR, *Circuit Theory with Computer Methods*, Holt, Rinehart & Winston, Inc., New York, 1972, 529 pp.

Appendix F: References

F-1 ELEMENTARY OR INTERMEDIATE TEXTBOOKS

ALLEY, CHARLES L., AND KENNETH W. ATWOOD, *Electronic Engineering*, 3rd ed., John Wiley & Sons, Inc., New York 1973, 864 pp.

ALLEY, CHARLES L., AND KENNETH W. ATWOOD, *Semiconductor Devices and Circuits*, John Wiley & Sons, Inc., New York, 1971, 490 pp.

ASELTINE, JOHN A., *Transform Method in Linear System Analysis*, McGraw-Hill, Inc., New York, 1958, 272 pp.

BALABANIAN, NORMAN, *Fundamentals of Circuit Theory*, Allyn and Bacon, Inc., Boston, 1961, 555 pp.

BALABANIAN, NORMAN, AND WILBUR R. LEPAGE, *Electrical Science: Book I: Resistive and Diode Networks*, McGraw-Hill Book Co., New York, 1970, 301 pp.

BALABANIAN, NORMAN, AND WILBUR R. LEPAGE, *Electrical Science: Book II: Dynamic Networks*, McGraw-Hill Book Co., New York, 1973, 640 pp.

CHAN, S. P., S. Y. CHAN, AND S. G. CHAN, *Analysis of Linear Networks and Systems—A Matrix-Oriented Approach with Computer Applications*, Addison-Wesley Publishing Company, Inc., Reading, Mass., 1972, 635 pp.

CHENG, DAVID K., *Analysis of Linear Systems*, Addison-Wesley Publishing Company, Inc., Reading, Mass., 1959, 431 pp.

CHIRLIAN, PAUL M., *Basic Network Theory*, McGraw-Hill Book Company, New York, 1969, 624 pp.

CHUA, LEON O., *Introduction to Nonlinear Network Theory*, McGraw-Hill Book Company, New York, 1969, 987 pp.

CLEMENT, PRESTON R. AND WALTER C. JOHNSON, *Electrical Engineering Science*, McGraw-Hill Book Company, New York, 1960, 588 pp.

CLOSE, CHARLES M., *The Analysis of Linear Circuits*, Harcourt, Brace & World, Inc., New York, 1966, 716 pp.

COOPER, G. R., AND C. D. McGILLEM, *Methods of Signal and System Analysis*, Holt, Rinehart & Winston, New York, 1965, 432 pp.

CRUZ, JOSE B., JR., AND M. E. VAN VALKENBURG, *Signals in Linear Circuits*, Houghton Mifflin Company, Boston, Mass., 1974, 596 pp.

DESOER, CHARLES A., AND ERNEST S. KUH, *Basic Circuit Theory*, McGraw-Hill Book Company, New York, 1969, 876 pp.

DIRECTOR, STEPHEN W., AND RONALD A. ROHRER, *An Introduction to System Theory*, McGraw-Hill Book Company, New York, 1972, 452 pp.

FRIEDLAND, B. J., OMAR WING, AND ROBERT ASH, *Principles of Linear Networks*, McGraw-Hill Book Company, New York, 1961, 270 pp.

GHAUSI, MOHAMMED S., *Electronic Circuits: Devices, Models, Functions, Analysis, and Design*, Van Nostrand Reinhold Company, New York, 1971, 731 pp.

GRAY, PAUL E., AND CAMPBELL L. SEARLE, *Electronic Principles: Physics, Models, and Circuits*, John Wiley & Sons, Inc., New York, 1969, 1016 pp.

GUILLEMIN, E. A., *Introductory Circuit Theory*, John Wiley & Sons, Inc., New York, 1953, 586 pp.

HAMMOND, S. B., AND D. K. GEHMLICH, *Electrical Engineering*, 2nd ed., McGraw-Hill Book Company, New York, 1971, 535 pp.

HAYT, WILLIAM H., JR., AND JACK E. KEMMERLY, *Engineering Circuit Analysis*, 2nd ed., McGraw-Hill Book Company, New York, 647 pp.

HAYT, W. J., JR., AND G. W. HUGHES, *Introduction to Electrical Engineering*, McGraw-Hill Book Company, New York, 1968, 443 pp.

HUANG, THOMAS S., AND RONALD R. PARKER, *Network Theory: An Introductory Course*, Addison-Wesley Publishing Company, Inc., Reading, Mass., 1971, 653 pp.

HUELSMAN, L. P., *Circuits, Matrices, and Linear Vector Spaces*, McGraw-Hill Book Company, New York, 1963, 281 pp.

HUGGINS, W. H., AND DORIS R. ENTWISLE, *Introductory Systems and Design*, Ginn/Blaisdell Publishing Company, Waltham, Mass., 1968, 683 pp.

KARNI, SHLOMO, *Network Theory: Analysis and Synthesis*, Allyn and Bacon, Inc., Boston, 1966, 483 pp.

KARNI, SHLOMO, *Intermediate Network Analysis*, Allyn and Bacon, Inc., Boston, 1971, 377 pp.

KIM, WAN H., AND HENRY E. MEADOWS, JR., *Modern Network Analysis*, John Wiley & Sons, New York, 1971, 431 pp.

KUO, FRANKLIN F., *Network Analysis and Synthesis*, 2nd ed., John Wiley and Sons, Inc., New York, 1966, 515 pp.

LATHI, B. P., *Signals, Systems, and Communication*, John Wiley & Sons, Inc., New York, 1965, 607 pp.

LEON, BENJAMIN J., *Lumped Systems*, Holt, Rinehart & Winston, Inc., New York, 1968, 223 pp.

LEON, BENJAMIN J., AND PAUL A. WINTZ, *Basic Linear Networks for Electrical and Electronics Engineers*, Holt, Rinehart & Winston, Inc., New York, 1970, 479 pp.

MANNING, LAURENCE A., *Electrical Circuits*, McGraw-Hill Book Company, New York, 1965, 567 pp.

MASON, S. J., AND HENRY J. ZIMMERMANN, *Electronic Circuits, Signals and Systems*, John Wiley & Sons, Inc., New York, 1960.

MERRIAM, C. W., III, *Analysis of Lumped Electrical Systems*, John Wiley & Sons, Inc., New York, 1969, 580 pp.

MILLIMAN, JACOB, AND CHRISTOS C. HALKIAS, *Integrated Electronics: Analog and Digital Circuits and Systems*, McGraw-Hill Book Company, New York, 1972, 911 pp.

MURDOCH, JOSEPH B., *Network Theory*, McGraw-Hill Book Company, New York, 1970, 525 pp.

PEDERSON, D. O., J. R. WHINNERY, AND J. J. STUDER, *Introduction to Electronic Systems, Circuits and Devices*, McGraw-Hill Book Company, New York, 1966, 457 pp.

ROHRER, RONALD A., *Circuit Theory: An Introduction to the State Variable Approach*, McGraw-Hill Book Company, New York, 1970, 314 pp.

RYDER, JOHN D., *Introduction to Circuit Analysis,* Prentice-Hall, Inc., Englewood Cliffs, N.J., 1973, 369 pp.

SCOTT, RONALD E., *Linear Circuits*, Addison-Wesley Publishing Company, Reading, Mass., 1960, 2 Vols., 922 pp.

SMITH, RALPH J., *Circuits, Devices and Systems*, 2nd ed., John Wiley & Sons, Inc., New York, 1971, 743 pp.

SMITH, RALPH J., *Electronics: Circuits and Devices*, John Wiley & Sons, Inc., New York, 1973, 459 pp.

STRUM, ROBERT D., AND JOHN R. WARD, *Laplace Transform Solution of Differential Equations* (a programmed text), Prentice-Hall, Inc., Englewood Cliffs, N.J., 1968, 197 pp.

TRUXAL, JOHN G., *Introductory System Engineering,* McGraw-Hill Book Co., New York, 1972, 596 pp.

TUTTLE, DAVID F., JR., *Electric Networks, Analysis and Synthesis*, McGraw-Hill Book Company, New York, 1965, 327 pp.

WARD, JOHN R., AND ROBERT D. STRUM, *State Variable Analysis* (a programmed text), Prentice-Hall, Inc., Englewood Cliffs, N.J., 1970, 270 pp.

WOLF, HELLMUTH, *Linear Systems and Networks*, Springer-Verlag, New York, 1971, 268 pp.

ZIMMERMAN, HENRY J., AND S. J. MASON, *Electronic Circuit Theory: Devices, Models and Circuits*, John Wiley & Sons, Inc., New York, 1959, 564 pp.

F-2 ADVANCED TEXTBOOKS ON CIRCUIT THEORY

BALABANIAN, NORMAN, THEODORE A. BICKART, AND SUNDARAM SESHU, *Electrical Network Theory*, John Wiley & Sons, Inc., New York, 1969, 931 pp.

BODE, H. W., *Network Analysis and Feedback Amplifier Design*, Van Nostrand Reinhold, Company, New York, 1945. 551 pp.

CHAN, SIIU-PARK, *Introductory Topological Analysis of Electrical Networks*, Holt, Rinehart & Winston, Inc., 1969, 482 pp.

CHEN, W. K., *Applied Graph Theory*, North-Holland Publishing Company, Amsterdam, 1971, 484 pp.

CHUA, LEON O., *Introduction to Nonlinear Network Theory*, McGraw-Hill Book Company, New York, 1969, 987 pp.

FRANK, HOWARD, AND IVAN T. FRISCH, *Communication, Transmission, and Transportation Networks*, Addison-Wesley Publishing Co., Reading, Mass., 1971, 496 pp.

FRANKS, L. E., *Signal Theory*, Prentice-Hall, Inc., Englewood Cliffs, N.J. 1969, 317 pp.

GABEL, ROBERT A., AND RICHARD A. ROBERTS, *Signals and Linear Systems*, John Wiley & Sons, Inc., New York, 1973, 415 pp.

GARDNER, MURRAY F., AND J. L. BARNES, *Transients in Linear Systems*, John Wiley & Sons, Inc., New York, 1942, 382 pp.

GHAUSI, MOHAMMED S., AND JOHN J. KELLY, *Introduction to Distributed-Parameter Networks*, Holt, Rinehart & Winston, New York, 1968, 331 pp.

HAZONY, DOV, *Elements of Network Synthesis*, Reinhold Publishing Corp., New York, 1963, 352 pp.

HUELSMAN, LAWRENCE P., *Theory and Design of Active RC Circuits*, McGraw-Hill Book Company, New York, 1968, 297 pp.

HUMPHERYS, DEVERL S., *The Analysis, Design and Synthesis of Electrical Filters*, Prentice-Hall, Inc., Englewood Cliffs, N.J., 1970, 675 pp.

KALMAN, R. E., AND N. DeCLARIS, eds., *Aspects of Network and System Theory*, Holt, Rinehart & Winston, Inc., New York, 1971, 648 pp.

KIM, WAN HEE, AND ROBERT T. CHIEN, *Topological Analysis and Synthesis of Communication Networks*, Columbia University Press, New York, 1962, 310 pp.

KUH, ERNEST S., AND R. A. ROHRER, *Theory of Linear Active Networks*, Holden-Day, Inc., San Francisco, 1967, 650 pp.

MAYEDA, WATARU, *Graph Theory*, John Wiley & Sons, Inc., New York, 1972, 588 pp.

MITRA, SANJIT K., *Analysis and Synthesis of Linear Active Networks*, John Wiley & Sons, Inc., New York, 1969, 565 pp.

NEWCOMB, ROBERT W., *Linear Multiport Synthesis*, McGraw-Hill Book Company, New York, 1966, 397 pp.

PENFIELD, PAUL JR., ROBERT SPENCE AND SIMON DUINKER, *Tellegen's Theorem and Electrical Networks*, The M.I.T. Press, Cambridge, Mass., 1970, 141 pp.

REZA, FAZLOLLAH, *Linear Spaces in Engineering*, Ginn/Xerox, Waltham, Mass., 1971, 416 pp.

SAEKS, RICHARD, *Generalized Networks*, Holt, Rinehart & Winston, New York, 1972, 433 pp.

SESHU, SUNDARAM, AND N. BALABANIAN, *Linear Network Analysis*, John Wiley & Sons, Inc., New York, 1959. 571 pp.

SESHU, SUNDARAM, AND MYRIL B. REED, *Linear Graphs and Electrical Networks*, Addison-Wesley Publishing Co., Reading, Mass., 1961, 315 pp.

SKWIRZYNSKI, J. K., *Design Theory and Data for Electrical Filters*, Van Nostrand Reinhold Co., New York, 1965, 701 pp.

SPENCE, ROBERT, *Linear Active Networks*, John Wiley & Sons, Inc., New York, 1970, 359 pp.

VAN VALKENBURG, M. E., *Introduction to Modern Network Synthesis*, John Wiley & Sons, Inc., New York, 1960, 498 pp.

VLACH, JIRI, *Computerized Approximation and Synthesis of Linear Networks*, John Wiley & Sons, Inc., New York, 1969, 477 pp.

WEINBERG, LOUIS, *Network Analysis and Synthesis*, McGraw-Hill Book Company, New York, 1962, 692 pp.

F-3 COMPUTER ORIENTED CIRCUIT THEORY TEXTBOOKS

BOWERS, JAMES C., AND STEPHEN R. SEDORE, *SCEPTRE: A Computer Program for Circuit and Systems Analysis*, Prentice-Hall, Inc., Englewood Cliffs, N.J., 1971, 455 pp.

CALAHAN, DONALD A., *Computer-Aided Network Design*, McGraw-Hill Book Company, New York, 1972, 350 pp.

CALAHAN, DONALD A., ALAN B. MACNEE AND E. L. McMAHON, *Computer-Oriented Circuit Analysis*, Holt, Rinehart & Winston, Inc., New York, 1974. To be published.

COMER, D. J., *Computer Analysis of Circuits*, Intext Educational Publishers, Scranton, Pa., 1971, 356 pp.

DERTOUZOS, MICHAEL L., MICHAEL ATHANS, RICHARD N. SPANN, AND SAMUEL J. MASON, *Systems, Networks, and Computation: Basic Concepts*, McGraw-Hill Book Company, New York, 1972, 514 pp.

DIRECTOR, S. W., *Circuit Theory—The Computational Approach*, John Wiley & Sons, Inc., New York, 1974. To be published.

GUPTA, SOMESHWAR C., JON W. BAYLESS, AND BEHROUZ PEIKARI, *Circuit Analysis with Computer Applications to Problem Solving*, Intext Educational Publishers, Scranton, Pa., 1972, 546 pp.

HERSKOWITZ, GERALD J., *Computer-Aided Integrated Circuit Design*, McGraw-Hill Book Company, New York, 1968, 432 pp.

HUELSMAN, LAWRENCE P., *Digital Computations in Basic Circuit Theory*, McGraw-Hill Book Company, New York, 1968, 203 pp.

HUELSMAN, LAWRENCE P., *Basic Circuit Theory with Digital Computations*, Prentice-Hall, Inc., Englewood Cliffs, N.J., 1972, 856 pp.

KINARIWALA, B. K., FRANKLIN F. KUO, AND NAI-KUAN TSAO, *Linear Circuits and Computation*, John Wiley & Sons, Inc., New York, 1973, 656 pp.

KUO, FRANKLIN F., AND JAMES F. KAISER, eds., *System Analysis by Digital Computer*, John Wiley & Sons, Inc., New York, 1966, 438 pp.

KUO, FRANKLIN F., AND WALDO G. MAGNUSON, JR., eds., *Computer Oriented Circuit Design*, Prentice-Hall, Inc., Englewood Cliffs, N.J., 1969, 561 pp.

LEY, B. JAMES, *Computer Aided Analysis and Design for Electrical Engineers*, Holt, Rinehart & Winston, Inc., New York, 1970, 852 pp.

PRINCE, M. DAVID, *Interactive Graphics for Computer-Aided Design*, Addison-Wesley Publishing Co., Reading, Mass., 1971, 301 pp.

RAMEY, ROBERT L., AND EDWARD J. WHITE, *Matrices and Computers in Electronic Circuit Analysis*, McGraw-Hill Book Company, New York, 1971, 390 pp.

SEELY, SAMUEL, NORMAN H. TARNOFF AND DAVID HOLSTEIN, *Digital Computers in Engineering*, Holt, Rinehart & Winston, Inc., New York, 1970, 329 pp.

STEIGLITZ, KENNETH, *An Introduction to Discrete Systems*, John Wiley & Sons, Inc., New York, 1973, 000 pp.

WING, OMAR, *Circuit Theory with Computer Methods*, Holt, Rinehart & Winston, Inc., New York, 1972, 529 pp.

WOLFENDALE, E., ed., *Computer-Aided Design Techniques*, Daniel Davey & Co., Inc., Hartford, Conn., 1970, 319 pp.

F-4 DEVICE MODELING

GRAY, PAUL E., AND CAMPBELL L. SEARLE, *Electronic Principles*, John Wiley & Sons, Inc., New York, 1969, 1016 pp.

HAMILTON, D. J., F. A. LINDOLM, AND A. H. MARSHAK, *Principles and Applications of Semiconductor Device Modeling*, Holt, Rinehart & Winston, Inc., New York, 1971, 485 pp.

MILLMAN, JACOB, *Electronic Devices and Models*, McGraw-Hill Book Company, New York, in preparation.

MILLMAN, JACOB, AND CHRISTOS C. HALKIAS, *Integrated Electronics: Analog and Digital Circuits and Systems*, McGraw-Hill Book Company, New York, 1972, 911 pp.

WEDLOCK, BRUCE D., AND JAMES K. ROBERGE, *Electronic Components and Measurements*, Prentice-Hall, Inc., Englewood Cliffs, N.J., 1969, 338 pp.

F-5 SYSTEMS

DERUSSO, P. M., R. J. ROY, AND C. M. CLOSE, *State Variables for Engineers*, John Wiley & Sons, Inc., New York, 1965, 608 pp.

FREDERICK, DEAN K., AND A. BRUCE CARLSON, *Linear Systems in Communication and Control*, John Wiley & Sons, Inc., New York, 1971, 575 pp.

KIRK, DONALD E., *Optimal Control Theory: An Introduction*, Prentice-Hall, Inc., Englewood Cliffs, N.J., 1971, 452 pp.

KUO, BENJAMIN C., *Linear Networks and Systems*, McGraw-Hill Book Company, New York, 1967, 411 pp.

MERRIAM, C. W. III, *Analysis of Lumped Electrical Systems*, John Wiley & Sons, Inc., New York, 1969, 580 pp.

PERKINS, WILLIAM R., AND JOSÉ B. CRUZ, JR., *Engineering of Dynamic Systems*, John Wiley & Sons, Inc., New York, 1969, 568 pp.

SCHWARZ, R. J., AND B. FRIEDLAND, *Linear Systems*, McGraw-Hill Book Company, New York, 1965, 521 pp.

TIMOTHY, LAMAR K., AND BLAIR E. BONA, *State Space Analysis: An Introduction*, McGraw-Hill Book Company, New York, 1968, 406 pp.

TRUXAL, JOHN G., *Introductory Systems Engineering*, McGraw-Hill Book Company, New York, 1972, 596 pp.

F-6 BACKGROUND MATERIAL OR MATERIAL OF HISTORICAL INTEREST

GANTMACHER, F. R., *The Theory of Matrices*, Vols, I and II, Chelsea Publishing Company, New York, 1959. 276 and 374 pp.

GANTMACHER, F. R., *Applications of the Theory of Matrices*, Interscience Publishers, Inc., New York, 1959, 317 pp.

GREENBERG, MICHAEL D., *Application of Green's Functions in Science and Engineering*, Prentice-Hall, Inc., Englewood Cliffs, N.J., 1971, 150 pp.

JORDAN EDWARD C., AND KEITH G. BALMAIN, *Electromagnetic Waves and Radiating Systems*, 2nd ed., Prentice-Hall, Inc., Englewood Cliffs, N.J., 1968, 753 pp.

REED, MYRIL B., AND GEORGIA B. REED, *Electric Network Theory: Laplace Transform Technique*, International Textbook Company, Scranton, Pa., 1968, 235 pp.

STEINMETZ, CHARLES PROTEUS, *Lectures on Electrical Engineering*, Dover Publications, Inc., New York, 1971, three volumes.

VAN VALKENBURG, M. E., *Classics of Circuit Theory,* Dowden, Hutchinson & Ross, Inc., Stroudsburg, Pa., 1974, 453 pp.

Appendix G: Answers to Selected Problems

Chapter 1

1-3. 2 C. **1-5.** $i = \omega C_0 V_0 (\cos \omega t - \cos 2\omega t)$ amp. **1-7.** 7.2×10^{-8} F. **1-9.** $C = 3\pi\epsilon R^2/d$ F. **1-15.** $W_C = q^4/4K^3$ J. **1-16**(b). $W = (2S_0/a^2) \sinh^2 (aq/2)$ J. **1-17.** 72×10^{-6} J. **1-19.** (a) i has the waveform of Fig. 1-11 and maximum amplitude CV_{max}. **1-21.** (a) 1 weber-turn, (c) 0.5 C. **1-24.** (a) 0.37 amp/sec, (b) 0.63 weber, (d) 0.37 V, (e) 0.198 J. (g) 0.233 W. **1-25.** (d) 0.37 V, (g) 0.137 W. **1-31.** $v_C = 4(t - 1)$, $1 \le t \le 2$; $v_C = 4 - 6(t - 2)$, $2 \le t \le 3$; $v_C = -2V.$, $t \ge 3$. **1-38.** $v_L = (\frac{1}{2}) \cos t$, $0 \le t \le \pi$.

Chapter 2

2-3. i intercept $= V/R_b$, slope $= -1/R_b$. **2-4.** See Fig. G-1. **2-8.** (a) $L_{eq} = L_1 + L_2 + 2M$ H. **2-9.** (b) $200e^{-t}$ V. **2-11.** All are planar. **2-13.** a-1, b-4, c-3, d-2. **2-15.** (a) 3, (b) 6.

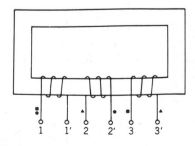

Fig. G-1.

Chapter 3

3-1. (a) $C_{eq} = C_1 C_2/(C_1 + C_2)$ F. **3-2.** (b) $L_{eq} = L_1 + L_2 - 2M$ H. **3-3.**
(b) $L_{eq} = (L_1 L_2 - M^2)/(L_1 + L_2 + 2M)$H. **3-6.** 2.5 amps, 5 amps/sec.
3-7. C = 0.15 F. **3-8.** (a) $v_1(t)$ is composed of straight-line segments and is
double-valued as shown by the following table:

t	0	1	2	3	4	5	—
$v(t-)$	0	0	5	4	14	12	—
$v(t+)$	0	1	4	6	12	12	—

3-12. (b) See Fig. G-2(a). (f) See Fig. G-2(b). **3-29.** (c) The node equations
are:

$$2C\frac{dv_a}{dt} + G_1 v_a + C\frac{dv_c}{dt} = C\frac{dv}{dt}$$

$$C_1\frac{dv_b}{dt} + 2Gv_b - Gv_c = Gv$$

$$-C\frac{dv_a}{dt} - Gv_b + C\frac{dv_c}{dt} + (G + G_L)v_c = 0$$

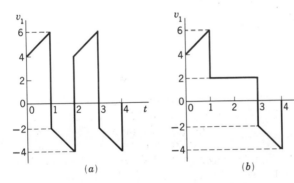

Fig. G-2.

3-30. Some terms: $b_{11} = 4 + (\frac{1}{2})d/dt$, $b_{33} = (\frac{3}{2})d/dt$, $b_{12} = -2$. **3-32.**
$i_1 = 405/159$, $i_2 = 210/159$, $i_3 = 270/159$. **3-34.** (a) 9, (b) 113. **3-35.** (b)
The equation is

$$\begin{bmatrix} 1.5 & -.25 & 0 & -1 \\ -.25 & 1.5 & -.25 & -1 \\ 0 & -.25 & 1.5 & -1 \\ -1 & -1 & -1 & -4 \end{bmatrix} \begin{bmatrix} v_a \\ v_b \\ v_c \\ v_d \end{bmatrix} = \begin{bmatrix} 0 \\ 0 \\ 0 \\ I \end{bmatrix}$$

3-40. $v_a = -0.533$, $v_b = -0.2$, $v_c = 4.03$, $v_0 = -0.267$ V. **3-58.** $L_{eq} = R_0^2 C$, $R_{eq} = R_0^2/R$. **3-61.** Make use of $L_1 C_1 = L_2 C_2 = L_3 C_3 = 1$ and
$Z_a = Z_b$ to get $C_1 = \frac{1}{2}$, $L_2 = 3$, $L_3 = 6$, $C_3 = \frac{1}{6}$.

Chapter 4

4-1. $(V/R_1)e^{-(R_1+R_2)t/L}$. **4-3.** (e) $v_2 = \frac{5}{3} - (\frac{2}{3})e^{-3t}$. **4-4.** $v_2 = -(\frac{1}{2})e^{-3t/4}$.
4-6. $0.6 - 0.1e^{-100t}$. **4-7.** $v_2 = e^{-t} - e^{-3t}$. **4-12.** $[V/R_1(R_1 + R_2)](R_1 +$

$R_2 - R_2 e^{-R_1 t/L}$). **4-13.** (b) $i = (Vt/L)e^{-Rt/L}$. **4-16.** (h)

$$i = \frac{[(a-1)\sin 2t - 2\cos t)e^{-t} + 2e^{-at}]}{(a-1)^2 + 4}$$

4-17. $i_L = 0.6 + 0.067e^{-3.57t}$.   **4-20.** (a) $R = 10^4\,\Omega$, (b) $C = 2.5\,\mu\text{F}$.
4-21. (b) $i = 0.01(1 - e^{-16t})$ amp. **4-22.** $w_C/w_T = (\frac{1}{2})(1 - e^{-t/RC})$.

Chapter 5

5-1. 0.1, -100, 100,000.   **5-3.** $d^2i/dt^2(0+) = -9 \times 10^5$ amp/sec.^{2} **5-5.**
$-8\ V/\text{sec}^2$, 26 V/sec^3. **5-6.** 0 V/sec^2. **5-8.** 100, -10^4, 10^6. **5-13.** (b) 0,
$R_2 V/(R_1 + R_2)$, (c) V/CR_1, 0, (d) $R_2 V/R_1 LC$. **5-15.** $d^2v_1/dt^2(0+) = V_0/R$.
5-18. 1.406×10^{-5} amp/sec.^{2} **5-20.** (b) 66.7 V, (c) 3.33, 1.67 amps, (d)
33.3, $-83,300$, **5-21.** $R_1 i(0+)$, 0, $R_1[(di/dt)(0+) - Ri(0+)/L]$, 0. **5-25.** (a)
VR_1/R_2, (b) $(1/C) - R_1$. **5-27.** (a) 0, $v(0+)/R_1 C_1$, (b) 0, 0, (c) 0, 0. **5-28.** (a)
(a) $v_K(0+) = VR_3/(R_1 + R_2 + R_3)$.

Chapter 6

6-3. (f) $i = (A\cos\sqrt{1.75}t + B\sin\sqrt{1.75}t)e^{-t/2}$. (h) $i = Ae^{-2t} + Bte^{-2t}$.
6-5. (a) $i = 2e^{-t} - e^{-2t}$. (b) $i = 3e^{-2t} - 2e^{-3t}$. (f) $i = e^{-0.5t}(\cos 1.32t +$
$0.379\sin 1.32t)$. (h) $i = e^{-2t} + te^{-2t}$. **6-9.** $i = Ae^{-.418t} + e^{-1.12t}(B\cos 1.06t$
$+ C\sin 1.06t)$. **6-11.** $t_1 = 1/\alpha$. **6-14.** $i = e^{-140t}\sin 1570t$. **6-15.** $i_L =$
$10\cos 316t$. **6-18.** (b) $i = K_1 e^{-t} + K_2 e^{-2t} - \frac{15}{4} + (5t/2)$, (e) $v = (K_1$
$+ t)e^{-2t} + (K_2 - 5t)e^{-3t}$. **6-23.** $i = 5t^2 e^{-t}$. **6-24.** $i = 10^{-4}e^{-t}(\cos t - \sin t)$
$- 10^{-4}e^{-10^4 t}$. **6-28.** $i = 10^{-3}(20.3\cos 377t + 14.8\sin 1414t - 0.8\cos 1414t)$
amp. **6-29.** $s(500) = -250 \pm j968$, $s(1000) = -500 \pm j866$, $s(3000) =$
$-2,620$, -380, $s(5000) = -4,790$, -210. **6-31.** Characteristic equation:
$(5 - K_1)s^2 + (6 - 2K_1)s + (2 - K_1) = 0$; the system is stable for $K_1 <$
2. **6-33.** (d) $w_S = w_R + w_L = (V^2/R)[t + (L/R)(e^{-Rt/L} - 1)]$.

Chapter 7

7-9. $F_1(s) = 2/[s(s^2 + 4)]$.   **7-16.** $F_8(s) = s^2/[(s^2 + a^2)^2]$.   **7-22.**
$(V/R_1)e^{-(R_1+R_2)t/L}$. **7-24.** $(V/R)\cos(t/\sqrt{LC})$. **7-30.** (a) $K_0 = \frac{3}{8}$, $K_2 = \frac{1}{4}$,
$K_4 = \frac{3}{8}$. **7-35.** $i = K_1 e^t + K_2 e^{-t} - 25 + (\frac{1}{3})e^{2t}$. **7-37.** $i = (\frac{1}{3})t^3 + 2e^{-t} + 2$.
7-42. $v_a = 13(e^{-0.0215t} - e^{-0.1285t}$. **7-43.** 16.7 sec. **7-44.** 14 V in 29.2 sec.
7-45. (a) $3.33 + 1.21e^{-6.35t} - 4.5e^{-23.6t}$.   **7-46.** $1020(e^{-0.155t} - e^{-0.645t})$.
7-49. $v_a = 4 + e^{-0.75t}(-1.5\cos 0.25t + 0.5\sin 0.25t)$.   **7-50.** $i_2 = -5 -$
$1.34e^{-0.707t} + 16.3e^{0.707t}$.

Chapter 8

8-1. $u(t-2) - u(t-3) + u(t-7) - u(t-8)$.   **8-9.** $v = K_0[t^2 u(t) -$
$4(t-1)u(t-1) - (t-2)^2 u(t-2)]$. **8-10.** Partial answer: $v = be^{[\ln(b/a)]t/c}$
$\cdot u(t) - \dots$ **8-14.** (c) $i = [5 \times 10^5 t - 1.25 + 1.25e^{-4\times 10^5 t}]u(t) - [5 \times$
$10^5(t - 10^{-6}) - 1.25 + 1.25e^{-4\times 10^5(t-10^{-6})}]u(t - 10^{-6})$. **8-17.** $i(t) = [1 -$
$e^{-(t-1)}]u(t-1) + [1 - e^{(t-2)}]u(t-2) + [1 - e^{-(t-3)}]u(t-3) + [1 - e^{-(t-4)}]$
$\cdot u(t-4) - 4[1 - e^{-(t-5)}]u(t-5)$. **8-19.** $F_1(s) = K[(1/s^2) + (t_0/s)]$. **8-26.**
$V(s) = (1/s)[e^{-s} + e^{-2s} + e^{-3s} + e^{-4s} - 4e^{-5s}]$.   **8-31.** $F(s) = 1/[s(1 -$
$e^{-st_0})]$.   **8-33.** $F(s) = \{1/[(s + \frac{1}{2})(1 + e^{-(0.5+s)})]\}$.   **8-34.** (a) $f(t) = te^{at}$.

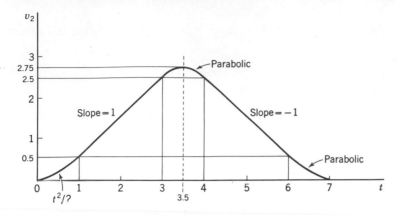

Fig. G-3.

8-36. See Fig. G-3.   **8-68.** $v = (4e^t - 3)u(t)$.   **8-69.** $h_3 = (t^2/2)u(t) - [3(t-1)^2/2]u(t-1) + [3(t-2)^2/2]u(t-2) - [(t-3)^2/2]u(t-3)$.   **8-70.** $v_2 = 2(t - 1 + e^{-t})u(t) - 4[1 - e^{-(t-1)}]u(t-1) - 2[t - 3 + e^{-(t-2)}u(t-2)]$.

Chapter 9

9-1. (a) $-2 \pm j1$, -3, (b) $-1 \pm j3$, -1, -4. **9-6.** See Fig. G-4. **9-7.** $Z = L_{eq}s$, $L_{eq} = (L_1L_2 + L_2L_3 + L_3L_1)/(L_2 + L_3)$. **9-8.** $Z(s) = (s + 1)/(s^2 + 3s + 1)$. **9-9.** $Z(s) = [(s^2 + 1)(s^2 + 9)]/[s(s^2 + 4)]$. **9-11.** $Y(s) = 1$. **9-27.** $I_3(s) = 4 \times 10^2/[s(s + 10)(s + 30)]$.   **9-29.** Amplitude $- R_3/(R_1 + R_3)$, duration $= T$ sec.   **9-30.** $V_{Th} = 4s/[(s^2 + 4)(s + 2)]$. $Z_{Th} = 1$.   **9-31.** (b) $V_{Th} = (2s + 1)/s[(5 - K_1)s + 3 - K_1]$, $Z_{Th} = [(3 - K_1)s + 2 - K_1]/[(5 - K_1)s + 3 - K_1]$. **9-32.** $Z_{Th} = R_1 + R_2 + R_3 - \mu_1 R_1 - \mu_2(R_1 + R_2) + \mu_1\mu_2 R_1$, $v_{Th} = (\mu_1 + \mu_2 - \mu_1\mu_2)v_1$.

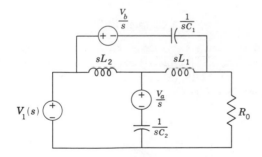

Fig. G-4.

Chapter 10

10-1. $Z_{12} = 1/[C(s + 1/RC)]$. **10-3.** $Z_{in} = 1$, $G_{12} = (s + 1)/(s^2 + 3s + 1)$. **10-4.** (a) $G_{12} = \frac{1}{41}$; (b) $G_{12} = \frac{1}{56}$.   **10-9.** $K = -3$.   **10-10.** $V_2/V_1 = (s^2 + 1)^2/(5s^4 + 5s^2 + 1)$.   **10-12.** $V_2/V_1 = 1/16s^4 + 12s^2 + 1)$.   **10-14.** $V_2/V_1 = 1/(s^4 + 3s^2 + 1)$ **10-16.** $R = 1$, $L = \frac{1}{3}$, $C = \frac{1}{10}$. **10-20.** (a) 0,0,3, (b) 0,0,3, (c) 2,0,1. **10-21.** (a) $K > \frac{1}{5}$, (b) When $K = \frac{1}{5}$, the real part vanishes. **10-23.** (a) 0,0,5, (b) 0,0,5, (c) 1,0,4, (d) 2,2,1, (e) 2,2,2. **10-29.** $K < (R_1 + R_2)/$

R_2 for stable amplifier, $K = (R_1 + R_2)/R_2$ causes oscillation at frequency $\omega = 1/\sqrt{LC}$.

Chapter 11

11-3.

$$z = \begin{bmatrix} \frac{13}{7} & \frac{2}{7} \\ \frac{2}{7} & \frac{3}{7} \end{bmatrix}, \qquad y = \begin{bmatrix} -\frac{3}{5} & -\frac{2}{5} \\ -\frac{2}{5} & \frac{13}{5} \end{bmatrix}$$

11-4.

$$z = \begin{bmatrix} -0.4 & 0.4 \\ -3.2 & 1.2 \end{bmatrix}, \qquad y = \begin{bmatrix} 1.5 & -0.5 \\ 4 & -0.5 \end{bmatrix}$$

11-10. Let $D = 2(s + 6)$. $y_{11} = (s^2 + 8s + 8)/D$, $y_{12} = -(s^2 + 6s + 8)/D$, $y_{21} = -(s^2 + 6s + 4)/D$, $y_{22} = (s^2 + 10s + 8)/D$. **11-13.** $h_{11} = R_1 + [(1 - \alpha)R_2R_3/(R_2 + R_3)]$, $h_{22} = 1/(R_2 + R_3)$, $h_{12} = R_2/(R_3 + R_3)$, $h_{21} = -(\alpha R_3 + R_2)/(R_2 + R_3)$. **11-26.** (a) $R_1 = \frac{1}{7}$, $R_2 = \frac{2}{7}$, $R_3 = \frac{11}{7}$. **11-27.** (b) $C_a = \frac{4}{5}$, $C_b = \frac{7}{5}$, $C_c = \frac{2}{5}$. **11-29.** $z_{11} = z_{22} = (\frac{1}{2})(Z_a + Z_b)$, $z_{12} = (\frac{1}{2})(Z_b - Z_a)$. **11-32.** $Z_{\text{series}} = R_0 Z_a/(2R_0 + Z_a)$, $Z_{\text{shunt}} = R_0 + 2Z_b$. **11-34.** See the bisected network in Fig. G-5.

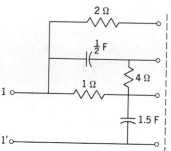

Fig. G-5.

Chapter 12

12-4. $A = 5$, $\phi = -\tan^{-1}\frac{3}{4} = -53°$. **12-7.** (a) $i_{ss} = 1/(2\sqrt{2})\sin(2t - 45°)$. **12-9.** $i_{ss} = (2/\sqrt{R^2 + (1/2C)^2})\cos(2t + \tan^{-1} 1/2RC)$. **12-16.** $v_a = 26.5\sin(10^6 t + 128°)$. **12-19.** (a) $v_1 = 2\sqrt{2}\sin(t/2 + 15°)$. **12-34.** $A = \frac{1}{3}$, $\phi = -90°$. **12-35.** See Fig. G-6.

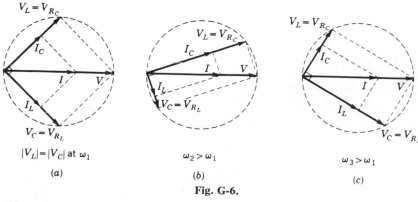

Fig. G-6.

Chapter 13

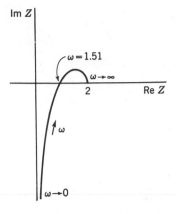

13-3. $G_{12}(j\omega) = (R^2/R^2 + \omega^2 L^2) - j(\omega RL/R^2 + \omega^2 L^2)$. **13-12.** (a) See Fig. G-7. **13-13.** (b) See Fig. G-8. **13-16.** Pole at $s = -220$. **13-17.** Poles at $s = -50 \pm j208$, zero at $s = 0$. **13-18.** (a) 4.04, (b) 0.124, (f) 0.5, (h) $LC = 0.0616$, $R = L$. **13-22.** Bandwidth is 2.1 times greater with the zero. **13-24.** $Q = 5$. **13-34.** $G(s) = [0.1s/(s/50 + 1)^3]$. **13-39.** $M_{max} = 20$ db, $\zeta = 0.05$, asymptotic slopes are 0 and -12 db/octave.

Fig. G-7.

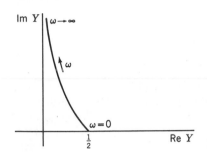

Fig. G-8.

Chapter 14

14-6. $V_{eff} = K_1/\sqrt{3}$. **14-11.** $V_{eff} = V_m/2$. **14-14.** $V_{eff} = (\sqrt{3}/4)V_0$. **14-16.** 12, 16, 7 or 25 V. **14-17.** $L = 1$ H. **14-26.** (b) 125 μF. **14-30.** $Z_x = R_x = 1\ \Omega$ **14-31.** (a) $\alpha = 10 \log (1 + \omega^6)$ db, (b) $\alpha = 10 \log (1 + \omega^8)$ db, (e) See Fig. G-9.

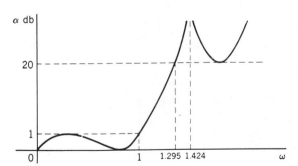

Fig. G-9.

Chapter 15

15-8. For $a = T/2$, $v = K[\frac{1}{2} + (2/\pi) \sin \omega_0 t - (2/3\pi) \sin 3\omega_0 t + \ldots]$. **15-9.** $v = K_1[\frac{1}{2} - (1/\pi) \sin \omega_0 t - (1/2\pi) \sin 2\omega_0 t - \ldots)]$. **15-10.** $a_0 = K_1/2$, $a_n = -(4K_2)/\pi^2 n^2)$, n odd. **15-12.** $|c_n| = [K_4\sqrt{(1 - \cos 2\pi\ nk_2)}/[\sqrt{2}\ \pi^2 n^2 k_2 (1 - k_2)]$, Angle of $c_n = \tan^{-1} (-\sin 2\pi nk_2)/(\cos 2\pi nk_2 - 1]$. **15-13.** $v = (4V_m/\pi)(\frac{1}{2} + \frac{1}{3} \cos \omega_0 t - \frac{1}{15} \cos 2\omega_0 t + \frac{1}{35} \cos 3\omega_0 t - \ldots)$. **15-14.** See solution for Prob. 15-32. **15-22.** (c) See Fig. G-10. **15-26.** (a) $\bar{c}_n = -jK/\pi n$, n odd, $\bar{c}_0 = K/2$. **15-32.** $\bar{c}_0 = V_m/\pi$, $\bar{c}_1 = V_m/4$, $\bar{c}_2 = V_m/3$, $\ldots$. **15-41.** $P_{av} = 7.07$ W. **15-42.**(a) $I_1 = 15e^{j(\omega_0 t - 30°)}$, $I_2 = 0.5e^{j(3\omega_0 t - 45°)}$. (b) $P_1 = 1080$ W, $P_3 = 1.77$ W, (c) 1081.77 W.

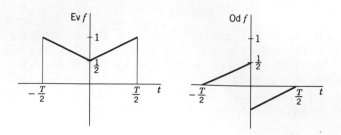

Fig. G-10.

Chapter 16

16-1. $|F(j\omega)| = |(2V_0/\omega)\sin(\omega a/2)|$.   **16-3.** $[2V_m/(1 - \omega^2)]\cos\omega(\pi/2)$.
16-4. $F(j\omega) = 4V_0/\omega(4 - \omega^2)]\sin(\omega\pi/2)$.   **16-5.** $F(j\omega) = V_0a\{(1 - \cos\omega a)/(a^2\omega^2) + j[(\sin\omega a) - (\omega a)]/(a^2\omega^2)\}$.   **16-17.** $\omega_0 V_0/(\alpha^2 + \omega_0^2 - \omega^2 + 2j\alpha\omega)$. **16-18.** $|F(j\omega)| = (V_0t_0)/(1 + \omega^2t_0^2)$. Angle $= -2\tan^{-1}\omega t_0$.
16-19. $F(j\omega) = \sqrt{\pi}\,V_0 t_0 e^{-\omega^2 t_0^2/4}$. **16-30.** Use bandwidth to first zero crossing, duration as π. **16-31.** First zero when $\omega_1 a/3 = 2\pi$, product becomes $\omega_1 a = 4\pi$. **16-35.** Let rise time be $t_r = T$ and half-power frequency is $\omega_1 = 1/T$ so that $t_r\omega_1 = 1$. **16-36.** (b) Overshoot $= v_{2m} - v(\infty) = e^{-(\zeta\pi)/(\sqrt{1-\zeta^2})}$.

Appendix H: Historical Notes

In the study of the fundamentals of network analysis, we encounter many names of persons who have contributed to our understanding of the subject through their contributions. The table that follows gives a brief introduction to a few of these.

Name	Year of Birth-Death	Occupation	Contribution Cited in Textbook
André Marie Ampère	1775—1836	Professor of Mathematics, Paris	Fundamental laws of electrodynamics (unit of current)
Hendrik Wade Bode	1905—	Bell Telephone Laboratories; Professor, Harvard University	Magnitude and phase coordinate system
Charles Augustin Coulomb	1738—1806	Experimentalist, France	Force relationship for charged bodies (unit of charge)
Peter Gustav Lejeune Dirichlet	1805—1859	Professor of Mathematics, Breslau, later Göttingen	Conditions for a Fourier series to exist
Leonhard Euler	1707—1783	Mathematician, Berlin Academy, later St. Petersburg Academy	Complex number relationship

Name	Year of Birth-Death	Occupation	Contribution Cited in Textbook
Michael Faraday	1791—1867	Assistant, Royal Institute, London	Electromagnetic induction (unit of capacitance)
Joseph Fourier	1768—1830	Professor of Mathematics, Ecole Polytechnique, Paris	Trigonometric series and integral
Benjamin Franklin	1706—1790	Printer-Philosopher, Philadelphia	Early conceptual scheme for electricity, direction of current.
Karl Friedrich Gauss	1777—1855	Professor of Mathematics, Göttingen	Laws on magnetism; scheme for matrix manipulation
Ernst Adolph Guillemin	1898—1970	Professor of Electrical Engineering, M.I.T.	Foundations of modern circuit theory; RC networks
Oliver Heaviside	1850—1925	Electrical Engineer, England	Operational mathematics, scheme for evaluating residues; modern concept of impedance
Hermann L. F. von Helmholtz	1821—1894	Professor of Physics, Heidelberg	Theorem for network reduction (cf. Thévenin)
Joseph Henry	1797—1878	Professor of Natural Philosophy, Princeton University	Electromagnetism (unit of inductance)
Heinrich Hertz	1857—1894	Professor of Physics, Bonn	Electromagnetic waves (unit of frequency)
A. Hurwitz	1862—1909	Mathematician, Germany	Location of zeros of polynomials (cf. Routh)
James P. Joule	1818—1889	Experimentalist, England	Law of heating (unit of energy)
Gustav Kirchhoff	1824—1887	Professor of Physics, Heidelberg	Conservation of voltage and current
Pierre Simon Laplace	1749—1825	Professor of Mathematics, Military School, Paris	Transform equation
James Clerk Maxwell	1831—1879	Professor of Mathematics, London and Cambridge	Electromagnetic laws

Name	Year of Birth-Death	Occupation	Contribution Cited in Textbook
Edward Lawry Norton	1898—	Electrical Engineer, Bell Telephone Laboratories	Theorem for network reduction (cf. Thévenin)
Harry Nyquist	1884—	Electrical Engineer, Bell Telephone Laboratories	Stability criterion
Hans Christian Oersted	1777—1851	Professor of Physics, University of Copenhagen	Discovery of electromagnetism
George Simon Ohm	1789—1854	Teacher, Koln	Law relating voltage and current (unit of resistance)
Edward John Routh	1831—1907	Professor of Mechanics, University of London	Location of zeros of polynomials (stability)
Charles Proteus Steinmetz	1865—1923	Electrical Engineer, General Electric Co.	Use of complex numbers for network analysis in the sinusoidal steady state
Bernard D. H. Tellegen	1900—	Philips Research Laboratories, Netherlands, and Technological University, Delft	Theorem relating to product of voltages and currents of a network
Leon Charles Thévenin	1857—1926	Telegraph Engineer, France	Theorem for network reduction
Alessandro Volta	1745—1827	Professor of Physics, Pavia	Electric generating cell (unit of voltage)
James Watt	1738—1819	Scottish engineer and inventor	(unit of power)
Charles Wheatstone	1802—1875	Professor of Physics, Kings College, London	Bridge form of network, used in measurements

Index